Sustainability of Olive Oil System

Sustainability of Olive Oil System

Editors

Cristina Alamprese
Emma Chiavaro
Francesco Caponio

MDPI • Basel • Beijing • Wuhan • Barcelona • Belgrade • Manchester • Tokyo • Cluj • Tianjin

Editors

Cristina Alamprese
University of Milan
Italy

Emma Chiavaro
University of Parma
Italy

Francesco Caponio
University of Bari "Aldo Moro"
Italy

Editorial Office
MDPI
St. Alban-Anlage 66
4052 Basel, Switzerland

This is a reprint of articles from the Special Issue published online in the open access journal *Foods* (ISSN 2304-8158) (available at: https://www.mdpi.com/journal/foods/special_issues/sustainability_olive_oil).

For citation purposes, cite each article independently as indicated on the article page online and as indicated below:

LastName, A.A.; LastName, B.B.; LastName, C.C. Article Title. *Journal Name* **Year**, *Volume Number*, Page Range.

ISBN 978-3-0365-3369-8 (Hbk)
ISBN 978-3-0365-3370-4 (PDF)

Cover image courtesy of Maria Paciulli.

© 2022 by the authors. Articles in this book are Open Access and distributed under the Creative Commons Attribution (CC BY) license, which allows users to download, copy and build upon published articles, as long as the author and publisher are properly credited, which ensures maximum dissemination and a wider impact of our publications.

The book as a whole is distributed by MDPI under the terms and conditions of the Creative Commons license CC BY-NC-ND.

Contents

About the Editors ix

Cristina Alamprese, Francesco Caponio and Emma Chiavaro
Sustainability of the Olive Oil System
Reprinted from: *Foods* 2021, *10*, 1730, doi:10.3390/foods10081730 1

Annalisa Rotondi, Lucia Morrone, Osvaldo Facini, Barbara Faccini, Giacomo Ferretti and Massimo Coltorti
Distinct Particle Films Impacts on Olive Leaf Optical Properties and Plant Physiology
Reprinted from: *Foods* 2021, *10*, 1291, doi:10.3390/foods10061291 5

Ana I. Carrapiso, Aránzazu Rubio, Jacinto Sánchez-Casas, Lourdes Martín, Manuel Martínez-Cañas and Concha de Miguel
Effect of the Organic Production and the Harvesting Method on the Chemical Quality and the Volatile Compounds of Virgin Olive Oil over the Harvesting Season
Reprinted from: *Foods* 2020, *9*, 1766, doi:10.3390/foods9121766 21

Amalia Piscopo, Rocco Mafrica, Alessandra De Bruno, Rosa Romeo, Simone Santacaterina and Marco Poiana
Characterization of Olive Oils Obtained from Minor Accessions in Calabria (Southern Italy)
Reprinted from: *Foods* 2021, *10*, 305, doi:10.3390/foods10020305 39

Giacomo Squeo, Roccangelo Silletti, Giacomo Mangini, Carmine Summo and Francesco Caponio
The Potential of Apulian Olive Biodiversity: The Case of Oliva Rossa Virgin Olive Oil
Reprinted from: *Foods* 2021, *10*, 369, doi:10.3390/foods10020369 49

Wilma Sabetta, Isabella Mascio, Giacomo Squeo, Susanna Gadaleta, Federica Flamminii, Paola Conte, Carla Daniela Di Mattia, Antonio Piga, Francesco Caponio and Cinzia Montemurro
Bioactive Potential of Minor Italian Olive Genotypes from Apulia, Sardinia and Abruzzo
Reprinted from: *Foods* 2021, *10*, 1371, doi:10.3390/foods10061371 63

Annalisa Rotondi, Lucia Morrone, Gianpaolo Bertazza and Luisa Neri
Effect of Duration of Olive Storage on Chemical and Sensory Quality of Extra Virgin Olive Oils
Reprinted from: *Foods* 2021, *10*, 2296, doi:10.3390/foods10102296 81

Stefano Farris, Susanna Buratti, Simona Benedetti, Cesare Rovera, Ernestina Casiraghi and Cristina Alamprese
Influence of Two Innovative Packaging Materials on Quality Parameters and Aromatic Fingerprint of Extra-Virgin Olive Oils
Reprinted from: *Foods* 2021, *10*, 929, doi:10.3390/foods10050929 91

Silvia Grassi, Olusola Samuel Jolayemi, Valentina Giovenzana, Alessio Tugnolo, Giacomo Squeo, Paola Conte, Alessandra De Bruno, Federica Flamminii, Ernestina Casiraghi and Cristina Alamprese
Near Infrared Spectroscopy as a Green Technology for the Quality Prediction of Intact Olives
Reprinted from: *Foods* 2021, *10*, 1042, doi:10.3390/foods10051042 105

Frederick Lia, Benjamin Vella, Marion Zammit Mangion and Claude Farrugia
Application of ^{1}H and ^{13}C NMR Fingerprinting as a Tool for the Authentication of Maltese Extra Virgin Olive Oil
Reprinted from: Foods **2020**, *9*, 689, doi:10.3390/foods9060689 . . . . **117**

Maria Paciulli, Graziana Difonzo, Paola Conte, Federica Flamminii, Amalia Piscopo and Emma Chiavaro
Physical and Thermal Evaluation of Olive Oils from Minor Italian Cultivars
Reprinted from: Foods **2021**, *10*, 1004, doi:10.3390/foods10051004 . . . . **131**

Vito Michele Paradiso, Francesco Longobardi, Stefania Fortunato, Pasqua Rotondi, Maria Bellumori, Lorenzo Cecchi, Pinalysa Cosma, Nadia Mulinacci and Francesco Caponio
Paving the Way to Food Grade Analytical Chemistry: Use of a Natural Deep Eutectic Solvent to Determine Total Hydroxytyrosol and Tyrosol in Extra Virgin Olive Oils
Reprinted from: Foods **2021**, *10*, 677, doi:10.3390/foods10030677 . . . . **153**

Federica Flamminii, Carla Daniela Di Mattia, Giampiero Sacchetti, Lilia Neri, Dino Mastrocola and Paola Pittia
Physical and Sensory Properties of Mayonnaise Enriched with Encapsulated Olive Leaf Phenolic Extracts
Reprinted from: Foods **2020**, *9*, 997, doi:10.3390/foods9080997 . . . . **167**

Paola Conte, Simone Pulina, Alessandra Del Caro, Costantino Fadda, Pietro Paolo Urgeghe, Alessandra De Bruno, Graziana Difonzo, Francesco Caponio, Rosa Romeo and Antonio Piga
Gluten-Free Breadsticks Fortified with Phenolic-Rich Extracts from Olive Leaves and Olive Mill Wastewater
Reprinted from: Foods **2021**, *10*, 923, doi:10.3390/foods10050923 . . . . **179**

Maria Angela Perito, Silvia Coderoni and Carlo Russo
Consumer Attitudes towards Local and Organic Food with Upcycled Ingredients: An Italian Case Study for Olive Leaves
Reprinted from: Foods **2020**, *9*, 1325, doi:10.3390/foods9091325 . . . . **195**

Matteo Carzedda, Gianluigi Gallenti, Stefania Troiano, Marta Cosmina, Francesco Marangon, Patrizia de Luca, Giovanna Pegan and Federico Nassivera
Consumer Preferences for Origin and Organic Attributes of Extra Virgin Olive Oil: A Choice Experiment in the Italian Market
Reprinted from: Foods **2021**, *10*, 994, doi:10.3390/foods10050994 . . . . **213**

Anna Maria Posadino, Annalisa Cossu, Roberta Giordo, Amalia Piscopo, Wael M. Abdel-Rahman, Antonio Piga and Gianfranco Pintus
Antioxidant Properties of Olive Mill Wastewater Polyphenolic Extracts on Human Endothelial and Vascular Smooth Muscle Cells
Reprinted from: Foods **2021**, *10*, 800, doi:10.3390/foods10040800 . . . . **231**

Mariangela Centrone, Mariagrazia D'Agostino, Graziana Difonzo, Alessandra De Bruno, Annarita Di Mise, Marianna Ranieri, Cinzia Montemurro, Giovanna Valenti, Marco Poiana, Francesco Caponio and Grazia Tamma
Antioxidant Efficacy of Olive By-Product Extracts in Human Colon HCT8 Cells
Reprinted from: Foods **2021**, *10*, 11, doi:10.3390/foods10010011 . . . . **243**

Souraya Benalia, Giacomo Falcone, Teodora Stillitano, Anna Irene De Luca, Alfio Strano, Giovanni Gulisano, Giuseppe Zimbalatti and Bruno Bernardi
Increasing the Content of Olive Mill Wastewater in Biogas Reactors for a Sustainable Recovery: Methane Productivity and Life Cycle Analyses of the Process
Reprinted from: *Foods* **2021**, *10*, 1029, doi:10.3390/foods10051029 . . . . **255**

Amin Nikkhah, Saeed Firouzi, Keyvan Dadaei and Sam Van Haute
Measuring Circularity in Food Supply Chain Using Life Cycle Assessment; Refining Oil from Olive Kernel
Reprinted from: *Foods* **2021**, *10*, 590, doi:10.3390/foods10030590 . . . . **277**

Alessia Pampuri, Andrea Casson, Cristina Alamprese, Carla Daniela Di Mattia, Amalia Piscopo, Graziana Difonzo, Paola Conte, Maria Paciulli, Alessio Tugnolo, Roberto Beghi, Ernestina Casiraghi, Riccardo Guidetti and Valentina Giovenzana
Environmental Impact of Food Preparations Enriched with Phenolic Extracts from Olive Oil Mill Waste
Reprinted from: *Foods* **2021**, *10*, 980, doi:10.3390/foods10050980 . . . . **293**

About the Editors

Cristina Alamprese (Associate Professor) obtained her M.Sc. in Food Science and Technology and her Ph.D. in Food Biotechnology at the University of Milan (Italy). Since 2016, she has served as an Associate Professor in Food Science and Technology at the Department of Food, Environmental and Nutritional Sciences (DeFENS) of the University of Milan. She has taught different courses in the area of Food Science and Technology since 2006. She is a Principal Investigator and participant in many national and international granted projects. Her research activity mainly covers the optimization of food formulations and processing conditions; the technological, rheological, mechanical, and ultrastructural characterization of foods; the food functionality–structure relationships; the application of NIR and MIR technology in the food sector, especially for food authentication and process control; the modelling of phenomena using Design of Experiments techniques and uni- and multivariate statistical methodologies. She is the author of more than 170 publications in international and national journals as well as in the proceedings of national and international conferences.

Emma Chiavaro (Associate Professor) obtained her M.Sc. in Pharmaceutical Chemistry and Technology at the University "LA SAPIENZA" in Rome and her Ph.D. in Animal Food Inspection at the Federico II University of Napoli. Since 2014, she has served as an Associate Professor in Food Science and Technology at the University of Parma, Department of Food and Drug. She has taught several Food Technology courses and acts as a supervisor of several theses for BSc and MSc courses in Food Science and Technology. She is the author of more than 180 publications in international and national peer-reviewed journals, as well as in the proceedings of national and international conferences. She is a participant in and principal investigator of granted national and international research projects. Her research activity covers the applicability of differential scanning calorimetry to the evaluation of the thermal properties of food, particularly extra-virgin olive oil and other vegetable oils. The technology of oils and fats, and the effect of processing and non-conventional preservation techniques on food formulation and shelf-life, are also topics of her investigations.

Francesco Caponio (Full Professor) graduated in Agrarian Sciences at the University of Bari (Italy), and since 2017 has served as a Full Professor in Food Science and Technology at the University of Bari, Department of Soil, Plant and Food Science (DISSPA). He is the principal investigator of several competitive research projects and was an invited speaker in many conferences, with national and regional diffusion. He is the author of over 230 publications in referenced international journals. Currently, he serves as the Coordinator of the Bachelor and Master Degrees in "Food Science and Technology" at the University of Bari Aldo Moro, and is a member of the Ph.D. School in "Soil and Food Sciences". His research activity covers olive-oil technology and the set-up of analytical techniques to ensure the quality and genuineness of foods; the extraction, characterization, and valorization of bioactive compounds from processing wastes and their use in foods in order to increase the shelf-life; the formulation of innovative and functional foods.

Editorial

Sustainability of the Olive Oil System

Cristina Alamprese [1,*], Francesco Caponio [2] and Emma Chiavaro [3]

1. Department of Food, Environmental, Nutritional Sciences (DeFENS), Università degli Studi di Milano, via G. Celoria 2, 20133 Milan, Italy
2. Department of Soil, Plant and Food Science (DISSPA), Università degli Studi di Bari Aldo Moro, via Amendola, 165/A, 70126 Bari, Italy; francesco.caponio@uniba.it
3. Department of Food and Drug, University of Parma, Parco Area delle Scienze 27/A, 43124 Parma, Italy; emma.chiavaro@unipr.it
* Correspondence: cristina.alamprese@unimi.it; Tel.: +39-0250319187

Citation: Alamprese, C.; Caponio, F.; Chiavaro, E. Sustainability of the Olive Oil System. *Foods* **2021**, *10*, 1730. https://doi.org/10.3390/foods10081730

Received: 21 July 2021
Accepted: 23 July 2021
Published: 27 July 2021

Publisher's Note: MDPI stays neutral with regard to jurisdictional claims in published maps and institutional affiliations.

Copyright: © 2021 by the authors. Licensee MDPI, Basel, Switzerland. This article is an open access article distributed under the terms and conditions of the Creative Commons Attribution (CC BY) license (https://creativecommons.org/licenses/by/4.0/).

Sustainability is a widely accepted goal across many sectors of our society and, according to new concepts, it includes resilience and adaptive capacity. Resilience is important for a system to guarantee the maintenance of functions and structures, including when subjected to shocks. Adaptability is fundamental to face unpredictability and unforeseen changes. Both aspects must be considered in the sustainable development of the olive oil system, including environmental, economic, and social issues, but also the improvement of the well-established historical tradition by maximizing the process efficiency and the quality of the end product. Therefore, this Special Issue about "Sustainability of Olive Oil System" intends to give an overview of several aspects related to sustainability of the olive oil processing chain, in order to open minds to new "sustainable thinking".

Starting from olive production, sustainability can be improved by using chemical-free alternatives to pesticides and organic production. Rotondi et al. [1] explored the effectiveness of kaolin-based and zeolitite-based particle films for hindering the attacks of the olive fruit fly (*Bactrocera oleae*), evaluating leaf gas exchanges and leaf optical properties. The zeolitite-based film showed the best performance, exerting a protective effect against olive fruit fly attacks without altering the leaf gas exchanges. Moreover, olive oils obtained from zeolitite-based particle film treatment showed intensities of gustatory and olfactory pleasant flavors higher than those of oils produced from kaolin and untreated olives. Carrapiso et al. [2] evaluated the effects on virgin olive oil characteristics of organic production without irrigation, traditional harvesting methods (tree vs. ground picked fruits), and harvesting time (over a six-week period). Organic production affected physical-chemical parameters and volatile compounds less than the harvesting method. Otherwise, a higher content in total phenols was found in the organic oils than in the conventional ones, probably explaining the increase in oil stability and the differences in the volatile compounds.

The valorization of minor olive accessions could represent a good way to improve the qualitative production of a specific territory while protecting biodiversity, an important aspect of sustainability. Four minor Italian cultivars were exploited by Piscopo et al. [3] to improve extra virgin olive oil (EVOO) production in the Calabria region; they obtained in most cases good quality oil in terms of free acidity, peroxides, spectrophotometric indexes, fatty acid composition, and bioactive compounds. Squeo et al. [4] investigated the cultivar "Oliva Rossa", which represents an old landrace belonging to the autochthon Apulian olive germplasm. The authors showed that the extracted virgin olive oils had a medium to high level of oleic acid. With colder temperatures, a higher content of monounsaturated fatty acids and antioxidants was observed, as well as a higher oleic/linoleic ratio. The phenolic profile was dominated by secoiridoid derivatives, which might indicate a product with remarkable pungent and bitter notes. Similarly, the volatile profile was dominated by the compounds arising from the lipoxygenase pathway. The recovery and valorization of other minor Italian olive cultivars were further investigated by Sabetta et al. [5]. A pattern of nine minor genotypes cultivated in three Italian regions was molecularly fingerprinted with 12 nuclear microsatellites that were able to unequivocally identify all genotypes. In addition, the monovarietal oils were evaluated for the principal phenolic compounds and the expression levels of related genes at different fruit development stages were investigated.

An important key factor to address emerging challenges of sustainable food consumption is the reduction of the environmental footprint of packed food. Thus, the performance of two innovative packaging materials in protecting EVOO from oxidation phenomena was investigated by Farris et al. [6]. In particular, a transparent plastic film loaded with a UV-blocker and a metallized material were compared to brown-amber glass during accelerated shelf-life tests at 40 and 60 °C. The transparent film emerged as the best-performing material to preserve EVOO quality.

A key role in sustainability is played by green chemistry, which can provide online techniques for automatic evaluation of food quality and optimization of food processes, while minimizing the use of hazardous materials. In this context, fingerprint techniques are valuable tools for both quality assessment and authentication issues. Grassi et al. [7] demonstrated that near-infrared (NIR) spectroscopy can be used in the field or at the mill entrance for a quick classification of the intact olive drupes as a function of their chemical parameters (moisture, oil content, soluble solids, total phenolic content, and antioxidant activity), in order to better design the olive oil quality. Lia et al. [8] developed a nuclear magnetic resonance (NMR) method for the discrimination of Maltese and non-Maltese EVOO, showing a higher effectiveness of ^{13}C NMR rather than ^{1}H NMR. Physical and thermal analyses were proposed by Paciulli et al. [9] as fast and green techniques to identify botanical and geographical origin of EVOO. In particular, thirteen EVOO samples obtained from minor olive cultivars, harvested at three different ripening stages in four Italian regions (Abruzzo, Apulia, Sardinia, and Calabria), were investigated for thermal properties, viscosity, and color, as influenced by fatty acid composition and chlorophyll content. The most influential thermal parameters and fatty acids were used to identify possible sample clusters by means of principal component analysis; while a clear distribution of the samples based on their botanical and geographical origin was evident, no pattern was highlighted in terms of olive harvesting time. Moreover, Paradiso et al. [10] proposed a green method for the determination of hydroxytyrosol and tyrosol content in EVOO, based on the use of a natural deep eutectic solvent composed of lactic acid and glucose for the liquid/liquid extraction step, followed by UV-spectrophotometric analysis.

Maximization of the production process efficiency goes through the valorization of olive oil by-products, by using polyphenolic extracts derived from olive leaves and mill wastewater as food ingredients. The papers by Flamminii et al. [11] and Conte et al. [12] suggest the use of free or encapsulated polyphenolic extracts in mayonnaise and gluten-free breadsticks, demonstrating the possibility of developing healthy foods, with extended shelf life. The likelihood of consumers' acceptance of these kinds of foods obtained with upcycled ingredients of olive oil production was studied by Perito et al. [13]. The authors found that, despite the negative influence of food technophobia, a core of sustainability-minded consumers interested in organic or local products could also favor the uptake of these foods enriched with ingredients made from olive oil by-products. Development of organic or local food products with upcycled ingredients could potentially be the right way to increase the probability of consumers' acceptance. Indeed, Carzedda et al. [14] investigated Italian consumers' behavior towards EVOO organic production methods and geographical origin to quantify the willingness to pay for these two attributes. Findings showed positive preference for origin attributes, especially linked to local productions.

The olive oil by-products are attracting great interest also in the pharmaceutical and renewable energy fields. Posadino et al. [15] produced a phenolic-rich extract from olive mill wastewater to assess the protection against oxidative cell death in human vascular cells. The tested extract protected cells from oxidative stress-induced cell death, failing indeed to interfere with cell viability and even with the metabolism, except for the highest tested concentration. Centrone et al. [16] explored the biological actions of extracts deriving from different olive by-products, including olive pomace, olive wastewater, and olive leaf, on human colorectal carcinoma HCT8 cells. Different effects on reactive oxygen species' generation and cell viability were found: the extract obtained from the olive mill wastewater showed higher antioxidant ability compared with the extracts derived from olive pomace and olive leaves. These biological effects may be related to the different phenolic composition of the extracts. Actually, the olive mill wastewater extract contained the highest amounts of hydroxytyrosol and tyrosol, which are considered potent antioxidant compounds. The advantages of using farm and food industry by-products to produce renewable energy as well as organic fertilizers, which could be used in situ to enhance farm sustainability, were demonstrated by Benalia et al. [17]. The authors explored the anaerobic co-digestion of olive mill wastewater to produce biogas and biomethane. Different mixtures of olive mill wastewater were tested under mesophilic conditions. By applying the life-cycle assessment (LCA) approach, it was demonstrated that a good biogas ecoprofile and a high process profitability can be obtained using 20% (v/v) olive mill wastewater.

Life-cycle-based methodologies are very powerful and reliable tools to quantify the impact generated from a product/service along the entire production process and throughout its whole duration. Nikkhah et al. [18] applied LCA to measure the circularity of refining oil from olive kernel, a common source of waste in olive fruit processing systems. The authors reported that the global warming potential of 1 kg oil produced from olive kernel was 1.37 kg CO_2eq, while the calculated damage of 1 kg oil production to human health, ecosystem quality, and resource depletion was 5.29×10^{-7} disability-adjusted life years (DALY), 0.12 PDF m^2 year., and 24.40 MJ, respectively. Pampuri et al. [19] quantified the environmental impact of four lab-scale food preparations (vegan mayonnaise, salad dressing, biscuits, and gluten-free breadsticks) enriched with phenolic extracts from olive oil by-products (i.e., mill wastewater and olive leaves), considering technological and nutritional parameters. The authors concluded that the phenolic extraction and encapsulation, even if characterized by low production yields, energy-intensive operations, and the partial use of chemical reagents, made a non-negligible environmental impact contribution to the food preparation. The addition of phenolic extracts to food products led to an enhanced environmental impact of the production process, but also to improved technological and nutritional performances. Impacts could be reduced through a scale-up process.

In summary, this Special Issue provides evidence that olive oil system sustainability can be improved in different ways, from the enhancement of biodiversity to the exploitation of waste and by-products for food and health-related purposes in a circular economy perspective.

Author Contributions: C.A., F.C., and E.C. contributed equally to the writing and editing of the editorial note. All authors have read and agreed to the published version of the manuscript.

Funding: This work did not receive external funding.

Conflicts of Interest: The authors declare no conflict of interest.

References

1. Rotondi, A.; Morrone, L.; Facini, O.; Faccini, B.; Ferretti, G.; Coltorti, M. Distinct particle films impacts on olive leaf optical properties and plant physiology. *Foods* **2021**, *10*, 1291. [CrossRef] [PubMed]
2. Carrapiso, A.; Rubio, A.; Sánchez-Casas, J.; Martín, L.; Martínez-Cañas, M.; de Miguel, C. Effect of the organic production and the harvesting method on the chemical quality and the volatile compounds of virgin olive oil over the harvesting season. *Foods* **2020**, *9*, 1766. [CrossRef] [PubMed]
3. Piscopo, A.; Mafrica, R.; De Bruno, A.; Romeo, R.; Santacaterina, S.; Poiana, M. Characterization of olive oils obtained from minor accessions in Calabria (Southern Italy). *Foods* **2021**, *10*, 305. [CrossRef] [PubMed]
4. Squeo, G.; Silletti, R.; Mangini, G.; Summo, C.; Caponio, F. The potential of Apulian olive biodiversity: The case of Oliva Rossa virgin olive oil. *Foods* **2021**, *10*, 369. [CrossRef] [PubMed]
5. Sabetta, W.; Mascio, I.; Squeo, G.; Gadaleta, S.; Flamminii, F.; Conte, P.; Di Mattia, C.; Piga, A.; Caponio, F.; Montemurro, C. Bioactive potential of minor Italian olive genotypes from Apulia, Sardinia and Abruzzo. *Foods* **2021**, *10*, 1371. [CrossRef] [PubMed]

6. Farris, S.; Buratti, S.; Benedetti, S.; Rovera, C.; Casiraghi, E.; Alamprese, C. Influence of two innovative packaging materials on quality parameters and aromatic fingerprint of extra-virgin olive oils. *Foods* **2021**, *10*, 929. [CrossRef] [PubMed]
7. Grassi, S.; Jolayemi, O.; Giovenzana, V.; Tugnolo, A.; Squeo, G.; Conte, P.; De Bruno, A.; Flamminii, F.; Casiraghi, E.; Alamprese, C. Near infrared spectroscopy as a green technology for the quality prediction of intact olives. *Foods* **2021**, *10*, 1042. [CrossRef] [PubMed]
8. Lia, F.; Vella, B.; Zammit Mangion, M.; Farrugia, C. Application of 1H and 13C NMR fingerprinting as a tool for the authentication of Maltese extra virgin olive oil. *Foods* **2020**, *9*, 689. [CrossRef]
9. Paciulli, M.; Difonzo, G.; Conte, P.; Flamminii, F.; Piscopo, A.; Chiavaro, E. Physical and thermal evaluation of olive oils from minor Italian cultivars. *Foods* **2021**, *10*, 1004. [CrossRef]
10. Paradiso, V.; Longobardi, F.; Fortunato, S.; Rotondi, P.; Bellumori, M.; Cecchi, L.; Cosma, P.; Mulinacci, N.; Caponio, F. Paving the way to food grade analytical chemistry: Use of a natural deep eutectic solvent to determine total hydroxytyrosol and tyrosol in extra virgin olive oils. *Foods* **2021**, *10*, 677. [CrossRef] [PubMed]
11. Flamminii, F.; Di Mattia, C.; Sacchetti, G.; Neri, L.; Mastrocola, D.; Pittia, P. Physical and sensory properties of mayonnaise enriched with encapsulated olive leaf phenolic extracts. *Foods* **2020**, *9*, 997. [CrossRef] [PubMed]
12. Conte, P.; Pulina, S.; Del Caro, A.; Fadda, C.; Urgeghe, P.; De Bruno, A.; Difonzo, G.; Caponio, F.; Romeo, R.; Piga, A. Gluten-free breadsticks fortified with phenolic-rich extracts from olive leaves and olive mill wastewater. *Foods* **2021**, *10*, 923. [CrossRef] [PubMed]
13. Perito, M.; Coderoni, S.; Russo, C. Consumer attitudes towards local and organic food with upcycled ingredients: An Italian case study for olive leaves. *Foods* **2020**, *9*, 1325. [CrossRef] [PubMed]
14. Carzedda, M.; Gallenti, G.; Troiano, S.; Cosmina, M.; Marangon, F.; de Luca, P.; Pegan, G.; Nassivera, F. Consumer preferences for origin and organic attributes of extra virgin olive oil: A choice experiment in the Italian market. *Foods* **2021**, *10*, 994. [CrossRef] [PubMed]
15. Posadino, A.; Cossu, A.; Giordo, R.; Piscopo, A.; Abdel-Rahman, W.; Piga, A.; Pintus, G. Antioxidant properties of olive mill wastewater polyphenolic extracts on human endothelial and vascular smooth muscle cells. *Foods* **2021**, *10*, 800. [CrossRef]
16. Centrone, M.; D'Agostino, M.; Difonzo, G.; De Bruno, A.; Di Mise, A.; Ranieri, M.; Montemurro, C.; Valenti, G.; Poiana, M.; Caponio, F.; et al. Antioxidant efficacy of olive by-product extracts in human colon HCT8 cells. *Foods* **2021**, *10*, 11. [CrossRef]
17. Benalia, S.; Falcone, G.; Stillitano, T.; De Luca, A.; Strano, A.; Gulisano, G.; Zimbalatti, G.; Bernardi, B. Increasing the content of olive mill wastewater in biogas reactors for a sustainable recovery: Methane productivity and Life Cycle Analyses of the process. *Foods* **2021**, *10*, 1029. [CrossRef] [PubMed]
18. Nikkhah, A.; Firouzi, S.; Dadaei, K.; Van Haute, S. Measuring circularity in food supply chain using Life Cycle Assessment; refining oil from olive kernel. *Foods* **2021**, *10*, 590. [CrossRef]
19. Pampuri, A.; Casson, A.; Alamprese, C.; Di Mattia, C.; Piscopo, A.; Difonzo, G.; Conte, P.; Paciulli, M.; Tugnolo, A.; Beghi, R.; et al. Environmental impact of food preparations enriched with phenolic extracts from olive oil mill waste. *Foods* **2021**, *10*, 980. [CrossRef]

Article

Distinct Particle Films Impacts on Olive Leaf Optical Properties and Plant Physiology

Annalisa Rotondi [1], Lucia Morrone [1,*], Osvaldo Facini [1], Barbara Faccini [2], Giacomo Ferretti [2] and Massimo Coltorti [2]

[1] Institute for the Bioeconomy, Italian National Research Council, via P. Gobetti 101, 40129 Bologna, Italy; annalisa.rotondi@ibe.cnr.it (A.R.); osvaldo.facini@ibe.cnr.it (O.F.)
[2] Department of Physics and Earth Science, University of Ferrara, via Saragat 1, 44122 Ferrara, Italy; fccbbr@unife.it (B.F.); frrgcm@unife.it (G.F.); massimo.coltorti@unife.it (M.C.)
* Correspondence: lucia.morrone@ibe.cnr.it

Citation: Rotondi, A.; Morrone, L.; Facini, O.; Faccini, B.; Ferretti, G.; Coltorti, M. Distinct Particle Films Impacts on Olive Leaf Optical Properties and Plant Physiology. *Foods* **2021**, *10*, 1291. https://doi.org/10.3390/foods10061291

Academic Editors: Cristina Alamprese, Emma Chiavaro and Francesco Caponio

Received: 27 April 2021
Accepted: 31 May 2021
Published: 4 June 2021

Publisher's Note: MDPI stays neutral with regard to jurisdictional claims in published maps and institutional affiliations.

Copyright: © 2021 by the authors. Licensee MDPI, Basel, Switzerland. This article is an open access article distributed under the terms and conditions of the Creative Commons Attribution (CC BY) license (https:// creativecommons.org/licenses/by/ 4.0/).

Abstract: The olive fruit fly is worldwide considered a major harmful pest of the olive agroecosystem. In Italy, the fruit fly infestation is traditionally countered by spraying chemical insecticides (e.g., dimethoate), but due to the recent ban of dimethoate by the Reg EU2019/1090 and the increasing awareness of consumers of food sustainability, the interest in developing chemical-free alternatives to pesticides, such as the use of particle-films, is rising. A field experiment was conducted to assess the effect of different particle films (kaolin-base and zeolitite-base) on leaf gas exchanges and leaf optical properties. Results showed that with the dust accumulation on the leaves' surface, photosynthesis, stomatal conductance, transpiration and water use efficiency were significantly lower in kaolin-treated olive trees compared to those treated with zeolitite and to the control, while olive trees treated with zeolitite showed physiological parameters similar to the untreated plants. Microstructural differences of different particle film on the leaf and olive surfaces emerged by ESEM observations also influenced leaf optical properties. Oils produced by zeolitite-treated plants show higher intensities of gustatory and olfactory secondary flavors compared to kaolin and test oils.

Keywords: *Olea europaea*; kaolin; zeolitite; foliar treatments; sustainable agriculture; crop defense

1. Introduction

The olive fruit fly (*Bactrocera oleae*) is worldwide considered a major harmful pest of the olive agroecosystem. Under certain environmental conditions (high humidity and precipitations and temperature below 28–30 °C), the fruit fly is responsible for large infestations that seriously compromise olive yield and oil quality [1].

The many olives strongly attacked by flies produce oxidized oils with a reduced quantity of phenolic substances, which therefore are unlikely to live up to the EFSA health claim [2]. In Italy, the fruit fly infestation is traditionally countered by spraying chemical insecticides such as dimethoate (in integrated regime) or by applying organic formulations (organic farming) [3]. However, taking into account the recent ban of dimethoate [4] and the increasing awareness of consumers of food sustainability, the interest in developing natural and chemical-free alternatives to pesticides, such as organic agrochemicals or the use of geologic material as particle film, is rising [5].

Agronomic practices are also one of the keys to allow the development of extra virgin olive oil (EVOO) market niches, guaranteeing high and constant quality standards [6].

The spraying of "rock dust" (e.g., kaolin) as foliar treatment in organic agriculture to reduce the negative impact of environmental stresses and to protect fruits from insect pests is a well-established approach [7]. Kaolinite ($Al_2Si_2O_5(OH)_4$) is an aluminium–silicate clay mineral composed of a layered silicon-oxygen tetrahedron and a layered aluminium–oxygen octahedron [8,9]; the commercial term "kaolin" refers to a rock whose percentage of kaolinite is higher than 50% [10]. In kaolin, kaolinite is often associated with other minerals such as quartz, feldspar and various phyllosilicates (such as muscovite and illite) [11]. Contrary to other clay minerals, such as smectites, kaolinite is characterized by a relatively low cation exchange capacity (CEC) (0.38 meq/g) [12]. The size of kaolinite particles can reach a colloidal level after milling and grinding during mineral processing [13].

Similarly, natural zeolites represent another geologic material that can be used as particle films for crop protection [14].

Zeolites are crystalline aluminosilicates composed of a 3D framework of linked $[SiO_4]^{4-}$ and $[AlO_4]^{5-}$. The framework delimits open cavities in the form of channels and cages in which H_2O molecules and extra-framework cations can be reversibly exchanged. The most important properties of zeolite minerals are (i) high cation-exchange capacity, (ii) reversible dehydration and (iii) molecular sieve. Nowadays, more than sixty types of natural zeolites have been described by researchers (http://www.iza-online.org/natural/default.htm (accessed on 15 March 2021)), each differing in terms of framework structure, mineral chemistry and ion exchange capacity, but only a few occur in sufficient amounts and purity to be considered as exploitable natural resources [15]. Among them, clinoptilolite is the most frequent and abundant sedimentary zeolite in nature, followed, in the order, by mordenite > chabazite > phillipsite > erionite [16]. Natural zeolites are often constituents of volcanic tuffs [17]; thus, the term "natural zeolites" is inappropriate from a geological perspective and it should be substituted by rocks or tuffs rich in zeolite. Analogously to kaolin, if the zeolite content is greater than 50%, the rock can be classified as "zeolitite", specifying the main zeolite constituent (e.g., chabazitic-zeolitite) [16].

Chabazite zeolite (CHA), although less abundant than clinoptilolite, is particularly attractive for agricultural and industrial applications because of its very high CEC (3.84 meq/g) and easiness in sorption and subsequent release of NH_4^+ ions [18,19]. The "honeycomb" framework of zeolite minerals, together with their carbon dioxide sorption and heat stress reduction capacity, makes them suitable as leaf coating products. Furthermore, their reversible dehydration makes them effective against fungal disease and insect pests [14]. Zeolite tuffs are most commonly used in agricultural practices as a soil amendment and for improving the nitrogen use efficiency (NUE) by crops because of their high affinity with NH_4^+ ions [20,21]. Recently, Italian CHA-zeolitite was used as a soil amendment in a long-term field experiment [22–24]. Laboratory incubations highlighted the positive effects of CHA-zeolitite on soil N and C gaseous emissions and microbial biomass [25,26]. The same rock was used for removing N and Na from animal liquid manure and low-quality irrigation waters, with promising results [27–29].

Studies on the use of powders for contrasting olive fly are fairly recent, and showed that kaolin application on the olive fruit fly significantly reduced the percentage of infested olives [30,31].

Rumbos et al. [32] studied the insecticide potential of zeolite formulations against stored grain insects but, to the best of our knowledge, zeolite tuffs have not yet been studied as a defense tool against the olive fly.

Regions characterized by arid climate and low rainfall regimes are the most suitable for this technology due to the reduced temperature of the leaves and the wash-off risk for the particle films. High rainfall regimes may lead to the necessity of multiple applications, increasing the costs (for materials and manpower) and hence significantly decreasing the attractiveness of the methodology [33].

Besides the effectiveness against the fruit fly, it is crucial to understand if the particle films interfere with the physiological activity of the plants, as the literature shows contrasting evidence on this subject. Some authors reported that kaolin film causes a reduction in leaf temperature, transpiration and water use efficiency (WUE) in soybean plants [34], as well as in apple leaves [35]. Contrarily, Jifon and Syvertsen [36] reported that the WUE of the kaolin-treated citrus leaf was higher than untreated leaves because photosynthesis was increased without an increase in leaf transpiration. In apple trees, the lower leaf temperature of kaolin-treated plants increased photosynthesis and stomatal conductance [37].

As mentioned above, the effect of zeolitite particle film on plant physiology is mostly unknown due to its recent application in agriculture. Besides the reduction of heat stress, zeolitites may also be used to reduce water stress. The adsorption selectivity of zeolites for water is greater than any other minerals [38], leading to an adsorption capacity that may reach up to 30% of the zeolite weight without any volume modification, depending on the zeolite type [39]. Thanks to these properties, together with the relatively low-cost and high abundance, zeolite attractiveness for agricultural utilizations has recently risen, overcoming that of kaolin.

According to Reddy et al. [40], the application of particle films over the stomata is known to increase resistance to water vapor losses. Moreover, particulate sprays modify the leaf optical properties, increasing foliage reflectivity and modifying plant physiological processes such as photosynthesis, morphogenesis and water balance [41].

The olive leaves are covered by trichomes, which may directly influence the diffusion boundary layer of the leaf surface, increase leaf reflectance for all wavelengths of solar radiation between 400 and 300 nm and restrict radiation absorbance, resulting in a reduction of the leaf load [42].

The experiments presented here were carried out in order to test the effectiveness of different particle films in a cold and humid environment, typical of northern Italy, where the olive fly attack is increasingly worrying. Here, small-scale, high quality olive oil production is carried out on the Emilia-Romagna Appennine hillsides.

In addition, this study aims to evaluate and compare the effects of two different particle films (kaolin and zeolitite) on leaf optical properties, leaf gas exchange and on the incidence of the olive fruit fly attack. ESEM observations allowed us to investigate the microstructural differences of the particle film on leaf and olive surfaces. Olive fruit analyses and sensory characterization of olive oils produced by the different treatments were also performed, in order to establish if the influence of foliar application on the ecophysiological parameters could affect harvest quality.

2. Materials and Methods

2.1. Treatments and Sampling

The study was carried out in 15-year-old commercial olive (*Olea europeae*) cv Correggiolo plants located in Bologna hills (Italy). One third of the olive orchard was submitted to kaolin treatment (K), 1/3 to zeolitite treatment (Z) and in the last 1/3 of orchard no applications were made (T). Two olive trees for each thesis were chosen, four branches for each tree were marked in different cardinal points, and for each branch three leaves were sampled. Twenty-four leaves for each thesis were considered for physiological, optical, ESEM and color leaf measurements. The tested treatments were:

(1) K: foliar application of kaolin at a dosage of 3.0 kg/100 L of H_2O;
(2) Z: foliar application of CHA-zeolitite at a dosage of 0.6 kg/100 L of H_2O;
(3) T: control (untreated).

The kaolin and the CHA-zeolitite were supplied by Balco s.p.a company. The mineralogical composition of both products is reported in Supplementary Table S1.

The tested application dosages were chosen according to the guidelines provided by the producer. Kaolin and CHA-zeolitite were applied by covering the total foliage using a mounted sprayer (flow max 50 L/min, capacity 200 l-Idromeccanica Bertolini-Reggio Emilia Italy) equipped with a handgun sprayer and testing different nozzle diameters. The average particle size of both kaolin and CHA-zeolitite was 6–10 µm.

The foliar applications started at the beginning of the summer, when olive fruits were developed enough to be attacked by *Bactrocera oleae*, and applications were repeated approximately every 20 days (13 June, 3 July, 21 July, 17 August, 5–12–19–29 September 2019), the applications were repeated after heavy precipitations (September) to guarantee sufficient coverage until the end of the growing season. Conventional orchard agronomic practises, pruning and winter treatment based on Bordeaux mixture, were applied for all thesis. Environmental temperatures and rainfalls were monitored using a weather station IRDAM WST 7000 C (IRDAM SA, Yverdon-les-Bains, Suisse).

50 g of leaves were randomly sampled from each olive plant to carry out elemental and isotopic analysis of C and N to check for possible differences in C-N composition between the studied plants. Once in the laboratory, the leaves were washed with deionized water, dried for three days at 60 °C and then ground to a fine powder. Additionally, to gain information on the soil environment, soil samples from the first 0.3 m depth were collected using a manual auger (Eijkelkamp). To address spatial variability, three logs per plant were mixed to form a global sample; each one was then sieved at 5 mm and air-dried before further analyses.

2.2. Environmental Electronic Microscope (ESEM) Observations

Leaf and fruit samples treated with different particle films were collected during the study according to the methodology reported by Lanza and Di Serio [43]. Samples were observed by ESEM (Zeiss, EVO LS 10, Oberkochen, Germany).

To assure a homogeneous distribution of the particle films on the olive surface, preliminary observations were carried out by ESEM. Generally, obtaining good coverage is mandatory when using non-systemic products, such as zeolitite or kaolin. This is because only the "covered" areas of the canopy surface are protected [44]. To this aim, the droplet size distribution during atomization is very important because it affects the biological activity and the spray drift [45]. Study by Skuterud et al. [46] showed that, when applying contact products such as zeolitite, it is important to use fine (60 µm) or medium-sized (60–200 µm) droplets. The final coverage is also affected by the spray type: high application volumes can result in product run-off, which leads to considerable losses. On the other hand, low volume spraying leads to very poor coverage of the leaf surface and hence loss of efficacy [45]. Considering also the lower concentration of zeolite compared to the concentration of kaolin it was necessary to identify the right diameter of nozzles to guarantee a homogenous coating. This was achieved through several ESEM observations and measurements of the distance between the crystals (Figure 1A). These observations and measurements have confirmed that good coverage was achieved when spraying CHA-zeolitite utilizing a handgun sprayer with 0.2 mm diameter nozzles. These nozzles cause a dispersion of the product characterized by a distance among crystals under tenths of millimetres which is far smaller than the area interested by oviposition puncture of *Bactrocera oleae* (triangular slot of 1–1.5 millimetres long).

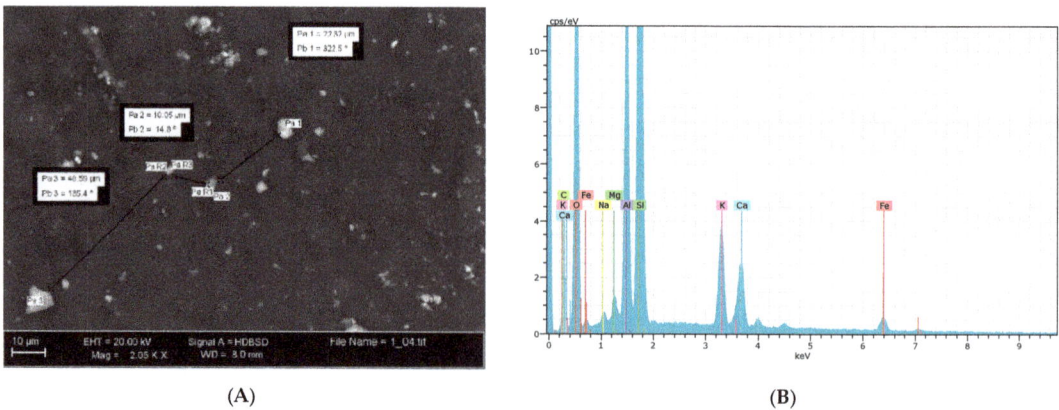

Figure 1. ESEM observations of CHA-zeolitite leaf coating to check the distribution protocol. (**A**) Measurements of the distance between CHA-zeolitite particles; (**B**) CHA-zeolitite particles' composition by EDS-microanalysis.

To establish the exact nature of the observed particles, semi-quantitative EDS (energy dispersive spectroscopy) microanalysis systems were carried out randomly to determine particles' composition (Figure 1B).

2.3. Chemical Analysis on Leaves and Soil Samples

A total of 50 g of leaves and 500 g of soil samples were analysed for total C and N and the relative isotopic signature (δ^{13}C and δ^{15}N) with an Elementar Vario Micro Cube Elemental Analyzer (EA) in line with an ISOPRIME 100 Isotopic RatioMass Spectrometer (IRMS) operating in continuous-flow mode (Elementar Analysensysteme GmbH, Langenselbold, Germany). Soil samples were additionally processed for X-ray fluorescence (XRF) analysis on powder pellets, using a wavelength-dispersive automated ARL Advant'X spectrometer (Thermo Electron SA, Ecublens, Switzerland). The organic matter of soil samples was measured by quantifying the weight loss after combustion at 550 °C.

2.4. Ecophyisiological, Optical Properties and Color Leaf Measurements

Leaf gas exchange measurements: photosynthesis (A), stomatal conductance (g), intercellular CO_2 concentration (Ci), transpiration rate (E) and intrinsic water use efficiency calculated as the ratio of photosynthesis rate to transpiration rate (WUE), were measured during a clear sky using a Li-Cor portable photosynthesis system (LiCor 6400, Lincoln, NE, USA) operating at 400 μmol m^{-2} s^{-1} flow rate. Measurements were taken in the morning (10:00 a.m to 12:00 p.m.), according to the protocols of Denaxa et al. [47] and Jifon and Syvertsen [36], on undamaged mature sun leaves located at the central part of the one-year-old shoot of the marked branches, according to Larbi et al. [48].

Total directional-hemispherical reflectance of the upper and lower leaf surface was measured with a calibrated spectroradiometer LiCor 1800 (Li-Cor, Nebr, Lincoln, NE, USA) able to scan from 300 to 1100 nm connected to a Li-Cor 1800-12 integrating sphere. To prevent spectral changes due to water losses and metabolic modification, spectral measurements were made immediately after the leaves were picked, according to Baldini et al. [42].

Leaves' colour was measured on the upper surface of one-year leaf using a Konica Minolta CR-400 Chroma Meter (Konica Minolta, Inc., Osaka, Japan) calibrated with a standard white plate at room temperature. The data collected were L* (lightness) and a* (red-green scale) recorded at three random locations on each leaf on twenty leaves collected from the olive trees submitted to different treatment (T, Z and K).

All leaf f measurements (ecophyisiological, optical properties and color surface) were carried out on 8 July, 24 August and 20 September 2019.

2.5. Olive Analyses and Olive Oils Sensory Evaluation

Considering that the optimal ripening index (RI) for the Correggiolo cultivar is included in the range 2–2.5 of the Jaén index [49], the RI was monitored for each treatment according to the method developed by the Agronomic Station of Jaén defining the RI as a function of fruit colour in both skin and pulp [50]. On the same samples, each olive fruit was examined for the presence of *Bactrocera oleae* infestation, dissecting the fruits to determine the percentage of total infestation (egg, larva or pupa, sting scar, exit holes). Olive water content was gravimetrically determined placing olive samples in oven at 60 °C for 8 days. Olive firmness was determined using a *penetrometer* (PCE-FM 200, PCE Group, Lucca Italy); it was measured at two points on each fruit, and the average readings were reported in g/mm^2 as exerted pressure.

The total production of the selected trees for each treatment was handpicked; an amount of 50 kg was transformed into oil. Olives were defoliated, washed and milled using a low scale continuous mill (Oliomio®; Toscana Enologica Mori, Firenze, Italy) equipped with blade crusher, horizontal malaxator and a two-phase decanter. Olive samples were processed within 24 h of harvest. For each sample the technological settings (temperature (below 27 °C) and the time of malaxation (20 min), the speed of the decanter (4200 rpm) and the flux of water in the separator (0.8 L h^{-1})) were standardized in order to minimize the variability due to the extraction procedures. Oil samples were filtered through cotton filters, poured into dark glass bottles, keeping the headspace to a minimum, and stored in a temperature-controlled cupboard set at 15 ± 1 °C until analysis.

Sensory analyses were carried out by a fully-trained analytical taste panel recognized by the International Olive Oil Council (IOOC) of Madrid and by the Italian Ministry of Agricultural, Food and Forestry Policies. The panel evaluated all oil samples following an incomplete randomized block design. Olive oil samples were placed in blue tasting glasses and the temperature of samples was kept at 15–18 °C. A panel test was established for the present study using a standard profile sheet (IOOC/T20) modified by IBIMET-CNR [51] that allows the obtaining of a more complete description of the organoleptic properties of the oils. The tasters evaluated direct or retronasal aromatic olfactory sensations (olive fruity, green/leaf and secondary positive flavours), gustatory sensations (olive fruity, bitterness and secondary positive flavours) and tactile/kinesthetic sensation (pungency). The tasters had to rate the intensity of the different descriptors on a continuous 0–10 cm scale. Values of the median of sensory data and robust standard deviation were calculated.

2.6. Statistical Analysis

The data collected were elaborated using Microsoft® Excel 2007/XLSTAT© (Version 2009.3.02, Addinsoft, Inc., Brooklyn, NY, USA). The significant differences among means at a 5% level were determined by ANOVA followed by a Tukey's Honestly Significant Difference (HSD) test. Principal component analysis (PCA) has been performed to explore data distribution patterns using physiological data.

3. Results and Discussion

3.1. ESEM Observations

Particles of both treatments (K and Z) were more homogenously distributed on the leaves' surface rather than on the surface of the olive. This higher attachment onto the leaves' surface is due to their peculiar morphology, characterized by overlapped stellar trichomes, particularly frequent on the lower surface (Figure 2A).

Since the first foliar application, a good distribution of both K and Z products was observed on the upper surface of the leaves, compared to the test which lacked particles on its surface (Figure 2B). In K treatment, kaolin appeared as a continuous layer and it was not possible to recognize the underlying star hairs (Figure 2C), while in Z treatment, the CHA-zeolitite film was more discontinuous and star hairs were still recognizable (Figure 2D). The same differences were also noted on the lower surface of the leaves.

This difference in the surface coverage is attributable to both higher amounts of kaolin sprayed at each application compared to the CHA-zeolitite and to the different morphology of kaolin and CHA-zeolitite particles (lamellar vs. pseudo-cubic).

Due to the particular morphology of the olive, characterized by a smooth and curved surface, the adhesion of the particles was less uniform than that observed on the leaves. This difference was observed from the first application and increased in subsequent applications, thanks to the accumulation of the deposited kaolin and CHA-zeolitite particles.

On the surface of the untreated olives (T), the epicuticular waxes arranged in crystalloid structures (membranous platelets) were well recognizable (Figure 3A), as observed in Carboncella olives by Lanza and Di Serio [43]. Micro-changes of epicuticular waxes, which occur with the progressing of ripening [43], were well visible in the T olives while in K and Z olives these micro-changes were hidden by particle accumulation, especially at the end of the experimentation.

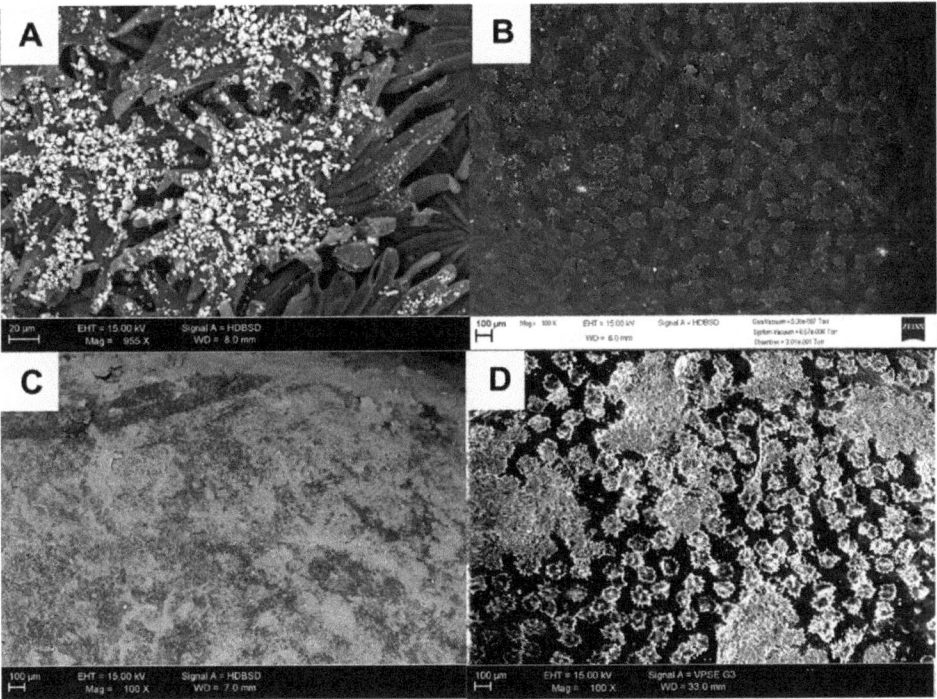

Figure 2. ESEM observations of treated and untreated olive leaves. (**A**) Accumulation of CHA-zeolitite on stellar trichomes that cover the lower surface of the olive leaves; (**B**) Upper surface of test olive leaf; (**C**) Olive leaf treated with kaolin; (**D**) Olive leaf treated with CHA-zeolitite.

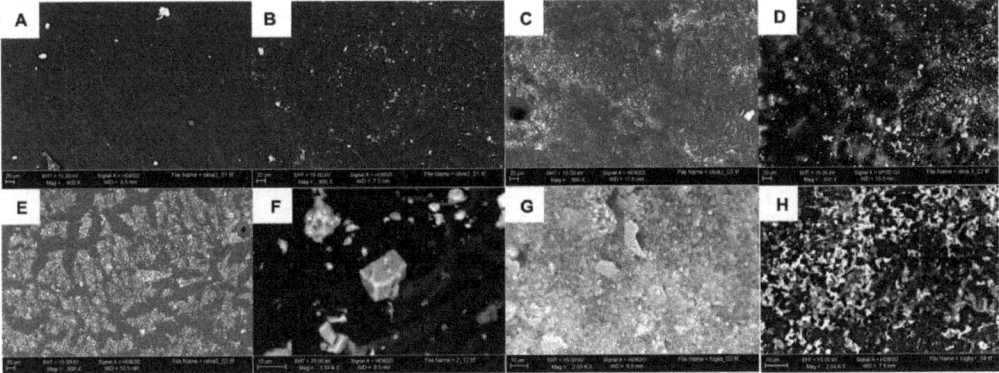

Figure 3. ESEM observations of treated and untreated olives. (**A**) Untreated (T) olive surface; (**B**) Olive surface treated with CHA-zeolitite (Z treatment, first application); (**C**) Olive surface treated with kaolin (K treatment, first application); (**D**) Olive surface treated with CHA-zeolitite (Z treatment, last application); (**E**) Olive surface treated with kaolin (K treatment, last application); (**F**) Morphology of CHA-zeolitite particles; (**G**) Morphology of kaolin particles; (**H**) Olive surface where kaolin appears to be incorporated by waxes.

As reported for the leaves, olives coverage was greater and more homogeneous in K treatment compared to Z treatment (Figure 3B,C); these differences were accentuated due to the accumulation of particles as the experiment progressed (Figure 3D).

ESEM observations after light rain events highlighted the tendency of K microaggregates to disperse and to form a continuous layer (macroscopically visible) on the surface of leaves and olives. With the growth of the olive tissues, the continuous K layer tends to fissure, leaving some areas uncovered (Figure 3E). The different aspect of K and Z films is linked to the specific morphology of the kaolin and CHA-zeolitite particles. CHA-zeolitite particles are mainly pseudo-cubic (Figure 3F) [52] while Kaolin is shaped as sheets/lamellae/irregular flakes (Figure 3G) [53].

Sample observations after heavy rain events showed that both K and Z coatings were well preserved, with the difference that the CHA-zeolitite particles kept their original shape and "anchored" themselves to the waxes, whereas those of kaolin appeared incorporated in waxes (Figure 3H); the same results were observed in apples treated with kaolin [54].

3.2. Chemical Analysis on Leaves and Soil Samples

The chemical composition of the soil between the various treatments was very similar in terms of soil organic matter, total N and C, major and trace elements (Supplementary Tables S2 and S3). The study area can be thus considered homogeneous in terms of soil chemistry and N availability to plants. Also, no significant differences were accounted in terms of total C, N and relative isotopic signature of the leaves at the end of the experimentation (Table 1).

Table 1. Results of leaves analysis (EA-IRMS) from each experimental plant treated with kaolin (K), CHA-zeolitite (Z) and the control (T). TN and TC are the total nitrogen and carbon content measured by EA analysis, $\delta^{15}N$ and $\delta^{13}C$ are the isotopic signatures expressed as delta notation by IRMS. Values are expressed as the mean of three replicates ± standard deviation. The same letters in the same column express no significant differences ($p > 0.05$) as results of ANOVA and Tukey's (HSD) tests.

Treatment	TN (%)	TC (%)	$\delta^{15}N$ (‰)	$\delta^{13}C$ (‰)
T	1.47 ± 0.02 a	45.51 ± 1.12 a	−3.71 ± 2.06 a	−27.76 ± 0.28 a
K	1.69 ± 0.17 a	47.71 ± 2.03 a	−0.53 ± 1.49 a	−28.59 ± 0.42 a
Z	1.44 ± 0.36 a	47.05 ± 3.39 a	−2.06 ± 1.71 a	−28.43 ± 0.37 a

3.3. Ecophysiological Parameters and Optical Properties

After the first two foliar applications, no significant changes in the photosynthetic rate (A) were observed between the treatments (Table 2). After the 7th application (20th September), a significant decrease of A and stomatal conductance (g) in K plants was observed (Table 2): K plants showed photosynthesis and stomatal conductance values 27% and 55% lower than those of the test plants, respectively. Similar results were found in bean plants by Tworkoski [55], whereas Jifon and Syvertsen [36] observed that the increasing leaf whiteness after kaolin sprays on grapefruit reduced the leaf temperature and increased stomatal conductance and net CO_2 assimilation rates. At first measurements (8th July), plants belonging to K, Z and T treatments did not show any difference in leaf transpiration (E) but, as the treatments continued (increasing particle accumulation), E decreased significantly in K plants and consequently, the WUE was significantly higher (Table 2). The effect of the kaolin accumulation on physiological parameters is possible to see in the PCA analysis where a clustering of the last date of the kaolin treatment occurs (Supplementary Figure S1). Similar results were observed by Jifon and Syvertsen [36] where WUE in kaolin sprayed leaves of grapefruits was 25% higher than that of control leaves.

Table 2. Ecophysiological parameters measured after each foliar application of K (kaolin), Z (CHA-zeolitite), and T (control). Data are presented as mean ± standard deviation. Different letters (a,b,c) indicate significant differences according to ANOVA and Tukey's HSD test ($p < 0.05$) at each application date.

Application Date	Treatment	A [1] µmol CO_2 $m^{-2} s^{-1}$	G [2] mmol $m^{-2} s^{-1}$	Ci [3] µmol CO_2 mol air	E [4] mol H_2O $m^{-2} s^{-1}$	WUE [5]
8 July	K	13.03 ± 0.75 a	0.39 ± 0.02 a	321.98 ± 4.05 a	9.68 ± 0.46 a	1.38 ± 0.09 a
	T	13.14 ± 1.05 a	0.32 ± 0.02 a	309.49 ± 4.89 a	9.82 ± 0.48 a	1.34 ± 0.09 a
	Z	12.14 ± 0.85 a	0.33 ± 0.04 a	309.36 ± 6.20 a	9.06 ± 0.71 a	1.42 ± 0.11 a
24 August	K	9.98 ± 0.64 a	0.20 ± 0.02 b	287.65 ± 5.02 b	8.01 ± 0.61 b	1.28 ± 0.07 a
	T	11.93 ± 0.78 a	0.28 ± 0.03 a,b	295.35 ± 4.57 b	10.28 ± 0.67 a	1.18 ± 0.06 a
	Z	12.35 ± 0.59 a	0.34 ± 0.02 a	312.77 ± 3.51 a	9.46 ± 0.58 a,b	1.37 ± 0.1 a
20 September	K	9.5 ± 0.55 b	0.12 ± 0.01 c	248.86 ± 6.94 c	2.91 ± 0.23 b	3.36 ± 0.15 a
	T	13.03 ± 0.49 a	0.28 ± 0.01 b	299.98 ± 2.06 b	5.55 ± 0.21 a	2.36 ± 0.06 b
	Z	12.19 ± 0.76 a	0.33 ± 0.01 a	317.69 ± 2.59 a	5.99 ± 0.20 a	2.02 ± 0.08 b

[1] A is the net photosynthetic rate; [2] g is stomatal conductance; [3] Ci is the intercellular CO_2 concentration; [4] E is the transpiration; [5] WUE is the water use efficiency calculated as the ratio of photosynthesis rate to transpiration rate.

No differences were observed in E and WUE between Z and T plants, while g was higher after the last two applications in the Z treatment (22 and 19%, respectively) without, however, influencing the photosynthetic rate (Table 2). Similar results were observed in soybean plants coated with kaolin, where the net radiation was reduced by 8% and short-wave irradiation was reduced by 20%, suggesting a potential reduction in transpiration and water use [34]. Also, Le Grange [56] reported a reduction in photosynthetic rates in kaolin sprayed leaves attributable to increased reflection and absorption of light reduced by 20–40%. Some authors [57,58] reported that kaolin treatment did not reduce the photosynthesis of single leaves but increased the photosynthesis of the whole canopy and therefore the productivity. In rainfed olive trees, Brito et al. [59] demonstrated that kaolin treatment counteracted the effect of water shortage and high light intensity on leaf sclerophyll and on stomatal density. Still in rainfed olive orchards, kaolin application contributed to keep a better water status by creating a specific microclimate around the leaves; moreover, it alleviated the adverse effect of summer stress through distinct physiological and biochemical responses [59].

In our study, the positive effect of kaolin was not observed because the olive trees are grown in environmental conditions (high rainfall and low temperatures) that do not lead to stress conditions; on the contrary, the abundant covering of the kaolin film had a negative effect on photosynthesis, that decreased during the delicate ripening phase of the olive fruits. Stomatal conductance and transpiration were also significantly reduced in K trees at the end of the experiment. This was probably the result of the abundant accumulation of kaolin on the leaf surfaces, leading to obstruction of stomata, with an alteration of leaf gas exchanges.

The authors are aware of the fact that, in these environmental conditions, a lower amount of kaolin or less frequent applications would have been sufficient (the concentration of kaolin was five-fold higher than that of CHA-zeolitite), but we aimed at reproducing the operative protocols commonly adopted for olive fruit fly defense. In the several Italian regions where olive cultivation is practiced, indeed, standardized protocols for protection from the olive fruit fly are used, regardless of the different climate conditions. A differentiation for kaolin-based treatments would be necessary and specific protocols should be developed for each different cultivation environment. These protocols must guarantee an adequate level of defense against fly attacks without significantly altering the physiological parameters of the plant.

Since the 4th application, a significant increase in Ci (CO_2 inside the lamina) was observed in Z plants compared to the other treatments. It has been reported that zeolites can adsorb carbon dioxide molecules and release them slowly into the environment; also, it has been suggested that when zeolites are spread on plant leaves, they may increase the amounts of CO_2 near the stomata, concomitantly increasing the photosynthesis rate [60]. In our experiment, however, we have observed no significant effect on the photosynthesis rate in Z plants. On the contrary, K leaves showed lower Ci that is in agreement with the observed decrease in A. Farquhar and Sharkey [61] indeed asserted that where CO_2 diffusion limits A, a decrease in Ci would also occur.

In our study, the upper and lower sides of K leaves showed a significant increase in reflectivity compared to the other treatments at all dates (Figure 4). The reflectance is the ability to reflect part of the incident light on a given surface and its effectiveness in reflecting radiant energy. Similar results were observed in grapefruit leaves coated with kaolin, which showed a higher reflectance compared to control leaves [36].

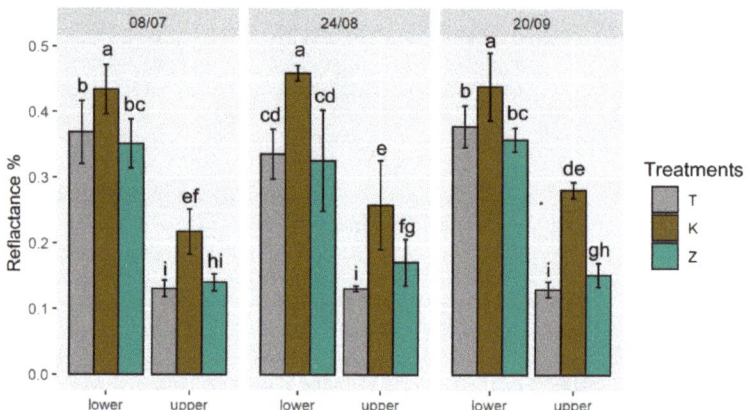

Figure 4. Mean reflectance between 400 and 700 nm measured on the lower and upper part of the leaves of T (control), K (kaolin) and Z (CHA-zeolitite) treatments. Error bars represent standard deviation. Different letters above the bars for each group of histograms indicate significant differences according to ANOVA and Tukey's HSD test ($p < 0.05$).

In Z leaves, the reflectance was similar to the T during the summer; only after the last application (29th September) did it increase (Figure 4). The different reflective capacity of the two films is attributable both to the different colour of the powders, white for kaolin (higher light reflectance) and light brown for CHA-zeolitite (lower light reflectance) and to the morphology of the particles, lamellar for kaolin (higher reflectance) vs. pseudo-cubic for CHA-zeolitite (lower reflectance). Furthermore, leaf reflectance data (Figure 4) showed that there is no difference when the number of treatments increases.

Colour measurement carried out on leaves treated with Kaolin showed a greater lightness (L*) compared to the test and CHA-zeolitite leaves at all dates (Figure 5). This difference in L* value between kaolin and CHA-zeolitite is due to the different conformation and colour of the kaolin (phyllosilicate) and natural zeolite (tectosilicate). After the 4th foliar application (24th August), the L* value of K leaves was greater by 22% and 19% with respect to those of T and Z leaves, respectively. Colour measurements performed on K leaves after the 3rd application showed lower L* values than the previous measurements, while no differences were accounted in T and Z treatments. The lower L* values observed in K treatment were probably due to kaolin leaching due to rainfall occurring during the previous days (Supplementary Figure S1). At the last measurement (performed on the 20th September), L* values were higher in both K and Z treatments compared to the T, suggesting that an accumulation of both kaolin and CHA-zeolitite on leaves occurred. Our data agrees with Jifon and Syvertsen's [36] measurements on grapefruit leaves treated with kaolin. Regarding the a* value, in the first two measurements T leaves showed higher values than those recorded in K and Z treatments, that were similar (Figure 6). It is interesting to note that contrary to L*, the a* value was not affected by the rainfall. After the 3rd measurement, a* decreased in all the treatments; this was probably caused by the leaf seasonality, since a reduction in the chlorophyll content in the leaf is expected at the end of the hot season [62].

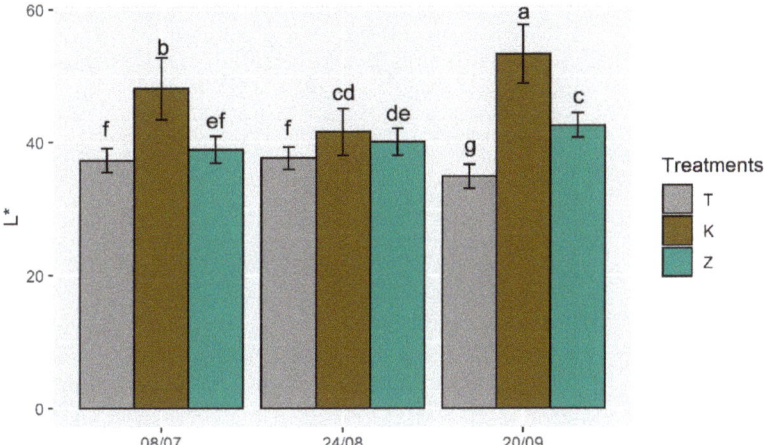

Figure 5. Changes in L*color values measured on the upper part of the leaves of T (control), K (kaolin) and Z (CHA-zeolitite) treatments. Error bars represent standard deviation. Different letters above the bars for each group of histograms indicate significant differences according to ANOVA and Tukey's HSD test ($p < 0.05$).

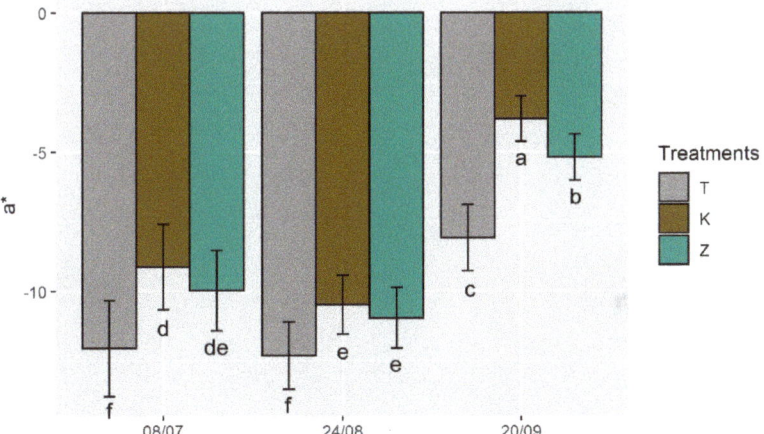

Figure 6. Changes in a*colour values measured on the upper part of the leaves of T (control), K (kaolin) and Z (CHA-zeolitite) treatments. Error bars represent standard deviation. Different letters above the bars for each group of histograms indicate significant differences according to ANOVA and Tukey's HSD test ($p < 0.05$).

3.4. Olive Analyses and Olive Oil Sensory Evaluation

In a year characterized by lower temperature (2019) (Supplementary Figure S2) with a high risk posed by the olive fruit fly, zeolite and kaolin sprays have significantly reduced the incidence of *Bactrocera oleae*; in fact, olives produced by Z and K olive trees present a decrease (over 40%) of infestation compared to control (Table 3). Water content of olive treated with kaolin was higher than the water content of olive treated with zeolitite while the olive from untreated trees showed similar value to both treatments (Table 3). The olive firmness did not statistically differ within the treatments, and the same results were observed in pear fruits treated with kaolin [63].

Table 3. Ripening index (RI), percentage of olive fruit fly infestation, water content and fruit firmness in olive from trees treated with K (kaolin), Z (CHA-zeolitite), and T (control). Data are presented as mean ± standard deviation. Different letters in the same column (a,b,c) indicate significant differences according to ANOVA and Tukey's HSD test ($p < 0.05$).

Treatment	RI	% Infestation	H$_2$O (%)	Firmness [1]
K	2.6	26	46.5 ± 1.4 a	55.0 ± 28.9
T	2.58	70	43.2 ± 0.6 b	52.7 ± 29.9
Z	2.48	34	44.7 ± 0.4 a,b	48.7 ± 28.8
p-value	/	/	0.038	ns

[1] express as g/mm^2.

The sensory profiles of olive oils extracted from plants treated with Kaolin (K), CHA-zeolitite (Z) and control (T) are shown in Figure 7. On a sensory level, the differences found in the oils were slight: Z and T olive oils showed a higher intensity of olfactory olive fruity than K olive oil. For the hint of bitterness K and T olive oil showed higher intensity than Z while for the hint of pungency test olive oil had higher intensity compared to K and Z olive oil. Test oil showed a lower intensity in olfactory secondary flavours while Z oil had a higher intensity in both olfactory and gustatory secondary flavours. Detailed examination of the pleasant flavours (Figure 8) revealed that, at the olfactory level, oils produced from both treatments had an artichoke scent and were perceived as fresher with respect to the oils produced by the test, that smelled of ripe tomato.

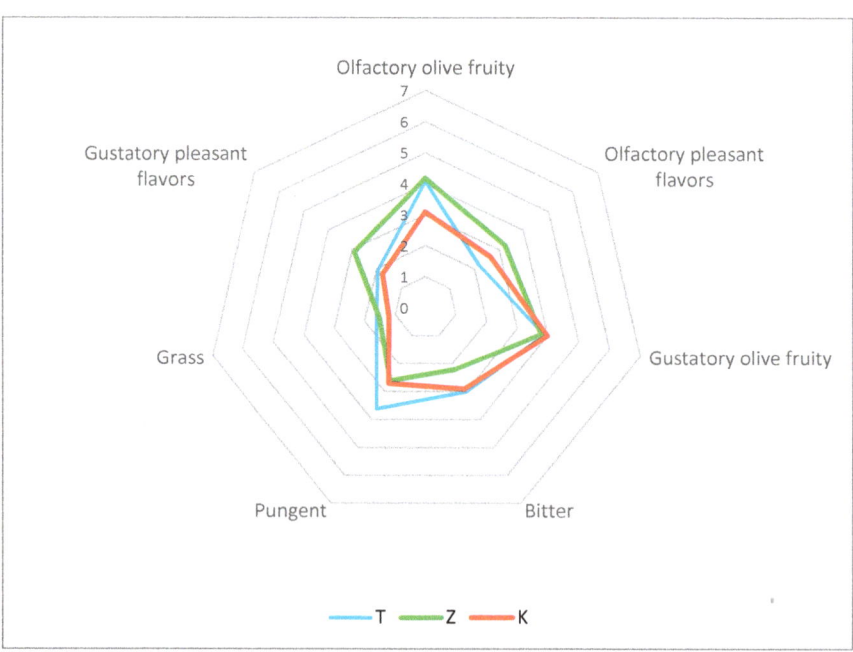

Figure 7. Sensory profiles of olive oils produced by plants treated with CHA-zeolitite (Z), Kaolin (K) and untreated plants (T).

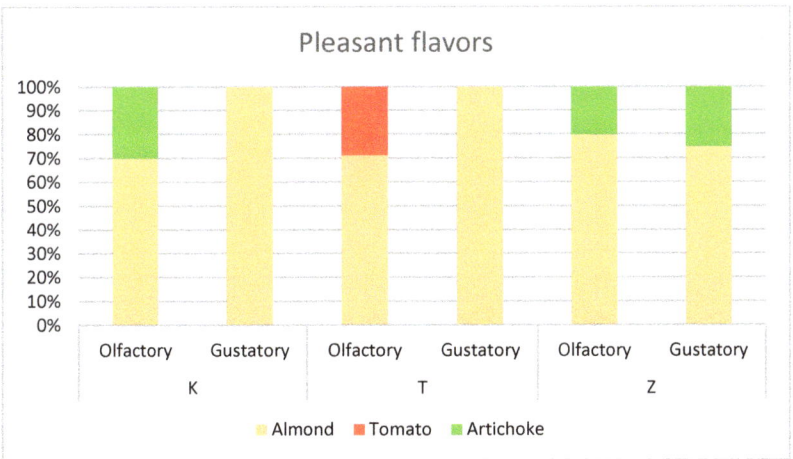

Figure 8. Pleasant flavors of olive oils produced by plants treated with CHA-zeolitite (Z), Kaolin (K) and untreated plants (T).

At the gustatory level, no differences were observed between the T and K oils, for which the tasters only perceived the hint of green almond. On the other hand, it is interesting to underline that the oil produced from olive trees treated with zeolite presented, in addition to the prevailing hint of almond, a note of artichoke which gave the oil a note of freshness compared to others.

De la Roca [30] found that kaolin application against the olive fruit fly significantly reduced the percentage of infested olives. Saour and Makee [31] showed that a kaolin-based particle film formulation significantly reduced fruit infestation levels; the authors hypothesized that adult flies may fail to recognize kaolin sprayed olive trees, and the gravid females are repelled due to the tactile unsuitable surface texture of particle film-treated olives.

4. Conclusions

In the scenario of sustainable and environmentally friendly olive oil production, both treatments represent valid alternatives to chemical insecticide. From an economic point of view, CHA-zeolitite represents an advantage because the recommended application rate is five times lower than that commonly used for kaolin. Moreover, CHA-zeolitite volcanic rocks abound in Central Italy and are already exploited for several purposes, including the production of micronized powder from the granular material resulting from building block cutting in quarries. CHA-zeolite supplying is thus relatively less impactful from an environmental point of view, with respect to other types of powders that are quarried and manufactured in foreign countries.

ESEM observation performed on leaf and olive surfaces highlighted microstructural differences between the two tested particle films which influenced some ecophysiological parameters. The intercellular CO_2 concentration was positively influenced by CHA-zeolitite application while kaolin application decreased photosynthesis, stomatal conductance and transpiration rates compared to the other foliar treatments. Therefore, in hot environments, the use of kaolin has the dual function of protecting the olive tree both from high temperatures and from the olive fly but the resulting impactful coating caused a reduction of photosynthesis that can, however, be compensated by an increase in WUE due to the reduced transpiration. The continuous layer of kaolin on leaf surface has also significantly influenced the leaf reflectance thanks to its crystal morphology, colour and application rate.

In a cold and humid environment (such as our experimental conditions), CHA-zeolitite was found to be the ideal compound because it exerted a protective effect against olive fruit fly attack, similar to kaolin, but left the leaf gas exchanges unaltered. Moreover, oils obtained from CHA-zeolitites showed higher intensities of gustatory and olfactory pleasant flavours than olive oils produced from kaolin and untreated trees, thus enhancing the quality and sustainability characteristics of this product.

Supplementary Materials: The following are available online at https://www.mdpi.com/article/10.3390/foods10061291/s1, Supplementary Table S1: Mineralogical composition (XRPD Rietveld-RIR method) of the kaolin and CHA-zeolitite supplied by Balco s.p.a and used in the experimentation. Data from the product's technical sheet supplied by the company; Supplementary Table S2: Results of soil analysis (oven-combustion, EA-IRMS) from each experimental plant treated with kaolin (K), CHA-zeolitite (Z) and the control (T); Supplementary Table S3: Results of soil analysis (2 replicates) by X-Ray Fluorescence (XRF) from each experimental plant treated with kaolin (K 1 and 2), CHA-zeolitite (Z 1 and 2) and the control (T 1 and 2); Supplementary Figure S1: PCA of the ecophysiological parameters measured after the foliar applications of K (kaolin), Z (CHA-zeolitite), and T (control); Supplementary Figure S2: Minimum, mean and maximum temperature (°C) and rainfall (mm) recorded in the period 1st Julay-31st October.

Author Contributions: Conceptualization, A.R.; methodology, L.M. and O.F.; formal analysis, L.M.; investigation, L.M., A.R. and O.F.; writing—original draft preparation, A.R.; writing—review and editing, L.M. and G.F.; supervision, B.F. and M.C.; project administration, A.R. All authors have read and agreed to the published version of the manuscript.

Funding: This research received no external funding.

Acknowledgments: The authors gratefully thank Matteo Mari for technical support, Barbara Alfei and Panel of ASSAM Marche for sensory analysis, the CNR-ISAC and CNR-ISP staff for the meteorological data and a special acknowledgment to Franco Corticelli for ESEM-EDX analysis support.

Conflicts of Interest: The authors declare no conflict of interest.

References

1. Haniotakis, G.E. Olive pest control: Present status and prospects. In Proceedings of the IOBC/WPRS Conference on Integrated Protection of Olive Crops, Chania, Greece, 29–31 May 2003.
2. European Commission. Regulation EC No. 432/2012 establishing a list of permitted health claims made on foods, other than those referring to the reduction of disease risk and to children's development and health. *Off. J. Eur. Union* **2012**, *L136*, 1–40.
3. Volpi, I.; Guidotti, D.; Mammini, M.; Petacchi, R.; Marchi, S. Managing complex datasets to predict *Bactrocera oleae* infestation at the regional scale. *Comput. Electron. Agric.* **2020**, *179*, 105867. [CrossRef]
4. European Commission. Commission Regulation (EU) No 1090/2019 of 26 June 2019. *Off. J. Eur. Union* **2019**, *173*, 39–41.
5. Daane, K.M.; Sime, K.R.; Wang, X.G.; Nadel, H.; Johnson, M.W.; Walton, V.M.; Kirk, A.; Pickett, C.H. *Psyttalia lounsburyi* (Hymenoptera: Braconidae), potential biological control agent for the olive fruit fly in California. *Biol. Control* **2008**, *44*, 79–89. [CrossRef]
6. Roselli, L.; Clodoveo, M.L.; Corbo, F.; De Gennaro, B. Are health claims a usefull tool to segment the category of extra-virgin olive oil? Threats and opportunities for the Italian olive oil supply chain. *Trends Food Sci. Technol.* **2017**, *68*, 176–181. [CrossRef]
7. Godfrey, L.; Grafton-Cardwell, E.; Kaya, H.; Chaney, W. Microorganisms and their byproducts, nematodes, oils and particle films have important agricultural uses. *Calif. Agric.* **2005**, *59*, 35–40. [CrossRef]
8. Brindley, G.W.; Robinson, K. Structure of kaolinite. *Nature* **1945**, *156*, 661–662. [CrossRef]
9. Zhu, X.; Zhu, Z.; Lei, X.; Yan, C. Defects in structure as the sources of the surface charges of kaolinite. *Appl. Clay Sci.* **2016**, *124*, 127–136. [CrossRef]
10. Dombrowsky, T. The origins of kaolinite—Implication for utlization. In *Science of Whiteware*; Carty, W., Ed.; Wiley: New York, NY, USA, 2000; pp. 3–12.
11. Andreola, F.; Castellini, E.; Manfredini, T.; Romagnoli, M. The role of sodium hexametaphosphate in the dissolution process of kaolinite and kaolin. *J. Eur. Ceram. Soc.* **2004**, *24*, 2113–2124. [CrossRef]
12. Kahr, G.; Madsen, F.T. Determination of the cation exchange capacity and the surface area of bentonite, illite and kaolinite by methylene blue adsorption. *Appl. Clay Sci.* **1995**, *9*, 327–336. [CrossRef]
13. Hu, Y.; Yang, Q.; Kou, J.; Sun, C.; Li, H. Aggregation mechanism of colloidal kaolinite in aqueous solutions with electrolyte and surfactants. *PLoS ONE* **2020**, *15*, e0238350. [CrossRef] [PubMed]
14. De Smedt, C.; Ferrer, F.; Leus, K.; Spanoghe, P. Removal of pesticides from aqueous solutions by adsorption on zeolites as solid adsorbents. *Adsorp. Sci. Technol.* **2015**, *33*, 457–485. [CrossRef]
15. Kesraoui-Ouki, S.; Cheeseman, C.R.; Perry, R. Natural zeolite utilisation in pollution control: A review of applications to metals' effluents. *J. Chem. Technol. Biotechnol.* **1994**, *59*, 121–126. [CrossRef]

16. Galli, E.; Passaglia, E. Natural zeolites in environmental engineering. In *Zeolites in Chemical Engineering*; Holzapfel, H., Ed.; Process Engineering GmbH: Vienna, Austria, 2007; pp. 392–416.
17. Delkash, M.; Bakhshayesh, B.E.; Kazemian, H. Using zeolitic adsorbents to cleanup special wastewater streams: A review. *Micropor. Mesopor. Mater.* **2015**, *214*, 224–241. [CrossRef]
18. Gualtieri, A.F.; Passaglia, E. Rietveld structure refinement of NH_4-exchanged natural chabazite. *Eur. J. Mineral* **2006**, *18*, 351–359. [CrossRef]
19. Mumpton, F.A. La roca magica: Uses of natural zeolites in agriculture and industry. *Proc. Natl. Acad. Sci. USA* **1999**, *96*, 3463–3470. [CrossRef]
20. Ferretti, G.; Di Giuseppe, D.; Natali, C.; Faccini, B.; Bianchini, G.; Coltorti, M. CN elemental and isotopic investigation in agricultural soils: Insights on the effects of zeolitite amendments. *Geochemistry* **2017**, *77*, 45–52. [CrossRef]
21. Gholamhoseini, M.; Ghalavand, A.; Khodaei-Joghan, A.; Dolatabadian, A.; Zakikhani, H.; Farmanbar, E. Zeolite-amended cattle manure effects on sunflower yield, seed quality, water use efficiency and nutrient leaching. *Soil Tillage Res.* **2013**, *126*, 193–202. [CrossRef]
22. Colombani, N.; Di Giuseppe, D.; Faccini, B.; Ferretti, G.; Mastrocicco, M.; Coltorti, M. Estimated water savings in an agricultural field amended with natural zeolites. *Environ. Process.* **2016**, *3*, 617–628. [CrossRef]
23. Faccini, B.; Di Giuseppe, D.; Ferretti, G.; Coltorti, M.; Colombani, N.; Mastrocicco, M. Natural and NH_4^+-enriched zeolitite amendment effects on nitrate leaching from a reclaimed agricultural soil (Ferrara Province, Italy). *Nutr. Cycling Agroecosyst.* **2018**, *110*, 327–341. [CrossRef]
24. Ferretti, G.; Galamini, G.; Deltedesco, E.; Gorfer, M.; Faccini, B.; Zechmeister-Boltenstern, S.; Coltorti, M.; Keiblinger, K.M. Effects of natural and NH_4^+-charged zeolite amendments and their combination with 3, 4-dimethylpyrazole phosphate (DMPP) on soil gross ammonification and nitrification rates. In *Geophysical Research Abstracts*; Copernicus publication: Munich, Germany, 2019; Volume 21.
25. Ferretti, G.; Keiblinger, K.M.; Zimmermann, M.; Di Giuseppe, D.; Faccini, B.; Colombani, N.; Mentlerb, A.; Zechmeister-Boltensternb, S.; Coltorti, M.; Mastrocicco, M. High resolution short-term investigation of soil CO_2, N_2O, NO_x and NH_3 emissions after different chabazite zeolite amendments. *Appl. Soil Ecol.* **2017**, *119*, 138–144. [CrossRef]
26. Ferretti, G.; Keiblinger, K.M.; Di Giuseppe, D.; Faccini, B.; Colombani, N.; Zechmeister-Boltenstern, S.; Coltorti, M.; Mastrocicco, M. Short-Term Response of Soil Microbial Biomass to Different Chabazite Zeolite Amendments. *Pedosphere* **2018**, *28*, 277–287. [CrossRef]
27. Ferretti, G.; Galamini, G.; Medoro, V.; Coltorti, M.; Di Giuseppe, D.; Faccini, B. Impact of Sequential Treatments with Natural and Na-Exchanged Chabazite Zeolite-Rich Tuff on Pig-Slurry Chemical Composition. *Water* **2020**, *12*, 310. [CrossRef]
28. Ferretti, G.; Di Giuseppe, D.; Faccini, B.; Coltorti, M. Mitigation of sodiumrisk in a sandy agricultural soil by the use of natural zeolites. *Environ. Monit. Assess.* **2018**, *190*, 646. [CrossRef]
29. Galamini, G.; Ferretti, G.; Medoro, V.; Tescaro, N.; Faccini, B.; Coltorti, M. Isotherms, Kinetics, and Thermodynamics of NH_4^+ Adsorption in Raw Liquid Manure by Using Natural Chabazite Zeolite-Rich Tuff. *Water* **2020**, *12*, 2944. [CrossRef]
30. De la Roca, M. Surround® Crop Protectant: La capa protectora natural para cultivos como el olivar. *Phytoma España* **2003**, *148*, 82–85.
31. Saour, G.; Makee, H. A kaolin-based particle film for suppression of the olive fruit fly *Bactrocera olCeae* Gmelin (Dip., Tephritidae) in olive groves. *J. Appl. Entomol.* **2003**, *128*, 28–31. [CrossRef]
32. Rumbos, C.I.; Sakka, M.; Berillis, P.; Athanassiou, C.G. Insecticidal potential of zeolite formulations against three stored-grain insects, particle size effect, adherence to kernels and influence on test weight of grains. *J. Stored Prod. Res.* **2016**, *68*, 93–101. [CrossRef]
33. McBride, J. Whitewashing agriculture. *Agric. Res.* **2000**, *48*, 14–17.
34. Doraiswamy, P.C.; Rosenberg, N.J. Reflectant Induced Modification of Soybean Canopy Radiation Balance. I. Preliminary Tests with a Kaolinite Reflectant 1. *Agron. J.* **1974**, *66*, 224–228. [CrossRef]
35. Schupp, J.R.; Fallahi, E.; Chun, I.J. Effect of particle film on fruit sunburn, maturity and quality of Fuji' and Honey crisp apples. *HortTechnology* **2002**, *12*, 87–90. [CrossRef]
36. Jifon, J.L.; Syvertsen, J.P. Moderate shade can increase net gas exchange and reduce photoinhibition in citrus leaves. *Tree Physiol.* **2003**, *23*, 119–127. [CrossRef] [PubMed]
37. Glenn, D.M. The mechanisms of plant stress mitigation by kaolin-based particle films and applications in horticultural and agricultural crops. *HortScience* **2012**, *47*, 710–711. [CrossRef]
38. Lalancette, N.; Belding, R.D.; Shearer, P.W.; Frecon, J.L.; Tietjen, W.H. Evaluation of hydrophobic and hydrophilic kaolin particle films for peach crop, arthropod and disease management. *Pest Manag. Sci.* **2005**, *61*, 25–39. [CrossRef] [PubMed]
39. Eriksson, H. Controlled release of preservatives using dealuminated zeolite Y. *J. Biochem. Bioph. Meth.* **2008**, *70*, 1139–1144. [CrossRef] [PubMed]
40. Reddy, P. Disguising the leaf surface. In *Recent Advances in Crop Protection*; Reddy, P., Ed.; Springer: New York, NY, USA, 2012; pp. 91–102.
41. Carter, G.A. Primary and secondary effect of water content on spectral reflectance of leaves. *Am. J. Bot.* **1991**, *78*, 916–924. [CrossRef]

42. Baldini, E.; Facini, O.; Nerozzi, F.; Rossi, F.; Rotondi, A. Leaf characteristics and optical properties of different woody species. *Trees* **1997**, *12*, 73–81. [CrossRef]
43. Lanza, B.; Di Serio, M.G. SEM characterization of olive (*Olea europaea* L.) fruit epicuticular waxes and epicarp. *Sci. Hortic.* **2015**, *191*, 49–56. [CrossRef]
44. Spanoghe, P.; De Schampheleire, M.; Van der Meeren, P.; Steurbaut, W. Influence of agricultural adjuvants on droplet spectra. *Pest Manag. Sci.* **2007**, *63*, 4–16. [CrossRef]
45. Gaskin, R.E.; Steele, K.D.; Forster, W.A. Characterising plant surfaces for spray adhesion and retention. *N. Zealand Plant Prot.* **2005**, *58*, 179–183. [CrossRef]
46. Skuterud, R.; Bjugstad, N.; Tyldum, A.; Tørresen, K.S. Effect of herbicides applied at different times of the day. *Crop Prot.* **1998**, *17*, 41–46. [CrossRef]
47. Denaxa, N.K.; Roussos, P.A.; Damvakaris, T.; Stournaras, V. Comparative effects of exogenous glycine betaine, kaolin clay particles and Ambiol on photosynthesis, leaf sclerophylly indexes and heat load of olive cv. Chondrolia Chalkidikis under drought. *Sci. Hortic.* **2012**, *137*, 87–94. [CrossRef]
48. Larbi, A.; Vázquez, S.; El-Jendoubi, H.; Msallem, M.; Abadía, J.; Abadía, A.; Morales, F. Canopy light heterogeneity drives leaf anatomical, eco-physiological, and photosynthetic changes in olive trees grown in a high-density plantation. *Photosynth. Res.* **2015**, *123*, 141–155. [CrossRef] [PubMed]
49. Rotondi, A.; Magli, M. Ripening of olives var. correggiolo: Modification of oxidative stability of oils during fruit ripening and oil storage. *J. Food Agric. Environ.* **2004**, *2*, 193–199.
50. Uceda, M.; Hermoso, M. La calidad del aceite de oliva. In *El Cultivo del Olivo*; Barranco, D., Fernàndez-Escobar, R., Rallo, L., Eds.; Junta de Andalucía Ediciones Mundi-Prensa: Madrid, Spain, 2001; pp. 589–614.
51. Rotondi, A.; Bendini, A.; Cerretani, L.; Mari, M.; Lercker, G.; Gallina Toschi, T. Effect of olive ripening degree on oxidative stability and organoleptic properties of Cv. Nostrana di Brisighella extra virgin olive oil. *J. Agric. Food Chem.* **2004**, *53*, 3649–3654. [CrossRef] [PubMed]
52. Sakamoto, Y.; Kaneda, M.; Terasaki, O.; Zhao, D.Y.; Kim, J.M.; Stucky, G.; Shin, H.J.; Ryoo, R. Direct imaging of the pores and cages of three-dimensional mesoporous materials. *Nature* **2000**, *408*, 449–453. [CrossRef]
53. Hu, P.; Yang, H. Insight into the physicochemical aspects of kaolins with different morphologies. *Appl. Clay Sci.* **2013**, *74*, 58–65. [CrossRef]
54. Curry, E.; Baer, D.; Young, J. X-Ray microanalysis of apples treated with kaolin indicates wax-Embedded Particulate in the Cuticle. In Proceedings of the XXVI International Horticultural Congress: Key Processes in the Growth and Cropping of Deciduous Fruit and Nut Trees, Toronto, ON, Canada, 11–17 August 2002; pp. 497–503.
55. Tworkoski, T.J.; Michael Glenn, D.; Puterka, G.J. Response of bean to applications of hydrophobic mineral particles. *Can. J. Plant Sci.* **2002**, *82*, 217–219. [CrossRef]
56. Le Grange, M.; Wand, S.J.E.; Theron, K.I. Effect of kaolin applications on apple fruit quality and gas exchange of apple leaves. In Proceedings of the XXVI International Horticultural Congress: Key Processes in the Growth and Cropping of Deciduous Fruit and Nut Trees, Toronto, ON, Canada, 11–17 August 2002; pp. 545–550.
57. Rosati, A.; Metcalf, S.G.; Buchner, R.P.; Fulton, A.E.; Lampinen, B.D. Effects of kaolin application on light absorption and distribution, radiation use efficiency and photosynthesis of almond and walnut canopies. *Ann. Bot.* **2007**, *99*, 255–263. [CrossRef] [PubMed]
58. Wünsche, J.N.; Lombardini, L.; Greer, D.H. 'Surround' Particle Film Applications-Effects on Whole Canopy Physiology of Apple. In Proceedings of the XXVI International Horticultural Congress: Key Processes in the Growth and Cropping of Deciduous Fruit and Nut Trees, Toronto, ON, Canada, 11–17 August 2002; pp. 565–571.
59. Brito, C.; Dinis, L.T.; Moutinho-Pereira, J.; Correia, C. Kaolin, an emerging tool to alleviate the effects of abiotic stresses on crop performance. *Sci. Hortic.* **2019**, *250*, 310–316. [CrossRef]
60. De Smedt, C.; Someus, E.; Spanoghe, P. Potential and actual uses of zeolites in crop protection. *Pest Manag. Sci.* **2015**, *71*, 1355–1367. [CrossRef] [PubMed]
61. Farquhar, G.D.; Sharkey, T.D. Stomatal conductance and photosynthesis. *Annu. Rev. Plant Physiol.* **1982**, *33*, 317–345. [CrossRef]
62. Proietti, P.; Famiani, F. Diurnal and seasonal changes in photosynthetic characteristics in different olive (*Olea europaea* L.) cultivars. *Photosynthetica* **2002**, *40*, 171–176. [CrossRef]
63. Colavita, G.M.; Blackhall, V.; Valdez, S. Effect of kaolin particle films on the temperature and solar injury of pear fruits. In Proceedings of the XI International Pear Symposium, General Roca, Rio Negro, Argentina, 23–26 November 2010; pp. 609–615.

Article

Effect of the Organic Production and the Harvesting Method on the Chemical Quality and the Volatile Compounds of Virgin Olive Oil over the Harvesting Season

Ana I. Carrapiso [1,*], Aránzazu Rubio [1], Jacinto Sánchez-Casas [2], Lourdes Martín [1], Manuel Martínez-Cañas [2] and Concha de Miguel [1]

[1] Escuela de Ingenierías Agrarias, Universidad de Extremadura, Av. Adolfo Suárez s/n, 06007 Badajoz, Spain; arubiosa@alumnos.unex.es (A.R.); martinlu@unex.es (L.M.); conchademiguelgordillo@gmail.com (C.d.M.)
[2] Centro de Investigaciones Científicas y Tecnológicas de Extremadura (CICYTEX), Instituto Tecnológico Agroalimentario de Extremadura (INTAEX), Av. Adolfo Suárez s/n, 06007 Badajoz, Spain; jacintojesus.sanchez@juntaex.es (J.S.-C.); manuel.martinez@juntaex.es (M.M.-C.)
* Correspondence: acarrapi@unex.es

Received: 8 November 2020; Accepted: 26 November 2020; Published: 28 November 2020

Abstract: Organic production has increasing importance in the food industry. However, its effect on the olive oil characteristics remains unclear. The purpose of this study was to research into the effect of organic production without irrigation, the traditional harvesting methods (tree vs. ground picked fruits), and the harvesting time (over a six-week period) on the oil characteristics. Free acidity, peroxide value, K_{232}, K_{270}, ΔK, total phenols, oxidative stability and the volatile compound profile (by SPME extraction, gas chromatography and mass detection) of olive oils from the Verdial de Badajoz cultivar were analysed. The organic production affected the peroxide value, total phenols, oxidative stability and 34 out of 145 volatile compounds. Its effect was much less strong than that of the harvesting method, which affected severely all the chemical and physical-chemical parameters and 105 out of 145 volatile compounds. Conversely, the harvesting time was revealed as a factor with little repercussion, on the chemical and physical-chemical parameters (only peroxide value was influenced), although it affected 83 out of 145 volatile compounds. The larger content in total phenols in the organic oils than in the conventional ones could explain the increase in oil stability and the differences in the volatile compounds.

Keywords: virgin olive oil; organic production; harvesting method; harvesting time; volatile compounds

---

1. Introduction

Virgin olive oil (VOO) is a valuable product obtained mechanically without any refining processes, so it keeps olive fruit compounds such as antioxidants [1] as well as compounds responsible for its typical colour and flavour. Olive oil flavour depends on the content in bitter-tasting compounds, such as phenolic compounds, but also on the volatile compounds, which are responsible for the typical odour notes and potential defects.

The most important VOO volatile compounds are formed through the lipoxygenase (LOX) pathway [2], C5 and C6 LOX compounds being the major contributors to the essential green sensory attribute [3]. When olives are released from the trees (either by falling down spontaneously or by harvesting them), progressive cell disruption takes place, which triggers the LOX pathway [3] and, therefore, the generation of C5 and C6 compounds and the development of the typical olive oil flavour. Factors affecting the activity of the enzymes involved in the LOX pathway, such as the fruit cultivar and the agronomic and processing conditions, may influence the volatile compound profile [2,3], and thus the flavour traits [4]. In addition, fruit decay favours chemical oxidation and microbial enzyme activity, which cause an increase in fermentation compounds, volatile phenols and chemical oxidation compounds [4], which are involved in most of the defective flavours of olive oil [3,5].

Over the last decades, organic agriculture has increased in importance. Regulation 834/2007 [6] (soon to be replaced) defines it as a system combining the best environmental practices, a high biodiversity level, the preservation of natural resources, animal welfare, and production methods based on natural substances and processes. Each State within the European Union have set up a control system to ensure that the organic agriculture comply with this regulation. It has been reported not to improve the sensory quality of food [7]. In the case of olive oil, results are not consistent. At small scale, it has been shown that restrictions in the use of chemicals (e.g., fertilisers) result in changes in the phenol content, some volatile compounds and sensory traits [8]. However, at large scale results do not seem to indicate a clear trend, which might be due to further differences in the agronomic practices [9]. In addition, little information about the effect of the organic practices on oil from unirrigated orchards and its volatile compounds is available [9], even though unirrigated and traditionally farmed systems tend to be based on more sustainable practices. In traditionally farmed orchards, harvesting is usually performed manually from the trees. Although not advisable for the production of high quality oil [10], harvesting the fruits from the ground is still resorted to in traditional systems when other methods are not feasible. It is currently performed when hand-picking is not convenient (then, the fruits are knocked down with a pole, e.g., when the fruits are scarce, scattered or difficult to access), or when the fruits fall down due to over-ripening or after climatic events such as strong winds. As ground fruits yield poor quality oils that usually cannot be marketed as virgin oil and need chemical refining, they are managed separately, and often undergo longer storage times, which favour further decay. It would be convenient to know if differences in the quality and the volatile compounds of oil when comparing both harvesting methods are consistent over the harvesting season. The harvesting time is of great interest for the industry. Early harvesting facilitates high prices in the market. However, as fruit ripeness increases, oil content rises, although in contrast oil quality worsens [11,12]. Little information is available about the effect of harvesting time on the volatile compounds, especially in the case of organic production [13].

About 1,800,000 tons of olive oil were produced in Spain in the 2018–2019 session [14]. The more abundant cultivars in Spain are Picual (over a million ha) and Hojiblanca (about 270,000 ha). Verdial de Badajoz is also among the main Spanish olive cultivars, with about 30,000 ha [15], being farmed mainly in the west (Extremadura region).

To date, most research about the effect of organic production and harvesting conditions has been focused on oil from irrigated orchards. More information on how these factors affect oil characteristics throughout the harvesting season is advisable to ensure the highest oil quality when there is no irrigation, which is more environmentally sustainable. Therefore, the aim of this study was to research the effect of organic production under traditional agronomic practices, the traditional harvesting methods (tree-picked vs. ground-picked fruits), and the harvesting time (over a six-week period) on the chemical and physical-chemical parameters and on the volatile compound profile of the olive oils produced from olives from the Verdial de Badajoz cultivar in a commercial olive mill.

2. Materials and Methods

2.1. Experimental Design and Samples

Olive fruits (*Olea europaea* L., Verdial de Badajoz cultivar) were collected from unirrigated trees grown in either organic or conventional orchards in the Lácara region (Badajoz, Spain). In the conventional production system, the fertiliser was a nitrogen-phosphorus-potassium-boron complex (20-8-14-0.1 B), as it is common practice in the region. In the organic production, the fertiliser was made up of weeds cut in springtime added to composted olive mill and pruning waste and hay. When mature (maturity index in the orchard ranging between 1 (fruits with green yellowish skin) and 5 (fruits with black skin and <50% purple flesh)), the fruits from the organic orchards were collected from the trees (Organic), whereas the ones from the conventional orchards were collected either from the trees (Conventional) or from the ground (Ground). They were mechanically processed, separately (in different days after proper cleaning to avoid cross-contamination), into oil under the same conditions in a local factory over the harvesting session. The organic production was subjected to the official control established in Spain according to Regulation 834/2007. Oil samples were taken from a tank filled during a week (20.000 L) for the Organic and Conventional oils, or directly from the production line for the Ground oil (which was produced once per week) once a week from the beginning of November to mid-January. Then, the eighteen (three types of oil x six weeks) virgin olive oils were kept at 6 °C and analysed.

2.2. Chemical and Physical-Chemical Analyses of Oil

The so-called quality parameters (free acidity, peroxide value, K_{232}, K_{270}, and ΔK extinction coefficients), which are taken into account to establish the olive oil categories within the European Union according to the Commission Regulation 2568/91 [16] and subsequent amendments, were determined [16].

The total polar phenol content was determined using the Folin-Ciocalteau colorimetric method [17]. The results were expressed as caffeic acid equivalent in mg·kg^{-1} oil.

The oxidative stability index (induction time expressed in hours) was determined by using an eight-channel 743 Rancimat instrument (Metrohm, Herisau, Switzerland), heating the oil samples (2.5 g) at 100 °C under an air flow of 10 L h^{-1} [18].

2.3. Volatile Compound Analysis

The virgin olive oil samples (5 g) were introduced into glass screw top vials, with laminated Teflon-rubber disks in the caps. The vials were left in a water bath at 40 °C for 10 min to equilibrate the volatile compounds in the headspace. Then, a solid-phase microextraction (SPME) needle was inserted through the disk, and a 1-cm 50/30 µm thickness DVB/Carboxen/PDMS fibre (Supelco, Bellefonte, PA, USA) was exposed to the headspace for 40 min while the vial was kept in the 40 °C-water bath. Later, the fibre was transferred to the gas-chromatograph inlet (splitless mode, 250 °C).

The chromatographic separation of the compounds was carried out using a HB-5 (50 m × 0.32 mm i.d, 1.05 µm) column (Agilent, Avondale, AZ, USA) placed into a gas-chromatograph (Agilent 6890 series) equipped with a mass spectrum detector (Agilent 5973). The oven temperature was held at 40 °C for 10 min and risen at 3 °C min^{-1} to a temperature of 160 °C, and then at 15 °C min^{-1} to a final temperature of 220 °C, where it was held for 10 min (total run time: 64 min). Mass spectra were generated by electronic impact at 70 eV, with a multiplier voltage of 1756 V. Data were collected at a rate of 1 scan s^{-1} over the 30–300 m/z range. The transfer line to the mass spectrometer was maintained at 280 °C. The Agilent MSD Chemstation software was used. n-alkanes (C5-C18) were analysed under the same conditions to calculate the linear retention indices (LRI).

The identification was performed by matching mass spectra (MS) and LRI with those of reference compounds analysed under the same conditions (a total of 62 Sigma-Aldrich reference compounds were used), or with those included in the Flavornet (www.flavornet.org) or NIST [19] databases. Two samples of each oil batch were analysed, and results were expressed as total area counts.

2.4. Data Analyses

A three-way (organic production, harvesting method, and harvesting time) Analysis of Variance (ANOVA) was performed on the data. When a significant effect was found, the Tukey test was carried out to compare the means. A Principal Component Analysis was performed on the mean values for each sample to evaluate the relations among variables and samples [20]. The statistical analyses were performed by means of the SPSS version 22.0.

3. Results and Discussion

3.1. Chemical and Physical-Chemical Parameters

Table 1 shows the results from the three-way ANOVA performed on the data from the chemical and physical-chemical analyses carried out on the Verdial de Badajoz virgin olive oils. The effect of the organic production was moderate (three out of seven parameters were affected), the harvesting method greatly influenced the parameters (all of them were affected), and the effect of the harvesting time was weak (only one out of seven parameters were affected) (Table 1).

Table 1. *p*-Values from a three-way ANOVA (type of production, harvesting method and harvesting time) performed on the chemical and physical-chemical data of Verdial de Badajoz olive oil.

	Type of Production	Harvesting Method	Harvesting Time
Free acidity (%)	0.953	<0.001	0.388
PV (mEq O_2 kg^{-1})	0.026	<0.001	0.006
K_{232}	0.965	0.016	0.488
K_{270}	0.823	0.024	0.506
ΔK	0.738	<0.001	0.276
Total phenols (mg kg−1)	0.039	<0.001	0.650
Oxidative stability (h)	0.006	<0.001	0.146

3.1.1. Effect of the Organic Production (Organic vs. Conventional)

In the case of the type of production system (Organic vs. Conventional), the effect was significant on the peroxide value (PV), the total phenols and the oxidative stability of oils (Table 1). The values averaged over the six week period for the Organic and Conventional oils are shown in Figure 1. Compared with the Conventional group, the Organic one had lower values for PV (9.63 ± 1.93 vs. 11.26 ± 2.12 mEq O_2 kg^{-1}), and higher for the total phenols (166.7 ± 15.0 vs. 149.0 ± 12.2 mg kg^{-1}) and the oxidative stability (26.3 ± 2.1 vs. 22.9 ± 2.0 h). However, most quality parameters (free acidity and the extinction coefficients) were not affected ($p > 0.730$ for all of them) (Table 1).

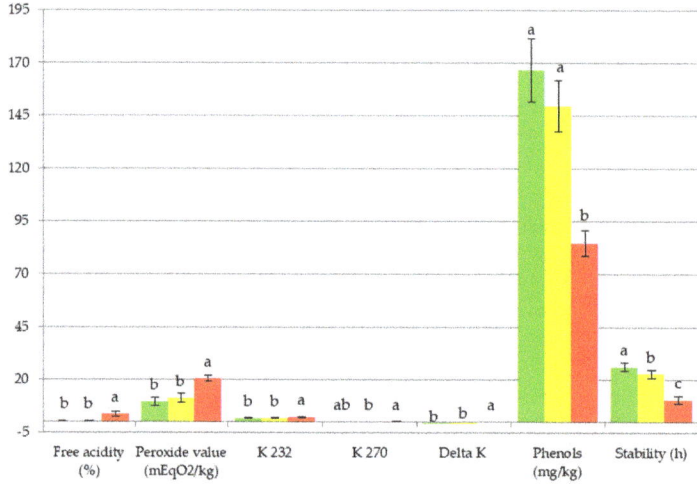

Figure 1. Mean values and standard deviation for the chemical and physical-chemical parameters over the six week period for the Organic (□), Conventional (□) and Ground (■) oils. Different letters indicate statistical differences (Tukey test's $p < 0.05$)

The lack of a marked effect on most of the quality parameters might be expected as the agronomic practices were very similar (no irrigation in both cases, and little vs. no use of phytosanitary chemicals) except for fertilisation. Currently, there is no general agreement on what the effect of the organic practices on the virgin olive oil quality is, apparently because of the difficulty of dealing with a relatively weak effect without excluding other environmental sources of variation. In this sense, a three-year study on the Leccino and Frantoio olive cultivars showed non-consistent differences in the oil quality parameters and phenol content due to the organic practices, suggesting that genotype and year-to-year variations in climate have a stronger effect [9]. However, both a decrease and an increase in oil quality have been reported in other studies. On the one hand, a decrease in oil quality (free acidity and K_{270}) was reported and attributed to infestation and fungal infection in the organic fruits as a consequence of the absence of pesticides [21]. Likewise, lower phenol contents have been reported in organic Picual and Hojiblanca oils than in the conventional ones [13]. On the other hand, an increase in oil quality or/and in phenol content was also reported as a result of the organic practices. A study on Picual oil showed lower values for PV, greater stability and higher phenol content in the organic oil than in the conventional one [22], which is in line with our results for Verdial de Badajoz oil. More recent research on Leccio and Frantoio olive oils found no effect on the quality parameters but showed an increase in the phenol content, which was attributed to the decreased availability of soil nitrogen when using organic practices, as phenol content increases with decreased soil nitrogen availability [23], suggesting that, under controlled environmental conditions, the effect of the agronomic practices on plant metabolism is clear [8]. Similarly, organic practices on Kolovi olive orchards resulted in an increase in phenols, including luteolin, which was proposed as a marker for organic production [24]. Our results seem to confirm those results and show that the increase in phenols also happens in Verdial de Badajoz oil from unirrigated orchards. The increase in the total phenols (which have antioxidant activity) could explain the increase in oil stability and the decrease in PV. In any case, the moderate effect of the organic production on these parameters might indicate that no large differences might be expected in the volatile compounds.

3.1.2. Effect of the Harvesting Method

The harvesting method affected all the parameters included in Table 1. Not only did the Ground oils (from ground-picked fruits) reach the highest and therefore worst values in all the quality parameters, but also the lowest in the total phenol content and oxidative stability (Figure 1). All the Ground oils, taken over the six-week period, exceeded considerably the 0.8% limit for free acidity for the extra virgin olive oil stated in European Commission Regulation (61/2011) [25], values being in the 2.4–5.3 range, and moreover all except the oil produced in week 2 exceeded the maximum allowance at least in one more quality parameter. Conversely, the two types of oils from tree-collected fruits (Conventional and Organic oils) were within the allowance limits for all the quality parameters. These results match those reported in Picual samples [26], with all the oils from ground-picked fruits exceeding the limit for free acidity and all the ones from tree-picked olives being below it.

With respect to the total phenol content, ground-picked olives, which are more exposed to microbiological infection than tree-picked olives, might have yielded lower phenol concentrations due to the phenol decline that the microbiological activity causes, in particular in oleuropein derivatives [27]. Furthermore, since phenols protect oil from oxidation [28], a decrease in them would facilitate a decline in the oxidative stability of oil. These results show a considerable repercussion of the harvesting method on the Verdial de Badajoz oil quality over, regardless of the harvesting time, and confirm the fact that fruits collected from the ground result in poor quality olive oils [10,26]. Thus, clear differences in the volatile compound profile between the Ground oils and the others might be expected.

3.1.3. Effect of the Harvesting Time

With respect to the harvesting time, the ANOVA revealed a slight effect, PV being the only parameter significant affected (Table 1). PV showed significant fluctuations throughout the six-week period instead of a steady trend (data not shown). Previous studies have reported that increased ripeness causes an undesirable increase in the values of the quality parameters, although with significant fluctuations, and a decrease in the oxidative stability and phenol content [12,29]. However, in our study harvesting was adjusted to real circumstances, where orchards management is set to harvest first the orchards which maturate first, as it is usually done in the commercial mills to achieve the best overall results, which could explain the lack of a clear trend. Our results show that in real conditions the harvesting time itself is not a critical quality factor for the Verdial de Badajoz oil as long as harvesting management is adequately set. The slight effect of harvesting time may anticipate slight changes in the volatile compound profile.

3.2. Volatile Compounds

A total of 145 volatile compounds were identified or tentatively identified in the headspace of the Verdial de Badajoz olive oils: 26 aldehydes, 13 ketones, 30 alcohols, 12 acids, 24 esters, 19 acyclic hydrocarbons, 13 cyclic hydrocarbons, four ethers and four other compounds (Table 2). The results from the three-way ANOVA show that few compounds were significantly affected by the organic production (34 out of 145), most of them (105 out of 145) were affected by the harvesting method and over half of them (83 out of 145) by the harvesting time (Table 2). Differences appeared not only in the lipoxygenase (LOX)-derived compounds (Table 3) but also in the fermentation (Table 4) and oxidation (Table 5) ones, as well as in all the chemical families of compounds (Table 2).

Table 2. Significance levels from a three-way ANOVA performed on the volatile compounds data * from the oil extracted from Verdial de Badajoz fruits farmed organically or conventionally (Prod.) and collected from the threes or from the ground (Meth.) over a six-week period (Week).

	p				p		
	Prod.	Meth.	Week		Prod.	Meth.	Week
Aldehydes				Alcohols			
Acetaldehyde [c]	0.044	<0.001	0.138	**Ethanol** [c]	<0.001	<0.001	<0.001
2-Methylpropanal [a]	0.404	<0.001	<0.001	Propan-1-ol [a]	0.497	<0.001	0.013
2-Methylprop-2-enal [b]	0.096	0.067	0.006	**2-methylpropan-1-ol** [b]	0.534	<0.001	0.001
Butanal [a]	0.181	0.644	0.709	**Butan-1-ol** [a]	0.667	<0.001	0.011
But-2-enal [a]	0.400	0.071	<0.001	**Pent-1-en-3-ol** [a]	0.008	0.001	<0.001
3-Methylbutanal [a]	0.549	<0.001	<0.001	Pentan-3-ol [b]	0.279	0.750	0.443
2-Methylbutanal [a]	0.472	0.003	0.191	3-Methylbut-3-en-1-ol [b]	0.311	<0.001	<0.001
(Z)-Pent-2-enal [b]	0.006	<0.001	0.046	**3-Methylbutan-1-ol** [a]	0.332	<0.001	<0.001
(E)-Pent-2-enal [a]	0.004	<0.001	0.001	**2-Methylbutan-1-ol** [a]	0.287	<0.001	<0.001
(Z)-Hex-3-enal [b]	0.003	<0.001	<0.001	Pentan-1-ol [a]	0.779	0.002	0.203
Hexanal [a]	<0.001	0.797	<0.001	(Z)-Pent-2-en-1-ol [b]	0.082	0.001	<0.001
(E)-Hex-2-enal [a]	<0.001	<0.001	0.001	**(Z)-Hex-3-en-1-ol** [a]	0.009	<0.001	0.001
Heptanal [a]	0.556	<0.001	0.058	**(E)-Hex-2-en-1-ol** [b]	0.237	0.004	0.003
Hexa-2,4-dienal [a]	0.025	<0.001	0.001	**Hexan-1-ol** [a]	0.642	<0.001	0.092
(E)-Hept-2-enal [a]	0.001	0.002	0.001	Heptan-2-ol [a]	0.685	0.096	0.284
Benzaldehyde [a]	0.688	<0.001	0.006	Phenol [b]	0.336	<0.001	0.065
Octanal [a]	0.449	<0.001	0.709	Heptan-1-ol [a]	0.974	<0.001	0.090
(2E)-Oct-2-enal [a]	0.011	0.001	0.826	Oct-1-en-3-ol [a]	0.478	<0.001	0.063
(3E)-Non-3-enal [b]	0.548	<0.001	0.723	Octan-3-ol [b]	0.001	<0.001	<0.001
Nonanal [a]	0.048	0.037	0.182	Octan-2-ol [a]	0.017	<0.001	<0.001
(E)-Non-2-enal [a]	0.357	0.001	0.011	2-Ethylhexan-1-ol [b]	0.681	0.081	0.044
Nona-2,4-dienal [b]	<0.001	<0.001	0.266	Phenylmethanol [b]	0.933	<0.001	0.150
(E)-Dec-2-enal [b]	0.249	0.226	0.370	(Z)-Oct-2-en-1-ol [b]	0.971	0.004	0.570
(E,Z)-Deca-2,4-dienal [b]	0.721	0.442	0.245	Octan-1-ol [a]	0.555	<0.001	0.542
(E,E)-Deca-2,4-dienal [b]	0.790	0.567	0.491	Phenylethanol [b]	0.959	<0.001	0.008
Undecenal [b]	0.274	0.657	0.549	(Z)-Non-3-en-1-ol [b]	0.694	<0.001	<0.001

Table 2. Cont.

	p				p		
	Prod.	Meth.	Week		Prod.	Meth.	Week
Ketones				2-Ethylphenol [b]	0.928	<0.001	0.047
Propan-2-one [c]	0.004	0.490	0.038	Nonan-1-ol [a]	0.502	<0.001	0.246
Butan-2-one [a]	0.755	<0.001	0.158	Decan-1-ol [b]	0.470	0.012	0.165
Pent-1-en-3-one [a]	0.031	0.045	0.066	4-Ethyl-2-methoxyphenol [b]	0.767	<0.001	0.100
Pentan-2-one [a]	0.675	0.196	0.320	*Acids*			
Pentan-3-one [b]	0.102	0.022	<0.001	**Acetic acid** [a]	0.576	0.020	<0.001
3-Hydroxybutan-2-one [b]	0.410	0.001	<0.001	2-Methylpropanoic acid [a]	0.916	0.001	0.317
4-Methylpentan-2-one [b]	0.556	0.026	0.005	**Butanoic acid** [a]	0.630	<0.001	0.024
2-Methylpentan-3-one [b]	0.971	<0.001	0.325	**2-Methylbutanoic acid** [a]	0.671	<0.001	0.014
Hexan-2-one [a]	0.851	0.002	0.338	Hexanoic acid [a]	0.570	<0.001	0.094
Heptan-2-one [a]	0.731	<0.001	0.236	Heptanoic acid [a]	0.482	0.318	0.511
Octan-3-one [a]	0.595	<0.001	0.002	Octanoic acid [a]	0.376	0.259	0.040
Octan-2-one [b]	0.728	<0.001	0.213	Nonanoic acid [a]	0.779	0.215	0.607
Nonan-2-one [b]	0.550	<0.001	0.044	Decanoic acid [a]	0.272	0.388	0.525
Esters				Undecanoic acid [b]	0.402	0.085	0.659
Methyl acetate [b]	0.003	0.043	<0.001	Dodecanoic acid [b]	0.681	0.391	0.632
Ethyl acetate [b]	0.001	<0.001	<0.001	Tridecanoic acid [b]	0.084	0.647	0.701
Methyl propanoate [b]	0.086	<0.001	<0.001	*Acyclic hydrocarbons*			
Ethyl propanoate [b]	0.508	<0.001	0.018	Pentane [a]	0.015	<0.001	<0.001
Methyl butanoate [a]	0.048	<0.001	0.001	2-Methylbut-2-ene [b]	0.806	0.597	0.002
Ethyl 2-methylpropanoate [a]	0.565	<0.001	<0.001	Penta-2,3-diene [b]	0.015	<0.001	<0.001
Methyl 2-methylbutanoate [a]	0.748	<0.001	0.017	Hexane [a]	<0.001	0.603	0.026
Methyl pentanoate [a]	0.055	0.016	0.002	2-Methylpent-2-ene [b]	0.724	<0.001	0.006
Ethyl 2-methylbutanoate [a]	0.205	0.085	<0.001	2-methylpenta-1,3-diene [b]	0.967	<0.001	0.003
3-Methylbutyl acetate [b]	0.119	<0.001	0.020	2-Methylhexane [b]	0.194	0.239	0.050
2-Methylbutyl acetate [b]	0.040	<0.001	0.011	5-methylhex-1-ene [b]	0.021	0.981	0.685
2-Methylpropyl 2-methylpropanoate [b]	0.145	<0.001	0.006	3-Methylhexane [b]	0.013	0.382	0.004
Methyl hexanoate [a]	0.003	0.001	0.041	Hept-1-ene [b]	0.522	<0.001	0.074
Methyl Hex-4-enoate [b]	0.867	0.639	0.455	Heptane [a]	0.511	0.001	0.068
Ethyl hexanoate [a]	0.001	<0.001	<0.001	Oct-1-ene [b]	0.706	<0.001	0.006
(E)-Hex-3-enyl acetate [b]	<0.001	0.026	0.023	Octane [a]	0.335	<0.001	0.003
Hexyl acetate [b]	0.799	<0.001	0.001	(Z)-Oct-2-ene [b]	0.441	<0.001	0.004
3-Methylbutyl 2-methylpropanoate [b]	0.348	<0.001	0.022	(E)-Oct-2-ene [b]	0.477	<0.001	0.001
Methyl heptanoate [b]	0.091	<0.001	0.554	(E)-β-Ocimene [b]	0.118	0.549	0.160
Methyl benzoate [b]	0.224	0.008	0.777	Alloocimene [b]	0.112	0.497	0.001
Methyl octanoate [b]	0.263	0.001	0.022	Dodecane [a]	0.617	0.751	<0.001
Ethyl benzoate [b]	0.301	<0.001	0.369	(E,E)-α-farnesene [b]	0.029	0.513	<0.001
Ethyl octanoate [b]	0.812	0.010	0.352	*Cyclic hydrocarbons*			
Methyl 2-methoxybenzoate [b]	0.531	0.334	0.372	Toluene [b]	0.934	<0.001	0.095
Others				1,4-Dimethylbenzene [b]	0.129	<0.001	0.028
Diethyl ether [a]	0.556	0.191	0.031	Styrene [b]	0.147	<0.001	0.001
2-Ethoxy-2-methylpropane [b]	0.819	0.001	0.039	1,3-Dimethylbenzene [b]	0.892	0.001	0.032
1-Methoxyhexane [b]	0.293	<0.001	0.071	1-Ethyl-3-methylbenzene [b]	0.026	0.200	0.186
Methoxybenzene [b]	0.015	0.228	0.126	1,2,4-Trimethylbenzene [b]	0.581	0.095	0.071
Dimethyl sulfide [b]	0.708	<0.001	0.009	1,3-Diethylbenzene [b]	0.100	0.030	<0.001
4-Methyl-2,3-dihydrofuran [b]	0.372	0.482	0.362	l-Limonene [a]	0.691	<0.001	0.010
Γ-Hexalactone [b]	0.001	<0.001	0.016	3-(4-Methylpent-3-enyl)furan [b]	0.290	<0.001	0.011
δ-Octalactone [b]	0.153	0.202	0.012	α-Copaene [b]	0.976	0.359	0.354
				γ-Cadinene [b]	0.907	0.561	0.283
				Eremophilene [b]	0.251	<0.001	0.224
				α-Muurolene [b]	0.843	0.037	0.391

* The compound was identified by comparing it with: [a] the MS and LRI of the reference compound; [b] MS and LRI from literature; [c] the MS and retention time of the reference compound. Bold: compounds included on Tables 3–5.

Table 3. Results (means and significance *) for the most representative C5 and C6 LOX volatile compounds of the Verdial de Badajoz oil headspace significantly affected by the organic production, harvesting method and/or harvesting time according to Table 2.

		Week 1	Week 2	Week 3	Week 4	Week 5	Week 6
(Z)-Pent-2-enal	Organic	0.42	0.45a	0.21	0.50a	0.20b	0.26ab
	Conventional	AB0.55	C0.3a	BC0.39	AB0.57a	A0.61a	ABC0.48a
	Ground	A0.31	AB0.06b	AB0.07	AB0.06b	B0.04c	AB0.05b
(E)-Pent-2-enal	Organic	A2.86ab	A2.46a	B0.77b	A2.48ab	B0.92b	B1.39b
	Conventional	AB2.97a	C1.65ab	BC2.24a	A3.20a	ABC2.51a	ABC2.42a
	Ground	A2.04b	ABC1.19b	A2.03a	AB1.61b	C0.42c	BC0.89c
Pent-1-en-3-one	Organic	A3.69	A4.4	B0.78	A4.82	B1.61	A4.04
	Conventional	5.50	1.26	5.44	7.33	5.15	7.46
	Ground	1.77	2.18	0.86	5.86	5.42	4.24
Pentan-3-one	Organic	A21.37	C7.92b	BC11.40b	AB17.42	C6.64b	C6.40ab
	Conventional	B13.65	B14.48a	A19.85a	B15.94	C9.78a	C9.34a
	Ground	AB13.38	AB11.57a	A15.80ab	AB12.68	BC7.49ab	C4.99b
Pent-1-en-3-ol	Organic	A5.62	B1.91	B2.06c	AB3.16ab	B1.14c	B1.49b
	Conventional	AB3.65	B3.12	A4.30a	A4.58a	B2.63a	B2.78a
	Ground	A3.37	BC2.35	AB3.21b	CD2.14b	DE1.43b	E0.81c
Pent-2-en-1-ol	Organic	A7.49	AB3.48	B2.97	AB3.95ab	B1.82	B2.65ab
	Conventional	4.76	4.11	4.94	5.40a	3.61	4.18a
	Ground	4.67	3.52	2.85	3.14b	1.67	1.62b
(Z)-Hex-3-enal	Organic	A10.42ab	A11.29a	BC5.63a	B7.29a	C5.14b	BC6.75b
	Conventional	A16.88a	B8.08a	B5.31a	B9.65a	AB11.07a	AB11.28a
	Ground	A3.51b	B1.06b	B0.72b	B0.74b	B0.07b	B0.20c
Hexanal	Organic	AB28.56	BC20.80b	C11.60b	A35.69b	BC18.16b	AB26.3b
	Conventional	BC31.70	C28.62ab	BC32.26a	A43.36a	ABC36.47a	AB38.49a
	Ground	AB41.49	ABC33.82a	ABC35.81a	A43.29a	C24.69b	BC28.51b
(E)-Hex-2-enal	Organic	A63.33b	A61.55a	B23.52c	A56.35b	B24.73b	B33.97b
	Conventional	AB82.37a	B68.07a	AB85.02a	A100.50a	AB83.06a	AB83.54a
	Ground	A74.71ab	B41.51b	B38.57b	B42.47c	C6.32c	C13.78c
(Z)-Hex-3-en-1-ol	Organic	AB73.07a	D44.12	BC67.15a	A85.86a	CD55.31a	BC63.62a
	Conventional	54.92b	56.64	64.12a	59.07b	47.84a	49.89b
	Ground	A55.38b	AB46.14	AB44.90b	BC42.64c	CD31.74b	D27.31c
(E)-Hex-2-en-1-ol	Organic	A22.64b	C4.86c	B12.95b	B12.53	C4.23c	C4.09c
	Conventional	B10.04a	A17.57b	A23.03a	B10.69	B8.78b	B9.49a
	Ground	B27.97b	A36.20a	C21.11a	D12.87	C20.85a	D7.37b
1-Hexanol	Organic	35.13a	17.65b	29.85b	35.39ab	24.9b	28.71b
	Conventional	24.03b	33.22a	35.56b	26.40b	22.30b	23.05b
	Ground	35.13a	39.87a	52.68a	41.82a	52.49a	36.18a
Hexyl acetate	Organic	0.33	0.31	0.40b	0.56b	0.35b	0.51b
	Conventional	B0.26	AB0.49	A0.72b	AB0.44b	B0.30b	AB0.34b
	Ground	C0.72	BC0.81	A1.58a	AB1.35a	ABC1.06a	ABC1.20a

* Different letters in the same row (A–C) or the same column (a–c) indicate statistical differences (Tukey test's $p < 0.05$). Results are expressed as AU $\times 10^{-6}$.

Table 4. Results (means and significance *) for the most representative fermentation compounds of the Verdial de Badajoz oil headspace significantly affected by the organic production, harvesting method and/or harvesting time according to Table 2.

		Week 1	Week 2	Week 3	Week 4	Week 5	Week 6
2-Methylpropanal	Organic	CD0.56	D0.31	CD0.66b	A1.72	BC1.03	AB1.45a
	Conventional	B0.46	B0.42	AB1.09b	A1.35	AB0.74	AB0.64b
	Ground	B1.35	B1.09	A3.65a	AB2.63	AB1.70	AB1.90a
2-Methylprop-2-enal	Organic	AB 0.38b	AB 0.35	B0.07c	AB 0.32b	AB0.18b	A0.48
	Conventional	AB0.46b	B0.26	AB0.40b	A0.66b	AB0.47a	AB0.45
	Ground	AB1.06a	C0.46	BC0.59a	A1.17a	C0.05b	C0.40
3-Methylbutanal	Organic	C0.99b	C0.65b	C0.79b	A4.67b	B2.43b	B2.99a
	Conventional	BC1.43b	C1.27b	A3.90a	A3.52c	B1.96c	BC1.88b
	Ground	BC4.13a	CD2.60a	AB5.40a	A6.37a	CD3.30a	D1.89b
2-Methylbutanal	Organic	2.68	0.91	2.69	4.81	3.79	6.41
	Conventional	2.62	2.79	3.95	2.19	3.06	3.43
	Ground	4.45	3.83	7.05	8.13	5.65	3.56
Ethanol	Organic	D59.14a	E43.79a	C91.84a	B112.40a	AB121.59a	A126.34b
	Conventional	D15.30	D21.38b	C52.41b	A89.89b	A80.90b	B69.13c
	Ground	C60.28a	C49.84a	A120.13a	B84.87b	C55.72a	A139.93a
2-Methylpropan-1-ol	Organic	B1.06b	B0.53b	A2.64b	A2.81b	A2.05b	A2.25b
	Conventional	C0.80b	C1.02b	A2.71b	B1.90c	C1.01c	C1.05c
	Ground	C6.05a	C6.39a	A16.34a	AB11.92a	BC8.54a	C5.69a
Butan-1-ol	Organic	0.43	0.73	0.35	0.49	1.17	0.96
	Conventional	0.35	0.22	0.76	0.74	0.87	0.70
	Ground	0.60	0.98	1.71	1.26	2.45	1.14
3-Methylbutan-1-ol	Organic	C3.36b	C1.96b	AB9.87b	A11.85b	B8.47b	AB10.13b
	Conventional	C3.00b	C4.14b	A10.42b	B7.09b	C4.14c	C4.45c
	Ground	B19.83a	B20.58a	A50.92a	A40.31a	B25.83a	B26.33a
2-Methylbutan-1-ol	Organic	D1.54b	D0.79b	C4.01b	A6.04b	BC4.60b	AB5.29b
	Conventional	C1.12b	C1.89b	A5.80b	B3.18c	C1.78c	C1.93c
	Ground	C8.53a	C9.58a	A23.3a	AB17.81a	BC14.48a	C10.87a
Acetic acid	Organic	C3.25	B17.76a	A64.99a	AB9.35	AB6.6a	AB9.45a
	Conventional	B24.18	B19.54a	A38.92b	C4.26	C3.9b	C5.94b
	Ground	2.04	3.98b	8.43c	5.77	4.21b	8.43ab
2-Methylpropanoic acid	Organic	0.16	0.11b	0.22b	0.24b	0.10	0.24b
	Conventional	0.12	0.19b	0.31b	0.05b	0.00	0.12c
	Ground	0.51	0.62a	1.16a	0.80a	4.12	2.74a
Butanoic acid	Organic	0.29	0.04b	0.28b	0.39b	0.09b	0.94
	Conventional	B0.00	B0.19b	A0.77b	B0.23b	B0.05b	B0.18
	Ground	B0.44	B1.31a	A2.89a	AB1.66a	B1.28a	B0.66
2-Methylbutanoic acid	Organic	B0.00b	B0.01b	A0.29b	B0.00	AB0.12b	AB0.11b
	Conventional	0.08b	0.00b	0.00b	0.11	0.07b	0.13b
	Ground	B0.37a	AB0.59a	A1.01a	AB0.76	AB0.78a	AB0.66a
Phenylmethanol	Organic	1.10b	1.14	1.09	1.89	0.63b	2.17b
	Conventional	1.32b	1.26	2.35	1.14	0.82b	0.85b
	Ground	3.81a	4.24	8.06	6.97	5.31a	5.29a
2-Phenylethanol	Organic	0.44b	0.35b	0.66b	0.9b	0.66b	1.03b
	Conventional	B0.18b	B0.64b	A1.35b	AB0.8b	B0.38b	B0.57b
	Ground	AB6.58a	AB7.37a	A9.54b	AB8.25a	B5.08a	AB7.55a
Phenol	Organic	0.15b	0.07b	0.06	0.02	0.00b	0.03c
	Conventional	0.16b	0.05b	0.09	0.02	0.07b	0.12b
	Ground	BC0.26a	BC0.32a	B0.34	C0.15	A0.52a	B0.33a
2-Ethylphenol	Organic	0.00b	0.41b	0.00b	1.87b	0.84b	0.03b
	Conventional	B0.31b	B0.04b	A2.06b	B0.58b	B0.45b	B0.28b
	Ground	6.52a	9.38a	18.42a	14.04a	8.1a	10.47a
4-Ethyl-2-methoxyphenol	Organic	0.00b	0.00b	0.05b	0.06b	0.04b	0.20b
	Conventional	0.00b	0.05b	0.00b	0.02b	0.09b	0.03b
	Ground	1.57a	2.23a	2.44a	2.38a	2.35a	2.26a

* Different letters in the same row (A–C) or the same column (a–c) indicate statistical differences (Tukey test's $p < 0.05$). Results are expressed as AU $\times 10^{-6}$.

Table 5. Results (means and significance *) for the most representative oxidation compounds of the Verdial de Badajoz oil headspace significantly affected by the organic production, harvesting method and/or harvesting time according to Table 2.

		Week 1	Week 2	Week 3	Week 4	Week 5	Week 6
Heptanal	Organic	1.90b	1.41	1.62	2.45ab	2.36	2.69
	Conventional	1.81b	2.34	1.77	2.23b	2.31	2.81
	Ground	3.66a	2.48	3.80	4.38a	4.14	3.05
Hexa-2,4-dienal	Organic	A4.95a	BC3.68a	D1.52b	AB4.28a	D1.20b	C2.83a
	Conventional	B3.44ab	B3.33a	A5.20a	A5.34a	B3.13a	B3.04a
	Ground	AB2.02b	ABC1.16b	A2.33b	A2.11b	C0.60b	BC0.81b
(E)-Hept-2-enal	Organic	AB6.64b	AB6.54b	D2.60c	A7.96b	C4.40b	BC5.91b
	Conventional	A9.16ab	B6.62b	AB7.96b	A8.87b	AB7.69a	AB8.56a
	Ground	B11.69a	C9.62a	B12.54a	A14.92a	D4.97b	C9.43a
Octanal	Organic	2.52b	1.66	2.09c	3.09	2.51b	3.91
	Conventional	2.02b	4.31	3.27b	2.77	2.29b	3.14
	Ground	4.94a	3.14	4.54a	4.49	6.43a	4.65
(E)-Oct-2-enal	Organic	0.79	0.65	0.40c	0.85b	0.69	0.69b
	Conventional	0.92	1.27	1.00b	0.76b	0.90	0.83b
	Ground	1.23	1.16	1.42a	1.53a	1.03	1.50a
Nonanal	Organic	8.75b	6.90	7.90	10.52	11.86	13.04
	Conventional	10.44b	15.90	9.14	12.88	10.73	16.36
	Ground	20.33a	12.45	12.61	15.34	16.57	15.60
Nona-2,4-dienal	Organic	0.84b	0.95	0.35b	1.06a	0.72b	0.91b
	Conventional	1.10a	1.11	1.32a	1.29a	1.15a	1.45a
	Ground	0.45c	0.27	0.11b	0.12b	0.13c	0.03c

* Different letters in the same row (A–C) or the same column (a–c) indicate statistical differences (Tukey test's $p < 0.05$). Results are expressed as $AU \times 10^{-6}$.

Regarding the type of production (Organic vs. Conventional), the modest effect found (Table 2) might be expected since the oils were just moderately different in the chemical and physical-chemical parameters (Table 1). Previous work on the effect of organic practices on the volatile compounds of olive oil has reported no consistent differences in a three-year study [9], but also a general rise [13,21], which could indicate that there might be further agronomic factors influencing the results.

Conversely, in the case of the harvesting method, the noticeable effect on the volatile compounds might be expected since the oils were markedly different in the chemical and physical-chemical parameters (Table 1). For most compounds, the largest abundances appeared in the Ground oils (Tables 3–5), which could be due to the mechanical damage caused by the drop of the fruits. The damage accelerates the decay process and makes easier the access of microorganisms to the fruits, whose infection results in changes in the volatile compound profile [27].

With regard to the harvesting time, the effect (Table 2) was stronger than expected taking into account the weak influence on the chemical and physical-chemical parameters (Table 1). In any case, a weak effect of the harvesting time was reported for Picual and Hojiblanca oils from irrigated orchards [13]. Our results for oil from unirrigated orchards suggest that the volatile compounds might be more affected by the harvesting time than what the chemical analyses may reveal.

3.2.1. LOX-Derived Volatile Compounds

Table 3 shows the results for the most representative C5 and C6 LOX volatile compounds affected by the organic production, the harvesting method, and/or the harvesting time. Those compounds were among the most abundant ones in the Verdial de Badajoz olive oil headspace, as it was previously reported for the oil from this cultivar [30] and others [3,4,10]. Most of those compounds have low odour-thresholds [3] and, therefore, could take part in oil flavour. In fact, C5 and C6 LOX compounds seem to contribute to the positive traits of olive oil [3,10]. The most abundant LOX compounds were (E)-hex-2-enal and (Z)-hex-3-en-1-ol, followed by hexan-1-ol and hexanal. It should be noted that hexanal, besides the LOX pathway, can be generated through oxidation reactions on linoleic acid [3], being involved in the rancid note of food when it appears at high concentrations.

According to the ANOVA results (Table 2), the organic production affected eight out of the 13 LOX compounds included in Table 3, the harvesting method 12 out of 13 (all except hexanal), and the harvesting time 12 out of 13 (all except hexan-1-ol). Some LOX compounds were not significantly affected by any factors, such as pentan-2-one and 3-pentanol (Table 2).

Effect of the Organic Production (Organic vs. Conventional)

The effect of the organic production was significant on four out of the six C5 compounds included in Table 3, and on four out of the seven C6 ones, according to the ANOVA results (Table 2). The effect was stronger than expected taking into account the relatively slight influence on the quality parameters (Table 1). Values for the C5 LOX compounds were generally lower in the Organic oils than in the Conventional ones, although in any case the Tukey test revealed only slight differences, especially on week 1. Different trends were found for important C6 compounds: (Z)-hex-3-en-1-ol tended to be more abundant in the Organic oils than in the Conventional ones (differences were significant in week 1, 4 and 6), whereas (Z)-hex-3-enal (weeks 1, 5 and 6), hexanal (weeks 3, 4, 5, 6) and (E)-hex-2-enal (weeks 1, 3, 4, 5, 6) showed the opposite trend.

To date, the effect of the organic practices on the volatile compounds has been scarcely studied, and results are not completely consistent. In this sense, our results, from unirrigated Verdial de Badajoz orchards, show a similar trend for hexanal to results for oil from the Leccino and Frantoio cultivars also farmed in unirrigated orchards, although for the other compounds no clear trends were reported [9]. Conversely, higher abundances in hexanal in Organic than in Conventional oils from Picual and Hojiblanca olives from irrigated orchards have been also reported [13]. Therefore, our data might confirm that there is an effect of the organic production on some compounds, such as hexanal, but this effect might depend on other factors, such as irrigation or the olive cultivar.

Effect of the Harvesting Method

The significant effect of the harvesting method (Table 2) on all the C5 LOX compounds and six (all except hexanal) out of the seven C6 ones included in Table 3 matches the substantial effect found on the chemical and physical-chemical parameters (Table 1). The C5 LOX compounds were generally more abundant in the Conventional oils (from tree-picked fruits) than in the Ground ones. However, for the C6 LOX compounds a mixed trend was found. (Z)-hex-3-enal, (E)-hex-2-enal and (Z)-hex-3-en-1-ol, which have been related to the green attribute [10], were significantly more abundant in the Conventional group than in the Ground one. Conversely, hexan-1-ol and hexyl acetate tended to be more abundant over time in the Ground oils, and (E)-hex-2-en-1-ol did not show a steady trend. Hexan-1-ol is considered to elicit a no agreeable odour in oil [10] and, therefore, its increase might have a detrimental effect on oil quality. Hexyl acetate, which contributes to the fruity note, is an indicator of ripeness [3], and its precursor (E)-hex-2-en-1-ol [3] has been related to some defects [10,31]. The differences between the Conventional and Ground groups (Table 3) in the LOX compounds increased over time, the Tukey test revealing that it was on week 6 when the most C5 and C6 LOX compounds were influenced by the harvesting method (Table 3). (Z)-hex-3-enal and (E)-hex-2-enal were the compounds most affected, differences being significant in the Tukey test on all the weeks of sampling (Table 3). It should be noted that hexanal was not affected by the harvesting method. This result did not match a previous study reporting an increase in it in oil from ground-picked fruits [26]. However, hexanal content depends on the LOX pathway but also on oxidation reactions, and thus the lack of effect in our study (Table 2) may be explained by a counteracting effect of both pathways.

Effect of the Harvesting Time

According to the ANOVA results (Table 2), the effect of the harvesting time was significant on five out of the six C5 LOX compounds and six out of the seven C6 ones included in Table 3. It affected all the oil groups to a similar extent (Table 3). The effect was stronger on these compounds than on the chemical and physical-chemical parameters (Table 1). Most C5 LOX compounds fluctuated over time, without a consistent trend, although pentan-3-one and pent-1-en-3-ol decreased significantly as the season went on (Table 3). A similar pattern was reported for C5 LOX compounds in Arbequina and Chéttoui olive oils [4]. A general decrease was also found for the C6 compounds over time, especially for (Z)-hex-3-enal, (E)-hex-2-enal and (E)-hex-2-en-1-ol (Table 3), which are related to positive flavour traits [10]. These results for Verdial de Badajoz olive oil match previous results on other cultivars [4,29], although it has been pointed out that the decrease in C6 LOX compounds might not affect all cultivars [3]. The decrease over time was more marked in the Ground oils, which might indicate that harvesting late would add to the detrimental effect of harvesting from the ground.

3.2.2. Fermentation Compounds

According to the ANOVA results (Table 2), the most important compounds related to the microbial activity were hardly affected by the organic practices (only ethanol was affected), but they were greatly influenced by the harvesting method (all the compounds included in Table 4 except 2-methylprop-2-enal) and the harvesting time (13 out of 18).

Effect of the Organic Production (Organic vs. Conventional).

Except for ethanol, neither the non-phenolic fermentation compounds (including short-chain acids and alcohols and branched C3 and C4 compounds) nor the volatile phenols were affected by the organic practices (Table 2). This result suggests that differences in the agronomic practices such as fertilisation do not affect to a considerable extent the degradation reactions in which microorganisms can be involved once the fruits are released from the trees, which is in line with the moderate effect on the chemical and physical-chemical parameters (Table 1). Scarce information about the effect of the organic practices on the fermentation compounds is available, since most attention has been devoted to the LOX compounds [9], and no information is available about oils from unirrigated trees. For oil from irrigated orchards, a slight effect on the fermentation compounds was also reported for Hojiblanca (only 3-methylbut-2-en-1-ol was affected, without a consistent trend over time) and Picual oil (only methanol, 2-methylbutanal and 3-methylbut-2-en-1-ol were affected, with larger abundances in the organic oil) [13]. A larger content in 2-methylpropan-1-ol was reported in a group of organic oils than in the conventional ones, but also no differences were found in other fermentation compounds [21]. Our results show that the organic practices have not a noticeable effect on the fermentation compounds of Verdial de Badajoz olive oil from unirrigated orchards, which is partly in line with previous studies on other cultivars and irrigated orchards.

Effect of the Harvesting Method

Regarding the harvesting method, both the non-phenolic and phenolic fermentation compounds were markedly affected (12 out of 13, and all the phenols, respectively), the compounds being generally more abundant in the Ground oils than in the Conventional ones (Table 4).

Almost all the non-phenolic compounds included in Table 4 (all except 2-methylprop-2-enal) were greatly affected by the harvesting method. Most of them were more abundant in the Ground oils than in the Conventional ones, although acetic acid followed the opposite trend. The generally higher values might be caused by the increased mechanical damage in the ground fruits and the subsequent opportunity for microbiological contamination and fermentation to occur. These results are in line with previous work reporting that fermentation compounds such as 2-methylbutan-1-ol and butan-1-ol were more abundant in oil from ground-picked olives than from tree-picked fruits [26]. Most of these compounds have low odour-thresholds [3], and 3-methylbutan-1-ol and short-chain acids have been related to the winey-vinegary and fusty defects [31].

With regard to the volatile phenols included in Table 4, they were all affected by the harvesting method (Table 2), all phenols being more abundant in the Ground oils. For three compounds

(2-phenylethanol, 2-ethylphenol, and 4-ethyl-2-methoxyphenol) the differences were significant in the six weekly samplings (Table 4). Our results for Verdial de Badajoz oil are in line with the increase in the volatile phenols reported in Picual oils [26]. The volatile phenols are markers of fruit degradation [32] and, in fact, it has been suggested that 4-ethylphenol is a microbial metabolite from hydroxycinnamic acids [26]. They are abundant in oils with strong fusty, musty and muddy defects [31,33]. In addition to their relatively low odour-thresholds, phenols affect the release of some volatile compounds during consumption [34].

Effect of the Harvesting Time

The harvesting time had a significant effect (Table 2) on most non-phenolic (11 out of 13) and some phenolic (two out of five) fermentation compounds included in Table 4. Most of the non-phenolic compounds (all except 2-methylbutanal and 2-methylpropanoic acid) were affected by the harvesting time (Table 2). These compounds tended to fluctuate from week to week, although a general increase in ethanol, 2-methylpropanal and 3-methylbutanal was found. For most compounds the highest values tended to appear in weeks 3 and 4. The increase in ethanol over the harvesting season matches a rise in this compound found throughout fruit ripening [13]. This compound, which arises from fruit sugar fermentation and has been proposed as a marker of oil deterioration [13], has been related to the winey-vinegary defect [31]. Likewise, the branched aldehydes increased over time (Table 4). Although no increase was found in Hojiblanca and Picual oils [13], an increase in these undesirable compounds during fruit storage has been reported [4]. These compounds generally possess low odour-thresholds [3] and, in fact, they and their corresponding alcohols and acids are related to the fusty defect [10].

With regard to the volatile phenols included in Table 4, only 2-phenylethanol and 2-ethylphenol were affected according to the the ANOVA results (Table 2), with significant fluctuations over the six weeks consisting of an increase in week 3 and a subsequent decrease (Table 4). To our knowledge no information about the effect of either the harvesting time or fruit ripening on the volatile phenols of olive oil is available. As mentioned above, volatile phenols are related to oil degradation [4]. It was suggested that its formation may depend on the resistance of olives to microbial decay, and they have been related to the free acidity values [35]. In our study there was not a clear change in the quality parameters over time (Table 1) and the only parameter affected (PV) also showed fluctuations instead of a steady increase. Therefore, results for the volatile phenols (Table 4), which fit those for the quality parameters, could confirm that there was not a clear quality loss as the harvesting season elapsed.

3.2.3. Oxidation-Derived Volatile Compounds

According to the ANOVA results (Table 2), most oxidation compounds included in Table 5 were significantly affected by the organic production (five out of the seven), all of them by the harvesting method, and two out of seven by the harvesting time. Some important oxidation compounds, such as (E)-dec-2-enal and the deca-2,4-dienal isomers, were not affected by any of the researched factors.

Effect of the Organic Production (Organic vs. Conventional)

Regarding the organic production, the differences found in all the compounds except for heptanal and octanal were larger than expected considering the relatively modest effect found on the chemical and physical-chemical parameters (Table 1). (E)-hept-2-enal and nona-2,4-dienal, which possess low odour-thresholds [3], tended to be more abundant in the Conventional than in Organic oils over time. Nonetheless, there was not a steady trend in the other compounds, without significant differences between the Organic and Conventional oils most of the weeks (Table 5), which suggests that other environmental factors might have modulated the effect of the organic practices on them. Previous studies on organic practices have not paid attention to its effect on the volatile oxidation compounds, apart from hexanal, which is also a well-known LOX compound. In any case, (E)-hept-2-enal and nona-2,4-dienal possess low odour thresholds [3].

Effect of the Harvesting Method

The effect of the harvesting method was significant on all the oxidation compounds included in Table 5, according to the ANOVA results (Table 2). Heptanal, (E)-hept-2-enal, (E)-oct-2-enal, octanal and nonanal were generally more abundant in the oils from ground-picked fruits than in the ones from tree-picked fruits for all the sampling weeks (Table 4). Conversely, hexa-2,4-dienal and nona-2,4-dienal were less abundant in the Ground oils.

Most of those compounds possess low odour-threshold [3]. (E)-hept-2-enal and (E)-2-octenal are among the main contributors to the rancid flavour in oil, and octanal and nonanal are also involved in this sensory defect [31]. In fact, octanal, nonanal and (E)-hept-2-enal are indicators of oxidative degradation [4]. Although oxidation compounds typically arise from oxidation reactions during oil storage, they are also formed as a consequence of fruit microbial activity [10,27], which is favoured when the fruits are collected from the ground. In fact, a higher content in octanal in oils from ground-picked fruits than from tree-picked ones was reported [26], although no information is available about the other compounds. Our results for the oxidation volatile compounds are in line with the marked effect found in the chemical and physical-chemical parameters (Table 1), and confirm for Verdial de Badajoz oil from unirrigated orchards the general rise in oxidation compounds when fruits are ground-collected.

Effect of the Harvesting Time

Regarding the harvesting time, only two out of the seven oxidation compounds included in Table 5 were affected, according to the ANOVA results (Table 2). There were significant fluctuations and also a slight decrease in hexa-2,4-dienal and (E)-hept-2-enal over the harvesting time regardless of the oil type (Table 5). However, most oxidation markers were not affected, harvesting over a six-week period having a slight influence on the oxidative volatile compounds of the Verdial de Badajoz oils, which could indicate that the harvesting time was suitably scheduled according to orchards ripening. Previous studies on the effect of harvesting during different periods did not included oxidation volatile compounds [13]. The slight effect is consistent with the results for the chemical and physical-chemical parameters, which were hardly affected (Table 1). Therefore, for unirrigated orchards, when timing is adequately set, only slight differences in the oxidation compounds are expected, the organic production and harvesting method having a much more noticeable effect.

3.3. Principal Component Analysis (PCA)

A Principal Component Analysis (PCA) was performed on the variables included in Tables 1 and 3–5 to explore the relationships among them and the general effect of the organic practices, harvesting method and harvesting time.

Results show a different distribution of the samples according to the organic practices and harvesting method. The Ground oils were plotted in the positive PC1 semiaxis (Figure 2a), where the quality parameters and most of the fermentation and oxidation compounds had large loadings (Figure 2b). Conversely, the oils from fruits collected from the trees (both the Conventional and Organic oils) appeared in the negative PC1 semiaxis, where the total phenols, oil stability and most LOX compounds reached large absolute loadings (Figure 2b). Figure 2b shows that most LOX compounds were positively related to the total phenol content and oil stability (most of them had negative loadings, generally under −0.5) and negatively related to the quality parameters and most of the fermentation and oxidation compounds (most of them had positive loadings, generally above 0.5). Therefore, factors which favour fermentation or oxidation (for example, collecting the fruits from the ground) seem to hinder the LOX-pathway compounds.

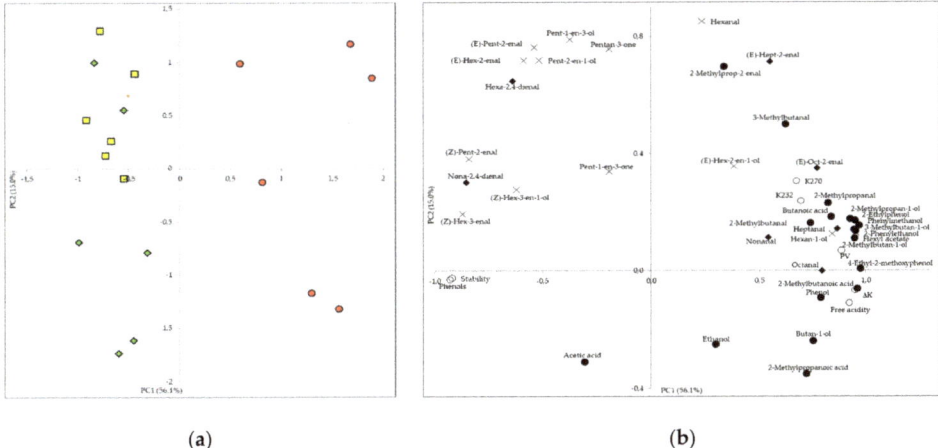

Figure 2. Projection of the oil samples (**a**) and variables from Tables 1 and 3–5 (**b**) onto the space defined by the first two principal components (PC1/PC2). ◆: Organic; ☐ Conventional; ⭕ Ground. o: Chemical and physical-chemical parameters; ×: LOX-derived volatile compounds; •: Fermentation compounds; ♦ Oxidation compounds.

In addition, the Organic oils tended to reach negative loadings in the PC2, whereas the Conventional ones tended to be displayed in the positive PC2 semiaxis. The variables with the largest absolute loadings in the PC2 were mostly LOX compounds, two oxidation compounds ((E)-hepten-2-al and hexa-2,4-dienal) and two fermentation compounds (2-methylprop-2-enal and 3-methylbutanal), all of them with positive values (Figure 2b). Therefore, these compounds seem to be more related to the Conventional than in the Organic oil according to the variability explained by the PC model.

Regarding the harvesting time, samples from the first weeks generally reached lower scores in the PC1 axis and higher in the PC2 axis, whereas the ones from the last weeks tended to follow the opposite trend. Therefore, at the beginning of the harvesting period, the oils tended to have higher scores in the variables positively related to oil quality, such as stability and phenol content, whereas at the end there was a slightly stronger relation to variables considered detrimental to oil quality (quality parameters, and fermentation and oxidation compounds).

4. Conclusions

Results show that the organic practices in unirrigated orchards had a noticeable yet commercially modest effect on the chemical and physical-chemical parameters and the volatile compound profile. The effect was much less strong than that of the harvesting method, which affected severely the chemical and physical-chemical parameters, including the quality parameters (which are used in the official oil grading), and the volatile compounds. Conversely, the harvesting time in real conditions was revealed to be a factor with little repercussion on the oil quality parameters, which might be due to a suitable harvesting time schedule, although it had a noticeable effect on some important volatile compounds. Previous studies have not shown a consistent effect of the organic production on the olive oil characteristics, partly because of the difficulty of controlling other sources of variation. Our results, for oil obtained in a commercial mill, reveal an increase in the total phenols in the organic oil, which was previously reported at laboratory scale and attributed to the decreased availability of nitrogen due to the lack of chemical fertilisation. This result, obtained under controlled conditions (e.g., the same mill, the same area, the same harvesting time and method) could explain the increase in oil stability and the changes in the volatile compound profile. Considering that some studies involving irrigated orchards have not shown an increase in phenols, it would be advisable to perform further studies under controlled conditions to shed light on how irrigation may affect the effect of the organic practices, and on whether or not the organic fertilisation currently used in commercially exploited orchards causes a consistent decrease in soil nitrogen availability. That information would help understand the real defect of the organic production and how to deal with it.

Author Contributions: Conceptualization and methodology, A.I.C., J.S.-C., L.M. and C.d.M.; formal analysis, A.I.C.; investigation, A.I.C., A.R., J.S.-C. and M.M.-C.; resources, A.I.C., J.S.-C. and L.M.; writing—original draft preparation, A.I.C. and A.R.; writing—review and editing, A.I.C., L.M., J.S.-C., M.M.-C. and C.d.M.; supervision and project administration, A.I.C., J.S.-C. and C.d.M.; funding acquisition, A.I.C., J.S.-C. and L.M. All authors have read and agreed to the published version of the manuscript.

Funding: This research was partially funded by Junta de Extremadura and FEDER, project GR18147.

Conflicts of Interest: The authors declare no conflict of interest.

References

1. Carrapiso, A.; Garcia, A.; Petron, M.; Martin, L. Effect of talc and water addition on olive oil quality and antioxidants. *Eur. J. Lipid Sci. Technol.* **2013**, *115*, 583–588. [CrossRef]
2. Sanchez-Ortiz, A.; Bejaoui, M.; Quintero-Flores, A.; Jimenez, A.; Beltran, G. Biosynthesis of volatile compounds by hydroperoxide lyase enzymatic activity during virgin olive oil extraction process. *Food Res. Int.* **2018**, *111*, 220–228. [CrossRef] [PubMed]
3. Kalua, C.M.; Allen, M.S.; Bedgood, D.R., Jr.; Bishop, A.G.; Prenzler, P.D.; Robards, K. Olive oil volatile compounds, flavour development and quality: A critical review. *Food Chem.* **2007**, *100*, 273–286. [CrossRef]
4. Hbaieb, R.; Kotti, F.; Gargouri, M.; Msallem, M.; Vichi, S. Ripening and storage conditions of Chetoui and Arbequina olives: Part, I. Effect on olive oils volatiles profile. *Food Chem.* **2016**, *203*, 548–558. [CrossRef]
5. Angerosa, F. Influence of volatile compounds on virgin olive oil quality evaluated by analytical approaches and sensor panels. *Eur. J. Lipid Sci. Technol.* **2002**, *104*, 639–660. [CrossRef]
6. Council Regulation (EC) 834/2007 of 28 June 2007 on Organic Production and Labelling of Organic Products and Repealing Regulation (EEC) No 2092/91. Available online: https://eur-lex.europa.eu/legal-content/EN/TXT/?uri=CELEX%3A32007R0834 (accessed on 22 November 2020).
7. Lima, G.P.P.; Vianello, F. Review on the main differences between organic and conventional plant-based foods. *Int. J. Food Sci. Technol.* **2011**, *46*, 1–13. [CrossRef]
8. Rosati, A.; Cafiero, C.; Paoletti, A.; Alfei, B.; Caporali, S.; Casciani, L.; Valentini, M. Effect of agronomical practices on carpology, fruit and oil composition, and oil sensory properties, in olive (*Olea europaea* L.). *Food Chem.* **2014**, *159*, 236–243. [CrossRef]
9. Ninfali, P.; Bacchiocca, M.; Biagiotti, E.; Esposto, S.; Servili, M.; Rosati, A.; Montedoro, G. A 3-year study on quality, nutritional and organoleptic evaluation of organic and conventional extra-virgin olive oils. *J. Am. Oil Chem. Soc.* **2008**, *85*, 151–158. [CrossRef]
10. Angerosa, F.; Servili, M.; Selvaggini, R.; Taticchi, A.; Esposto, S.; Montedoro, G. Volatile compounds in virgin olive oil: Occurrence and their relationship with the quality. *J. Chromatogr. A* **2004**, *1054*, 17–31. [CrossRef]
11. Salvador, M.; Aranda, F.; Fregapane, G. Influence of fruit ripening on 'Cornicabra' virgin olive oil quality—A study of four successive crop seasons. *Food Chem.* **2001**, *73*, 45–53. [CrossRef]
12. Sonmez, A.; Ozdikicierler, O.; Gumuskesen, A. Evaluation of olive oil quality during the ripening of the organic cultivated olives and multivariate discrimination of the variety with a chemometric approach. *Riv. Ital. Delle Sostanze Grasse* **2018**, *95*, 173–181.
13. Jimenez, B.; Rivas, A.; Lorenzo, M.; Sanchez-Ortiz, A. Chemosensory characterization of virgin olive oils obtained from organic and conventional practices during fruit ripening. *Flavour Fragr. J.* **2017**, *32*, 294–304. [CrossRef]
14. MAPA (Ministerio de Agricultura, Pesca y Alimentación). *Anuario de Estadística 2019*; MAPA (Ministerio de Agricultura, Pesca y Alimentación): Madrid, Spain, 2020.

15. Uceda, M.; Aguilera, M.P.; Jiménez, A.; Beltrán, G. Chapter 4: Variedades de olivo y aceituna. Tipos de Aceites. In *El Aceite de Oliva Virgen: Tesoro de Andalucía*; Gutiérrez, A.F., Carretero, A.S., Eds.; Unicaja Foundation Publishing: Málaga, Spain, 2009; pp. 107–137.
16. Commission Regulation 2568/91 of 11 July 1991 on the Characteristics of Olive Oil and Olive-Residue Oil and on the Relevant Methods of Analysis. Available online: https://eur-lex.europa.eu/legal-content/EN/TXT/?uri=CELEX%3A01991R2568-20151016 (accessed on 22 November 2020).
17. Montedoro, G.; Servili, M.; Baldioli, M.; Miniati, E. Simple and hydrolyzable phenolic-compounds in virgin olive oil. 1. Their extraction, separation, and quantitative and semiquantitative evaluation by HPLC. *J. Agric. Food Chem.* **1992**, *40*, 1571–1576. [CrossRef]
18. Franco, M.; Galeano-Diaz, T.; Sanchez, J.; De Miguel, C.; Martin-Vertedor, D. Antioxidant capacity of the phenolic fraction and its effect on the oxidative stability of olive oil varieties grown in the southwest of Spain. *Grasas Y Aceites* **2014**, *65*. [CrossRef]
19. Linstrom, P.J.; Mallard, W.G. *NIST Chemistry WebBook, NIST Standard Reference Database Number 69*; National Institute of Standards and Technology: Gaithersburg, MD, USA, 2020.
20. Hair, J.; Anderson, R.; Tatham, R.; Black, W. *Multivariate Data Analysis*, 5th ed.; Prentice Hall: Upper Saddle River, NJ, USA, 1998.
21. Garcia-Gonzalez, D.L.; Aparicio-Ruiz, R.; Morales, M.T. Chemical characterization of organic and non-organic virgin olive oils. *OCL Oilseeds Fats Crop. Lipids* **2014**, *21*, 501–506. [CrossRef]
22. Gutierrez, F.; Arnaud, T.; Albi, M.A. Influence of ecological cultivation on virgin olive oil quality. *J. Am. Oil Chem. Soc.* **1999**, *76*, 617–621. [CrossRef]
23. Fernandez-Escobar, R.; Beltran, G.; Sanchez-Zamora, M.; Garcia-Novelo, J.; Aguilera, M.; Uceda, M. Olive oil quality decreases with nitrogen over-fertilization. *Hortscience* **2006**, *41*, 215–219. [CrossRef]
24. Kalogiouri, N.; Aalizadeh, R.; Thomaidis, N. Investigating the organic and conventional production type of olive oil with target and suspect screening by LC-QTOF-MS, a novel semi-quantification method using chemical similarity and advanced chemometrics. *Anal. Bioanal. Chem.* **2017**, *409*, 5413–5426. [CrossRef]
25. Commission Regulation 61/2011 of 24 January 2011 Amending Regulation 2568/91 on the Characteristics of Olive Oil and Olive-Residue Oil and on the Relevant Methods of Analysis. Available online: https://eur-lex.europa.eu/legal-content/EN/TXT/?uri=CELEX%3A32011R0061 (accessed on 22 November 2020).
26. Jimenez, A.; Aguilera, M.; Beltran, G.; Uceda, M. Application of solid-phase microextraction to virgin olive oil quality control. *J. Chromatogr. A* **2006**, *1121*, 140–144. [CrossRef]
27. Vichi, S.; Romero, A.; Tous, J.; Caixach, J. The activity of healthy olive microbiota during virgin olive oil extraction influences oil chemical composition. *J. Agric. Food Chem.* **2011**, *59*, 4705–4714. [CrossRef]
28. Bendini, A.; Cerretani, L.; Carrasco-Pancorbo, A.; Gomez-Caravaca, A.; Segura-Carretero, A.; Fernandez-Gutierrez, A.; Lercker, G. Phenolic molecules in virgin olive oils: A survey of their sensory properties, health effects, antioxidant activity and analytical methods. An overview of the last decade. *Molecules* **2007**, *12*, 1679–1719. [CrossRef] [PubMed]
29. Gomez-Rico, A.; Salvador, M.D.; La Greca, M.; Fregapane, G. Phenolic and volatile compounds of extra virgin olive oil (Olea europaea L. cv. Cornicabra) with regard to fruit ripening and irrigation management. *J. Agric. Food Chem.* **2006**, *54*, 7130–7136. [CrossRef] [PubMed]
30. Luaces, P.; Perez, A.; Sanz, C. The effect of olive fruit stoning on virgin olive oil aroma. *Grasas Y Aceites* **2004**, *55*, 174–179.
31. Morales, M.; Luna, G.; Aparicio, R. Comparative study of virgin olive oil sensory defects. *Food Chem.* **2005**, *91*, 293–301. [CrossRef]
32. Vichi, S.; Romero, A.; Gallardo-Chacon, J.; Tous, J.; Lopez-Tamames, E.; Buxaderas, S. Influence of Olives' Storage Conditions on the Formation of Volatile Phenols and Their Role in Off-Odor Formation in the Oil. *J. Agric. Food Chem.* **2009**, *57*, 1449–1455. [CrossRef]
33. Vichi, S.; Romero, A.; Tous, J.; Tamames, E.; Buxaderas, S. Determination of volatile phenols in virgin olive oils and their sensory significance. *J. Chromatogr. A* **2008**, *1211*, 1–7. [CrossRef]
34. Genovese, A.; Caporaso, N.; Villani, V.; Paduano, A.; Sacchi, R. Olive oil phenolic compounds affect the release of aroma compounds. *Food Chem.* **2015**, *181*, 284–294. [CrossRef]

35. Vichi, S.; Romero, A.; Gallardo-Chacon, J.; Tous, J.; Lopez-Tamames, E.; Buxaderas, S. Volatile phenols in virgin olive oils: Influence of olive variety on their formation during fruits storage. *Food Chem.* **2009**, *116*, 651–656. [CrossRef]

Publisher's Note: MDPI stays neutral with regard to jurisdictional claims in published maps and institutional affiliations.

© 2020 by the authors. Licensee MDPI, Basel, Switzerland. This article is an open access article distributed under the terms and conditions of the Creative Commons Attribution (CC BY) license (http://creativecommons.org/licenses/by/4.0/).

Article

Characterization of Olive Oils Obtained from Minor Accessions in Calabria (Southern Italy)

Amalia Piscopo, Rocco Mafrica, Alessandra De Bruno, Rosa Romeo, Simone Santacaterina and Marco Poiana *

Department of AGRARIA, University Mediterranea of Reggio Calabria, 89124 Vito, Reggio Calabria, Italy; amalia.piscopo@unirc.it (A.P.); rocco.mafrica@unirc.it (R.M.); alessandra.debruno@unirc.it (A.D.B.); rosa.romeo@unirc.it (R.R.); simone.santacaterina@unirc.it (S.S.)
* Correspondence: mpoiana@unirc.it; Tel.: +39-0965-1694367

Abstract: The valorization of minor accessions of olive is potentially a good way to improve the qualitative production of a specific territory. Olive oils of four minor accessions (Ciciarello, Tonda di Filogaso, and Ottobratica Calipa and Ottobratica Cannavà clones) produced in the same area of the Calabria region were characterized for the principal qualitative analyses at two drupe harvesting periods (October and November). Good quality in terms of free acidity, peroxides, spectrophotometric indexes, and fatty acid composition was observed in olive oils produced at both drupe harvesting times, with the exception of those of Tonda di Filogaso, which showed a free acidity level over the legal limit for extra virgin olive oil in the second harvesting time. All of the olive oils possessed at both production periods averagely abundant total polyphenols (460–778 mg/kg) and tocopherols (224–595 mg/kg), and the amounts changed in the experimental years for expected different environmental variations. Ottobratica Cannavà and Ottobratica Calipa clones showed some peculiar qualitative characteristics (free acidity, peroxides, fatty acid composition, and total polyphenols), distancing themselves from the principal variety of reference, Ottobratica.

Keywords: clones; minor accessions; olive oil; quality

1. Introduction

In view of the recognized importance of the right lifestyle, primarily resulting in healthy eating, the daily consumption of olive oil is highly recommended for its dotation in monounsaturated fatty acids, in particular oleic acid, and antioxidant compounds, proven to reduce the incidence of cardiovascular and age-associated diseases [1]. Olive variety has a remarkable impact on absolute and relative concentrations of oil components, such as fatty acids, triacylglycerols, and sterols [2–4], and sensorial characteristics [5] and antioxidant compounds, such as polyphenols, tocopherols [6,7], and squalene [8,9]. Nowadays, studies on minor olive cultivars, also called neglected, have sparked interest in different countries for the topic of biodiversity protection and the possibility to improve, enrich, and diversify local olive oil productions [10–13].

The Italian olive heritage contains over 500 varieties; many of these are in Calabria [14], a region located in the southern Italy, particularly due to favorable geographic area, climate, and soil conditions that promote the diffusion of cultivars (about 33) to a different extent. Some of these are largely present along the Calabria region, such as Carolea cv. [15], some others grow in more specific areas, such as Grossa di Gerace, Ottobratica, and Sinopolese cv. [16,17], and others grow in limited towns, such as Roggianella [18]. In previous works, it was evidenced that the cultivation in the different areas of Calabria, where different microclimates are present, significantly impacts the diversification and typical characterization of productions, both from different varieties [16] and from the same cultivar [19]. Correlated with these results, the authors have conducted with this study a first investigation on qualitative parameters of olive oils obtained from four minor olive accessions, Ciciarello, Tonda di Filogaso, Ottobratica Calipa, and Ottobratica Cannavà, that are grown in the same area of Calabria.

This paper aims to investigate for the first time the chemical characteristics of olive oils from four minor olive accessions, Ciciarello, Tonda di Filogaso, Ottobratica Calipa, and Ottobratica Cannavà, present in the Tyrrenian Southern area of Calabria. Ottobratica Calipa and Ottobratica Cannavà are in particular two genotypes selected within the Ottobratica population variety in the last decades by the olive growers of this specific territory of the Calabria region [20]. The study focused on olive trees cultivated in the same area of Calabria. This approach was considered to exclude possible different effects of climatic conditions among the varieties, except those linked to the annual trend that occurred similarly for all four varieties. This research represents an interesting opportunity for olive oil production in Calabria. Despite their low diffusion in the whole region as a result of past selections, the minor olive accessions must be studied because, being autochthonous, they possess various characteristics of rusticity and adaptability to the microclimate. This study can also contribute to the protection of olive biodiversity in the Calabria region and its valorization at the same time. Moreover, the chemical characterization of obtained olive oils gives new knowledge, and can be considered as a valid instrument to improve and strengthen qualitative olive productions in Calabria.

2. Materials and Methods

2.1. Sampling

The studied olive oils were obtained from four olive accessions (Ciciarello, Tonda di Filogaso, Ottobratica Calipa, and Ottobratica Cannavà) in a fifteen-year-old olive grove located in Gioia Tauro Plain, an important olive growing area located on the Tyrrhenian side of the Calabria region (southern Italy). Ottobratica cultivar was also submitted to the research as a reference for its related clones. Previous morphological and molecular characterization studies conducted on the two clones of Ottobratica [20] averted the risks of cases of synonymy or homonymy, both between the two clones and with the most widespread type of Ottobratica (used in this study as a reference element). The orchard was characterized by homogeneous trees, in good vegetative and productive condition, trained according to the open-center training system, spaced 6.0 × 6.0 m, and grown under rain-fed conditions. The soil of the olive orchard was deep, without a skeleton, had a medium texture, was non-calcareous, and with a sub-acid reaction. During the three years of trials, 2017, 2018, and 2019, the average annual temperature and rainfall were, respectively, 15 °C and 1427 mm. The fertilization was carried out at the end of winter with the controlled release fertilizer (N:P:K 21:5:9 with microelements) at three kilograms per tree. In order to ensure the health integrity of trees and fruits, continuous monitoring for the main olive parasites was carried out, using pest control treatments when necessary and according to the principles of integrated pest management.

The experiment was carried out considering three blocks of the four accessions, each composed of three olive trees. About 10–12 kg of drupes were sampled from each block in two harvesting times: October (O) and November (N) of 2017, 2018, and 2019.

2.2. Analytical Methods

The oil yield (% oil dry weight) was determined in drupes after stone removing by extraction with petroleum ether in a Soxhlet apparatus (Bicasa s.r.l., Bernareggio, MI, Italy). For the olive oil extraction, about 15 kg of drupes were milled with a hammer mill. The obtained paste was mixed at a temperature below 20–25 °C for 30 min and pressed using a hydraulic press (pressure up to 200 bar) in a small olive oil press mill Mini 30 system (Agrimec Valpesana, Firenze, Italy). After centrifugation and filtration through paper, olive oils were then stored in dark glass bottles at room temperature and analyzed for the total free acidity value, peroxide index, and UV light absorption coefficients according to EC regulations [21,22]. Pigments were extracted from the oil samples (5 mL of oil and 5 mL of cyclohexane) following the method reported by Minguez-Mosquera et al. [23], and the total contents of chlorophylls and carotenoids were determined spectrophotometrically (670 nm and 470 nm, respectively). Total tocopherols were evaluated according to Bakre et al. [24]. The oil samples were diluted in isopropanol (1:10) and filtered (0.45 µ pore size). An aliquot of 5 µL of samples was injected in an ultra-high performance liquid chromatography (UHPLC) system (UHPLC PLATINblue, Knauer, Germany), coupled with a fluorescence detector RF-20A/RF-20Axs model (Shimadzu Corporation, Kyoto, Japan) and analyzed (flow rate of 0.4 mL min^{-1}) through a mobile phase of methanol/acetonitrile (50:50). The detector was set at a 290 nm excitation wavelength and 330 nm emission wavelength. The identification and quantification were performed by calibration curve, using pure α-tocopherol as the standard and concentrations ranging from 10 to 500 mg kg^{-1}. Results were expressed as mg kg^{-1}. Determination of total polyphenols was performed following Baiano et al. [25]. Two mL of methanol/water (70:30, v/v) and 2 mL of hexane were added to 5 g of oil samples and mixed with a vortex (10 min). The hydro-alcoholic phase containing phenols was separated from the oil phase by several centrifugations; 100 µL of phenolic extract were mixed with 100 µL of Folin–Ciocalteau reagent (2N) and, after 4 min, with 800 µL of an aqueous solution of Na_2CO_3 (5%). The mixture was heated in a 40 °C water bath for 20 min, and the total phenol content was determined calorimetrically at 750 nm. The total phenolic content was expressed as milligrams of gallic acid equivalents per kilogram of oil. The total antioxidant activity of the olive oils was detected by 2,2′-azino-bis (3-ethylbenzothiazoline-6-sulfonic acid) (ABTS)/Trolox equivalent antioxidant capacity (TEAC), according to Re et al. [26], and 2,2-diphenyl-1-picrylhydrazyl (DPPH), following the opportunely modified method of Brand-Williams et al. [27]. Fatty acid composition was determined as methyl esters (FAME) following the official method [22].

2.3. Statistical Data Elaboration

The results of the analyses were elaborated as mean ± standard deviations of three sampling years for two harvesting times. Significant differences ($p < 0.05$) were obtained by one-way analysis of variance (ANOVA) and multivariate analysis. Pearson's coefficient was used to study the correlation among qualitative parameters of olive oils. SPSS Software (Version 15.0, SPSS Inc., Chicago, IL, USA) was used for statistical elaboration.

3. Results

Mean results of the olive oil yield during the three years of study for the minor olive varieties are reported in Figure 1. Drupes of Tonda di Filogaso and Ottobratica Calipa possessed similar oil content at the first sampling (27–29% d.m.), whereas Ciciarello and Ottobratica Cannavà differed for less abundant oiliness (18–19%). In the following harvesting period, the oil content remained significantly similar in Tonda di Filogaso cv, whereas it tended to increase with the highest result in Ottobratica Calipa (44%). The ripening index varied among varieties and harvesting months, as Supplementary Materials shows (Table S1).

The results of principal qualitative parameters of oils, as three-year means, are illustrated in Tables 1 and 2.

During the three years and at the second production in particular, that is November, a large variability in free acidity was observed in olive oils from the same accession, except in the Ottobratica Cannavà clone (0.55 ± 0.13). The range observed in oils produced in October was 0.31–0.57%; at November, it tended to increase in all of the samples, exceeding the 0.8% in some years, except for Ottobratica Cannavà oils. The other productions were affected probably by a different varietal response to some negative environmental factors linked to a specific year (Tables S2–S6): 2018 for Ottobratica Calipa and Ciciarello (total acidity >1%), and 2019 for Tonda di Filogaso (total acidity near 2.5%), as evidenced by the values of the standard deviations. The observed low quality of Ottobratica olive oil produced in November (Table 2) was confirmed by previous works [28,29], and reflected the origin of its name, strictly linked to its optimal ripening in the month of October. It is interesting to note that for one of the two Ottobratica-related clones, Ottobratica Cannavà, the free acidity inside the legal limit of 0.8% in that period expressed a positive result of the performed new genetic duplication.

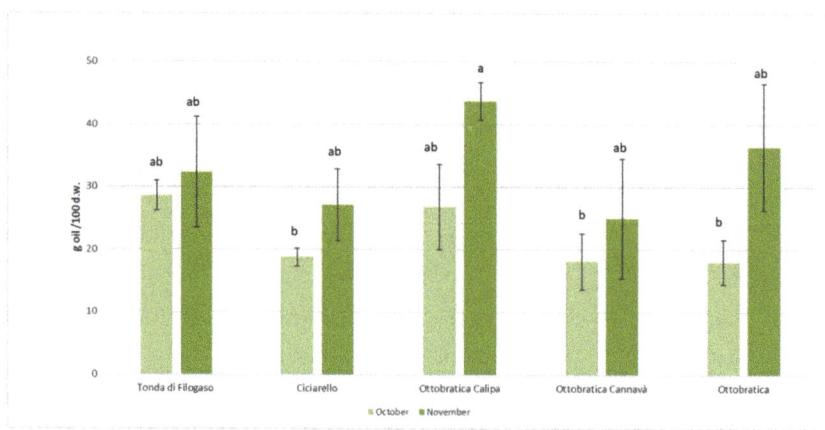

Figure 1. Olive oil yield at two harvesting times of the four studied minor varieties, with Ottobratica cv used as the reference for Calipa and Cannavà clones. Values are the means of 2017, 2018, and 2019. Different letters show significant differences at $p < 0.05$ by Tukey's post hoc test.

Table 1. Principal chemical parameters of olive oils of Tonda di Filogaso (TF), Ciciarello (C), Ottobratica Calipa. (O. CLP) Ottobratica Cannavà (O. CNV) and Ottobratica (O) accessions.

Qualitative Parameters	Accessions	Harvesting Times		Sign.
		O	N	
FA (oleic acid %)	TF	0.55 ± 0.18a	1.47 ± 1.11	*
	C	0.31 ± 0.06b	0.69 ± 0.39	*
	O. CLP	0.57 ± 0.24a	0.74 ± 0.54	n.s.
	O. CNV	0.53 ± 0.13ab	0.55 ± 0.13	n.s
	O	0.40 ± 0.09ab	1.18 ± 1.08	*
Sign.		*	n.s.	
PV (mEq O_2/kg)	TF	3.57 ± 1.05ab	3.59 ± 1.94	n.s.
	C	2.50 ± 0.52b	3.13 ± 1.53	n.s.
	O. CLP	2.71 ± 0.99ab	3.14 ± 1.70	n.s
	O. CNV	3.88 ± 1.27a	5.58 ± 1.12	*
	O	2.30 ± 0.64b	6.63 ± 3.88	**
Sign.		**	n.s.	
K_{232}	TF	1.94 ± 0.15	1.62 ± 0.22	**
	C	1.58 ± 0.55	1.51 ± 0.20	n.s
	O. CLP	2.08 ± 0.39	1.82 ± 0.25	n.s
	O. CNV	1.82 ± 0.47	1.74 ± 0.40	n.s
	O	1.81 ± 0.26	1.76 ± 0.51	n.s.
Sign.		n.s.	n.s.	
K_{270}	TF	0.22 ± 0.04	0.15 ± 0.05	*
	C	0.19 ± 0.07	0.19 ± 0.03	n.s
	O. CLP	0.23 ± 0.09	0.22 ± 0.08	n.s
	O. CNV	0.18 ± 0.14	0.13 ± 0.09	n.s
	O	0.19 ± 0.06	0.19 ± 0.07	n.s.
Sign.		n.s.	n.s.	
ΔK	TF	0.00 ± 0.00	0.00 ± 0.00	n.s.
	C	0.00 ± 0.00	0.00 ± 0.00	n.s
	O. CLP	0.00 ± 0.00	0.00 ± 0.00	n.s
	O. CNV	0.00 ± 0.00	0.00 ± 0.00	n.s
	O	0.00 ± 0.00	0.00 ± 0.00	n.s
Sign.		n.s.	n.s.	

The data are presented as means ± standard deviations. ** Significance at $p < 0.01$; * significance at $p < 0.05$; n.s., not significant; a, ab, b see Figure 1.

Table 2. Fatty acid compositions of olive oils of Tonda di Filogaso (TF), Ciciarello (C), Ottobratica Calipa. (O. CLP) Ottobratica Cannavà (O. CNV) and Ottobratica (O) accessions.

	Accessions	Harvesting Times O	N	Sign.		Accessions	Harvesting Times O	N	Sign.
C16:0 (%)	TF	15.07 ± 1.87	15.11 ± 0.13	n.s.	C18:2 (%)	TF	8.55 ± 1.62ab	10.06 ± 1.59	n.s.
	C	13.23 ± 0.84	13.51 ± 0.93	n.s.		C	5.96 ± 1.78b	7.59 ± 2.90	n.s.
	O. CLP	13.86 ± 3.04	14.49 ± 1.42	n.s.		O. CLP	8.93 ± 3.35ab	8.92 ± 2.35	n.s.
	O. CNV	13.40 ± 2.37	13.42 ± 2.22	n.s.		O. CNV	7.78 ± 2.59ab	7.52 ± 2.31	n.s.
	O	15.74 ± 0.69	14.96 ± 1.21	n.s.		O	9.28 ± 1.19a	9.14 ± 1.95	n.s.
Sign.		n.s.	n.s.		Sign.		*	n.s..	
C16:1 (%)	TF	1.50 ± 0.23	1.63 ± 0.24	n.s.	C18:3 (%)	TF	0.83 ± 0.38	0.56 ± 0.09	n.s.
	C	0.99 ± 0.25	1.03 ± 0.19	n.s.		C	0.58 ± 0.05	0.52 ± 0.06	*
	O. CLP	1.63 ± 0.81	1.64 ± 0.43	n.s.		O. CLP	0.74 ± 0.11	0.64 ± 0.17	n.s.
	O. CNV	1.07 ± 0.33	1.40 ± 0.67	n.s.		O. CNV	0.74 ± 0.50	0.61 ± 0.17	n.s.
	O	1.59 ± 0.27	1.49 ± 0.41	n.s.		O	0.62 ± 0.08	0.63 ± 0.08	n.s.
Sign.		*	n.s.		Sign.		n.s..	n.s..	
C17:0 (%)	TF	0.09 ± 0.07ab	0.09 ± 0.06	n.s.	C20:0 (%)	TF	0.34 ± 0.12b	0.38 ± 0.04b	n.s.
	C	0.06 ± 0.04b	0.08 ± 0.05	n.s.		C	0.46 ± 0.05a	0.46 ± 0.02a	n.s.
	O. CLP	0.04 ± 0.02b	0.04 ± 0.03	n.s.		O. CLP	0.41 ± 0.10ab	0.37 ± 0.04b	n.s.
	O. CNV	0.09 ± 0.11ab	0.05 ± 0.01	n.s.		O. CNV	0.45 ± 0.02ab	0.41 ± 0.05ab	n.s.
	O	0.18 ± 0.14a	0.10 ± 0.07	n.s.		O	0.41 ± 0.05ab	0.42 ± 0.06ab	n.s.
Sign.		*	n.s.		Sign.		*	**	
C17:1 (%)	TF	0.17 ± 0.14	0.19 ± 0.11	n.s.	C20:1 (%)	TF	0.28 ± 0.06	0.23 ± 0.01b	n.s.
	C	0.10 ± 0.08	0.15 ± 0.08	n.s.		C	0.30 ± 0.03	0.29 ± 0.05a	n.s.
	O. CLP	0.09 ± 0.04	0.10 ± 0.05	n.s.		O. CLP	0.27 ± 0.09	0.30 ± 0.05a	n.s.
	O. CNV	0.13 ± 0.09	0.10 ± 0.03	n.s.		O. CNV	0.29 ± 0.05	0.28 ± 0.03ab	n.s.
	O	0.72 ± 1.43	0.19 ± 0.11	n.s.		O	0.26 ± 0.04	0.28 ± 0.02ab	n.s.
Sign.		n.s.	n.s.		Sign.		n.s.	*	
C18:0 (%)	TF	1.72 ± 0.76	3.44 ± 0.34	**	C22:0 (%)	TF	0.22 ± 0.28	0.12 ± 0.01ab	n.s.
	C	2.32 ± 1.11	2.56 ± 1.82	n.s.		C	0.15 ± 0.02	0.15 ± 0.02a	n.s.
	O. CLP	1.33 ± 0.63	2.06 ± 1.19	n.s.		O. CLP	0.12 ± 0.05	0.12 ± 0.02b	n.s.
	O. CNV	1.95 ± 0.72	1.84 ± 1.16	n.s.		O. CNV	0.15 ± 0.01	0.13 ± 0.02ab	n.s.
	O	2.49 ± 0.93	1.77 ± 0.61	n.s.		O	0.15 ± 0.01	0.14 ± 0.03ab	n.s.
Sign.		n.s.	n.s.		Sign.		n.s.	*	
C18:1 (%)	TF	71.11 ± 3.99ab	68.15 ± 1.25b	n.s.	MUFA/PUFA	TF	8.13 ± 1.78	6.77 ± 1.14	n.s.
	C	75.79 ± 2.82a	73.61 ± 4.36ab	n.s.		C	12.56 ± 2.84	10.47 ± 3.41	n.s.
	O. CLP	72.46 ± 4.48ab	71.26 ± 3.92ab	n.s.		O. CLP	8.78 ± 3.72	8.31 ± 2.82	n.s.
	O. CNV	73.87 ± 4.88a	74.18 ± 4.22a	n.s.		O. CNV	9.95 ± 3.88	10.24 ± 3.57	n.s.
	O	68.50 ± 1.26b	70.83 ± 3.01ab	*		O	7.29 ± 1.09	7.88 ± 2.50	n.s.
Sign.		**	*		Sign.		n.s.	n.s.	

The data are presented as means ± standard deviations. **, *, n.s. see Table 1; a, ab, b see Figure 1.

Peroxide values of oils were in the range of 2.50–5.58 mEq O_2/kg, with significant differences between the two harvesting times only in Ottobratica Cannavà oils. Higher peroxide values were noted in the oils of clones compared to those from Ottobratica cv in October, whereas an opposite result was detected for the productions of November (6.63 mEq O_2/kg in oils from Ottobratica). Spectrophotometric indices denoted olive oil productions of good quality at both harvesting times without significant differences, with the only exception of Tonda di Filogaso olive oils.

The major fatty acid in olive oil is oleic acid; in our study, its content varied with significance from 68.15% (Tonda di Filogaso oils in November) to 75.79% (Ciciarello oils in October). The followed principal detected fatty acids were palmitic acid (C16:0), quantified from 13.23% to 15.74%, and then linoleic acid (C18:2) that varied from 5.96 to 10.06%; both components were similar among minor varieties. The stearic acid (C18:0) was significantly higher in the oils of Tonda di Filogaso obtained in November (3.44%) than in the other samples. The other fatty acids that significantly varied among the samples were C16:1, among the unsaturated ones, from 0.99 to 1.64% and C20:0, among the saturated ones, from 0.34 to 0.46%, as evidenced by the Tukey's post hoc test ($p < 0.05$) elaboration (data not shown). Olive oils from Ciciarello showed the highest oleic/linoleic (11–13) and monounsaturated/polyunsaturated acid (MUFA/PUFA) (10–12) ratios, confirming the previously discussed results for fatty acid quantification. The antioxidant compositions of olive oil from minor accessions is reported in Table 3.

Table 3. Antioxidant composition and activity of Tonda di Filogaso (TF), Ciciarello (C), Ottobratica Calipa. (O. CLP) Ottobratica Cannavà (O. CNV) and Ottobratica (O) accessions.

		Accessions	Harvesting Times		Sign.
			O	N	
Antioxidant property	TChl	TF	10.21 ± 5.05	3.03 ± 0.90b	**
		C	10.95 ± 3.41	3.41 ± 1.02b	**
		O. CLP	7.00 ± 2.18	2.53 ± 1.23b	**
		O. CNV	6.35 ± 3.58	6.19 ± 3.84a	n.s
		O	8.76 ± 3.72	3.52 ± 1.17ab	**
	Sign.		n.s.	**	
	TCa	TF	5.97 ± 1.82	2.71 ± 0.64b	**
		C	7.16 ± 2.28	3.25 ± 1.15ab	**
		O. CLP	5.17 ± 0.92	2.26 ± 0.90b	**
		O. CNV	4.98 ± 2.69	5.46 ± 3.58a	n.s
		O	5.72 ± 1.58	3.32 ± 1.02ab	**
	Sign.		n.s.	*	
	TT	TF	227 ± 19c	224 ± 10c	n.s
		C	289 ± 18bc	242 ± 30bc	n.s
		O. CLP	324 ± 19b	309 ± 36a	n.s
		O. CNV	595 ± 66a	286 ± 32ab	**
		O	266 ± 44bc	238 ± 33bc	n.s.
	Sign.		**	**	
	TP	TF	615 ± 403	516 ± 130	n.s.
		C	460 ± 123	486 ± 196	n.s.
		O. CLP	617 ± 397	446 ± 279	n.s
		O. CNV	778 ± 235	695 ± 318	n.s
		O	560 ± 453	334 ± 245	n.s.
	Sign.		n.s.	n.s.	
Antioxidant activity	DPPH assay	TF	20.58 ± 13.62	14.09 ± 6.65b	n.s
		C	25.40 ± 13.70	13.80 ± 3.11b	n.s
		O. CLP	21.17 ± 11.50	20.50 ± 7.77ab	n.s
		O. CNV	34.71 ± 3.99	26.07 ± 2.70a	n.s
		O	21.97 ± 6.76	15.62 ± 12.33ab	n.s
	Sign.		n.s.	*	
	ABTS assay	TF	29.01 ± 3.67b	31.72 ± 12.11	n.s
		C	31.69 ± 10.26ab	33.79 ± 17.80	n.s
		O. CLP	23.29 ± 10.58b	24.76 ± 10.07	n.s
		O. CNV	45.67 ± 15.69a	37.52 ± 13.27	*
		O	25.49 ± 7.44b	32.07 ± 23.36	n.s
	Sign.		**	n.s.	

Amount expressed as mg/kg for TChl (total chlorophylls), TCa (total carotenoids), TT (total tocopherols), TP (total polyphenols), and % inhibition/mg for DPPH and ABTS assays. The data are presented as means ± standard deviations; **, *, n.s. see Table 1; a, ab, b, c see Figure 1.

Significant ($p < 0.01$) differences of pigment amounts were observed among the samples; the olive oils of Ciciarello and Tonda di Filogaso obtained in October were the richest in chlorophylls (10.95 ± 3.41 mg/kg and 10.21 ± 5.05 mg/kg, respectively). Ciciarello olive oils were even the richest in total carotenes (7.16 ± 2.28 mg/kg). The oils extracted in November showed reduced pigment amounts and, in particular, a major reduction was observed in Ottobratica Calipa olive oils (TCL: 2.53 ± 1.23 mg/kg and TCA: 2.26 ± 0.90 mg/kg). ANOVA data elaboration showed variations for pigments between harvesting times, except in the oils of the Ottobratica Cannavà clone (Table 3).

Chlorophyll and carotenoid amounts were also significantly ($p < 0.05$) higher than those resulted in the cultivar population of reference (Ottobratica cv).

Comparing the total mean amounts of polyphenols quantified at two harvesting times, no significant differences were noted.

Among productions in November, Ottobratica Cannavà oils possessed higher mean phenolic antioxidant amount than the other clone, Ottobratica Calipa, and Ottobratica cv. The total tocopherols detected in the oils from minor accessions were in the range of 225–595 mg/kg; Ottobratica Cannavà olive oils were the richest for this typology of antioxidants (with the only observed significant variation between harvesting times), followed by Ottobratica Calipa olive oils, whereas lower amounts were detected in those from Tonda di Filogaso (224–227 mg/kg) as confirmed by literature [30]. A significant decrease in TT content was observed only in oils from Ottobratica Cannavà extracted in November. In the other productions, the total tocopherols remained constant.

The antioxidant activity of the oils was analyzed by the reaction against two radicals, DPPH and ABTS. The obtained results denoted a higher response with the second antioxidant assay (23.29–45.67%) than the DPPH radical (13.80–34.71%). The largest differences among the varieties were significantly observed in the oils produced in October for ABTS assays (Ottobratica Cannavà > Ottobratica Calipa > Ciciarello = Tonda di Filogaso).

4. Discussion

The olive oil accumulation on fruits during ripening follows the triglyceride-forming biosynthesis pathway up to the achievement of full drupe maturation. The olive oil yield in fruits is influenced by the choice of the right harvesting time for each variety and by several growing conditions, such as water availability [31]. In our study, an evident effect of varietal characteristics was observed, and the two Ottobratica clones differed for mean oil production at both harvesting times. Free acidity is generally the first parameter discussed to evaluate the quality of olive oil production; it is well-known that oil-free acidity can be affected by many factors, including fruit handling harvesting mean, storage, and processing, but also by the harvesting time. All of the oils produced in October were inside the limits for the extra virgin olive category [21], and, in particular, those of Ciciarello possessed the lowest mean acidity among the other minor accessions ($p < 0.05$). It is interesting to note that good oils from Ottobratica clones could be obtained at both harvesting times, without significant variations among each other ($p > 0.05$) and, different from Ottobratica, their reference variety. The oils of this last production in November denoted the previously discussed variability during the years and, on average, poor quality (free acidity of 1.18 ± 1.08%). Peroxides and extinction coefficients complied with the regulation limits for the extra virgin olive oil [21]. Fatty acid composition of samples was inside the limits imposed by European regulation for the extra virgin category [32]. The fatty acid composition did not largely vary between harvesting times in oils from each minor accession; only the oils of Tonda di Filogaso were different for stearic acid (means of 1.72% in October and 3.44% in November). Oleic acid was particularly abundant in oils from Ciciarello and Ottobratica Cannavà; for this chemical parameter, oils becoming to this clone were of higher quality with respect to those from Ottobratica cv.

The molecules responsible for olive oil color are pigments belonging to chlorophyll and carotenoid compounds. Their quantification is important to determine not only the sensorial characters and consumer acceptability of olive oils, but also their antioxidant potentiality; olive oil chlorophylls react as radical scavengers in dark storage and as prooxidant (sensitizer pigments) in light. Carotenes instead protect cells against the light, with oxygen and sensitizer pigment effects having the ability to quench singlet oxygen and excited sensitizer molecules. Moreover, they can also react as antioxidants under conditions other than photosensitization [33]. The total pigment content varies among varieties, drupe ripening, or olive oil stocking before the extraction [34,35].

In particular, the total content can range from 2 to 40 mg/kg for chlorophylls and from zero to a few mg/kg for carotenoids [36,37]. Olive oil produced at the second harvesting time showed reduced amounts of both pigments in our study, according to the literature [38]. Among the oil samples from the different accessions, only those of Cannavà did not vary for total pigments. Total tocopherols were detected in oils from minor varieties at a higher content than those observed in oils obtained from other cultivars in Calabria [15]. The quantified total phenols were in the range of 460–778 mg/kg, according to Fabiani [39], manifesting a strong antioxidant potentiality. A positive correlation between the total phenol content and ABTS assay was indeed evidenced by a high Pearson coefficient (r = 0.7–0.9) in all of the oils, in particular those extracted in November, according to Sicari [40]. It confirms their usefulness in providing the minimum intake of 5 mg of hydroxytyrosol per serving of olive oil (total phenol content >250 mg/kg) that is required to manifest the antioxidant effect in a balanced diet.

Finally, a multivariate data analysis was performed to evidence the influence of varietal characteristics or drupe harvesting times to the olive oil quality (Table 4). Results showed an evident effect due to the olive origin, especially for the prevalent fatty acids ($p < 0.00$) in the olive oils. From this data elaboration that considered overall data for all of the three years of experimentation, it was noted that, among the other qualitative parameters, in particular the total carotene content was affected exclusively by the drupe ripening, as the literature [35] confirms.

Table 4. Multivariate analyses of qualitative characteristics of olive oils extracted from drupes produced from the four minor accessions and at two harvesting times.

Variables	Accession	Harvesting Time
FA	**	**
PV	**	**
ΔK	**	n.s.
TChl	*	**
TCa	n.s.	**
TT	**	**
TP	*	**
C16:0	**	n.s.
C18:1	**	n.s.
C18:2	**	n.s.

** Significance at $p < 0.01$; * significance at $p < 0.05$; n.s., not significant.

5. Conclusions

This study allowed the characterization of the oil productions from four minor olive accessions grown in the same area of Calabria, with the aim to compare the qualitative differences measured during three years of observations, excluding climatic variables due to different environmental conditions. For some of these (oils from Tonda di Filogaso and Ciciarello cv), harvesting times significantly affected the results for free acidity and total pigments. All of the oil productions obtained in October possessed the chemical parameters to be classified as extra virgin olive oils. Comparing all of the studied accessions, olive oils from Ciciarello cv and Calipa and Cannavà clones also showed good quality when extracted in November. This is interesting for growing practices in the same studied area and for new knowledge about the potentiality of the two new clones obtained from the Ottobratica cultivar. In particular, the oils of the Ottobratica Cannavà clone showed better quality for free fatty acidity at both harvesting times, with oleic acid content and total antioxidants (polyphenols and tocopherols) with respect to the cultivar of reference, Ottobratica, largely diffused in the considered territory.

Supplementary Materials: The following are available online at https://www.mdpi.com/2304-8158/10/2/305/s1, Table S1: Ripening index of the olives at harvest times, Table S2: Qualitative characterization of the Tonda di Filogaso oils, Table S3: Qualitative characterization of the Ciciarello oils, Table S4: Qualitative characterization of the Ottobratica Calipa oils, Table S5: Qualitative characterization of the Ottobratica Cannavà oils, Table S6: Qualitative characterization of the Ottobratica oils.

Author Contributions: Conceptualization, R.M. and M.P.; Data curation, A.P., M.P., and R.M.; Formal analysis, A.D.B., R.R., and S.S.; Funding acquisition, M.P.; Methodology, A.P., A.D.B., and R.R.; Resources, M.P.; Supervision, A.P. and M.P. All authors have read and agreed to the published version of the manuscript.

Funding: This research was funded by AGER 2 PROJECT, grant number 2016-0105.

Institutional Review Board Statement: Not applicable.

Informed Consent Statement: Not applicable.

Data Availability Statement: The datasets presented in this study are available as Supplementary materials.

Acknowledgments: The authors thank the Olearia San Giorgio firm for sample supplying and Antonino Tramontana for assistance in conducting olive oil extraction.

Conflicts of Interest: The authors declare no conflict of interest.

References

1. Nocella, C.; Cammisotto, V.; Fianchini, L.; D'Amico, A.; Novo, M.; Castellani, V.; Stefanini, L.; Violi, F.; Carnevale, R. Extra Virgin Olive Oil and Cardiovascular Diseases: Benefits for Human Health. *Endocr. Metab. Immune Disord. Drug Targets* **2018**, *18*, 4–13. [CrossRef]
2. Poiana, M.; Mincione, A. Fatty acids evolution and composition of olive oils extracted from different olive cultivars grown in Calabrian area. *Grasas Aceites* **2004**, *55*, 282–290. [CrossRef]
3. Servili, M.; Selvaggini, R.; Esposto, S.; Taticchi, A.; Montedoro, G.F.; Morozzi, G. Health and sensory properties of virgin olive oil hydrophilic phenols: Agronomic and technological aspects of production that affect their occurrence in the oil. *J. Chromatogr. A* **2004**, *1054*, 113–127. [CrossRef]
4. Giuffrè, A.M. Variation in triacylglycerols of olive oils produced in Calabria (Southern Italy) during olive ripening. *Riv. Ital. Sostanze Grasse* **2014**, *91*, 221–240.
5. Inglese, P.; Famiani, F.; Servili, M. I fattori di variabilità genetici, ambientali e colturali della composizione dell'olio di oliva. *Italus Hortus* **2009**, *16*, 67–81.
6. Lo Curto, S.; Dugo, G.; Mondello, L.; Errante, G.; Russo, M.T. Variation of tocopherol content in virgin italian olive oils. *J. Ital. Food Sci.* **2001**, 221–228.
7. Dugo, L.; Russo, M.; Cacciola, F.; Mandolfino, F.; Salafia, F.; Vilmercati, A.; Fanali, C.; Casale, M.; De Gara, L.; Dugo, P.; et al. Determination of the Phenol and Tocopherol Content in Italian High-Quality Extra-Virgin Olive Oils by Using LC-MS and Multivariate Data Analysis. *Food Anal. Methods* **2020**, *13*, 1027–1041. [CrossRef]
8. Beltran, G.; Bucheli, M.E.; Aguilera, M.P.; Belaj, A.; Jimenez, A. Squalene in virgin olive oil: Screening of variability in olive cultivars. *Eur. J. Lipid Sci. Technol.* **2016**, *118*, 1250–1253. [CrossRef]
9. Navas-López, J.F.; Cano, J.; de la Rosa, R.; Velasco, L.; León, L. Genotype by environment interaction for oil quality components in olive tree. *Eur. J. Agron.* **2020**, *119*, 126115. [CrossRef]
10. Conte, P.; Squeo, G.; Difonzo, G.; Caponio, F.; Fadda, C.; Del Caro, A.; Urgeghe, P.P.; Montanari, L.; Montinaro, A.; Piga, A. Change in Quality During Ripening of Olive Fruits and Related Oils Extracted from Three Minor Autochthonous Sardinian Cultivars. *Emir. J. Food Agric.* **2019**, *31*, 196–205. [CrossRef]
11. Mousavi, S.; Stanzione, V.; Mencuccini, M.; Baldoni, L.; Bufacchi, M.; Mariotti, R. Biochemical and molecular profiling of unknown olive genotypes from central Italy: Determination of major and minor components. *Eur. Food Res. Technol.* **2019**, *245*, 83–94. [CrossRef]
12. Omri, A.; Abdelhamid, S.; Ayadi, M.; Araouki, A.; Gharsallaoui, M.; Gouiaa, M.; Benincasa, C. The investigation of minor and rare Tunisian olive cultivars to enrich and diversify the olive genetic resources of the country. *J. Food Compos. Anal.* **2021**, *95*, 103657. [CrossRef]
13. Salazar-García, D.C.; Malheiro, R.; Pereira, J.A.; Lopéz-Cortés, I. Unexplored olive cultivars from the Valencian Community (Spain): Some chemical characteristics as a valorization strategy. *Eur. Food Res. Technol.* **2019**, *245*, 325–334. [CrossRef]
14. Marra, F.P.; Caruso, T.; Costa, F.; Di Vaio, C.; Mafrica, R.; Marchese, A. Genetic relationships, structure and parentage simulation among the olive tree (*Olea europaea* L. subsp. *europaea*) cultivated in Southern Italy revealed by SSR markers. *Tree Genet. Genomes* **2013**, *9*, 961–973. [CrossRef]

15. Piscopo, A.; De Bruno, A.; Zappia, A.; Ventre, C.; Poiana, M. Characterization of monovarietal olive oils obtained from mills of Calabria region (Southern Italy). *Food Chem.* **2016**, *213*, 313–318. [CrossRef]
16. Mafrica, R.; Piscopo, A.; De Bruno, A.; Pellegrino, P.; Zappia, A.; Zappia, R.; Poiana, M. Integrated Study of Qualitative Olive and Oil Production from Three Important Varieties Grown in Calabria (Southern Italy). *Eur. J. Lipid Sci. Technol.* **2019**, *121*, 1900147. [CrossRef]
17. Sicari, V.; Leporini, M.; Giuffré, A.M.; Aiello, F.; Falco, T.; Pagliuso, M.T.; Ruffolo, A.; Reitano, A.; Romeo, R.; Tundis, R.; et al. Quality parameters, chemical compositions and antioxidant activities of Calabrian (Italy) monovarietal extra virgin olive oils from autochthonous (Ottobratica) and allochthonous (Coratina, Leccino, and Nocellara Del Belice) varieties. *J. Food Meas. Charact.* **2020**, 1–13. [CrossRef]
18. Giuffrè, A.M.; Piscopo, A.; Sicari, V.; Poiana, M. The effects of harvesting on phenolic compounds and fatty acids content in virgin olive oil (cv Roggianella). *Riv. Ital. Sostanze Gr.* **2010**, *LXXXVII*, 14–23.
19. Piscopo, A.; De Bruno, A.; Zappia, A.; Ventre, C.; Poiana, M. Data on some qualitative parameters of Carolea olive oils obtained in different areas of Calabria (Southern Italy). *Data Brief.* **2016**, *9*, 78–80. [CrossRef]
20. Marra, F.P.; Marchese, A.; Campisi, G.; Guzzetta, G.; Caruso, T.; Mafrica, R.; Pangallo, S. Intra-cultivar diversity in Southern Italy olive cultivars depicted by morphological traits and SSR markers. *Acta Hortic.* **2014**, *1057*, 571–576. [CrossRef]
21. European Union Commission. Commission regulation No 61/2011 of 24 January 2011. *Off. J. Eur. Union* **2011**, *23*, 1–14.
22. European Union Commission. Commission Implementing Regulation No 348/2013 of 17 December 2013. *Off. J. Eur. Union* **2013**, *108*, 31–67.
23. Mínguez-Mosquera, M.I.; Rejano-Navarro, L.; Gandul-Rojas, B.; Sanchez Gomez, A.H.; Garrido-Fernandez, J. Color-Pigment Correlation in Virgin Olive Oil. *J. Am. Oil Chem. Soc.* **1991**, *68*, 332–336.
24. Bakre, S.M.; Gadmale, D.K.; Toche, R.B.; Gaikwad, V.B. Rapid determination of alpha tocopherol in olive oil adulterated with sunflower oil by reversed phase high-performance liquid chromatography. *J. Food Sci. Technol.* **2015**, *52*, 3093–3098. [CrossRef] [PubMed]
25. Baiano, A.; Gambacorta, G.; Terracone, C.; Previtali, M.A.; Lamacchia, C.; La Notte, E. Changes in phenolic content and antioxidant activity of italian extravirgin olive oils during storage. *J. Food Sci.* **2009**, *74*, 177–183. [CrossRef]
26. Re, R.; Pellegrini, N.; Proteggente, A.; Pannala, A.; Yang, M.; Rice-Evans, C. Antioxidant Activity applying an improbe ABTS radical cation decolorization assay. *Free Radic. Biol. Med.* **1999**, *26*, 9–10. [CrossRef]
27. Brand-Williams, W.; Cuvelier, M.E.; Berset, C. Use of free radical method to evaluate antioxidant activity. *Lebensm. Wiss. Technol.* **1995**, *28*, 25–30. [CrossRef]
28. Sicari, V.; Giuffrè, A.M.; Piscopo, A.; Poiana, M. Effect of "Ottobratica" variety ripening stage on the phenolic profile of the obtained olive oil. *Riv. Ital. Sostanze Gr.* **2009**, *86*, 215–219.
29. Piscopo, A.; Zappia, A.; De Bruno, A.; Poiana, M. Effect of the harvesting time on the quality of olive oils produced in Calabria. *Eur. J. Lipid Sci. Technol.* **2018**, *120*, 1700304. [CrossRef]
30. De Bruno, A.; Romeo, R.; Piscopo, A.; Poiana, M. Antioxidant quantification in different portions obtained during olive oil extraction process in an olive oil press mill. *J. Sci. Food Agric.* **2021**, *101*, 1119–1126. [CrossRef]
31. Dag, A.; Kerem, Z.; Yogev, N.; Zipori, I.; Lavee, S.; Ben-David, E. Influence of time of harvest and maturity index on olive oil yield and quality. *Sci. Hort.* **2011**, *127*, 358–366. [CrossRef]
32. European Union Commission. Commission Regulation No 1830/2015 of July 8 2015. *Off. J. Eur.* **2015**, *266*, 9–13.
33. Krinsky, N.I. Antioxidant functions of carotenoids. *Free Radic. Biol. Med.* **1989**, *7*, 617–635. [CrossRef]
34. Roca, M. and Mínguez-Mosquera, M.I. Changes in chloroplast pigments of olive varieties during fruit ripening. *J. Agric. Food Chem.* **2001**, *49*, 832–839. [CrossRef] [PubMed]
35. Piscopo, A.; De Bruno, A.; Zappia, A.; Gioffrè, G.; Grillone, N.; Mafrica, R.; Poiana, M. Effect of olive storage temperature on the quality of Carolea and Ottobratica oils. *Emir. J. Food Agric.* **2018**, 563–572. [CrossRef]
36. Giuliani, A.; Cerretani, L.; Cichelli, A. Chlorophylls in Olive and in Olive Oil: Chemistry and Occurrences. *Crit. Rev. Food Sci. Nutr.* **2011**, *51*, 678–690. [CrossRef]
37. Psomiadou, E.; Tsimidou, M. Pigments in Greek virgin olive oils: Occurrence and levels. *J. Sci. Food Agric.* **2001**, *81*, 640–647. [CrossRef]
38. Cerretani, L.; Motilva, M.J.; Romero, M.P.; Bendini, A.; Lercker, G. Pigment profile and chromatic parameter of monovarietal olive oils from different Italian cultivars. *Eur. Food Res. Technol.* **2008**, *226*, 1251–1258. [CrossRef]
39. Fabiani, R. Anti-cancer properties of olive oil secoridoid phenols: A systematic review on in vivo studies. *Food Funct.* **2016**, *7*, 4145–4159. [CrossRef]
40. Sicari, V. Antioxidant potential of extra virgin olive oils extracted from three different varieties cultivated in the Italian province of Reggio Calabria. *J. Appl. Bot. Food Qual.* **2017**, *90*, 76–82. [CrossRef]

Article

The Potential of Apulian Olive Biodiversity: The Case of Oliva Rossa Virgin Olive Oil

Giacomo Squeo [1,*], Roccangelo Silletti [1], Giacomo Mangini [2], Carmine Summo [1] and Francesco Caponio [1]

[1] Food Science and Technology Unit, Department of Soil Plant and Food Sciences, University of Bari "Aldo Moro", Via Amendola 165/A, 70126 Bari, Italy; roccangelo.silletti@uniba.it (R.S.); carmine.summo@uniba.it (C.S.); francesco.caponio@uniba.it (F.C.)

[2] Institute of Biosciences and Bioresources (IBBR), National Research Council (CNR), Via Amendola 165/A, 70126 Bari, Italy; giacomo.mangini@ibbr.cnr.it

* Correspondence: giacomo.squeo@uniba.it

Abstract: In this study, the drupes and virgin olive oils extracted from the Oliva Rossa landrace are characterized. Oliva Rossa is an old landrace part of the autochthonous Apulian olive germplasm for which only few data have been reported till now. During the study, the maturity patterns of the drupes had been followed. Four samplings per year were planned, one every 14 days starting from the middle of October. The pigmentation index, the oil content and the total phenolic content of the drupes were measured. Simultaneously, virgin olive oils were extracted at the lab scale and analyzed for the fatty acid composition, the basic quality parameters and the content of minor compounds. The pigmentation pattern of the drupes was different among the years and, despite this trend, at the third sampling time the stage of maximum oil accumulation was always over. The extracted virgin olive oils had a medium to high level of oleic acid. With colder temperatures, a higher level of monounsaturated fatty acids, oleic/linoleic ratio and antioxidants was observed. The phenolic profile was dominated by 3,4-DPHEA-EDA and *p*-HPEA-EDA while the volatile profile by (*E*)-2-hexenal and 3-ethyl-1,5-octadiene.

Keywords: olive landrace; ripening; harvest season; antioxidants; minor compounds; oil quality

Citation: Squeo, G.; Silletti, R.; Mangini, G.; Summo, C.; Caponio, F. The Potential of Apulian Olive Biodiversity: The Case of Oliva Rossa Virgin Olive Oil. *Foods* **2021**, *10*, 369. https://doi.org/10.3390/foods10020369

Academic Editor: Beatriz Gandul-Rojas

Received: 23 December 2020
Accepted: 6 February 2021
Published: 9 February 2021

Publisher's Note: MDPI stays neutral with regard to jurisdictional claims in published maps and institutional affiliations.

Copyright: © 2021 by the authors. Licensee MDPI, Basel, Switzerland. This article is an open access article distributed under the terms and conditions of the Creative Commons Attribution (CC BY) license (https://creativecommons.org/licenses/by/4.0/).

1. Introduction

Extra virgin olive oil (EVOO) is one of the most renown advocate of the Mediterranean diet all over the world, representing its principal source of fat [1–3]. Its worldwide appreciation is linked to key features that include its hedonistic aspects together with the nutritional and the healthy ones [4–8]. However, despite the great importance gained, quite often the discussion about EVOO is generic and the "extra virgin olive oil" class becomes a huge container in which the single features and particularities of different EVOOs are lost. This approach may lead unfamiliar and also traditional consumers to the misconception that all the EVOOs are the same and also increase the perception of EVOO as a commodity [9]. In last decades, this behavior has been even enhanced by the fact that the majority of the virgin olive oil (VOO) is produced from few cultivars together with the development of super-intensive cultivation [10], despite the huge degree of olive biodiversity at disposal worldwide. As proof of this, it is reported that around 9% of the total Spanish varieties accounted for 96% of the olive-growing surface area in Spain; in Italy, 80 varieties out of 538 accounted for 90% of the total area; and in Greece, only 3 cultivars accounted for about 90% of the total area [10].

However, in contrast with such trend, actions have been taken toward the characterization, differentiation and valorization of monovarietal olive oils and olive biodiversity. This is proven by the rise in studies aimed at the valorization of olive biodiversity at the national [11–14] or local scale [15–18].

The safeguard and valorization of biodiversity has become a crucial matter in national and international policies [19]. Biodiversity is the key for a resilient and sustainable agriculture and represents one of the most important heritage to be preserved. One of the most important tools in this sense, adopted by the EU about two decades ago, is the designation of origin (DO) of food products [20]. DOs represent and implicit link with specific territories and, indeed, with their genetic resources.

In recent years, an increasing role of the olive variety and of the place of origin in driving the marketing strategies has been observed in the Italian olive oil market, even if the differentiation among EVOOs is still vertical and the market segmentation strongly based on habits [21]. Nonetheless, Cacchiarelli and colleagues [21] observed that EVOOs from local cultivars still get a less premium price, probably as a consequence of a lack of information on the consumers' side. Hence, it appears clear that supporting biodiversity and local typical production goes through the knowledge and the communication to the consumers of the specificity of local products.

Italy has a large olive germplasm, estimated in over 500 accessions, including varieties, old local landraces and feral forms, which makes it the leading country for olive biodiversity [10,15,22]. In particular, old local landraces may represent a treasure at the disposal of farmers, producers, olive mills and sellers in the framework of a comprehensive valorization of the territory. Among these, "Oliva Rossa" represents an old landrace belonging to the autochthon Apulian olive germplasm [23].

Only few traces about the characteristics of this cultivar and of the derived VOOs could be found in the literature. According to the Italian National Review of Monovarietal Olive Oils [24] and the Olea database [25], Oliva Rossa is a synonym of "Oliastro" and is reported to be autochthonous to the Apulia region. Other synonyms of the landrace are "Lezze" in the area of Brindisi; "Olivastro del Gargano" in the area of Foggia; and "Olivasto di Conversano" in Bari province [23,25]. The old landrace is designed for oil extraction while no information about other purposes has been reported. The trees show a dense canopy and a medium vigor while the drupes are characterized by many lenticels, an elliptic shape and a low-to-medium weight and oil content [25]. Moreover, from the little information available, Oliva Rossa oils are characterized by almond, artichoke and cut-grass notes, among others [24], and the phenolic content is noteworthy, with a mean value of about 700 mg kg^{-1} (based on three samples from two different harvest crops). Today, Oliva Rossa could be considered basically an old olive landrace. It is planted only in small areas, in which it is still difficult to find representative plantations since the trees are dispersed in arable or in old orchards consociated with other tree fruits, such as almond or cherry.

The drupes are generally harvested together with modern olive cultivars for oil extraction. As a consequence, monovarietal oils from Oliva Rossa are more unique than rare.

Thus, in the framework of the valorization of the Apulian autochthonous olive germplasm, the aim of the study was to characterize and extend the knowledge about Oliva Rossa VOO. For the purpose, the maturity pattern of the drupes and the chemical characteristics of the relative VOOs were studied for four consecutive harvest seasons.

2. Materials and Methods

2.1. Plant Material, Sampling Plan and VOO Extraction

Drupes of the old olive landrace Oliva Rossa were harvested in four consecutive harvest seasons, from 2016/2017 to 2019/2020, in an olive tree field located in Putignano (Bari, Italy). Drupes were sampled from three different trees every 14 days (Sampling 1, S1; Sampling 2, S2; Sampling 3, S3; and Sampling 4, S4), starting around the middle of October (±one week) and, more in detail, two-weeks before the expected physiological maturity (defined as the half-veraison of the fruits). On the whole, around 10 kg of drupes were collected at each sampling time. For each sampling, the drupes were divided into 3 aliquots (around 3 kg each), representing the biological replicates. VOOs were extracted starting from the 2nd sampling (S2)—supposed to correspond to the physiological maturity—to the last (S4) by a lab scale plant made up of a hammer mill and a bask centrifuge [26]. On the whole, $n = 9$ VOOs were supposed to be extracted per each harvest season, 3 per each sampling point. Once extracted, the VOOs were sealed in glass bottles and stored in cold and dark conditions till the moment of analysis. Data about temperatures and rainfalls during the studied years were obtained from the Apulian civil protection [27] and reported in Figure S1. The weather station was about at 8 km from the orchard including the Oliva Rossa trees.

2.2. Drupes Analysis

The ripening degree of the fruits at each sampling time was calculated as the pigmentation index, Pi [28]. Pi ranges from 0 to 5 and is obtained by dividing the drupes into 5 classes taking into account the color of the skin and eventually the flash color. The moisture of the drupes was measured with an automatic moisture analyzer (Mod. MAC 110/NP, Radwag Wagi Elektroniczne, Radom, Poland) and used to express the results of the subsequent determinations on the basis of fruit dry weight. The total oil content was determined by using the Soxhlet apparatus and diethyl ether as a solvent. After extraction the solvent was removed, and the oil content was express as a percentage with respect to the fruit dry weight. For the extraction of total phenolic compounds, about 1 g of olive paste was mixed with 10 mL of a mixture of methanol/water (70/30) and 5 mL of hexane. Then the mixture was kept in agitation for 10 min and afterward centrifuged (SL 16R Centrifuge, Thermo Fisher Scientific Inc., Waltham, MA, USA) for 10 min at 4 °C at $3941 \times g$. The hydro alcoholic phase was recovered and centrifuged again at $8867 \times g$ for 5 min. Finally, the methanolic phase was filtered (0.45 µm, VWR International, Center Valley, PA, USA) into an amber glass vial. The phenolic extracts were used for the determination of the total phenolic content (TPC) by means of the Folin-Ciocalteu reagent, as previously reported [29,30]. In brief, 100 µL of extract was mixed with the same amount of Folin reactive (Sigma-Aldrich Co. LLC, St. Louis, MO, USA) and, after 4 min, 800 µL of Na_2CO_3 (5% w/v in water) (Carlo Erba Reagents S.r.l., Cornaredo, Italy) were added. The final solution was heated at 40 °C for 20 min and after 15 min of cooling the absorbance at 750 nm was measured (Agilent Cary 60 spectrophotometer, Agilent Technologies, Santa Clara, CA, USA). If needed, proper dilution of the extract was carried out. Quantitation was achieved by means of an external calibration curve of gallic acid ($R^2 = 0.998$) and the results expressed as mg of gallic acid equivalents (GAE) per kg of dry weight.

2.3. VOO Analysis

The determination of fatty acid composition was carried out after sample transesterification with KOH 2N in methanol [31] by a GC (Agilent 7890A gas chromatograph, Agilent Technologies, Santa Clara, CA, USA) equipped with an FID detector (set at 220 °C) and an SP2340 capillary column of 60 m × 0.25 mm (i.d.) × 0.2 mm film thickness (Supelco Park, Bellefonte, PA, USA). The identification of each fatty acid was carried out by comparing the retention time with that of the corresponding standard methyl ester (Sigma-Aldrich, St. Louis, MO, USA). The amount of single fatty acids was expressed as area % with respect to the total area. Basic quality analyses of the VOOs (free fatty acids, FFA; peroxide value, PV; K_{232}, K_{270}) were carried out according to the official regulations [31]. VOO hydrophilic antioxidants were extracted as previously described for drupes with slight modifications. Basically, about 1 g of oil was mixed with 5 mL of methanol/water mixture and 2 mL of hexane. Then the extraction protocol, the Folin assay and the quantitative determination were done the same as previously reported (see Section 2.2). Pigments (chlorophylls and carotenoids) were measured spectrophotometrically (Agilent Cary 60 spectrophotometer, Agilent Technologies, Santa Clara, CA, USA), as reported in [32,33], respectively. Tocopherols were determined by RP-UHPLC-FLD (ThermoScientific, Waltham, MA, USA), as reported in previous work [28]. For the analysis of single phenolic compounds, phenolic extracts were obtained according to the procedure already described for the Folin assay with minor modifications. In detail, 5 g of sample were used with 2 mL of methanol/water mixture and 2 mL of hexane. Then, 250 µL of a 100 mg L^{-1} solution of gallic acid was added as internal standard for quantitation. The separation was obtained by RP-UHPLC-DAD (ThermoScientific, Waltham, MA, USA), as described in [34]. The identification was carried out by comparison of the retention times with those of pure standards and, if not available, with the literature data. Quantification was carried out on the signal recorded at 280 nm and the results were expresses as mg of gallic acid equivalent (GAE) per kg of oil. The phenolic profile was reported only for the harvest season 2017/2018.

The analysis of volatile compounds was carried out by means of HS-SPME-GC-MS, as reported in a previous work [35]. Briefly, for the extraction of volatiles, about 1 g of sample was sealed in a 20 mL vial after the addition of 100 µL of a 60 mg kg^{-1} solution of 1-octanol in purified olive oil as internal standard for quantification. Then, it was first conditioned at 40 °C for 2 min and, thereafter, the SPME fiber (50/30 µm DVB/CAR/PDMS; Supelco, Bellefonte, PA, USA) was exposed in the vial headspace for 20 min. The desorption was carried out directly in the GC-MS (Agilent 6850 series gas chromatograph coupled to a mass spectrometer Agilent 5975 series; Agilent Technologies, Santa Clara, CA, USA) injector at 250 °C for 2 min. The stationary phase was an HP-Innovax polar column. Volatile compounds were identified by comparison of their mass spectra with those in the NIST library. Only those with a match quality above 70% were considered. Results were expressed as mg of 1-octanol equivalents (OE) per kg of oil. The volatile profile was reported only for the harvest season 2017/2018.

2.4. Statistical Analysis

Each analysis was carried out at least in duplicate per each biological replicate ($n = 3$). One-way and two-way analysis of variance were carried out by means of Minitab 17 (Minitab Inc., State College, PA, USA) and significant differences were highlighted by Fisher's LSD post-hoc test at $\alpha = 0.05$.

3. Results

3.1. Sampling Issues

Along the four years of the current study, two harvest seasons (2016/2017 and 2018/2019) have been characterized by intense pest attacks and/or other agro-climatic issues, such as late frost, which destroyed almost all the production [36]. In these seasons, only two samplings (S1 and S2) out of the four planned were carried out. As a consequence, the discussion of the results will be mostly based on those obtained during the harvest seasons 2017/2018 and 2019/2020 although all the available results are reported.

3.2. Drupes Characteristics

The ripening process of the drupes was measured as Pi and it is reported in Figure 1. As a first glance, it appears the maturity pattern was different among the harvest seasons. Depending on the year, the drupes started with different values of Pi. In 2018/2019 and 2019/2020, the drupes started totally green (Pi = 0) while they were slightly spotted in the case of 2016/2017 and 2017/2018 (Pi around 1, i.e., <50% black skin with white flash).

It is worthy to highlight that the first sampling (S1) fell in all the studied seasons around the middle of October. Following the evolution, the drupes reached a Pi of about 2 ($\geq$50% black skin with white flash) in all the years, with exception of the season 2018/2019. This stage (Pi = 2) correspond to the stage commonly considered the physiological maturity (half-veraison of the fruits) and expected to be the stage in which the oil accumulation reached its maximum (i.e., technological maturity).

Thereafter, an important gap among the years 2017/2018 and 2019/2020 was recorded at S3. In the latter, the full maturity seemed almost reached and, from this point on, only a slight increase in Pi was recorded. Differently in 2017/2018, the Pi remained practically unchanged between S2 and S3 while the most significant jump was observed from S3 to S4. On the whole, the ripening process proceeded in a more regular way during the harvest season 2019/2020.

Figure 1. Trends in the Pi (mean ± SD, n = 3) of the Oliva Rossa drupes during ripening in four consecutive harvest seasons. Different letters for each harvest year indicate significant differences according to one-way ANOVA followed by Fisher's LSD post-hoc test (α = 0.05). S1–S4: four subsequent samplings.

As evidenced by Camposeo et al. [37], the observed differences are likely related to the climatic conditions and crop load, which, among other factors, strongly influence drupe ripening. In this regard, the lower temperatures recorded during the ripening period in 2017 with respect to 2019 (±2–3 °C less, Figure S1) could be responsible for the observed delay in the pigmentation pattern.

From a technological point of view, it should be stressed how these differences were reflected in the drupes and in the oil characteristics. The evolution of the oil content during maturation is reported in Figure 2.

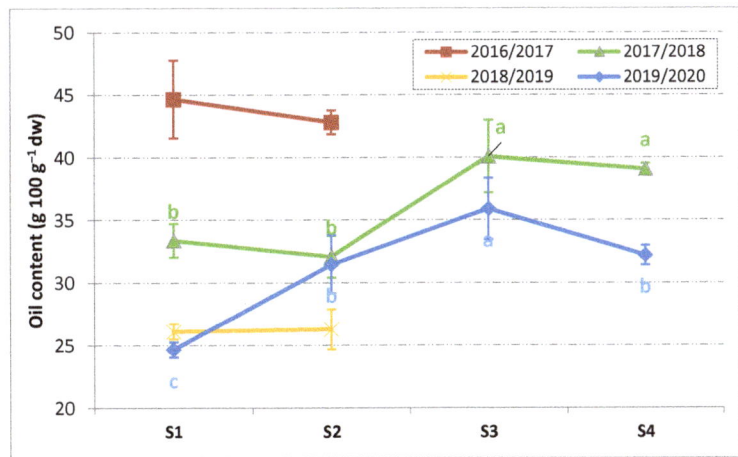

Figure 2. Oil content on dry weight (mean ± SD, n = 3) of the Oliva Rossa drupes during ripening in four consecutive harvest seasons. Different letters for each harvest year indicate significant differences according to one-way ANOVA followed by Fisher's LSD post-hoc test (α = 0.05). S1–S4: four subsequent samplings.

The oil content at S1 ranged between 25 and 26 g 100 g^{-1} on dry weight (2018/2019 and 2019/2020) and goes up to 45 g 100 g^{-1} in 2016/2017. The season 2017/2018 lied in between the others, showing an initial oil content of 33 g 100 g^{-1}. The very high oil content of the harvest season 2016/2017 should be considered as an exceptional case, confirming that such a season was an outlier with respect to the general pattern. In 2019/2020, the evolution of oil content was very similar to that of the Pi in Figure 1. Indeed, a regular accumulation of the oil was observed with a positive slope of about 5 g 100 g^{-1} each 14 days from S1 to S3. Thereafter, a small significant decrease in oil was reported. Such a slight decrease was likely linked to the natural variability of the sampled fruits among different trees and canopy positions. During 2017/2018, the most significant phase of oil accumulation occurred between S2 and S3, moving from 32 g 100 g^{-1} up to 40 g 100 g^{-1}. Such an increase in oil was not reflected in the Pi, which was unaffected when moving from S2 to S3 (Figure 1). Afterward, the oil content remained constant. As a most important finding, in both the harvest seasons for which the full maturity pattern was followed, S3 corresponded to the maximum oil content. Unfortunately, the Pi did not reflect directly such a conclusion. In fact, at the half-veraison of the drupes (i.e., Pi = 2), generally considered as the moment of physiological maturity, the fruits could still be in the stage of intense oil accumulation, as observed in the 2019/2020 crop year. Figure 3 shows the evolution of the olives TPC.

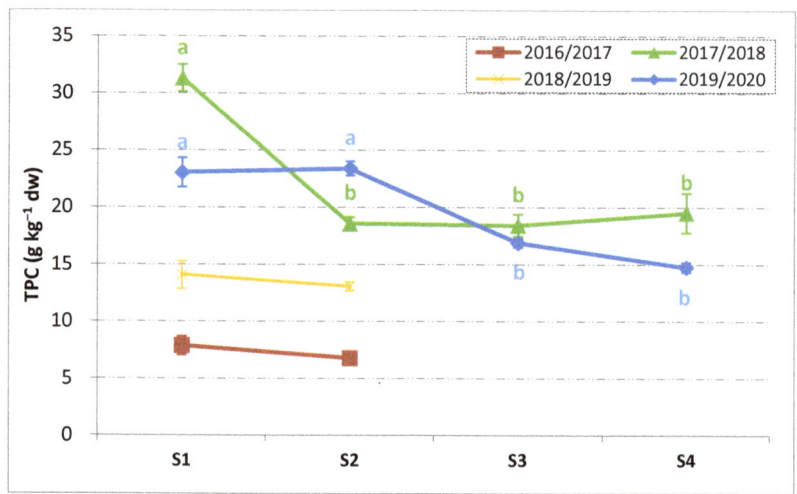

Figure 3. Total phenolic content (TPC) on dry weight (mean ± SD, n = 3) of the Oliva Rossa drupes during ripening in four consecutive harvest seasons. Different letters per each harvest year indicate significant differences according to one-way ANOVA followed by Fisher's LSD post-hoc test (α = 0.05). S1–S4: four subsequent samplings.

During fruit ripening the TPC decreased, as is already known [38]. Further, in the harvest seasons 2016/2017 and 2018/2019, the negative effect of the pest attacks and late frost on the phenolic content of the drupes was easily observed. A high TPC was observed in 2017/2018, which was around 30 g kg^{-1} and about 25% less than in 2019/2020. The TPC of the Oliva Rossa drupes was similar to that reported for the Frantoio cultivar [38], while less than what was reported for the Sardinian cultivars [16]. Despite the initial gap at S1 between 2017/2018 and 2019/2020, the TPC tended to become similar during ripening and in particular at S3 the differences were negligible. The sharp TPC decrease from S1 to S2 in 2017/2018 could be linked to the abundant rainfalls recorded in October (Figure S1). Indeed, the tree water status is reported to be inversely correlated with the phenolic content [39]. It is noteworthy pointing out that, at S3, the phase of the more intense oil accumulation was already over (Figure 2) and the fruits also had comparable TPC. Thus, based on the available results, it would be reasonable considering S3 as the best moment for harvesting. Nonetheless the main issue of a feasible and reliable way of determining such a moment is still of concern considering that S3 corresponded to the stage in which the difference in the Pi was the biggest among the considered harvest seasons (Figure 1). The phenolic content of the drupes represents an important feature for the future VOOs and it should be considered that only 1–2% of the olives phenols moves into the oily phase during extraction [40], which may, in turn, depend on the fruit ripening [41].

3.3. VOOs Characteristics

Different features influence the definition of VOO quality. From a consumer point of view, the hedonistic aspects are relevant although also the nutritional ones may be of concern. The main impact on the nutritional score of VOO is ascribable to the FA (fatty acid) composition and particularly to the amount of oleic acid [42]. The phenolic compounds, together with other bioactive ones, also play a key role, as proven by the specific health claim adopted by the European Union [43]. On the other hand, while phenolic compounds are responsible for the organoleptic features of VOOs [44], it is not true for oleic acid. Finally, from a producer and seller point of view, compliance with the mandatory quality parameters defined by the EU law is of primary concern [31]. Starting from these considerations, and in the view of the valorization of the oils from Oliva Rossa, the VOOs were characterized for their fatty acid composition as well as the basic quality parameters and the minor compounds. Given the sampling issues related to the crop years 2016/2017 and 2018/2019—which may bias the comparison—only the results of the years 2017/2018 and 2019/2020 are reported and discussed. Table 1 reports the FA composition of the Oliva Rossa VOOs.

Table 1. Fatty acid composition of the Oliva Rossa virgin olive oils (VOOs) extracted during fruit ripening in two different harvest seasons (mean ± SD, n = 3).

	2017/2018			2019/2020		
Pi *	1.92	1.93	3.12	2.43	3.57	3.83
Fatty acid	S2	S3	S4	S2	S3	S4
C14:0	0.01 ± 0.00 [b,c]	0.01 ± 0.00 [b,c]	0.01 ± 0.00 [c]	0.01 ± 0.01 [c]	0.02 ± 0.00 [a,b]	0.03 ± 0.01 [a]
C16:0	11.11 ± 0.28 [b]	10.24 ± 0.02 [b]	12.84 ± 0.61 [a]	12.54 ± 0.28 [a]	10.31 ± 0.03 [b]	10.75 ± 0.80 [b]
C16:1	0.48 ± 0.01 [c]	0.48 ± 0.00 [c]	0.55 ± 0.01 [c]	1.04 ± 0.07 [a]	0.78 ± 0.01 [b]	0.79 ± 0.10 [b]
C17:0	0.04 ± 0.00 [d]	0.05 ± 0.00 [c,d]	0.05 ± 0.00 [c,d]	0.10 ± 0.01 [a]	0.10 ± 0.01 [a,b]	0.07 ± 0.03 [b,c]
C17:1	0.07 ± 0.01 [a]	0.08 ± 0.00 [a]	0.07 ± 0.00 [a]	0.07 ± 0.01 [a]	0.05 ± 0.01 [b]	0.04 ± 0.01 [b]
C18:0	2.75 ± 0.01 [a]	2.56 ± 0.00 [b,c,d]	2.47 ± 0.04 [c,d]	2.46 ± 0.10 [d]	2.65 ± 0.03 [a,b]	2.59 ± 0.11 [b,c]
C18:1	73.58 ± 0.17 [c]	76.21 ± 0.01 [a]	74.38 ± 0.50 [b]	71.83 ± 0.32 [d]	70.84 ± 0.19 [e]	70.73 ± 0.50 [e]
C18:2	10.37 ± 0.11 [b]	8.80 ± 0.01 [c]	8.24 ± 0.13 [d]	10.63 ± 0.17 [b]	13.51 ± 0.11 [a]	13.20 ± 0.31 [a]
C18:3	0.77 ± 0.00 [a]	0.77 ± 0.00 [a]	0.65 ± 0.02 [b]	0.57 ± 0.02 [c]	0.61 ± 0.04 [b,c]	0.56 ± 0.06 [c]
C20:0	0.40 ± 0.00 [b]	0.40 ± 0.01 [b]	0.37 ± 0.00 [b]	0.31 ± 0.09 [b]	0.59 ± 0.07 [a]	0.60 ± 0.06 [a]
C20:1	0.38 ± 0.02 [c]	0.37 ± 0.02 [c]	0.36 ± 0.03 [c]	0.32 ± 0.02 [c]	0.46 ± 0.04 [b]	0.54 ± 0.03 [a]
C22:0	0.01 ± 0.00 [b]	0.01 ± 0.00 [b]	0.01 ± 0.00 [b]	0.06 ± 0.03 [a]	0.06 ± 0.03 [a]	0.06 ± 0.02 [a]
C24:0	0.01 ± 0.00 [b]	0.01 ± 0.00 [b]	0.01 ± 0.00 [b]	0.05 ± 0.02 [a]	0.03 ± 0.01 [b]	0.03 ± 0.01 [b]
SFA	14.34 ± 0.27 [b]	13.27 ± 0.03 [c]	15.75 ± 0.65 [a]	15.53 ± 0.30 [a]	13.76 ± 0.04 [b,c]	14.13 ± 0.75 [b,c]
MUFA	74.52 ± 0.16 [c]	77.15 ± 0.02 [a]	75.36 ± 0.54 [b]	73.27 ± 0.27 [d]	72.12 ± 0.16 [e]	72.10 ± 0.38 [e]
PUFA	11.14 ± 0.11 [b]	9.57 ± 0.01 [c]	8.89 ± 0.11 [d]	11.20 ± 0.15 [b]	14.12 ± 0.15 [a]	13.76 ± 0.37 [a]
O/L	7.09 ± 0.06 [c]	8.66 ± 0.01 [b]	9.03 ± 0.08 [a]	6.76 ± 0.12 [d]	5.24 ± 0.06 [e]	5.36 ± 0.09 [e]

* Pi (mean, n = 3) of the drupes at the moment of oil extraction; SFA, total saturated fatty acids; MUFA, total monounsaturated fatty acids; PUFA, total polyunsaturated fatty acids; O/L, oleic acid over linoleic acid ratio. Different letters for each parameter indicate significant differences according to two-way ANOVA followed by Fisher's LSD post-hoc test (α = 0.05). S2–S4: three subsequent samplings.

The palmitic (C16:0), stearic (C18:0), oleic (C18:1) and linoleic (C18:2) acids were the most abundant, as is typical in olive oils. Oleic acid, the main monounsaturated fatty acid, ranged from about 71% to just over 76%. The global mean value was roughly 73%, which is similar to the data reported elsewhere [24,25]. The crop year, the sampling time and the interaction among those factors showed an effect on the C18:1 content. The VOOs extracted during 2017/2018 had significantly higher values of C18:1 than those of 2019/2020. Further, while in 2017/2018 an increase in C18:1 was observed during ripening, the opposite was found in 2019/2020. The second most abundant fatty acid, linoleic acid, ranged between around 8% to 13.51%. A mean value of about 8% was reported in the Italian monovarietal oils databank [24].

Linoleic acid content was generally higher in 2019/2020 with respect to 2017/2018; also, a significant effect of the interaction between harvest year and sampling was observed. In particular, while in 2017/2018 a decrease in linoleic acid was reported during ripening, in 2019/2020 it was the opposite. The palmitic acid content showed a significant difference among the samples with an opposite trend in the studied harvest seasons. Indeed, the highest content was found at S4 in 2017/2018 while in S2 in 2019/2020. The stearic acid content was on average around 2.60% and it showed a regular decrease in content during 2017/2018, while again an opposite trend was observed in 2019/2020. Finally, other significant differences were observed for all the minor fatty acids. The content of some minor fatty acids (namely, miristic, arachic and gadoleic) in 2019/2020 oils deserves special attention, as they were close to the maximum value allowed by official regulations [31]. Altogether, it comes out that Oliva Rossa VOOs were richer in MUFA (total monounsaturated fatty acids) in 2017/2018 whilst richer in PUFA (total polyunsaturated fatty acids) in 2019/2020. Giving a clear explanation of the observed differences could be a difficult task considering the complex effect of single factors (genetic, environmental and agronomic), their interactions or independence, as reviewed by Inglese and others [39]. However, differences in the oleic, palmitic and linoleic acid content were mostly related to the course of temperatures during the year. In particular, lower temperatures could be correlated with a higher content of oleic acid and, on the opposite, with a lower content of palmitic and/or linoleic acids [39]. In the 2017/2018 crop season, the mean temperatures registered during fruits ripening were actually lower than those registered in the 2019/2020 season (Figure S1) and could justify the differences observed in terms of fatty acids composition.

From a technological point of view, it is a fact that the different composition in fatty acids could in turn influence the oxidative stability of the product. In particular, values of O/L equals or higher than 7 have been suggested as an indication of good stability [45]. From Table 1, it could be observed that the ratio was different between the years and, in 2017/2018, it was always higher than 7.

The results of the basic chemical analyses of the VOOs are reported in Table 2.

Table 2. Basic quality parameters of the Oliva Rossa VOOs extracted during fruit ripening in two different harvest seasons (mean ± SD, n = 3).

Harvest Season	Pi *	Sampling	FFA (g 100 g^{-1})	PV (mEq O$_2$ kg^{-1})	K$_{232}$	K$_{270}$
2017/2018	1.92	S2	0.61 ± 0.06 [a]	7.49 ± 0.28 [c]	1.94 ± 0.08 [c]	0.21 ± 0.03 [a,b]
	1.93	S3	0.58 ± 0.07 [a]	7.15 ± 0.23 [c]	1.80 ± 0.12 [d]	0.20 ± 0.06 [a,b]
	3.12	S4	0.58 ± 0.06 [a]	6.05 ± 0.28 [d]	1.76 ± 0.02 [d]	0.23 ± 0.04 [a]
2019/2020	2.43	S2	0.52 ± 0.04 [a,b]	8.92 ± 0.22 [b]	2.34 ± 0.05 [b]	0.17 ± 0.02 [a,b]
	3.57	S3	0.43 ± 0.06 [b,c]	11.02 ± 0.48 [a]	2.46 ± 0.10 [a,b]	0.18 ± 0.03 [a,b]
	3.83	S4	0.34 ± 0.07 [c]	11.03 ± 0.67 [a]	2.52 ± 0.04 [a]	0.16 ± 0.03 [b]

* Pi (mean, n = 3) of the drupes at the moment of oil extraction; FFA, free fatty acids; PV, peroxide value; K$_{232}$ and K$_{270}$, specific absorption at 232 and 270 nm, respectively. Different letters for each parameter indicate significant differences according to two-way ANOVA followed by Fisher's LSD post-hoc test (α = 0.05). S2–S4: three subsequent samplings.

FFA is commonly the benchmark parameter for the vertical differentiation of olive oils into a specific product class (extra, virgin and lampante) [9]. The mean oil acidity ranged from 0.61% to 0.34%, being on average higher in the harvest season 2017/2018. The observed values could be considered quite high for freshly extracted oils. During maturation, a significant decrease in FFA was observed in 2019/2020. This decreasing trend was also reported by other authors [16]. Moving to the oxidation assessment, PV and K_{232} are generally considered markers of primary oxidation, while K_{270} is a marker of secondary oxidation products. Differences were highlighted considering both the harvest season and the sampling time. A specular trend was observed between the years. The oils extracted in 2019/2020 were affected by a higher extent of primary oxidation (PV and K_{232}) but, on the average, by a less pronounced secondary oxidation (K_{270}). This might be linked to the significantly higher amount of PUFA in oils from 2019/2020 and to the significantly lower value of O/L (Table 1), indices of a higher susceptibility to oxidation.

On the whole, from Table 2, it could be stated that the Oliva Rossa VOOs still were classified as extra virgin olive oil (based on the reported parameters) till S3, which, in turn, corresponded to the maximum oil content (Figure 2). Obviously, these results cannot be generalized for all the VOOs obtained from this cultivar because the quality parameters are mostly influenced by the technological aspects of olive processing. Hence, it should be clear that such conclusions are related to the experimental conditions of the present study.

The profile of the minor compounds of the oils is reported in Table 3. Minor compounds are fundamental molecules of VOO, which, in turn, influence its stability to oxidation, its nutritional and healthy aspects and its organoleptic features [44,46]. The oils had a high-medium content of TPC, ranging from about 800 to 350 mg kg^{-1}. These results agree with the few available data in the literature [24,25]. The phenolic content decreased significantly during ripening, as is well known, in both the harvest seasons [37,47,48]. Moreover, the effect of the harvest season on the TPC was evident, with the 2017/2018 season having a general higher content with respect to the 2019/2020 season. The tocopherols content was also remarkable. The content of α-tocopherol decreased during ripening, without significant differences in 2017/2018. In 2019/2020, an outstanding content of α-tocopherol was reported at S2. The total amount of tocopherols followed the general trend reported for α-tocopherol considering that the latter represents roughly 90% of the total tocopherols in olive oils. The sum of the β- and γ-tocopherols ranged from about 6 to about 3 mg kg^{-1} with the highest significant value reached at S2 in 2019/2020.

Table 3. Minor compounds (mg kg^{-1}) of the Oliva Rossa VOOs extracted during fruit ripening in two different harvest seasons (mean ± SD, n = 3).

Harvest Season	Pi *	Sampling	TPC	α-Tocopherol	β- and γ-Tocopherols	Total Tocopherols	Carotenoids	Chlorophylls
2017/2018	1.92	S2	761.64 ± 20.36 [a]	246.20 ± 10.59 [b]	5.28 ± 0.49 [a,b]	251.47 ± 10.94 [b]	57.77 ± 0.94 [a]	54.68 ± 0.90 [b]
	1.93	S3	577.36 ± 15.33 [b]	237.21 ± 4.49 [b]	4.27 ± 0.82 [b,c]	241.47 ± 4.82 [b]	34.39 ± 0.30 [c]	26.96 ± 0.10 [d]
	3.12	S4	472.83 ± 12.60 [c]	230.48 ± 14.03 [b]	5.60 ± 0.73 [a,b]	235.82 ± 13.45 [b]	48.14 ± 0.95 [b]	40.23 ± 0.73 [c]
2019/2020	2.43	S2	592.86 ± 72.99 [b]	349.74 ± 5.05 [a]	6.32 ± 0.50 [a]	356.06 ± 4.60 [a]	47.27 ± 1.50 [b]	66.30 ± 1.84 [a]
	3.57	S3	379.60 ± 8.43 [d]	235.79 ± 34.25 [b]	3.39 ± 1.44 [c]	239.18 ± 35.67 [b]	23.42 ± 1.71 [d]	23.13 ± 0.44 [f]
	3.83	S4	358.67 ± 25.97 [d]	248.33 ± 11.82 [b]	4.34 ± 0.17 [b,c]	252.67 ± 11.99 [b]	25.22 ± 0.91 [d]	25.27 ± 0.52 [e]

* Pi (mean, n = 3) of the drupes at the moment of oil extraction; TPC, total phenolic compounds. Different letters for each parameter indicate significant differences according to two-way ANOVA followed by Fisher's LSD post-hoc test (α = 0.05). S2–S4: three subsequent samplings.

Pigments are important compounds of olive oils, which affect the product stability and also could give useful information about fruit ripening and authenticity [46,49]. During ripening, the pigment content decreased although we observed a parabolic trend, especially considering the harvest season 2017/2018. Ripening is generally reported as the main source of variability in pigment content despite other factors [48,50], even if technology could also have an impact [51]. The ratio between chlorophylls and carotenoids (data not shown) was always very close to unity, as reported in other studies and suggested for authenticity purposes [49].

The profile of the phenolic compounds by HPLC for the harvest season 2017/2018 is reported in Table 4. In Figure S2, the relative chromatograms can be observed.

Table 4. Identified phenolic compounds (mg GAE kg^{-1}) of the Oliva Rossa VOOs extracted during fruit ripening in the 2017/2018 harvest season (mean $\pm$ SD, n = 3).

Compound	S2	S3	S4
3,4-DHPEA	0.33 $\pm$ 0.05 [a,b]	0.42 $\pm$ 0.06 [a]	0.25 $\pm$ 0.07 [b]
p-HPEA	0.54 $\pm$ 0.03 [b]	0.54 $\pm$ 0.02 [b]	0.68 $\pm$ 0.05 [a]
Vanillic acid	0.19 $\pm$ 0.02 [a]	0.20 $\pm$ 0.02 [a]	0.21 $\pm$ 0.06 [a]
Syringic acid	0.44 $\pm$ 0.03 [a]	0.29 $\pm$ 0.07 [b]	0.36 $\pm$ 0.04 [a,b]
3,4-DHPEA-EDA	62.35 $\pm$ 3.29 [a]	25.81 $\pm$ 4.57 [b]	17.57 $\pm$ 1.77 [c]
3,4-DHPEA-EDA-CARB	1.64 $\pm$ 0.23 [b]	5.65 $\pm$ 1.04 [a]	1.31 $\pm$ 0.40 [b]
p-HPEA-EDA	30.87 $\pm$ 0.93 [a]	13.97 $\pm$ 0.78 [c]	20.76 $\pm$ 1.40 [b]
Pinoresinol	13.78 $\pm$ 0.28 [a]	9.90 $\pm$ 0.69 [c]	11.34 $\pm$ 0.51 [b]
Luteolin	2.02 $\pm$ 0.23 [b]	6.76 $\pm$ 0.62 [a]	2.35 $\pm$ 2.15 [b]
p-HPEA-EA	9.01 $\pm$ 0.94 [a]	8.63 $\pm$ 1.68 [a]	5.84 $\pm$ 1.27 [b]
Apigenin	3.21 $\pm$ 0.33 [a]	3.75 $\pm$ 1.27 [a]	4.31 $\pm$ 1.50 [a]
Total	124.38 $\pm$ 3.80 [a]	75.92 $\pm$ 7.40 [b]	64.98 $\pm$ 5.60 [b]

GAE, gallic acid equivalents. 3,4-DHPEA, hydroxytyrosol; p-HPEA, tyrosol; 3,4-DHPEA-EDA, decarboxymethyl oleuropein-aglycone di-aldehyde; 3,4-DHPEA-EDA-CARB, carboxymethyl oleuropein-aglycone di-aldehyde; p-HPEA-EDA, decarboxymethyl ligstroside-aglycone di-aldehyde; p-HPEA-EA, ligstroside-aglycon. Different letters for each parameter indicate significant differences according to one-way ANOVA followed by Fisher's LSD post-hoc test (α = 0.05). S2–S4: three subsequent samplings.

The identified compounds were those commonly found in VOOs, belonging to the classes of phenolic alcohols (3,4-DHPEA, p-HPEA), phenolic acids (vanillic and syringic acids), flavonoids (luteolin and apigenin), lignans (pinoresinol) and, most importantly, secoiridoid derivatives (3,4-DHPEA-EDA, p-HPEA-EDA and p-HPEA-EA). Phenolic acids were found in very little amounts, according to the literature data [44].

Similarly, a minor contribution of phenolic alcohols was observed. It is known that phenolic alcohols originate from the more complex secoiridoid moieties mostly during VOO storage [44]. The di-aldehydic forms of 3,4-DHPEA and p-HPEA were the most abundant phenols in all the maturity stages, followed by a remarkable amount of pinoresinol. At the first sampling (S2), the oils showed the highest significant content of almost all the identified phenolic compounds. During ripening, the profile changed with a sharp reduction in 3,4-DHPEA-EDA and a less pronounced decrease in p-HPEA-EDA. At S3, the oils were characterized on one hand by the significantly higher amount of luteolin and 3,4-DHPEA-EDA-CARB while on the other by the significant lower content of p-HPEA-EDA and pinoresinol. At S4, the lowest significant amount of phenolic compounds was observed. Some phenols showed a parabolic trend during ripening, having a maximum or minimum content in correspondence to the technological optimum (S3); it was also the case for 3,4-DHPEA-EDA-CARB, p-HPEA-EDA, pinoresinol and luteolin. Such a trend could be linked to the complex pattern of biochemical and chemical phenomena, which could affect the relative amount of phenols in the VOOs and, to a lesser extent, to the natural variability of the sampled material.

Although no organoleptic assessment of the oils was carried out in this study, the relationship between phenolic compounds and the sensory features of the product is well known and confirmed by numerous works [5,44,52]. p-HPEA-EDA, also known as oleocanthal, have been proved to be the major compound responsible for pungent notes while other secoiridoid derivatives are more strongly linked to bitterness [44]. Considering the TPC (Table 3), the phenolic profile (Table 4) and the similarities with the data reported for the Coratina cultivar olive oils [11,28,53], which are well known to be strongly bitter and pungent, it might be supposed that the VOOs from the Oliva Rossa landrace have noteworthy bitter and pungent notes, although such a hypothesis should be verified by an in-depth study. To date, the few results available in the literature seem to confirm such a hypothesis [24].

The headspace volatile profile of the oils extracted in 2017/2018 is reported in Table 5.

Table 5. Identified volatile compounds (mg OE kg^{-1}) of the Oliva Rossa VOOs extracted during fruit ripening in the 2017/2018 harvest season (mean ± SD, n = 3).

Compound	S2	S3	S4
Methyl acetate	2.28 ± 0.58 [b]	1.82 ± 0.89 [b]	5.25 ± 0.29 [a]
Ethyl acetate	0.40 ± 0.05 [b]	1.25 ± 0.49 [a]	1.47 ± 0.03 [a]
2-Methyl butanal	1.83 ± 0.43 [b]	1.65 ± 0.22 [b]	5.18 ± 0.54 [a]
3-Methyl butanal	2.53 ± 0.57 [b]	2.07 ± 0.18 [b]	7.84 ± 0.44 [a]
3-Pentanone	1.25 ± 0.29 [a]	1.64 ± 0.29 [a]	1.69 ± 0.20 [a]
Pentanal	1.07 ± 0.12 [c]	13.78 ± 1.73 [a]	4.00 ± 0.79 [b]
3-Ethyl-1,5-octadiene	15.81 ± 3.00 [b]	37.05 ± 9.52 [a]	16.60 ± 2.21 [b]
1-Penten-3-one	6.23 ± 0.67 [c]	16.34 ± 0.84 [a]	8.42 ± 0.78 [b]
Hexanal	2.72 ± 0.44 [b]	4.81 ± 0.42 [a]	4.29 ± 0.85 [a]
(E)-2-Pentenal	0.82 ± 0.19 [a,b]	1.32 ± 0.24 [a]	0.65 ± 0.32 [b]
1-Penten-3-ol	2.95 ± 0.34 [c]	4.85 ± 0.42 [b]	10.52 ± 0.15 [a]
(E)-2-Hexenal	70.37 ± 9.18 [b]	111.88 ± 14.16 [a]	86.69 ± 7.13 [a,b]
(Z)-3-Hexen-1-yl acetate	0.84 ± 0.20 [b]	0.10 ± 0.03 [b]	11.42 ± 1.11 [a]
(Z)-2-Penten-1-ol	4.51 ± 0.46 [c]	18.16 ± 1.30 [a]	9.94 ± 1.07 [b]
Acetic acid	3.69 ± 0.36 [b]	8.48 ± 1.19 [a]	3.59 ± 1.44 [b]
Hexan-1-ol	1.49 ± 0.92 [a]	1.10 ± 0.30 [a]	1.99 ± 0.69 [a]
(Z)-3-Hexen-1-ol	4.47 ± 0.31 [a]	1.71 ± 0.09 [c]	3.41 ± 0.44 [b]
Total	123.25 ± 9.81 [b]	228.00 ± 17.31 [a]	182.95 ± 7.97 [a]

OE, 1-octanol equivalents. Different letters for each parameter indicate significant differences according to one-way ANOVA followed by Fisher's LSD post-hoc test (α = 0.05). S2–S4: three subsequent samplings.

The volatile compounds found in Oliva Rossa oils were those usually found in good-quality VOOs [4]. The most abundant ones were C5 and C6 aldehydes, alcohols and esters deriving from the well-known lipoxygenase pathway (LOX) [4], together with a remarkable amount of 3-ethyl-1,5-octadiene.

Regardless of the ripening stage of the drupes, (E)-2-hexenal was the most abundant compound. (E)-2-hexenal was reported to be strongly correlated with the bitter, almond, green, green apple-like, fatty, almond-like and cut-grass notes of the VOOs [4]. In decreasing order, 3-ethyl-1,5-octadiene was the following one. It was already detected in VOOs, although generally in quite lower concentrations [35]. The remarkable amount found in Oliva Rossa oils might be a typical trait of this olive landrace. Indeed, 3-ethyl-1,5-octadiene was already identified as one of the markers for varietal discrimination in Turkish olive oils [54]. It is worthy to note that the results reported here refer to only one crop season and thus further studies are needed to corroborate such evidence.

During ripening, significant differences in the volatile profile were observed [4,5]. Less ripe drupes (S2) gave oils with less total volatile compounds, whose profile was dominated by (E)-2-hexenal, 3-ethyl-1,5-octadiene, C5 compounds (1-penten-3-one, (Z)-2-penten-1-ol, 1-penten-3-ol) and (Z)-3-hexen-1-ol. In particular, (Z)-3-hexen-1-ol, which was significantly higher at S2 with respect to the other stages, was already reported as one of the indicators of the early stage of ripeness [4].

At the technological optimum (i.e., S3), the highest amount of total volatile compounds was observed. In detail, the oils were characterized by the significantly higher content of pentanal, 3-ethyl-1,5-octadiene, 1-penten-3-one, (E)-2-hexenal and (Z)-2-penten-1-ol. It is reported that the maximum content of (E)-2-hexenal is reached when the drupes' pigmentation changes from green to purple and then decrease with ripening [4,5]. However, this trend could be even cultivar dependent.

At S3, a noticeable content of acetic acid was found, too. Acetic acid could originate from the fermentation process and might be linked to some organoleptic defects such as wine-vinegary [55]. However, the detection of such defects involves many other volatile compounds [55]. At S4, a decrease in the total volatile compounds was observed with respect to S3. The oils were characterized by an increase in the acetate esters, which was associated with a decrease in the acetic acid content. A significantly higher content of

2-methyl butanal, 3-methyl butanal and 1-penten-3-ol was reported, too. The correlation among the volatile compounds and sensory features of the VOOs is well known [4,5,56]. Based on this, and considering the data reported in Table 5, it could be supposed that Oliva Rossa oils have notes of fruity, green, cut-leaves and almond, which is in accordance with the few data available in the literature [24].

4. Conclusions

Oliva Rossa is an old olive landrace from the Apulia region mostly unknown and not yet valorized. Aiming at its valorization, some useful results have been reported. First, depending on the crop year, the evolution of the drupes' pigmentation may vary significantly and the half-varaison of the skin color could not match the technological optimum (i.e., the maximum oil content). At its optimum, the drupes had a remarkable amount of phenolic compounds. A medium to high content of oleic acid was observed although significant differences along the years were highlighted, likely due to the different climatic conditions. When a high content of PUFA was reported, there also was significant primary oxidation observed. Thus, with respect to the European limits concerning the primary oxidation markers, this aspect deserves attention, even at the technological step, which has not been considered in this study. The oils had a remarkable amount of minor compounds, which were affected by ripening. Depending on the crop year, the noteworthy levels of MUFA and TPC could lead to oils with a good stability. The phenolic profile in the considered crop year was dominated by secoiridoids derivatives, which might indicate a product with remarkable pungent and bitter notes. Similarly, based on the available data, the volatile profile was dominated by C6 and C5 compounds arising from the LOX pathway. A distinct trait seemed to be the high content of 3-ethyl-1,5-octadiene. Further investigations are needed to confirm the results obtained and to estimate the environmental effect on the oil composition of Oliva Rossa.

Supplementary Materials: The following are available online at https://www.mdpi.com/2304-8158/10/2/369/s1, Figure S1: Monthly average temperature (°C) and rainfall (mm) recorded at the nearest weather station of Turi (Bari, Italy) for four years (2016–2019). The field including Oliva Rossa trees is about at 8 km from the weather station. Figure S2: UHPLC-DAD phenolic profile of Oliva Rossa VOOs extracted at different sampling times in 2017/2018 harvest season. IS, internal standard; (1) 3,4-DHPEA; (2) p-HPEA; (3) Vanillic acid; (4) Syringic acid; (5) 3,4-DHPEA-EDA; (6) 3,4-DHPEA-EDA-CARB; (7) p-HPEA-EDA; (8) Pinoresinol; (9) Luteolin; (10) p-HPEA-EA; (11) Apigenin.

Author Contributions: Conceptualization, G.S., C.S. and F.C.; methodology, G.S., R.S. and G.M.; formal analysis, G.S.; investigation, G.S. and R.S.; data curation, G.S. and F.C.; writing—original draft preparation, G.S., G.M. and F.C.; writing—review and editing, G.S., R.S., G.M., C.S. and F.C.; funding acquisition, F.C. All authors have read and agreed to the published version of the manuscript.

Funding: This research was funded by the AGER 2 Project, grant no. 2016-0105.

Institutional Review Board Statement: Not applicable.

Informed Consent Statement: Not applicable.

Data Availability Statement: Not applicable.

Conflicts of Interest: The authors declare no conflict of interest.

References

1. Wahrburg, U.; Kratz, M.; Cullen, P. Mediterranean diet, olive oil and health. *Eur. J. Lipid Sci. Technol.* **2002**, *104*, 698–705. [CrossRef]
2. UNESCO. Intergovernmental Committee for the Safeguarding of the Intangible Cultural Heritage. ITH/13/8.COM/Decisions. Decision 8.COM 8.10. Paris. 2013. Available online: https://ich.unesco.org/en/8com (accessed on 10 December 2020).
3. CIHEAM/FAO. Mediterranean Food Consumption Patterns: Diet, Environment, Society, Economy and Health. Available online: http://www.fao.org/3/a-i4358e.pdf (accessed on 3 December 2020).
4. Kalua, C.M.; Allen, M.S.; Bedgood, D.R., Jr.; Bishop, A.G.; Prenzler, P.D.; Robards, K. Olive oil volatile compounds, flavour development and quality: A critical review. *Food Chem.* **2007**, *100*, 273–286. [CrossRef]

5. Campestre, C.; Angelini, G.; Gasbarri, C.; Angerosa, F. The compounds responsible for the sensory profile in monovarietal virgin olive oils. *Molecules* **2017**, *22*, 1833. [CrossRef] [PubMed]
6. Colomer, R.; Menéndez, J.A. Mediterranean diet, olive oil and cancer. *Clin. Transl. Oncol.* **2006**, *8*, 15–21. [CrossRef]
7. Covas, M.I. Olive oil and the cardiovascular system. *Pharmacol. Res.* **2007**, *55*, 175–186. [CrossRef]
8. Boronat, A.; Mateus, J.; Soldevila-Domenech, N.; Guerra, M.; Rodríguez-Morató, J.; Varon, C.; Muñoz, D.; Barbosa, F.; Morales, J.C.; Gaedigk, A.; et al. Cardiovascular benefits of tyrosol and its endogenous conversion into hydroxytyrosol in humans. A randomized, controlled trial. *Free Radic. Biol. Med.* **2019**, *143*, 471–481. [CrossRef]
9. Carbone, A.; Cacchiarelli, L.; Sabbatini, V. Exploring quality and its value in the Italian olive oil market: A panel data analysis. *Agric. Food Econ.* **2018**, *6*, 6. [CrossRef]
10. Ilarioni, L.; Proietti, P. Olive tree cultivars. In *The Extra-Virgin Olive Oil Handbook*, 1st ed.; Peri, C., Ed.; John Wiley & Sons, Ltd.: West Sussex, UK, 2014; pp. 59–67.
11. Rotondi, A.; Alfei, B.; Magli, M.; Pannelli, G. Influence of genetic matrix and crop year on chemical and sensory profiles of Italian monovarietal extra-virgin olive oils. *J. Sci. Food Agric.* **2010**, *90*, 2641–2648. [CrossRef]
12. Di Lecce, G.; Piochi, M.; Pacetti, D.; Frega, N.G.; Bartolucci, E.; Scortichini, S.; Fiorini, D. Eleven monovarietal extra virgin olive oils from olives grown and processed under the same conditions: Effect of the cultivar on the chemical composition and sensory traits. *Foods* **2020**, *9*, 904. [CrossRef] [PubMed]
13. Lukić, I.; Lukić, M.; Žanetić, M.; Krapac, M.; Godena, S.; Brkić Bubola, K. Inter-varietal diversity of typical volatile and phenolic profiles of croatian extra virgin olive oils as revealed by GC-IT-MS and UPLC-DAD analysis. *Foods* **2019**, *8*, 565. [CrossRef] [PubMed]
14. Luna, G.; Morales, M.T.; Aparicio, R. Characterisation of 39 varietal virgin olive oils by their volatile compositions. *Food Chem.* **2006**, *98*, 243–252. [CrossRef]
15. Miazzi, M.M.; di Rienzo, V.; Mascio, I.; Montemurro, C.; Sion, S.; Sabetta, W.; Vivaldi, G.A.; Camposeo, S.; Caponio, F.; Squeo, G.; et al. Re.Ger.OP: An integrated project for the recovery of ancient and rare olive germplasm. *Front. Plant. Sci.* **2020**, *11*, 73. [CrossRef]
16. Conte, P.; Squeo, G.; Difonzo, G.; Caponio, F.; Fadda, C.; Del Caro, A.; Urgeghe, P.P.; Montanari, L.; Montinaro, A.; Piga, A. Change in quality during ripening of olive fruits and related oils extracted from three minor autochthonous Sardinian cultivars. *Emir. J. Food Agric.* **2019**, *31*, 196–205. [CrossRef]
17. Squeo, G.; Difonzo, G.; Silletti, R.; Paradiso, V.M.; Summo, C.; Pasqualone, A.; Caponio, F.C. Bambina, una varietà minore pugliese: Profilo di maturazione, composizione delle drupe e caratterizzazione chimica dell'olio vergine. *Riv. Ital. Sostanze Grasse* **2019**, *96*, 143–149.
18. Piscopo, A.; De Bruno, A.; Zappia, A.; Ventre, C.; Poiana, M. Characterization of monovarietal olive oils obtained from mills of Calabria region (Southern Italy). *Food Chem.* **2016**, *213*, 313–318. [CrossRef] [PubMed]
19. FAO. *The State of the World's Biodiversity for Food and Agriculture*; Bélanger, J., Pilling, D., Eds.; FAO Commission on Genetic Resources for Food and Agriculture Assessments: Rome, Italy, 2019; Available online: http://www.fao.org/3/CA3129EN/CA3129EN.pdf (accessed on 3 December 2020).
20. Council of the European Communities. Council regulation (EEC) No 2081/92 of 14 July of 1992 and subsequent integrations and amendments. *O. J. Eur. Communities* **1992**, *208*, 1–8.
21. Cacchiarelli, L.; Carbone, A.; Laureti, T.; Sorrentino, A. The value of the certifications of origin: A comparison between the Italian olive oil and wine markets. *Br. Food J.* **2016**, *118*, 824–839. [CrossRef]
22. Rotondi, A.; Magli, M.; Morrone, L.; Alfei, B.; Pannelli, G. Italian National Database of Monovarietal Extra Virgin Olive Oils. In *The Mediterranean Genetic Code—Grapevine and Olive*; Poljuha, D., Sladonja, B., Eds.; IntechOpen: London, UK, 2013; pp. 179–200.
23. Muzzalupo, I.; Lombardo, N.; Musacchio, A.; Noce, M.E.; Pellegrino, G.; Perri, E.; Sajjad, A. DNA sequence analysis of microsatellite markers enhances their efficiency for germplasm management in an Italian olive collection. *J. Am. Soc. Hortic. Sci.* **2006**, *131*, 352–359. [CrossRef]
24. Oli Monovarietali Italiani by ASSAM Marche and CNR-IBE. Available online: http://www.olimonovarietali.it/ (accessed on 3 December 2020).
25. Oleadb: Worldwide Database for the Management of Genetic Resources of Olive (Olea europaea L.). Available online: http://www.oleadb.it/olivodb.html (accessed on 3 December 2020).
26. Squeo, G.; Difonzo, G.; Summo, C.; Crecchio, C.; Caponio, F. Study of the influence of technological coadjuvants on enzyme activities and phenolic and volatile compounds in virgin olive oil by a response surface methodology approach. *LWT Food Sci. Technol.* **2020**, *133*, 109887. [CrossRef]
27. Protezione Civile Regione Puglia. Available online: https://protezionecivile.puglia.it/ (accessed on 25 January 2021).
28. Difonzo, G.; Fortunato, S.; Tamborrino, A.; Squeo, G.; Bianchi, B.; Caponio, F. Development of a modified malaxer reel: Influence on mechanical characteristic and virgin olive oil quality and composition. *LWT Food Sci. Technol.* **2021**, *135*, 110290. [CrossRef]
29. Zago, L.; Squeo, G.; Bertoncini, E.I.; Difonzo, G.; Caponio, F. Chemical and sensory characterization of Brazilian virgin olive oils. *Food Res. Int.* **2019**, *126*, 108588. [CrossRef] [PubMed]
30. Squeo, G.; Caponio, F.; Paradiso, V.M.; Summo, C.; Pasqualone, A.; Khmelinskii, I.; Sikorska, E. Evaluation of total phenolic content in virgin olive oil using fluorescence excitation–emission spectroscopy coupled with chemometrics. *J. Sci. Food Agric.* **2019**, *99*, 2513–2520. [CrossRef] [PubMed]

31. Commission of the European Communities. Commission regulation (EEC) No 2568/91 of 11 July of 1991 and subsequent integrations and amendments. *Off. J. Eur. Communities* **1991**, *248*, 1–83.
32. Pokorny, J.; Kalinova, L.; Dysseler, P. Determination of chlorophyll pigments in crude vegetable oils: Results of a collaborative study and the standardized method (Technical Report). *Pure Appl. Chem.* **1995**, *67*, 1781–1787. [CrossRef]
33. Makhlouf, F.Z.; Squeo, G.; Barkat, M.; Trani, A.; Caponio, F. Antioxidant activity, tocopherols and polyphenols of acorn oil obtained from *Quercus* species grown in Algeria. *Food Res. Int.* **2018**, *114*, 208–213. [CrossRef]
34. Paradiso, V.M.; Squeo, G.; Pasqualone, A.; Caponio, F.; Summo, C. An easy and green tool for olive oils labelling according to the contents of hydroxytyrosol and tyrosol derivatives: Extraction with a natural deep eutectic solvent and direct spectrophotometric analysis. *Food Chem.* **2019**, *291*, 1–6. [CrossRef]
35. Caponio, F.; Leone, A.; Squeo, G.; Tamborrino, A.; Summo, C. Innovative technologies in virgin olive oil extraction process: Influence on volatile compounds and organoleptic characteristics. *J. Sci. Food Agric.* **2019**, *99*, 5594–5600. [CrossRef]
36. Tozzini, L. Pesano sulla produzione i danni della gelata. *Olivo e Olio* **2019**, *1*, 8–9. (In Italian)
37. Camposeo, S.; Vivaldi, G.A.; Gattullo, C.E. Ripening indices and harvesting times of different olive cultivars for continuous harvest. *Sci. Hortic.* **2013**, *151*, 1–10. [CrossRef]
38. Trapani, S.; Migliorini, M.; Cherubini, C.; Cecchi, L.; Canuti, V.; Fia, G.; Zanoni, B. Direct quantitative indices for ripening of olive oil fruits to predict harvest time. *Eur. J. Lipid Sci. Technol.* **2016**, *118*, 1202–1212. [CrossRef]
39. Inglese, P.; Famiani, F.; Galvano, F.; Servili, M.; Esposto, S.; Urbani, S. Factors Affecting Extra-Virgin Olive Oil Composition. In *Horticultural Reviews*; Janik, J., Ed.; John Wiley & Sons, Ltd.: West Sussex, UK, 2011; Volume 38, pp. 83–148.
40. Klen, T.J.; Vodopivec, B.M. The fate of olive fruit phenols during commercial olive oil processing: Traditional press versus continuous two-and three-phase centrifuge. *LWT Food Sci. Technol.* **2012**, *49*, 267–274. [CrossRef]
41. Artajo, L.S.; Romero, M.P.; Motilva, M.J. Transfer of phenolic compounds during olive oil extraction in relation to ripening stage of the fruit. *J. Sci. Food Agric.* **2006**, *86*, 518–527. [CrossRef]
42. Sales-Campos, H.; Reis de Souza, P.; Crema Peghini, B.; Santana da Silva, J.; Ribeiro Cardoso, C. An overview of the modulatory effects of oleic acid in health and disease. *Mini-Rev. Med. Chem.* **2013**, *13*, 201–210. [CrossRef] [PubMed]
43. European commission. Commission Regulation (EU) No 432/2012 of 16 May 2012. *Off. J. Eur. Union* **2012**, *136*, 1–40.
44. Bendini, A.; Cerretani, L.; Carrasco-Pancorbo, A.; Gómez-Caravaca, A.M.; Segura-Carretero, A.; Fernández-Gutiérrez, A.; Lercker, G. Phenolic molecules in virgin olive oils: A survey of their sensory properties, health effects, antioxidant activity and analytical methods. An overview of the last decade. *Molecules* **2007**, *12*, 1679–1719. [CrossRef]
45. Kiritsakis, A.K.; Nanos, G.D.; Polymenoupoulos, Z.; Thomai, T.; Sfakiotakis, E.Y. Effect of fruit storage conditions on olive oil quality. *J. Am. Oil Chem. Soc.* **1998**, *75*, 721–724. [CrossRef]
46. Choe, E.; Min, D.B. Mechanisms and factors for edible oil oxidation. *Compr. Rev. Food. Sci. Food Saf.* **2006**, *5*, 169–186. [CrossRef]
47. Caponio, F.; Gomes, T.; Pasqualone, A. Phenolic compounds in virgin olive oils: Influence of the degree of olive ripeness on organoleptic characteristics and shelf-life. *Eur. Food Res. Technol.* **2001**, *212*, 329–333. [CrossRef]
48. Beltrán, G.; Aguilera, M.P.; Del Rio, C.; Sanchez, S.; Martinez, L. Influence of fruit ripening process on the natural antioxidant content of Hojiblanca virgin olive oils. *Food Chem.* **2005**, *89*, 207–215. [CrossRef]
49. Gandul-Rojas, B.; Cepero, M.R.L.; Mínguez-Mosquera, M.I. Use of chlorophyll and carotenoid pigment composition to determine authenticity of virgin olive oil. *J. Am. Oil Chem. Soc.* **2000**, *77*, 853–858. [CrossRef]
50. Criado, M.N.; Romero, M.P.; Casanovas, M.; Motilva, M.J. Pigment profile and colour of monovarietal virgin olive oils from Arbequina cultivar obtained during two consecutive crop seasons. *Food Chem.* **2008**, *110*, 873–880. [CrossRef]
51. Roca, M.; Minguez-Mosquera, M.I. Change in the natural ratio between chlorophylls and carotenoids in olive fruit during processing for virgin olive oil. *J. Am. Oil Chem. Soc.* **2001**, *78*, 133–138. [CrossRef]
52. Pedan, V.; Popp, M.; Rohn, S.; Nyfeler, M.; Bongartz, A. Characterization of phenolic compounds and their contribution to sensory properties of olive oil. *Molecules* **2019**, *24*, 2041. [CrossRef]
53. Caponio, F.; Squeo, G.; Brunetti, L.; Pasqualone, A.; Summo, C.; Paradiso, V.M.; Catalano, P.; Bianchi, B. Influence of the feed pipe position of an industrial scale two-phase decanter on extraction efficiency and chemical-sensory characteristics of virgin olive oil. *J. Sci. Food Agric.* **2018**, *98*, 4279–4286. [CrossRef] [PubMed]
54. Ilyasoglu, H.; Ozcelik, B.; Van Hoed, V.; Verhe, R. Cultivar characterization of Aegean olive oils with respect to their volatile compounds. *Sci. Hortic.* **2011**, *129*, 279–282. [CrossRef]
55. Cayuela, J.A.; Gómez-Coca, R.B.; Moreda, W.; Pérez-Camino, M.D.C. Sensory defects of virgin olive oil from a microbiological perspective. *Trends Food Sci. Technol.* **2015**, *43*, 227–235. [CrossRef]
56. Cecchi, T.; Alfei, B. Volatile profiles of Italian monovarietal extra virgin olive oils via HS-SPME–GC–MS: Newly identified compounds, flavors molecular markers, and terpenic profile. *Food Chem.* **2013**, *141*, 2025–2035. [CrossRef] [PubMed]

Article

Bioactive Potential of Minor Italian Olive Genotypes from Apulia, Sardinia and Abruzzo

Wilma Sabetta [1,2,*], Isabella Mascio [3], Giacomo Squeo [3], Susanna Gadaleta [3], Federica Flamminii [4], Paola Conte [5], Carla Daniela Di Mattia [4], Antonio Piga [5], Francesco Caponio [3] and Cinzia Montemurro [2,3,6]

1. Institute of Biosciences and BioResources, National Research Council (IBBR-CNR), Via Amendola 165/A, 70125 Bari, Italy
2. Spin off Sinagri s.r.l., University of Bari Aldo Moro, Via Amendola 165/A, 70125 Bari, Italy; cinzia.montemurro@uniba.it
3. Department of Soil, Plant and Food Sciences, University of Bari Aldo Moro, Via Amendola 165/A, 70125 Bari, Italy; mascioisa@gmail.com (I.M.); giacomo.squeo@uniba.it (G.S.); sanna14@hotmail.it (S.G.); francesco.caponio@uniba.it (F.C.)
4. Faculty of Bioscience and Technology for Agriculture, Food and Environment, University of Teramo, Via Renato Balzarini 1, 64100 Teramo, Italy; fflamminii@unite.it (F.F.); cdimattia@unite.it (C.D.D.M.)
5. Dipartimento di Agraria, University of Sassari, Viale Italia 39/A, 07100 Sassari, Italy; pconte@uniss.it (P.C.); pigaa@uniss.it (A.P.)
6. Institute for Sustainable Plant Protection–Support Unit Bari, National Research Council (IPSP-CNR), Via Amendola 165/A, 70125 Bari, Italy
* Correspondence: wilma.sabetta@ibbr.cnr.it; Tel.: +39-080-5583400

Citation: Sabetta, W.; Mascio, I.; Squeo, G.; Gadaleta, S.; Flamminii, F.; Conte, P.; Di Mattia, C.D.; Piga, A.; Caponio, F.; Montemurro, C. Bioactive Potential of Minor Italian Olive Genotypes from Apulia, Sardinia and Abruzzo. *Foods* **2021**, *10*, 1371. https://doi.org/10.3390/foods10061371

Academic Editor: Alessandra Bendini

Received: 30 April 2021
Accepted: 7 June 2021
Published: 14 June 2021

Publisher's Note: MDPI stays neutral with regard to jurisdictional claims in published maps and institutional affiliations.

Copyright: © 2021 by the authors. Licensee MDPI, Basel, Switzerland. This article is an open access article distributed under the terms and conditions of the Creative Commons Attribution (CC BY) license (https://creativecommons.org/licenses/by/4.0/).

Abstract: This research focuses on the exploration, recovery and valorization of some minor Italian olive cultivars, about which little information is currently available. Autochthonous and unexplored germplasm has the potential to face unforeseen changes and thus to improve the sustainability of the whole olive system. A pattern of nine minor genotypes cultivated in three Italian regions has been molecularly fingerprinted with 12 nuclear microsatellites (SSRs), that were able to unequivocally identify all genotypes. Moreover, some of the principal phenolic compounds were determined and quantified in monovarietal oils and the expression levels of related genes were also investigated at different fruit developmental stages. Genotypes differed to the greatest extent in the content of oleacein (3,4-DHPEA-EDA) and total phenols. Thereby, minor local genotypes, characterized by stable production and resilience in a low-input agro-system, can provide a remarkable contribution to the improvement of the Italian olive production chain and can become very profitable from a socio-economic point of view.

Keywords: *Olea europaea*; autochthonous cultivars; molecular fingerprinting; polyphenol content; gene expression; fruit developmental stages

1. Introduction

Olive (*Olea europaea* ssp. *europaea*) is considered among the historically and traditionally most important crops in the world and especially in the Mediterranean basin, where it has been cultivated for centuries not only for nutrition but also for cultural and religious reasons. In particular, the strategic geographical position of Italy in a temperate area has favored olive cultivation and therefore the enrichment of its germplasm over time, which is estimated to include about 800 cultivars [1,2]. The great diffusion of olive trees in the Italian territory highlights its importance for production, economy and local traditions, as proven and documented by innumerable historical catalogues and archives up to the XIII century B.C. [3].

As a consequence of the general awareness about the loss of plant genetic diversity and the drastic climate change currently underway, the attention of the scientific community has been recently put toward more sustainable agriculture, as opposed to the intensive, mono-cultivar farming systems, and to the safeguard of plant biodiversity as source of new interesting traits. Similarly to other cultivations, also for olive, a valid and intriguing opportunity to help in overcoming these issues is offered by the unexploited or still poorly characterized germplasm, including local or minor genotypes, i.e., autochthonous landraces generally spread at regional level and well adapted to specific pedoclimatic conditions in traditional groves with very low agronomic input [4]. These genotypes could represent an interesting reservoir of useful traits, such as nutraceutical and antioxidant compounds and/or resistances to environmental stresses, that makes their diffusion in the market and their utilization in breeding programs extremely promising. In this perspective, the first step for the valorization of these genotypes is represented by their fine characterization that provides the genetic identity of both plant material and derived products (drupes and oils) and that guarantees an unequivocal varietal recognition. Microsatellite markers (SSRs) are among the most simple, fast and economic molecular tools widely used for olive varietal identification [5–8] and for food tracking and tracing [9–13].

The extra virgin olive oil is considered a fundamental element of the Mediterranean diet [14], thus representing one of the pillars of the Italian economy. Being the most important fat source in the human diet, olive oil is characterized by a distinctive high content of mono-unsaturated fatty acids (MUFA) and by a certain level of secondary metabolites with nutraceutical properties, such as polyphenols [15]. The most abundant polyphenol class in olive oil is represented by the secoiridoids, i.e., complex molecules including oleuropein, ligstroside and their derivates (oleacein, oleocanthal, an isomer of oleuropein aglycone called 3,4-DHPEA-EA and the ligstroside aglycone also known as p-HPEA-EA), the latter compounds being produced during olive oil mechanical extraction. Polyphenols also include two important phenolic alcohols worthy to be mentioned for their antioxidant power, i.e., tyrosol and hydroxy-tyrosol [16].

A huge literature has been dedicated to the healthy value of polyphenols [17–20]. Numerous research has demonstrated their positive effects in contrasting cardiovascular diseases thanks to their anti-inflammatory and -oxidative actions [21–24]. In particular, polyphenols are able to bind low-density lipoprotein (LDL), avoiding its oxidation [25], modulate angiogenic responses [26], protect against endothelial dysfunction [27] and contribute to decrease the blood pressure [28]. Moreover, the positive effect of tyrosol and hydroxy-tyrosol against reactive oxygen species has been deeply investigated [29–32].

Furthermore, some phenolic compounds, especially oleuropein, ligstroside aglycone and derived molecules, are responsible for the pungency and bitter taste of olive oil [33–36]. For this reason, in recent decades, breeders and researchers focused on polyphenols as new trait of interest to satisfy the increasing request of high-quality olive oil for nutritional and organoleptic aspects [37–39].

The biosynthetic pathway of olive secoiridoids has been studied and proposed by several authors, even though it is not completely clarified yet. Transcriptomic and metabolomic analysis of developing olive fruits and leaves in different cultivars have shed light on the enzymes involved in polyphenol metabolism [40–44].

The main aim of this work was the valorization of some minor olive genotypes and cultivars, autochthonous of Central and Southern Italy, that are still poorly studied, but interesting for both their resilience and adaptation to low-input agriculture and for the quality of olive oil. The application of SSR markers has been necessary to obtain a clear molecular fingerprinting of these genotypes, as a recognized tool of primary importance for certifying plant material production, varietal tracing and authenticity testing. Moreover, an integrated approach of oil biochemical characterization, with particular focus on tyrosol, hydroxy-tyrosol and oleacein, coupled with a genetic expression study of key enzymes involved in the first steps of their biosynthesis, has contributed to clarify the molecular mechanisms underlying the olive polyphenol biosynthetic pathway and to further valorize the Italian olive germplasm.

2. Materials and Methods

2.1. Plant Material and Olive Sampling

In total, nine olive cultivars were selected for this work; they were grown in groves located in three different Italian regions: Sardinia, Abruzzo and Apulia.

The Sardinian cultivars "Sivigliana da olio", Semidana and "Corsicana da olio" have been chosen for their suitability to be processed as table olives (Sivigliana da olio), for their resistance to some common diseases and high productivity (Corsicana da olio) and for the increasing interest of growers to introduce it in all Sardinian orchards (Semidana) [45,46]. Specialized non-irrigated orchards were located in the province of Sassari (Sardinia, Italy); olives of Sivigliana and Semidana cultivars were hand harvested in Ittiri (40°37′15.7″ NL 8°32′20.7″ EL), while samples of Corsicana were picked in Usini (40°39′39.0″ NL 8°31′32.7″ EL).

With regard to Abruzzo (Central Italy), the varieties Dritta, Tortiglione and Gentile dell'Aquila were selected as the most representative of the provinces of Pescara, Teramo and L'Aquila, respectively, also in consideration of their specific adaptation to the local pedoclimatic conditions correlated with peculiar agronomic and biochemical traits [47,48]. Dritta and Tortiglione were grown in orchards located in Notaresco, in the province of Teramo (42°38′54.6″ NL, 13°52′57.4″ EL), while Gentile dell'Aquila was grown in orchards from Vittorito, in the province of L'Aquila (42°08′16.6″ NL, 13°48′37.3″ EL).

In the Apulia region (Southern Italy), in addition to some well-known and notable cultivars for olive oil extraction, a great number of other varieties and/or landraces are present [3]. Among these, three have been considered in this study, specifically the varieties Bambina, Oliva Rossa and Cima di Melfi, for which few reports are currently available in the literature [49,50]. The Apulian genotype Bambina was cultivated in the orchard located in Gravina di Puglia (40°49′0″ NL, 16°25′0″ EL), while genotypes Cima di Melfi and Oliva Rossa were grown in the orchard of Putignano (40°51′0″ NL, 17°7′0″ EL), both in the province of Bari.

For the molecular characterization, young leaves of each cultivar were collected and stored at −20 °C before use. For gene expression analysis, drupes were harvested at different developmental stages of fruit, generally from mid-October to the beginning of November: T1, yellow-green olives; T2, turning olives; and T3, almost dark olives (Table 1). In order to minimize the effects of fruit asynchronous maturation within the same tree, drupes were harvested from the external parts of the canopy of trees. After harvesting, olives were immediately frozen in liquid nitrogen and stored at −80 °C until further processing. For two of the considered time-points (T2 and T3), one part of the collected drupes was immediately used for virgin olive oil extraction and biochemical characterization. An additional timepoint at 100% ripening (T4, fully dark olives, from the end of November to the beginning of December) was considered exclusively for oil production and biochemical characterization (Table 1). For each cultivar/genotype, three biological replicates were taken. Sampling was carried out for two consecutive years.

Table 1. Overview of the four drupe-collecting stages and type of analysis carried out in this study.

Sampling Time-Point	Drupe Status	Performed Analysis
T1	yellow-green	Gene expression
T2	turning	Gene expression + oil biochemical characterization
T3	almost dark	Gene expression + oil biochemical characterization
T4	fully dark	Oil biochemical characterization

2.2. Genetic Characterization

Leaf samples were collected from three plants per cultivar, lyophilized and finely pulverized before use. Total genomic DNA was extracted from 200 mg of dry tissue according to [51], checked for quantity and quality and normalized to 50 ng μL^{-1}. Olive genotypes were molecularly characterized by means of 12 nuclear SSRs (Table S1) [52–54], chosen on the basis of their suitability and reliability proven in several studies about olive variety identification [55–57]. Primer pairs were synthesized by Thermo Fisher Scientific (Waltham, MA, USA) and all forward primers were labeled with one of the following dyes: 6FAM™, NED™, VIC® and PET™. The amplification reactions were carried out in a final volume of 12.5 µL, using a T100 thermal cycler (Bio-Rad Laboratories, Segrate, MI, Italy) according to [9]. Two microliters of each PCR product was added to 0.5 µL of GeneScan™ 600 LIZ® Size Standard (Applied Biosystem, Foster City, CA, USA) and 9.5 µL of Hi-Di Formamide (Applied Biosystem, Foster City, CA, USA), and successively separated by capillary electrophoresis using an automatic sequencer ABI PRISM 3100 Avant Genetic Analyzer (Applied Biosystems, Foster City, CA, USA). Detection, sizing and data collection were carried out by means of the GeneMapper® genotyping software v.5.0 (Applied Biosystems, Foster City, CA, USA) as in [58]. To estimate the genetic distances among the considered genotypes, cluster analysis based on the Unweighted Neighbor Joining method was performed using DARWIN software v. 6.0.010 (http://darwin.cirad.fr, accessed on 26 April 2019), with 1000 bootstrap values for tree construction.

2.3. Oil Extraction and Characterization

Around 10 kg of drupes were collected from three trees per each sampling point and variety. Fruits were then divided into three aliquots representing the biological replicates. Oil extraction was performed within 5 h after harvesting at laboratory scale. The extraction system was made up of a semi-industrial scale hammer crusher (RETSCH GmbH 5657, Haan, Germany) working at 2850 rpm and a basket centrifuge with a bowl of 19 cm working at 2700 rpm (Marelli Motori S.p.A., Arzignano, VI, Italy) [59]. Briefly, about 1 kg of olives per sample and replicate was crushed and then the olive paste was transferred into the basket centrifuge for the oil recovery. Once extracted, the virgin olive oils ($n = 3$) were stored at 20 °C in 100 mL dark glass bottles until the analyses.

Two phenolic alcohols (i.e., hydroxyl-tyrosol and tyrosol) and the oleacein (3,4-DHPEA-EDA) were chosen as target phenolic compounds for HPLC analyses, being amongst the most abundant ones in olive oils and because of their powerful antioxidant activity and impact on oil sensorial feature [60]. These phenolic compounds were extracted by liquid–liquid extraction using a mixture of methanol/water (70/30 v/v) according to previous papers [61]. The extraction procedure was similar for both the Folin–Ciocalteu assay and for HPLC analysis, with the only difference being that in the latter case 250 µL of a 100 mg kg^{-1} solution of gallic acid as internal standard for quantification was added. Total phenolic compounds (TPC) were quantified spectrophotometrically [61] by means of a calibration curve of pure gallic acid and the results are expressed as gallic acid equivalent (GAE, mg kg^{-1}). HPLC-DAD analysis was carried out as previously reported [61] using a UHPLC binary system (Dionex Ultimate 3000 RSLC, Waltham, MA, USA). The identification of hydroxy-tyrosol, tyrosol and 3,4-DHPEA-EDA was performed by comparing the peak retention times with those obtained by the injection of pure standards and/or with data in the literature [62]. The results are expressed as gallic acid equivalent (GAE, mg kg^{-1}).

2.4. Polyphenolic Compound Gene Expression

The mesocarps of frozen olives were mechanically crushed by the use of a tissue lyser and total RNA was extracted from 100 mg of the obtained powder according to the manufacturer's instructions of Spectrum Plant Total RNA Kit (Sigma-Aldrich, St. Louis, MO, USA). An additional step for on column genomic DNA digestion was added. RNA quantity and quality were checked by spectrophotometric measurement using a Nanodrop 2000 spectrophotometer (Thermo Fisher Scientific, Waltham, MA, USA) and by electrophoresis on 1.2% Certified Molecular Biology Agarose gel (Bio-Rad Laboratories, Segrate, MI, Italy) in 1X TBE buffer (1 M Trizma base, 1 M Boric Acid, 20 mM EDTA, pH 8.3).

Quantitative real-time polymerase chain reactions (qRT-PCRs) were carried out on three genes involved in the polyphenolic biosynthesis, named TYRD for tyrosine/dopa decarboxylase, CuAO for copper amine oxidase, and ALDH for alcohol dehydrogenase.

The used primer pairs are listed in Supplementary Material (Table S2). The reverse transcription of 700 μg RNA samples was performed with the SuperScript™ VILO™ cDNA Synthesis Kit (Thermo Fisher Scientific, Waltham, MA, USA) according to the manufacturer's instructions. qRT-PCR reactions were carried out using the SsoAdvanced Universal SYBER® Green Supermix (Bio-Rad Laboratories, Segrate, MI, Italy) and the CFX96 Touch Real-Time PCR Detection System (Bio-Rad Laboratories, Segrate, MI, Italy). Thermal cycling parameters were: initial denaturation at 95 °C for 3 min, followed by 40 cycles of 95 °C for 10 s and 60 °C for 30 s. The specificity of the amplification product per each primer pair was confirmed by evaluating the melting curve through an increase of 0.2 °C every 5 s from 65 to 95 °C. For qRT-PCR assay, each amplification reaction was run in triplicates of three biological replicates. The elongation factor 1α (EF1α, AM946404) was selected as a reference gene for normalization and the comparative Ct method ($2^{-\Delta\Delta Ct}$ method) was used to analyze the expression levels of the selected genes [63].

2.5. Statistical Analysis

All values of chemical and genetic analysis are means ($n = 3$) ± standard deviation (sd). The statistical analysis was performed, for samples of each region, by one-way analysis of variance (ANOVA) with sampling time as the group factor and post-hoc Fisher's LSD test, using the software Statistica 10.0 for Windows. Differences were considered to be significant when $p < 0.05$.

3. Results

3.1. Genetic Characterization

The chosen nuclear microsatellites successfully amplified all the olive cultivars under analysis, with identical genetic profiles among the replicates of each cultivar. With the exception of missing data for only three loci related to the markers DCA09, DCA17 and GAPU101, unique alleles for all the used SSR were assigned to each cultivar, thus allowing an accurate varietal identification (Table 2 and Figure S1).

The Unweighted Neighbor Joining dendrogram allowed us to group the genotypes predominantly on the basis of their regional origin, except one. Indeed, as expected, three main clusters were identified (Figure 1). Cluster I included two of the three analyzed Abruzzo cultivars, i.e., Tortiglione and Dritta. The third Abruzzo cultivar, Gentile dell'Aquila, was included in Cluster II together with all the Sardinian cultivars (Semidana, Sivigliana and Corsicana). Finally, Cluster III encompassed cultivars from the Apulia region, i.e., Bambina, Cima di Melfi and Oliva Rossa.

Table 2. Molecular fingerprinting of the analyzed cultivars by means of 12 microsatellite markers (SA, AB and AP, respectively, indicate Sardinian, Abruzzo and Apulian varieties). Allele sizes are reported in base pair (bp).

Cultivar	Microsatellite Marker																							
	DCA03		DCA05		DCA09		DCA13		DCA15		DCA17		DCA18		GAPU45		GAPU71b		GAPU101		EMO90		EMOL	
Corsicana (SA)	232	253	198	208	162	182	120	124	246	257	109	115	177	183	181	181	127	144	192	200	186	188	190	198
Semidana (SA)	245	253	202	206	162	174	122	140	246	257	117	117	177	179	183	185	130	144	192	218	186	188	190	198
Sivigliana (SA)	243	253	206	208	162	174	120	120	246	257	113	117	179	183	181	181	127	144	198	200	188	194	190	198
Tortiglione (AB)	229	245	198	206	186	186	120	120	246	246	113	113	177	185	195	195	124	127	192	192	188	198	198	198
Dritta (AB)	229	229	208	212	186	194	120	120	246	257	113	113	177	177	181	185	127	127	190	206	186	190	192	192
Gentile dell'Aquila (AB)	243	249	206	206	162	162	120	120	246	246	-	-	179	181	183	183	144	144	200	206	188	194	198	198
Bambina (AP)	243	253	198	206	-	-	120	120	246	266	117	117	177	177	183	185	127	144	-	-	188	188	198	198
Cima di Melfi (AP)	243	253	198	206	184	184	120	120	246	246	143	143	177	181	181	185	124	144	182	182	190	190	192	198
Oliva Rossa (AP)	239	253	206	206	184	194	120	122	246	273	143	143	177	181	185	185	127	144	182	218	188	188	200	214

"-" indicates missing data.

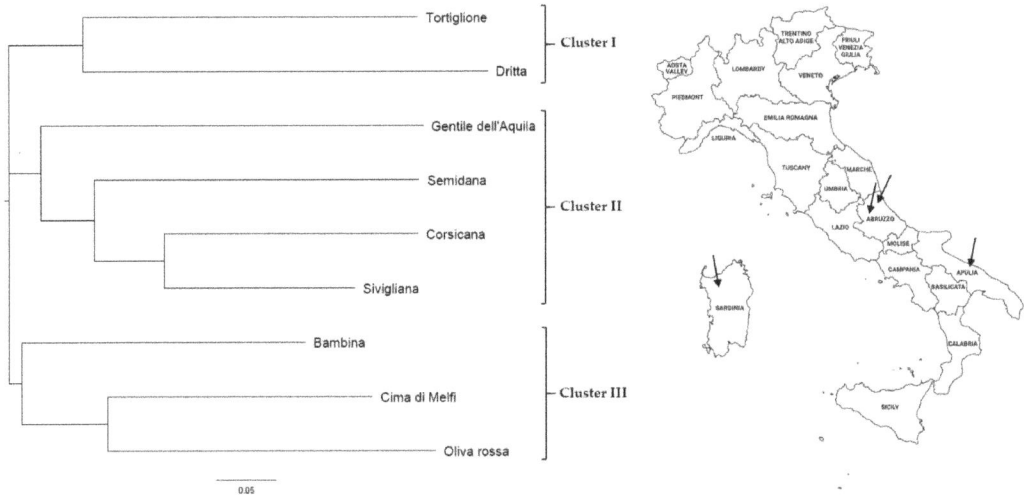

Figure 1. Unweighted Neighbor Joining dendrogram illustrating the clusterization of the olive cultivars under study, and a map of Italy with indication of the considered regions and the corresponding drupe sampling sites.

3.2. Monovarietal Olive Oil Phenolic Content

This research focused on three of the main active constituents of olive polyphenols, i.e., tyrosol, hydroxy-tyrosol and oleacein (3,4-DHPEA-EDA), and on the TPC of monovarietal oils. In general, the concentration and the main trend of these compounds varied among the analyzed cultivars. For example, tyrosol content was higher in cultivar Tortiglione than in Dritta and Gentile dell'Aquila with no significant changes ($p > 0.05$) during ripening, while it significantly increased in Sardinian cultivars and in two of the Apulian genotypes, i.e., Bambina and Oliva Rossa. An opposite trend was observed in cultivar Cima di Melfi, that showed a significant drop in tyrosol content at complete maturity of drupes (Table 3; Figure 2A).

Table 3. Phenolic compounds and total phenolic content (TPC) detected in VOOs of all the analyzed cultivars from fruits at different ripening stages (T2 = turning drupes, T3 = almost dark drupes, T4 = fully dark drupes).

Region	Genotype	Drupe Collecting Time-Point	Hydroxy-Tyrosol (3,4-DHPEA)	Tyrosol (p-HPEA)	Oleacein (3,4-DHPEA-EDA)	TPC
Sardinia	Corsicana	T2	0.14 ± 0.12 d	0.32 ± 0.02 f	33.95 ± 0.21 a	572 ± 3 a
	Corsicana	T3	0.27 ± 0.01 d	0.36 ± 0.02 f	30.06 ± 0.17 b	558 ± 1 a
	Corsicana	T4	0.13 ± 0.01 d	0.43 ± 0.01 ef	9.20 ± 0.43 e	511 ± 1 b
	Semidana	T2	0.81 ± 0.28 bc	1.18 ± 0.17 b	13.32 ± 3.95 d	479 ± 23 de
	Semidana	T3	0.74 ± 0.19 c	1.32 ± 0.08 b	9.69 ± 0.61 e	469 ± 9 ef
	Semidana	T4	0.91 ± 0.15 abc	1.55 ± 0.12 a	7.78 ± 0.67 e	453 ± 3 f
	Sivigliana	T2	1.07 ± 0.20 ab	0.65 ± 0.10 d	20.83 ± 1.46 c	503 ± 10 bc
	Sivigliana	T3	0.70 ± 0.15 c	0.55 ± 0.04 de	7.88 ± 0.55 e	487 ± 6 cd
	Sivigliana	T4	1.14 ± 0.29 a	0.83 ± 0.04 c	12.72 ± 1.46 d	477 ± 10 de

Table 3. Cont.

Region	Genotype	Drupe Collecting Time-Point	Hydroxy-Tyrosol (3,4-DHPEA)	Tyrosol (p-HPEA)	Oleacein (3,4-DHPEA-EDA)	TPC
Apulia	Bambina	T2	0.40 ± 0.06 a	1.51 ± 0.06 c	4.85 ± 0.52 b	358 ± 15 b
	Bambina	T3	0.36 ± 0.06 a	1.86 ± 0.07 b	10.75 ± 1.76 a	406 ± 10 a
	Bambina	T4	0.32 ± 0.09 a	2.35 ± 0.13 a	3.12 ± 1.83 b	392 ± 6 a
	Cima di Melfi	T2	0.75 ± 0.01 a	3.01 ± 0.01 a	17.49 ± 1.89 c	386 ± 5 c
	Cima di Melfi	T3	0.39 ± 0.06 b	1.36 ± 0.01 b	36.79 ± 2.76 b	502 ± 20 b
	Cima di Melfi	T4	0.41 ± 0.05 b	1.00 ± 0.05 c	65.52 ± 12.26 a	730 ± 11 a
	Oliva Rossa	T2	0.30 ± 0.05 ab	0.48 ± 0.02 b	55.34 ± 2.07 a	677 ± 47 a
	Oliva Rossa	T3	0.35 ± 0.05 a	0.45 ± 0.01 b	21.36 ± 3.46 b	478 ± 5 b
	Oliva Rossa	T4	0.22 ± 0.06 b	0.59 ± 0.04 a	15.46 ± 1.77 c	416 ± 19 c
Abruzzo	Dritta	T2	0.12 ± 0.02 c	0.25 ± 0.01 b	3.71 ± 0.24 de	129 ± 27 d
	Dritta	T3	0.03 ± 0.01 c	0.23 ± 0.02 b	0.50 ± 0.09 e	39 ± 2 f
	Dritta	T4	0.09 ± 0.03 c	0.26 ± 0.04 b	1.41 ± 0.08 e	71 ± 9 e
	Gentile dell'Aquila	T2	0.05 ± 0.01 c	0.17 ± 0.03 d	6.19 ± 1.37 d	123 ± 38 d
	Gentile dell'Aquila	T3	0.07 ± 0.02 c	0.13 ± 0.05 cd	1.09 ± 0.51 e	106 ± 26 de
	Gentile dell'Aquila	T4	0.09 ± 0.01 c	0.16 ± 0.02 cd	16.80 ± 1.35 c	176 ± 15 c
	Tortiglione	T2	0.59 ± 0.12 a	0.38 ± 0.07 a	42.37 ± 3.78 a	804 ± 68 a
	Tortiglione	T3	0.29 ± 0.08 b	0.21 ± 0.02 b	24.44 ± 3.72 b	377 ± 30 b
	Tortiglione	T4	0.28 ± 0.06 b	0.34 ± 0.01 a	18.34 ± 1.61 c	395 ± 6 b

Data are means (n = 3) ± standard deviation. Different letters for each parameter and region indicate significant differences according to one-way ANOVA followed by Fisher's LSD post-hoc test (p = 0.05).

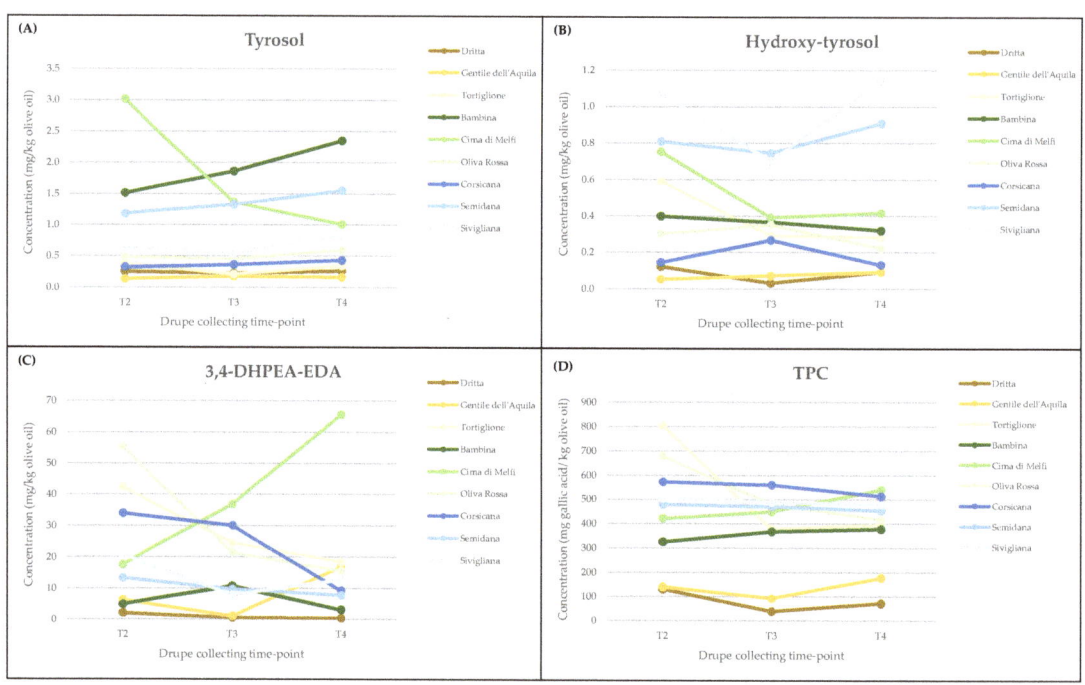

Figure 2. Trend of variation in some polyphenolic compound contents in VOOs of the analyzed genotypes, during three developmental stages of olive fruits (T2, T3, T4). (**A**) Tyrosol; (**B**) hydroxy-tyrosol; (**C**) oleacein (3,4-DHPEA-EDA); (**D**) total phenol content (TPC).

Hydroxy-tyrosol concentration was the lowest among the three studied compounds, generally showing its highest amount in two time-points, that is at T2 and T4. Among all the analyzed cultivars, two Sardinian cultivars (Semidana and Sivigliana) had the highest hydroxy-tyrosol content, with values of about 0.9 and 1.2 mg kg^{-1} olive oil, respectively. In particular, in Sivigliana cultivar, this compound first decreased when drupes were in advanced stage of ripening (T3) and then significantly increased at the end of fruit maturation. An exception to this behavior was recorded for cultivars Oliva Rossa and Corsicana, whose hydroxy-tyrosol contents were slightly higher at T3 (Table 3; Figure 2B).

The concentration of oleacein generally decreased with maturity stage in all the considered cultivars, with the only exception of Cima di Melfi and Gentile dell'Aquila, that showed a more evident and significant increase in this compound in fully ripe fruits (Table 3; Figure 2C). Cultivars Dritta and Bambina were the genotypes with the lowest amount of 3,4-DHPEA-EDA in their oils. Bambina VOO was already known for its particular phenolic pattern, which shows a significant contribution of flavonoids and a lesser amount of secoiridoid derivatives [49].

Finally, changes in the total phenol content (TPC) were generally recorded in most of the cultivars. Indeed, the concentration of total phenols was almost constant during drupe maturation or slightly increasing at the last stage of fruit ripening (T4), except for the Abruzzo cultivar Tortiglione and the Apulian cultivar Oliva Rossa, that showed a drastic and significant drop in the total phenol content in comparison with the first stage of fruit ripening (T4 vs. T2) (Table 3; Figure 2D). Sardinian cultivars showed, on the other hand, a slight but significant decrease in TPC at the T4 sampling. Anyway, the oils obtained from these cultivars, together with oils derived from most of the Apulian genotypes, were the ones with the highest and most stable amount of total phenols during all stages of fruit ripening.

3.3. Gene Expression Analysis

The expression profiles of the selected genes did not follow the same trend among the cultivars. In most genotypes, the mRNA levels of TYRD were quite low and remained almost constant as the fruit development proceeded (Figure 3A). An exception was represented by cultivars Sivigliana and Gentile dell'Aquila, in which this gene was highly expressed at the first drupe sampling (T1) and then drastically dropped at the second sampling, albeit with no significance. On the contrary, in cultivars Semidana and Tortiglione, TYRD mRNA levels remarkably increased from the second to the third sampling, thus reaching the highest values when drupes were almost dark and mature.

In some cultivars, the CuAO expression pattern did not notably change during fruit maturation, while in other genotypes, such as Semidana and Bambina, a slight increase was observed up to the third sampling time. Interestingly, the CuAO expression level was noteworthy in one of the Apulian cultivars, i.e., Cima di Melfi, not for its general trend but specifically for its abundance during the entire drupe maturation process, since it was extremely and significantly different (higher) in comparison to that detected in all other cultivars (Figure 3B).

In the majority of genotypes, ALDH was almost exclusively expressed during the T3 stage of fruit development, with a remarkable increase starting at the turning phase of drupes, in particular in the cases of Gentile dell'Aquila and Cima di Melfi, and with a less pronounced increment observed in the cases of Bambina, Semidana and Sivigliana. On the contrary, ALDH expression levels in the remaining cultivars were very low (Figure 3C).

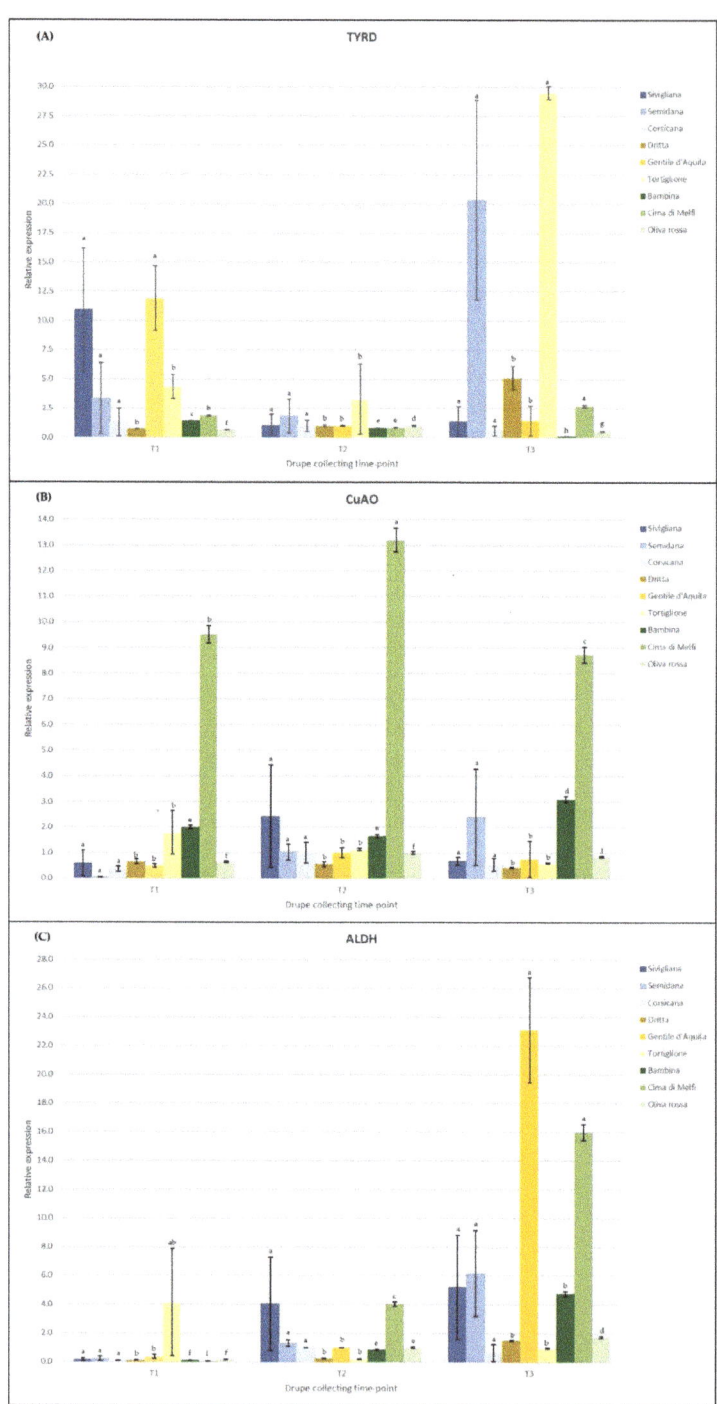

Figure 3. Expression profiles of the analyzed genes involved in the first step of the phenol biosynthetic pathway during fruit development (T1 = yellow-green drupes, T2 = turning drupes, T3 = almost dark

drupes). The mRNA levels were determined by qRT-PCR and relatively expressed as ΔΔCt. (**A**) TYRD = tyrosine/dopa decarboxylase; (**B**) CuAO = copper amine oxidase; (**C**) ALDH = alcohol dehydrogenase. Data are means ($n = 3$) ± standard deviation. Different letters indicate significant differences according to one-way ANOVA followed by Fisher's LSD post-hoc test ($p = 0.05$).

4. Discussion

The identification of minor olive cultivars with favorable agronomic and nutraceutical features among the available and still uncharacterized Italian germplasm, and their fine characterization with different approaches, could effectively contribute to both the valorization and the spread of these cultivars worldwide. The possibility of their introduction in the olive-growing sector is reinforced by the availability of a molecular fingerprint, that can provide a crucial tool for the identification of these genotypes and the traceability of their oils. The use of microsatellite markers in this study has allowed not only a fine characterization of the considered genotypes, but also clarification of the phylogenetic relationships among them and assessment of an identification key. The Unweighted Neighbor Joining dendrogram mostly separated the genotypes according to their geographic origin as reported in many other studies about the evaluation of the genetic diversity of the Italian olive germplasm [5,64]. Surprisingly, only one of the analyzed genotypes has escaped this kind of clustering; indeed, the Gentile dell'Aquila cultivar, belonging to the autochthonous cultivars of Abruzzo region, grouped with all the Sardinian genotypes instead of with the other cultivars from Central Italy (Figure 1). The explanation of this result can be traced back to the complex migration history of the olive. In 2015, Dìez and colleagues [65] demonstrated the existence of three different gene pools (Q1, Q2 and Q3) in the Mediterranean area and the presence of a broad 'mosaic' group as a mixture of Q1 and Q2. The Italian cultivars considered in that research mainly belonged to the Q2 group. Anyway, the authors postulated that the wide occurrence of the 'mosaic' cultivars might be indicative of admixture events. Thereby, both the outcrossing nature of *O. europaea* ssp. *europaea* as well as the exchanges related to human migration over the centuries [64,66] could explain a certain level of genetic admixture between Gentile dell'Aquila and the Sardinian cultivars or their origin from a common wild ancestor.

Over the last decades, virgin olive oil (VOO) became the symbol of good nutritional habits. This role has been mostly linked to some features such as the high content of MUFA and of phenolic compounds, which allow us to recognize VOO as a functional food [67]. As proof of that, the European Food Safety Authority (EFSA) has permitted the declaration "Olive oil polyphenols contribute to the protection of blood lipids from oxidative stress" when olive oil contains a minimum of "5 mg hydroxy-tyrosol and its derivatives per 20 g of olive oil" [68,69]. However, VOO phenolic content not only plays an important nutritional role but also has a great impact on the organoleptic features and on the stability of the product [60]. Given the wide spectrum of features directly linked to olive phenolics, it is clear how important the understanding of their synthesis and regulation is.

Polyphenol accumulation has been reported to be a complex process considerably varying among genotypes, tissues, developmental stages and in response to different agronomic and environmental conditions [70–72]. In accordance with other studies, there was a component of the variability of the phenolic composition observed among the analyzed monovarietal virgin oils that mostly depended on the genotype. On average, the Apulian cultivars generally showed the highest values of the detected compounds at the complete maturation stage of drupes, followed by the Sardinian cultivars. This could also be related to the different cultivation areas and environmental temperatures of the considered groves, since the climates of Apulia and Sardinia are generally similar to each other and usually milder than that of Central Italy (Abruzzo). As reported by [73], the pathways related to olive phenol compounds could be differentially modulated on the basis of altitude and temperatures, phenols biosynthesis being prolonged in temperate areas.

Quantitative differences in phenolic compounds were highlighted both among the analyzed cultivars and among fruits of the same cultivar at different ripening stages. With regard to total phenols (TPC), a sharp decrease is generally and mainly observed between June and September, as result of the oleuropein hydrolysis during drupe maturation (high β-glucosidase activity) [74,75]. Indeed, the very early stages of drupe formation are dominated by secoiridoids (above all oleuropein), while in the last ripening stages, simple phenols and flavonoids become the major components. In our study, the drupe collecting period between October and December did not allow us to highlight such remarkable differences in total phenol content, this period being characterized by a higher stability of phenol metabolism (Table 3; Figure 2D). Thereby, with the only exception of cultivars Tortiglione and Oliva Rossa, that showed a sharp decrease from T2 to T3 stage, the variation of TPC was minimal in all cultivars during the fruit maturation period considered in this study. An intermediate behavior was registered in oils obtained by Sardinian cultivars, that showed a slight, but significant, decrease only at the overripe (T4) stage.

On the contrary, the content of single compounds was very variable among cultivars and sampling times. Secoiridoid derivatives, i.e., aglycon forms of the secoiridoid glucosides usually formed during oil extraction by the enzymatic hydrolysis of oleuropein, demethyloleuropein and ligstroside [76], were reported to be the most abundant phenols in monovarietal olive oils [72]. Among those, oleacein (3,4-DHPEA-EDA) is an important compound, structurally similar to oleocanthal and with comparable pharmacological properties [77]. In our study, the oleacein levels considerably differed among all genotypes, progressively decreasing as fruit maturation proceeded (Table 3; Figure 2C). Indeed, the higher values were found at the T2 time-point, although with remarkable differences among cultivars (in the range between 2 and 63 mg/kg). As a result of an opposite behavior to this general trend, oils obtained by fully ripe drupes (T4) of the Apulian cultivar Cima di Melfi and of the Abruzzo cultivar Gentile dell'Aquila were those with the highest oleacein concentration. As already mentioned, variation in the VOOs' phenolic composition can be ascribable to several factors, including the technological steps of olive oil extraction, which is far from being merely a mechanical process. Indeed, starting from drupe crushing, several enzymes are set free whose activities are strongly influenced by the genotype, the maturity degree and the environmental conditions (time, temperature, atmosphere) [78,79]. Moreover, secoiridoid derivatives, such as 3,4-DHPEA-EDA, are among the most affected olive phenols during the extraction process. For example, in Tuscan olive oils, oleacein was found only a long time after milling, likely due to the activity of different esterases [80]. Thus, linearity between drupe genetic expression and VOO phenols could be difficult to see and the identification of clear markers of maturity is a complex task [81].

The content of minor phenols such as tyrosol and hydroxy-tyrosol was also evaluated in the considered genotypes, due to their relevance as bioactive molecules with beneficial properties on human health [16,82]. Hydroxy-tyrosol is a simple alcohol conjugated to form oleuropein derivatives, usually highly expressed in young olive fruits [42,43]. In accordance with other reports, the hydroxy-tyrosol amount decreased from semi-green to nearly ripe drupes (with the exception of Corsicana, Oliva Rossa and Gentile dell'Aquila), but then an interesting increase was again observed in the final developmental stage, i.e., in totally black fruits (Table 3; Figure 2B). This result is generally attributed to oleuropein catabolism rather than to new hydroxy-tyrosol biosynthesis, as a consequence of the recycling of all the valuable molecules of complex compounds and their conversion to simple phenols [42,44,83]. Moreover, the oleuropein catabolism in the mature fruit is necessary to reduce the bitter taste typically associated with this compound.

Changes in tyrosol content with ripening did not follow a clear pattern among the studied varieties. In some cases (such as Semidana and Cima di Melfi) an opposite trend with the oleacein content was found. These variations could be supposed to be linked to the genetic expression of the enzymes in the biosynthetic pathway (Figure 3). Indeed, tyrosol could be initially synthesized—and mostly found—in free form, while subsequently it is used as a precursor for oleacein production (Table 3; Figure 2A). However, as already reported [81], finding a straightforward relationship between gene expression and phenotype is very difficult, as the interaction with agronomic and pedoclimatic conditions should be also taken into account. Moreover, the content of phenolic compounds depends also on the activity of other enzymes such as the olive β-glucosidases (not considered in this study) which strongly affect their release in the oily phase. Different patterns of tyrosol accumulation with respect to the cultivar and the maturity degree have been already observed by other authors [84].

When this study started, the exact polyphenol biosynthetic pathway was still not clarified in detail; therefore, among genes annotated in phenol biosynthesis, for expression analysis our choice fell on three genes (TYRD, CuAO and ALDH) on the basis of the involvement of the corresponding encoded enzymes in the first steps of this biosynthetic pathway and in relation to the compounds detected in VOOs under characterization. Tyrosine is referred to as the precursor of the secoiridoid class [85]; in particular, the biosynthesis of tyrosol and hydroxy-tyrosol has been proposed to proceed through the formation of tyramine or, otherwise, it can follow the path including L-DOPA and dopamine as intermediates (Figure 4). Initially, an enzyme similar to a tyrosine/DOPA decarboxylase (TYRD) was recognized to be responsible for the conversion of both tyrosine in tyramine and L-DOPA in dopamine [40,86] (Figure 4A). Recently, these two conversions have been clarified to take place by the action of two different enzymes, i.e., a tyrosine decarboxylase (TDC) in the first way and a DOPA decarboxylase (DCC) in the parallel path (Figure 4B) [44]. Moreover, in the previous proposed version, the subsequent steps leading to the conversion of tyramine to tyrosol and dopamine to hydroxy-tyrosol required two enzymes called CuAO and ALDH [87]. Now, the presence of other intermediates in both paths and the action of an additional enzyme, that is, a phenylacetaldehyde reductase (PAR), have been recognized (Figure 4B) [44]. The reaction of the conversion of tyrosol to hydroxy-tyrosol and vice versa seems to involve a polyphenol oxidase (PPO) [42]. Since most of the genes responsible for secoiridoid formation have been identified in a restricted number of species, some steps of the biosynthesis of oleuropein are still unclear, even though the pathway from geranyl diphosphate to secologanin has been elucidated [42,44].

The relative expression of the selected genes was monitored during three stages of fruit maturation, at T1, T2 and T3 sampling times. The expression analysis showed the general tendency of these mRNA levels to decrease or remain more or less stable in semi-green fruits and then to increase, slightly or significantly in a few cases, in semi-dark drupes (Figure 3). These results are in line with the biochemical characterization of monovarietal oils at T2 and T3 stages, which proved that at these time-points the tyrosol and hydroxy-tyrosol contents were progressively diminishing, following the same pattern as the genetic expression data. Some exceptions to this trend have been surprisingly observed for few varieties and in particular for TYRD and ALDH genes. Indeed, these two genes were highly expressed, respectively, in the Tortiglione and Semidana genotypes, and in Cima di Melfi and Gentile dell'Aquila, with a remarkable increase in their transcript levels from T2 to T3 time-points. These unexpected but interesting results, not yet reported in other olive cultivars, to our knowledge, certainly require further investigation. In order to clarify the role of TYRD and ALDH genes in these cultivars, their expression profile should be investigated at more developmental stages of drupes during more consecutive years. Protein expression experiments as well as enzymatic activity assays could also help us to understand the involvement of these genes in polyphenol metabolism of these minor cultivars. At the moment, we could hypothesize that these genotypes are effectively characterized by a high level of gene expression even towards the end of drupe maturation, contrarily to that reported in this study and, more in general, in the literature. It is also plausible that the corresponding enzymes are conveyed to other metabolisms or that some post-translational control takes place and makes these enzymes unfunctional. Exceptions apart, the general agreement of qRT-PCR results with the biochemical outcomes confirmed that hydroxy-tyrosol biosynthesis is under a transcriptional control, as also reported by other authors [40,42–44].

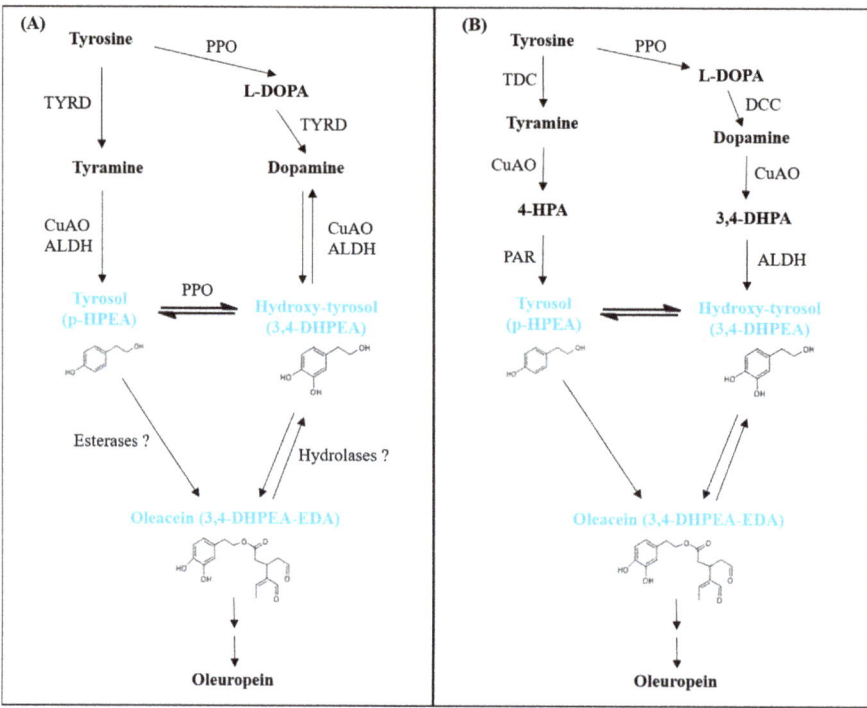

Figure 4. Simplified representation of the biosynthetic pathway of some important olive polyphenols. The previously identified enzymes and reactions (**A**) as well as the recently identified ones, with additional steps in the biosynthesis (**B**), are reported. Abbreviations: 3,4-DHPEA-EDA: elenolic acid linked to 3,4-dihydroxyphenyl ethanol (3,4-DHPEA); 4-HPA: 4-hydroxyphenylacetic acid; 3,4-DHPA: 3,4-dihydroxyphthalic acid; TYRD: tyrosine/DOPA decarboxylase; PPO: polyphenol oxidase; CuAO: copper amine oxidase; ALDH: alcohol dehydrogenase; DCC: DOPA decarboxylase; PAR: phenylacetaldehyde reductase. Adapted from [40,44].

5. Conclusions

Unveiling the genetic potentialities of minor cultivars/genotypes could open the doors to their valorization. Besides the safeguarding of local varieties, the information provided in this work could be useful for breeding programs with selection of genotypes that, while being resilient and better adapted to specific environmental conditions, could also exploit their potentiality in terms of accumulation of bioactive compounds with important consequences on the nutritional, sensory and stability properties of VOO. With a view to an eco-sustainable and low-input agriculture, the introduction of minor cultivars may offer interesting features that, together with the extraction technology, could address the needs of a worldwide market with growing interest towards VOOs. Thus, the results and information obtained in this study are encouraging for the valorization of the still poorly explored Italian germplasm, with important both economic and scientific consequences.

Supplementary Materials: The following are available online at https://www.mdpi.com/article/10.3390/foods10061371/s1, Figure S1: Example of SSR electropherograms, Table S1: Microsatellite markers used in this study, Table S2: List of primers utilized for qRT-PCR analysis.

Author Contributions: Conceptualization, W.S., C.M. and F.C.; investigation, S.G., I.M., G.S., F.F., P.C. and A.P.; writing—original draft preparation, W.S.; writing—review and editing, W.S., G.S., F.F., C.D.D.M., P.C., A.P., C.M. and F.C.; supervision, W.S., C.M. and F.C.; funding acquisition, F.C., W.S. and C.M. All authors have read and agreed to the published version of the manuscript.

Funding: This research was mainly funded by the Italian AGER 2 Project, grant no. 2016-0105, and it was in part supported by the CNR project NUTR-AGE (FOE-2019, DSB.AD004.271) and the Apulia Region Project "Approcci di Next Generation Sequencing per l'analisi di variabilità e di espressione genica in genotipi di olivo autoctoni pugliesi" L.R. 28 December 2018 n. 67, art. 37.

Conflicts of Interest: The authors declare that they have no competing financial or personal interests that could influence the work reported in this paper.

References

1. Muzzalupo, I. *Olive Germplasm–Italian Catalogue of Olive Varieties*; InTech: Rijeka, Croatia, 2012; p. 430. [CrossRef]
2. Bartolini, G.; Cerreti, S.; Briccoli Bati, C.; Stefani, F.; Zelasco, S.; Perri, E.; Petruccelli, R. Oleadb: Worldwide database for the management of genetic resources of olive (*Olea europaea* L.). In Proceedings of the X National Congress on Biodiversity, Rome, Italy, 3–5 September 2014; pp. 18–23.
3. Miazzi, M.M.; di Rienzo, V.; Mascio, I.; Montemurro, C.; Sion, S.; Sabetta, W.; Vivaldi, G.A.; Camposeo, S.; Caponio, F.; Squeo, G.; et al. REGEROP: An integrated project for the recovery of ancient and rare olive germplasm. *Front. Plant Sci.* **2020**, *11*, 73. [CrossRef]
4. Rotondi, A.; Cultrera, N.G.; Mariotti, R.; Baldoni, L. Genotyping and evaluation of local olive varieties of a climatically disfavoured region through molecular, morphological and oil quality parameters. *Sci. Hortic.* **2011**, *130*, 562–569. [CrossRef]
5. Muzzalupo, I.; Vendramin, G.G.; Chiappetta, A. Genetic biodiversity of Italian olives (*Olea europaea* L.) germplasm analyzed by SSR markers. *Sci. World J.* **2014**, *2014*. [CrossRef]
6. Mousavi, S.; Mariotti, R.; Regni, L.; Nasini, L.; Bufacchi, M.; Pandolfi, S.; Baldoni, L.; Proietti, P. The first molecular identification of an olive collection applying standard simple sequence repeats and novel expressed sequence tag markers. *Front. Plant Sci.* **2017**, *8*, 1283. [CrossRef] [PubMed]
7. Sebastiani, L.; Busconi, M. Recent developments in olive *Olea europaea* L.) genetics and genomics: Applications in taxonomy, varietal identification, traceability and breeding. *Plant Cell Rep.* **2017**, *36*, 1345–1360. [CrossRef] [PubMed]
8. El Bakkali, A.; Essalouh, L.; Tollon, C.; Rivallan, R.; Mournet, P.; Moukhli, A.; Hayat Zaher, H.; Mekkaoui, A.; Hadidou, A.; Sikaoui, L.; et al. Characterization of worldwide olive germplasm banks of Marrakech (Morocco) and Córdoba (Spain): Towards management and use of olive germplasm in breeding programs. *PLoS ONE* **2019**, *14*, e0223716. [CrossRef] [PubMed]
9. Montemurro, C.; Miazzi, M.M.; Pasqualone, A.; Fanelli, V.; Sabetta, W.; di Rienzo, V. Traceability of PDO olive oil "terra di Bari" using high resolution melting. *J. Chem.* **2015**, *2015*. [CrossRef]
10. Sabetta, W.; Miazzi, M.M.; di Rienzo, V.; Fanelli, V.; Pasqualone, A.; Montemurro, C. Development and application of protocols to certify the authenticity and the traceability of Apulian typical products in olive sector. *Riv. Ital. Sost. Grasse* **2017**, *94*, 37–43.
11. Xanthopoulou, A.; Ganopoulos, I.; Bosmali, I.; Tsaftaris, A.; Madesis, P. DNA fingerprinting as a novel tool for olive and olive oil authentication, traceability, and detection of functional compounds. In *Olives and Olive Oil as Functional Foods: Bioactivity, Chemistry and Processing*; Shahidi, F., Kiritsakis, A., Eds.; John Wiley & Sons Ltd.: Hoboken, NJ, USA, 2017; pp. 587–601.
12. Gomes, S.; Breia, R.; Carvalho, T.; Carnide, V.; Martins-Lopes, P. Microsatellite High-Resolution Melting (SSR-HRM) to track olive genotypes: From field to olive oil. *J. Food Sci.* **2018**, *8*, 2415–2423. [CrossRef]
13. Piarulli, L.; Savoia, M.A.; Taranto, F.; D'Agostino, N.; Sardaro, R.; Girone, S.; Gadaleta, S.; Fucilli, V.; De Giovanni, C.; Montemurro, C.; et al. A robust DNA isolation protocol from filtered commercial olive oil for PCR-based fingerprinting. *Foods* **2019**, *8*, 462. [CrossRef]
14. International Olive Council (IOC). 2019. Available online: https://www.internationaloliveoil.org (accessed on 5 April 2021).
15. Uylaşer, V.; Yildiz, G. The historical development and nutritional importance of olive and olive oil constituted an important part of the Mediterranean diet. *Crit. Rev. Food Sci. Nutr.* **2014**, *54*, 1092–1101. [CrossRef]
16. Tuck, K.L.; Hayball, P.J. Major phenolic compounds in olive oil: Metabolism and health effects. *J. Nutr. Biochem.* **2002**, *13*, 636–644. [CrossRef]
17. Harwood, J.L.; Yaqoob, P. Nutritional and health aspects of olive oil. *Eur. J. Lipid Sci. Technol.* **2002**, *104*, 685–697. [CrossRef]
18. Stark, A.H.; Madar, Z. Olive oil as a functional food: Epidemiology and nutritional approaches. *Nutr. Rev.* **2002**, *60*, 170–176. [CrossRef] [PubMed]
19. Piroddi, M.; Albini, A.; Fabiani, R.; Giovannelli, L.; Luceri, C.; Natella, F.; Rosignoli, P.; Rossi, T.; Taticchi, A.; Servili, M.; et al. Nutrigenomics of extra-virgin olive oil: A review. *BioFactors* **2017**, *43*, 17–41. [CrossRef]
20. Jimenez-Lopez, C.; Carpena, M.; Lourenço-Lopes, C.; Gallardo-Gomez, M.; Lorenzo, J.M.; Barba, F.J.; Prieto, M.A.; Simal-Gandara, J. Bioactive Compounds and Quality of Extra Virgin Olive Oil. *Foods* **2020**, *9*, 1014. [CrossRef] [PubMed]
21. Beauchamp, G.K.; Keast, R.S.; Morel, D.; Lin, J.; Pika, J.; Han, Q.; Lee, C.; Smith, A.B.; Breslin, P.A.S. Ibuprofen-like activity in extra-virgin olive oil. *Nature* **2005**, *437*, 45–46. [CrossRef] [PubMed]
22. Servili, M.; Esposto, S.; Fabiani, R.; Urbani, S.; Taticchi, A.; Mariucci, F.; Selvaggini, R.; Montedoro, G.F. Phenolic compounds in olive oil: Antioxidant, health and organoleptic activities according to their chemical structure. *Inflammopharmacology* **2009**, *17*, 1–9. [CrossRef]
23. Lucas, L.; Russell, A.; Russell, K. Molecular mechanisms of inflammation. Anti-inflammatory benefits of virgin olive oil and the phenolic compound oleocanthal. *Curr. Pharm. Des.* **2011**, *17*, 754–768. [CrossRef]

24. Yubero-Serrano, E.M.; Lopez-Moreno, J.; Gomez-Delgado, F.; Lopez-Miranda, J. Extra virgin olive oil: More than a healthy fat. *Eur. J. Clin. Nutr.* **2019**, *72*, 8–17. [CrossRef]
25. De la Torre-Carbot, K.; Chávez-Servín, J.L.; Jaúregui, O.; Castellote, A.I.; Lamuela-Raventós, R.M.; Nurmi, T.; Poulsen, H.E.; Gaddi, A.V.; Kaikkonen, J.; Zunft, H.F.; et al. Elevated circulating LDL phenol levels in men who consumed virgin rather than refined olive oil are associated with less oxidation of plasma LDL. *J. Nutr.* **2010**, *140*, 501–508. [CrossRef]
26. Calabriso, N.; Massaro, M.; Scoditti, E.; D'Amore, S.; Gnoni, A.; Pellegrino, M.; Storelli, C.; De Caterina, R.; Palasciano, G.; Carluccio, M.A. Extra virgin olive oil rich in polyphenols modulates VEGF-induced angiogenic responses by preventing NADPH oxidase activity and expression. *J. Nutr. Biochem.* **2016**, *28*, 19–29. [CrossRef]
27. Storniolo, C.E.; Rosello-Catafau, J.; Pintó, X.; Mitjavila, M.T.; Moreno, J.J. Polyphenol fraction of extra virgin olive oil protects against endothelial dysfunction induced by high glucose and free fatty acids through modulation of nitric oxide and endothelin-1. *Redox Biol.* **2014**, *2*, 971–977. [CrossRef]
28. Lockyer, S.; Rowland, I.; Spencer, J.P.E.; Yaqoob, P.; Stonehouse, W. Impact of phenolic-rich olive leaf extract on blood pressure, plasma lipids and inflammatory markers: A randomised controlled trial. *Eur. J. Nutr.* **2017**, *56*, 1421–1432. [CrossRef]
29. Moreno, J.J. Effect of olive oil minor components on oxidative stress and arachidonic acid mobilization and metabolism by macrophages RAW 264.7. *Free Radic. Biol. Med.* **2003**, *35*, 1073–1081. [CrossRef]
30. Martin, M.A.; Ramos, S.; Granado-Serrano, A.B.; Rodríguez-Ramiro, I.; Trujillo, M.; Bravo, L.; Goya, L. Hydroxy-tyrosol induces antioxidant/ detoxificant enzymes and Nrf2 translocation via extracellular regulated kinases and phosphatidylinositol-3-kinase/protein kinase B pathways in HepG2 cells. *Mol. Nutr. Food Res.* **2010**, *54*, 956–966. [CrossRef] [PubMed]
31. Giordano, E.; Davalos, A.; Visioli, F. Chronic hydroxy-tyrosol feeding modulates glutathione-mediated oxido-reduction pathways in adipose tissue: A nutrigenomic study. *Nutr. Metab. Cardiovasc. Dis.* **2014**, *24*, 1144–1150. [CrossRef] [PubMed]
32. Orak, H.H.; Karamac, M.; Amarowicz, R.; Orak, A.; Penkacik, K. Genotype-related differences in the phenolic compound profile and antioxidant activity of extracts from olive (*Olea europaea* L.) leaves. *Molecules* **2019**, *24*, 1130. [CrossRef] [PubMed]
33. Caponio, F.; Gomes, T.; Pasqualone, A. Phenolic compounds in virgin olive oils: Influence of the degree of olive ripeness on organ-oleptic characteristics and shelf-life. *Eur. Food Res. Technol.* **2001**, *212*, 329–333. [CrossRef]
34. Siliani, S.; Mattei, A.; Innocenti, L.B.; Zanoni, B. Bitter taste and phenolic compounds in extra virgin olive oil: An empirical relationship. *J. Food Qual.* **2006**, *29*, 431–441. [CrossRef]
35. Beltrán, G.; Ruano, M.T.; Jiménez, A.; Uceda, M.; Aguilera, M.P. Evaluation of virgin olive oil bitterness by total phenol content analysis. *Eur. J. Lipid Sci. Technol.* **2007**, *109*, 193–197. [CrossRef]
36. Pedan, V.; Popp, M.; Rohn, S.; Nyfeler, M.; Bongartz, A. Characterization of phenolic compounds and their contribution to sensory properties of olive oil. *Molecules* **2019**, *24*, 2041. [CrossRef]
37. Serrilli, A.M.; Padula, G.; Bianco, A.; Petrosino, L.; Ripa, V. Searching new olive (*Olea europaea* L.) cultivars: Analysis of phenolic fraction in olive selections derived from a breeding program. *Adv. Hort. Sci.* **2008**, *22*, 104–109.
38. Riachy, E.M.; Priego-Capote, F.; Rallo, L.; Luque-de Castro, M.D.; León, L. Phenolic composition of virgin olive oils from cross breeding segregating populations. *Eur. J. Lipid Sci. Tech.* **2012**, *114*, 542–551. [CrossRef]
39. Pérez Rubio, A.G.; León, L.; Sanz, C.; Rosa, R.D.L. Fruit Phenolic Profiling: A New Selection Criterion in Olive Breeding Programs. *Front. Plant Sci.* **2018**, *9*, 241. [CrossRef] [PubMed]
40. Alagna, F.; Mariotti, R.; Panara, F.; Caporali, S.; Urbani, S.; Veneziani, G.; Esposto, S.; Taticchi, A.; Rosati, A.; Rao, R.; et al. Olive phenolic compounds: Metabolic and transcriptional profiling during fruit development. *BMC Plant Biol.* **2012**, *12*, 162. [CrossRef]
41. Alagna, F.; Geu-Flores, F.; Kries, H.; Panara, F.; Baldoni, L.; O'Connor, S.E.; Osbourn, A. Identification and characterization of the iridoid synthase involved in oleuropein biosynthesis in olive (*Olea europaea* L.) fruits. *J. Biol. Chem.* **2015**, *291*, 5542–5554. [CrossRef] [PubMed]
42. Mougiou, N.; Trikka, F.; Trantas, E.; Ververidis, F.; Makris, A.; Argiriou, A.; Vlachonasios, K.E. Expression of hydroxy-tyrosol and oleuropein biosynthetic genes are correlated with metabolite accumulation during fruit development in olive, Olea europaea, cv Koroneiki. *Plant Physiol. Biochem.* **2018**, *128*, 41–49. [CrossRef] [PubMed]
43. Guodong, R.; Jianguo, Z.; Xiaoxia, L.; Ying, L. Identification of putative genes for polyphenol biosynthesis in olive fruits and leaves using full-length transcriptome sequencing. *Food Chem.* **2019**, *300*, 125246. [CrossRef]
44. Guodong, R.; Zhang, J.; Liu, X.; Li, X.; Wang, C. Combined Metabolome and Transcriptome Profiling Reveal Optimal Harvest Strategy Model Based on Different Production Purposes in Olive. *Foods* **2021**, *10*, 360. [CrossRef]
45. Bandino, G.; Mulas, M.; Sedda, P.; Moro, C. Le varietà di Olivo Della Sardegna; Consorzio Interprovinciale per la Frutticoltura di Cagliari, Oristano e Nuoro: Cagliari, Italy, 2001; p. 256.
46. Deiana, P.; Santona, M.; Dettori, S.; Molinu, M.G.; Dore, A.; Culeddu, N.; Azara, E.; Naziri, E.; Tsimidou, M.Z. Can all the Sardinian varieties support the PDO "Sardegna" virgin olive oil? *Eur. J. Lipid Sci. Technol.* **2019**, *121*, 1800135. [CrossRef]
47. Lombardo, N.; Alessandrino, M.; Godino, G.; Madeo, A.; Ciliberti, A.; Pellegrino, M. Caratterizzazione della crescita, produttività, inoliazione e qualità dell'olio di 10 cultivar di olivo autoctone abruzzesi. In Proceedings of the National Congress "Maturazione e Raccolta Delle Olive: Strategie e Tecnologie Per Aumentare la Competitività in Olivicoltura", ARSSA, Regione Abruzzo Alanno, Italy, Alanno (PE), Italy, 1 April 2006; p. 238.
48. Albertini, E.; Torricelli, R.; Bitocchi, E.; Raggi, L.; Marconi, G.; Pollastri, L.; Di Minco, G.; Battistini, A.; Papa, R.; Veronesi, F. Structure of genetic diversity in *Olea europaea* L. cultivars from central Italy. *Mol. Breed.* **2011**, *27*, 533–547. [CrossRef]

49. Squeo, G.; Difonzo, G.; Silletti, R.; Paradiso, V.M.; Summo, C.; Pasqualone, A.; Caponio, F.C. Bambina, una varietà minore pugliese: Profilo di maturazione, composizione delle drupe e caratterizzazione chimica dell'olio vergine. *Riv. Ital. Sost. Grasse* **2019**, *94*, 143–149.
50. Squeo, G.; Silletti, R.; Mangini, G.; Summo, C.; Caponio, F. The potential of Apulian olive biodiversity: The case of Oliva Rossa Virgin Olive Oil. *Foods* **2021**, *10*, 369. [CrossRef] [PubMed]
51. Spadoni, A.; Sion, S.; Gadaleta, S.; Savoia, M.A.; Piarulli, L.; Fanelli, V.; Di Rienzo, V.; Taranto, F.; Miazzi, M.M.; Montemurro, C.; et al. A simple and rapid method for genomic DNA extraction and microsatellite analysis in tree plants. *J. Agric. Sci. Technol.* **2019**, *21*, 1215–1226.
52. Sefc, K.M.; Lopes, M.S.; Mendonça, D.; Rodrigues Dos Santos, M.; Laimer Da Câmara Machado, M.; Da Câmara Machado, A. Identification of microsatellite loci in olive (*Olea europaea* L.) and their characterization in Italian and Iberian olive trees. *Mol. Ecol.* **2000**, *9*, 1171–1193. [CrossRef]
53. Carriero, F.; Fontanazza, G.; Cellini, F.; Giorio, G. Identification of simple sequence repeats (SSRs) in olive (*Olea europaea* L.). *Theor. Appl. Genet.* **2002**, *104*, 301–307. [CrossRef] [PubMed]
54. De La Rosa, R.; James, C.M.; Tobutt, K.R. Using microsatellites for paternity testing in olive progenies. *Hort. Sci.* **2004**, *39*, 351–354. [CrossRef]
55. Pasqualone, A.; Montemurro, C.; Summo, C.; Sabetta, W.; Caponio, F.; Blanco, A. Effectiveness of microsatellites DNA markers in checking the identity of PDO extra virgin olive oil. *J. Agric. Food Chem.* **2007**, *55*, 3857–3862. [CrossRef] [PubMed]
56. Baldoni, L.; Cultrera, N.G.; Mariotti, R.; Ricciolini, C.; Arcioni, S.; Vendramin, G.G.; Buonamici, A.; Porceddu, A.; Sarri, V.; Ojeda, M.A.; et al. A consensus list of microsatellite markers for olive genotyping. *Mol. Breed.* **2009**, *24*, 213–231. [CrossRef]
57. Boucheffa, S.; Miazzi, M.M.; di Rienzo, V.; Mangini, G.; Fanelli, V.; Tamendjari, A.; Pignone, D.; Montemurro, C. The coexistence of oleaster and traditional varieties affects genetic diversity and population structure in Algerian olive (*Olea europaea* L.) germplasm. *Gen. Res. Crop Evol.* **2017**, *64*, 379–390. [CrossRef]
58. Di Rienzo, V.; Sion, S.; Taranto, F.; D'Agostino, N.; Montemurro, C.; Fanelli, V.; Sabetta, W.; Boucheffa, S.; Tamendjari, A.; Pasqualone, A.; et al. Genetic flow among olive populations within the Mediterranean basin. *PeerJ* **2018**, *6*, e5260. [CrossRef] [PubMed]
59. Caponio, F.; Catalano, P. Hammer crushers vs disk crushers: The influence of working temperature on the quality and preservation of virgin olive oil. *Eur. Food Res. Technol.* **2001**, *213*, 219–224. [CrossRef]
60. Bendini, A.; Cerretani, L.; Carrasco-Pancorbo, A.; Gómez-Caravaca, A.M.; Segura-Carretero, A.; Fernández-Gutiérrez, A.; Lercker, G. Phenolic molecules in virgin olive oils: A survey of their sensory properties, health effects, antioxidant activity and analytical methods. An overview of the last decade Alessandra. *Molecules* **2007**, *12*, 1679–1719. [CrossRef] [PubMed]
61. Zago, L.; Squeo, G.; Bertoncini, E.I.; Difonzo, G.; Caponio, F. Chemical and sensory characterization of Brazilian virgin olive oils. *Food Res. Int.* **2019**, *126*, 108588. [CrossRef] [PubMed]
62. International Olive Council (IOC). *Determination of Biophenols in Olive Oils by HPLC*; COI/ T.20/Doc. No 29/Rev.1; International Olive Council: Madrid, Spain, 2017.
63. Livak, K.J.; Schmittgen, T.D. Analysis of Relative Gene Expression Data Using Real-Time Quantitative PCR and the $2^{-\Delta\Delta CT}$ Method. *Methods* **2001**, *25*, 402–408. [CrossRef] [PubMed]
64. Baldoni, L.; Tosti, N.; Ricciolini, C.; Belaj, A.; Arcioni, S.; Pannelli, G.; Germanà, M.A.; Mulas, M.; Porceddu, A. Genetic structure of wild and cultivated olives in the central Mediterranean basin. *Ann. Bot.* **2006**, *98*, 935–942. [CrossRef]
65. Díez, C.M.; Trujillo, I.; Martinez-Uriroz, N.; Barranco, D.; Rallo, L.; Marfil, P.; Gaut, B.S. Olive domestication and diversification in the Mediterranean basin. *New Phytol.* **2015**, *206*, 436–447. [CrossRef]
66. Erre, P.; Chessa, I.; Munoz-Diez, C.; Belaj, A.; Rallo, L.; Trujillo, I. Genetic diversity and relationships between wild and cultivated olives (*Olea europaea* L.) in Sardinia as assessed by SSR markers. *Genet. Resour. Crop Evol.* **2010**, *57*, 41–54. [CrossRef]
67. Covas, M.I.; Ruiz-Gutiérrez, V.; De La Torre, R.; Kafatos, A.; Lamuela-Raventós, R.M.; Osada, J.; Visioli, F. Minor components of olive oil: Evidence to date of health benefits in humans. *Nutr. Rev.* **2006**, *64* (Suppl. 4), S20–S30. [CrossRef]
68. The European Parliament and The Council of the European Union. Commission Regulation (EU) No 1924/2006 of 20 December 2006 on Nutrition and Health Claims Made on Foods. *Off. J. Eur. Union* **2006**, *1924*. Available online: https://eur-lex.europa.eu/eli/reg/2006/1924/oj#document1 (accessed on 5 April 2021).
69. Commission Regulation (EU) No 432/2012 of 16 May 2012 Establishing a List of Permitted Health Claims Made on Foods, Other than Those Referring to the Reduction of Disease Risk and to Children's Development and Health. Available online: https://eur-lex.europa.eu/eli/reg/2012/432/oj (accessed on 5 April 2021).
70. Malik, N.S.A.; Bradford, J.M. Recovery and stability of oleuropein and other phenolic compounds during extraction and processing of olive (*Olea europaea* L.) leaves. *J. Food Agric. Environ.* **2008**, *6*, 8–13.
71. Baiano, A.; Terracone, C.; Viggiani, I.; Del Nobile, M.A. Effects of cultivars and location on quality, phenolic content and antioxidant activity of extra-virgin olive oils. *J. Am. Oil Chem. Soc.* **2013**, *90*, 103–111. [CrossRef]
72. Miho, H.; Díez, C.M.; Mena-Bravo, A.; Sánchez de Medina, V.; Moral, J.; Melliou, E.; Magiatis, P.; Rallo, L.; Barranco, D.; Priego-Capote, F. Cultivar influence on variability in olive oil phenolic profiles determined through an extensive germplasm survey. *Food Chem.* **2018**, *266*, 192–199. [CrossRef]

73. Bruno, L.; Picardi, E.; Pacenza, M.; Chiappetta, A.; Muto, A.; Gagliardi, O.; Muzzalupo, I.; Pesole, G.; Bitonti, M.B. Changes in gene expression and metabolic profile of drupes of *Olea europaea* L. cv Carolea in relation to maturation stage and cultivation area. *BMC Plant Biol.* **2019**, *19*, 428. [CrossRef]
74. Talhaoui, N.; Gómez-Caravaca, A.M.; León, L.; De la Rosa, R.; Fernández-Gutiérrez, A.; Segura-Carretero, A. Pattern of Variation of Fruit Traits and Phenol Content in Olive Fruits from Six Different Cultivars. *J. Agric. Food Chem.* **2015**, *63*, 10466–10476. [CrossRef] [PubMed]
75. Velázquez-Palmero, D.; Romero-Segura, C.; García-Rodríguez, R.; Hernández, M.L.; Vaistij, F.E.; Graham, I.A.; Pérez, A.G.; Martínez-Rivas, J.M. An oleuropein b-glucosidase from olive fruit is involved in determining the phenolic composition of Virgin Olive Oil. *Front. Plant Sci.* **2017**, *8*, 1902. [CrossRef]
76. Servili, M.; Selvaggini, R.; Esposto, S.; Taticchi, A.; Montedoro, G.; Morozzi, G. Health and sensory properties of virgin olive oil hydrophilic phenols: Agronomic and technological aspects of production that affect their occurrence in the oil. *J. Chromatogr. A* **2004**, *1054*, 113–127. [CrossRef]
77. Paiva-Martins, F.; Fernandes, J.; Rocha, S.; Nascimento, H.; Vitorino, R.; Amado, F.; Santos-Silva, A. Effects of olive oil polyphenols on erythrocyte oxidative damage. *Mol. Nutr. Food Res.* **2009**, *53*, 609–616. [CrossRef]
78. Artajo, L.S.; Romero, M.P.; Motilva, M.J. Transfer of phenolic compounds during olive oil extraction in relation to ripening stage of the fruit. *J. Sci. Food Agric.* **2006**, *86*, 518–527. [CrossRef]
79. Squeo, G.; Difonzo, G.; Summo, C.; Crecchio, C.; Caponio, F. Study of the influence of technological coadjuvants on enzyme activities and phenolic and volatile compounds in virgin olive oil by a response surface methodology approach. *LWT* **2020**, *133*, 109887. [CrossRef]
80. Cecchi, L.; Migliorini, M.; Cherubini, C.; Giusti, M.; Zanoni, B.; Innocenti, M.; Mulinacci, N. Phenolic profiles, oil amount and sugar content during olive ripening of three typical Tuscan cultivars to detect the best harvesting time for oil production. *Food Res. Int.* **2013**, *54*, 1876–1884. [CrossRef]
81. Kalua, C.M.; Allen, M.S.; Bedgood, D.R.; Bishop, A.G.; Prenzler, P.D. Discrimination of olive oils and fruits into cultivars and maturity stages based on phenolic and volatile compounds. *J. Agric. Food Chem.* **2005**, *53*, 8054–8062. [CrossRef] [PubMed]
82. Cirilli, M.; Caruso, G.; Gennai, C.; Urbani, S.; Frioni, E.; Ruzzi, M.; Servili, M.; Gucci, R.; Poerio, E.; Muleo, R.M. The Role of Polyphenoloxidase, Peroxidase, and β-Glucosidase in Phenolics Accumulation in *Olea europaea* L. Fruits under Different Water Regimes. *Front. Plant Sci.* **2017**, *8*, 717. [CrossRef]
83. Ortega-Garcia, F.; Blanco, S.; Peinado, M.A.; Peragon, J. Polyphenol oxidase and its relationship with oleuropein concentration in fruits and leaves of olive (Olea europaea) cv. "Picual" trees during fruit ripening. *Tree Physiol.* **2008**, *28*, 45–54. [CrossRef] [PubMed]
84. Morelló, J.R.; Romero, M.P.; Ramo, T.; Motilva, M.J. Evaluation of L-phenylalanine ammonia-lyase activity and phenolic profile in olive drupe (*Olea europaea* L.) from fruit setting period to harvesting time. *Plant Sci.* **2005**, *168*, 65–72. [CrossRef]
85. Ryan, D.; Antolovich, M.; Herlt, T.; Prenzler, P.D.; Lavee, S.; Robards, K. Identification of phenolic compounds in tissues of the novel olive cultivar hardy's mammoth. *J. Agric. Food Chem.* **2002**, *50*, 6716–6724. [CrossRef]
86. Facchini, P.J.; De Luca, V. Differential and tissue-specific expression of a gene family for tyrosine/dopa decarboxylase in opium poppy. *J. Biol. Chem.* **1994**, *269*, 26684–26690. [CrossRef]
87. Saimaru, H.; Orihara, Y. Biosynthesis of acteoside in cultured cells of Olea europaea. *J. Nat. Med.* **2010**, *64*, 139–145. [CrossRef]

Article

Effect of Duration of Olive Storage on Chemical and Sensory Quality of Extra Virgin Olive Oils

Annalisa Rotondi, Lucia Morrone *, Gianpaolo Bertazza and Luisa Neri

Institute for the Bioeconomy, Italian National Research Council, via P. Gobetti 101, 40129 Bologna, Italy; annalisa.rotondi@ibe.cnr.it (A.R.); gianpaolo.bertazza@ibe.cnr.it (G.B.); luisa.neri@ibe.cnr.it (L.N.)
* Correspondence: lucia.morrone@ibe.cnr.it

Abstract: This work considered the influence of the duration of olive storage on the chemical and sensory properties of extra virgin olive oil. In total, 228 batches of olives collected during three successive crop seasons were sampled in seven industrial mills; information about olive batches (variety, harvest date) was collected, together with the produced oils. Four classes of storage times were considered: ≤24 h, 2–3 days, 4–6 days, ≥7 days. The oils' quality parameters free acidity, peroxide number and K232 increased significantly as storage duration increased, while phenolic content decreased significantly, with a resulting effect on oil stability. The fatty acid composition was not affected by the olive storage period, while α-tocopherol, lutein and β-carotene content decreased as storage duration lengthened. Finally, the main positive sensory attributes (olive fruity, green notes, bitter and pungency) underwent a statistically significant reduction with the increase in storage duration, while the intensity of defects increased, suggesting that the duration of olive storage has an important effect on the quality of the final oil.

Keywords: olive oil; olive storage duration; oil chemical composition; sensory properties

1. Introduction

Olive oil plays an important role in the diet in Mediterranean countries [1]. Extra virgin olive oil (EVOO) is the only vegetable oil that must be extracted only by mechanical means without any adjuvants [2]. EVOO is therefore, in effect, a fruit juice, hence the phytosanitary state of drupes is the main factor determining the quality of the extracted olive oil [3]. To best preserve the raw material before processing, post-harvest management is strategic to obtain extra virgin olive oils, since during this period oxidation of fat matrix and fermentation can occur [4].

However, in olive-producing regions such as Italy, Spain and Greece, because of the difficulty in synchronizing fruit harvesting and extraction of its oil, the olive sector is often forced to store the fruits piled up, in poor conditions and for periods of up to several weeks. During this period, the fruits suffer mechanical, physicochemical and physiological alterations that may eventually cause the breakdown of their cell structures [5,6]. During prolonged olive storage, anaerobiosis processes can occur in the lower portion of the olives kept inside the containers, and heat production from the respiratory activity may also accelerate fruit deterioration and eventually cause the breakdown of the cell structure [6]. Olive oils obtained from damaged olives present a characteristic high acidity, low oxidative stability and high level of oxidation, due to the increased peroxide value, and specific extinction coefficients at 232 and 270; they can also develop a high content of volatile acids (acetic or butyric) that cause a typical musty smell [7]. These processes will deteriorate the chemical and sensory quality of the resulting EVOO, so in order to better manage the postharvest period, several technological solutions have been proposed such as cold storage of olives [7], storage in a modified atmosphere [8], and other preservation conditions such as storage in sea water, brine or drinking water have also been investigated [9].

The importance of processing olives a short time after harvesting is also linked to the fact that most fruit is harvested mechanically and could, therefore, be internally damaged, more so than in the case of manual harvesting; however, allowing for proper storage conditions the fruits can be stored for several days maintaining the appropriate chemical and sensory quality standards of the final oil. Yousfi et al. [10], in fact, studied the quality of EVOO from mechanically harvested Arbequina olives under different storage conditions, and found that storage at 3 °C for a period of up to 10 days allowed the highest commercial level of oil quality to be maintained.

The problem of synchronization of harvest and transformation phases has not been widely considered in Italy, where this study was carried out in the past due to the production fragmentation, the structure of olive mills (small and widespread) and the presence of different olive cultivars, a factor broadening the collection window. However, the presence of numerous different cultivars on the Italian territory is a characteristic feature of Italian olive growing that increases its sustainability as the loss of biodiversity is an environmental threat. The production of monovarietal olive oils has increased to a great extent lately since the quality of olive oil depends on the olive variety from which it originates [11]. Nowadays, however, the structure of production is changing in Italy, due to the presence of an increasing number of intensive orchards that can exceed the processing capacity of the mills, and therefore synchronization between harvest and transformation should be considered.

The aim of this study was to assess the effect of the duration of olive storage on the chemical and sensory quality of the EVOO, identifying which parameters were most affected by olive storage; in particular, we focused on product parameters that are more easily illustrated to actors in the supply chain (mills, producers and consumers), thus making it easier to understand and assimilate the results.

2. Materials and Methods

2.1. Olive Fruit Analysis and Oil Sampling

Olive fruits and the corresponding oil samples (n = 228) were collected during 3 crop seasons, from seven industrial oil mills located in the Emilia-Romagna region in northern Italy, all equipped with hammer crusher, two-phase decanter, and centrifugation and filtration facilities. Data characterising olive oil samples, such as olive cultivar, harvesting method, and olive storage duration, were collected by interviewing olive growers. Only samples of healthy olives without signs of infection were considered after visual inspection.

Oil samples were poured into dark glass bottles, keeping headspace to a minimum, and stored in the dark in a temperature-controlled cupboard set at 15 ± 1 °C, until chemical and sensory analyses were carried out.

2.2. Chemical Analysis of Olive Oils

Free acidity, peroxide value, and UV-spectrophotometric indices (K_{232}, K_{270}) were evaluated in triplicate in line with official methods described in Regulation EC 2568/91 and subsequent amendments [12].

Analysis of fatty acids was carried out according to Regulation EC 2568/91 and subsequent amendments [12] using a Chrompack CP 9000 gas chromatograph with a flame ionization detector (FID), equipped with a capillary column (Stabilwax, Restek Corporation, Bellefonte, PA, USA) and helium as the carrier gas (flow rate = 1 mL min^{-1}; split ratio of 1:20, v:v). Chromatographic parameters were as follows: injection and detection temperature 250 °C; 230 °C; column oven temperature, 240 °C. All parameters were determined in triplicate for each sample.

The phenolic fraction was extracted in triplicate from 30 g of oil using 30 mL of methanol. The combined extract was brought to dryness through a rotary evaporator and then suspended in 2 mL 50% methanolic solution. Total phenol content was determined by the Folin–Ciocalteau spectrophotometric method at 750 nm [13] using a spectrophotometer (Jasco V-500, Jasco Corporation, Tokyo, Japan).

Quantitative analysis of tocopherols, lutein and β-carotene was carried through olive oil filtration on PTFE (Polytetrafluoroethylene) membrane filter of 25 mm, 0.2 μm pore size (GyroDisc, Orange Scientific, Waterloo, Belgium) and direct injection of 20 μL in HPLC (high-performance liquid chromatography) [14] equipment (LC-10ADvp, Shimadzu, Kyoto, Japan) with a degasser (Flow 154, Gastorr Flom, Tokyo, Japan), a low-pressure gradient unit (FCV-10ALvp, Shimadzu, Kyoto, Japan) and a column oven (CTO-10ASvp, Shimadzu, Kyoto, Japan). Analytes were separated on a C18 column 150 mm × 4.6 mm (Inertsil ODS-2 5U, Alltech, Deerfield, IL, USA); the flow rate was 1 mL min^{-1}, the injection volume was 20 μL and the column temperature was 25 °C. The eluent used was: A methanol: water 80:20 (v/v) and B methanol: tetrahydrofuran 20:80 (v/v). Quantification of analytes was carried out using their relative analytical standard's calibration curves all purchased from Merk (Deisenhofen, Germany). Tocopherol quantification was carried out at 295 nm, β-carotene and lutein at 450 nm using a photodiode array detector (UV6000, ThermoQuest, San Jose, CA, USA).

2.3. Oil Stability Determination

For determination of oil stability, an eight-channel Oxidative Stability Instrument (OSI) (Omnion, Decatur, IL, USA) was used; the instrument was set at 110 °C and at 120 mL min^{-1} (airflow) [15]. The OSI index was expressed as time (hours and hundredth of hours) and was reported as "OSI time".

2.4. Sensory Analysis

Sensory analysis was performed by the panel of Agency for Agrofood Sector Services of Marche region (ASSAM), a fully-trained analytical taste panel recognized by the International Olive Oil Council (IOC) of Madrid, Spain, and by the Italian Ministry for Agriculture, Food, and Forestry Policy. The panel was composed of 8 assessors, 50% male and 50% female. The method applied was QDA (Quantitative Descriptive Analysis). A profile sheet IOC method T20 n. 15 modified by IBIMET-CNR and ASSAM was used, allowing to obtain a QDA of the oils' sensory profile and more complete description of the organoleptic properties of the sampled oils: the sensory assessors evaluated direct or retronasal aromatic olfactory sensations (aroma of olive fruity and green notes), gustatory sensations (olive fruity and bitterness) and tactile/kinesthetic sensation (pungency), organoleptic defects (Supplementary Table S1) as well as overall judgment. The sensory assessors had to rate the intensity of the different descriptors on a continuous 0–10 cm scale. Values of median of sensory data were calculated.

2.5. Statistical Analysis

The significance of differences at a 5% level was determined by one-way ANOVA using Tukey's test with Microsoft® Excel 2007/XLSTAT© (Version 2009.3.02, Addinsoft, Inc., Brooklyn, NY, USA). Sensory data were also processed for Principal Component Analysis (PCA) to explore data distribution patterns and to visualize the "distance" between oils produced following the differing storage times.

3. Results and Discussion

After interviewing olive growers, it was recorded that the olive and the correspondent Extra virgin olive oil (EVOO) samples collected (n = 228) were mainly composed of mixed varieties (blends) (45%), while the remaining samples were monovarietal from cv. Nostrana di Brisighella (25%), cv. Correggiolo (16%), cv. Leccino (8%) and other minor cultivars (6%). Furthermore, it was assessed that the olives' storage method was the same for all the analyzed samples: fruits were stored in small plastic bins with holes to allow for ventilation, and never in stacks nor in plastic or jute bags.

Olive storage duration before technological transformation ranged from 0 to over 7 days: chemical and sensory data were thus processed dividing them into four classes of storage times: ≤24 h, 2–3 days, 4–6 days, ≥7 days. Only 39% of olive samples were processed within 24 h, while 23% and 20% of olive samples were stored between 2 and 3 days and between 4 and 6 days, respectively; finally, 18% of samples were processed after 7 days of storage.

The content of free acids is an important quality factor, extensively used as the major criterion for the classification of olive oil at various commercial grades [16]. The values of free acidity, peroxide number, and K232 increased significantly along with the increase in olive storage duration (Table 1). There was a free acidity increase from 0.30% to 0.56% during the olive storage period studied; peroxide number from 6.96 to 9.56 mEqO_2kg^{-1}, K232 from 1.48 to 1.66 while K270 was not affected by time of storage probably because indicates secondary oxidation. It is interesting to note that all these values fall within the legal limit of the classification of extra virgin olive oil. This indicates that, although the oxidation process starts to take place during the olives' storage time, the phytosanitary state and the integrity of the raw material affect the speed of this process. The total phenol content of oil samples suffered a progressive reduction as olive storage duration proceeded: as observed in Table 1, oils produced within 24 h from the olive collection had a phenolic content of 243 mg kg^{-1} of gallic acid while oils produced from olives stored for over 7 days presented 143.6 mg kg^{-1} of gallic acid, a decrease of 41%. This phenol loss could be attributed both to bacteria and fungi proliferation on cellular fluid exuding from damaged fruits [17] and to endogenous oxidoreductases [18] as suggested by Clodoveo et al. [5], results of which were consistent with the data presented here. This impoverishment in the phenol fraction also affected the oils' stability, with a reduction from 28 h in oils produced within a day to 19 h in oils obtained from olives stored for more than a week, with a decrease of 30% (Table 1). These results agree with the studies of Vichi et al. [19] and Youssef et al. [20]. Our results partially agree with Pereira et al. [21], which found a significant decrease in oil stability during the first period of storage (0–7 days), while for peroxide number, free acidity and K232 and K270 the values were not significantly affected by storage duration. As explained by Pereira himself [21], the verified decrease in stability was due to the consumption of minor compounds such as phenols and tocopherols, that hindered the formation of peroxides.

Table 1. Quality index of oil samples extracted after different olive storage duration. Values are mean ± standard deviation. Values followed by different letters in the same column (a, b, c) were significantly different according to Tukey's test ($p < 0.05$).

Storage Classes	Free Acidity	Peroxide Number	K232	K270	Total Phenol	OSI
<24 h	0.30 ± 0.09 b	6.96 ± 2.17 b	1.48 ± 0.18 b	0.08 ± 0.02	246.00 ± 104.02 a	28.28 ± 10.83 a
2–3 days	0.37 ± 0.17 b	9.03 ± 2.76 a	1.64 ± 0.23 a	0.09 ± 0.02	202.51 ± 90.14 b	22.60 ± 7.53 b
4–6 days	0.51 ± 0.36 a	8.67 ± 2.94 a	1.55 ± 0.25 ab	0.08 ± 0.02	179.22 ± 83.49 bc	16.90 ± 5.74 b
>7 days	0.56 ± 0.43 a	9.56 ± 2.64 a	1.66 ± 0.29 a	0.09 ± 0.02	143.60 ± 63.39 c	19.91 ± 10.08 b
p-value	<0.0001	<0.0001	<0.0001	0.394	<0.0001	<0.0001

Free acidity is expressed as g/100 g of oleic acid; peroxide number as mEq O_2 kg^{-1} oil; OSI, oxidative stability index, as hours; total phenol as mg kg^{-1} of gallic acid.

Fatty acid and sterolic profile can be used as an exceptional compositional marker for olive oil authenticity [11]. Fatty acid composition of all of the EVOO samples extracted from olives after different storage duration was characterized by the high oleic acid content (Table 2), coherent with the cold climate of the region; the relationship between fatty acid composition and climate is well known [22]. Several studies reported that fatty acid composition in oils did not show any change as the period for which olives were stored prior to crushing increased [10,23] and neither did they even when the olive storage period was very long, e.g., 45 days, as reported by Gutierrez et al. [24]. However, other studies [20,21] found differences in fatty acids content during storage, in agreement with our results. Specifically, we found differences in the content of C16:0, C18:1 and C18:2. In fact, the C16:0 content tended to increase as the olive storage period lengthened, while concentrations of C18:1 and C18:2 did not show a clear trend.

Table 2. Fatty acid composition of oils produced by olives after different olive storage duration. Values are mean ± standard deviation. Values followed by different letters in the same column (a, b) were significantly different according to Tukey's test ($p < 0.05$).

Storage Classes	C16	C16:1	C18	C18:1	C18:2	C18:3
<24 h	12.67 ± 0.96 b	1.25 ± 0.26	2.09 ± 0.23	75.39 ± 1.48 a	6.74 ± 0.98 b	0.74 ± 0.09
2–3 days	13.07 ± 0.73 ab	1.2 ± 0.18	2.03 ± 0.25	75.02 ± 1.53 ab	6.97 ± 1.07 ab	0.71 ± 0.07
4–6 days	13.17 ± 1.05 a	1.2 ± 0.17	2.04 ± 0.19	74.49 ± 1.80 b	7.31 ± 1.18 a	0.73 ± 0.08
>7 days	13.15 ± 0.76 a	1.15 ± 0.16	1.99 ± 0.16	75.17 ± 1.61 ab	6.75 ± 1.16 ab	0.73 ± 0.08
p-value	0.004	0.071	0.089	0.020	0.026	0.213

The content of α-tocopherol, the naturally occurring form of vitamin E assimilated by the human body, found in oils obtained from olives belonging to different conservation classes showed a statistically significant decreasing trend (Table 3). Vitamin E is an antioxidant, working as peroxyl radical scavenger that terminates chain reactions [25]. As it is well documented, oxidation phenomena are the main cause of tocopherol degradation [26]; the data here presented showed that milling olives within 24 h from the collection was the only way to protect the tocopherol fraction. In this study, the decrement in α-tocopherol content found after 7 days of olive storage was about 17%, consistent with the reduction of 22% of total tocopherol content found by Yousfi et al. [10] for cv. Arbequina and Pereira and colleagues [21] for cv. Verdeal Transmontana. An important and significant decrease in α-tocopherol after a short (48 h) olive storage period was as well found for cv. Nostrana di Brisighella oils while not for cv. Correggiolo oils [27].

Table 3. Tocopherol and carotenoid content in oil samples after different olive storage durations. Values are mean ± standard deviation. Compounds are expressed as mg of relative standard compound per kg of oil. Values followed by different letters in the same column (a, b) were significantly different according to Tukey's test ($p < 0.05$).

Storage Classes	α-Tocopherol	β+γ-Tocopherol	Lutein	β Carotene
<24 h	184.25 ± 36.59 a	8.29 ± 1.84	1.67 ± 0.62 a	1.05 ± 0.63 a
2–3 days	165.63 ± 32.8 b	8.31 ± 2.32	1.34 ± 0.94 b	0.80 ± 0.68 b
4–6 days	160.98 ± 32.85 b	8.33 ± 2.07	1.19 ± 0.42 b	0.64 ± 0.41 b
>7 days	157.77 ± 30.96 b	8.24 ± 1.80	1.46 ± 0.46 ab	0.82 ± 0.34 b
p-value	<0.0001	0.988	0.000	0.000

In the present study, carotenoid pigments decreased during the first three storage times analyzed, in agreement with other works [28,29]. At the last time of storage duration analyzed (>7 days), olive oils showed an increase in carotenoid content, in line with Yousfi and colleagues [10], who hypothesized a greater extractability of the pigments in olives during storage, due to the degradation of the chloroplast membranes; the degradation was found to be a consequence of the growing dehydration of the olives during storage [10].

A correlation analysis was carried out to quantify the intensity of the connection between EVOO properties and olive storage duration (Table 4). A positive correlation (Pearson) with storage duration was found for the parameter acidity, peroxide number, K232 and palmitic acid content, while palmitoleic and stearic acid, total phenol content and oil stability exhibited a negative correlation, diminishing with the increase in storage duration. In the case of fruits left for long periods before transformation, in the produced oils was observed, in addition to an increase in free acidity, even a gradual depletion in the content of oleic acid and total phenols, with the consequent reduction in stability during storage. It is also important to underline the correlation (r = 0.388, $p \leq 0.0001$) between oleic acid and OSI found in this work (data not shown). In fact, a high concentration of oleic acid enhances the oil stability and EVOO oxidative stability is mainly linked to its fatty acid composition, therefore the induction period is the result of the fatty acid composition and the simultaneous activity of various prooxidant and antioxidant factors endogenous in the oils [30].

Table 4. Pearson correlation between oil chemical parameters and olive storage duration.

Variable	r	p
Acidity	0.244	0.000
Peroxide number	0.312	<0.0001
Total phenols	−0.319	<0.0001
C 16	0.149	0.031
C 16:1	−0.184	0.007
C 18	−0.139	0.044
C 18:1	−0.009	0.894
C 18:2	−0.038	0.587
C 18:3	−0.021	0.757
OSI	−0.262	0.000
K_{232}	0.295	<0.0001
K_{270}	0.035	0.615

The sensory profile was modified according to the olive storage duration before processing, with the main positive attributes (olive fruity, green notes, bitterness and pungency) undergoing a statistically significant reduction with the increasing of olive storage days (Table 5). This result was in line with the previously recorded decrease in total phenols: the typical bitter taste and pungent note of fresh EVOO rich in total phenols decreased in intensity as olive storage duration lengthened. The intensity of the defects perceived by the sensory assessors increased with the progress of the days of olives storage (Table 5). Even in oils processed within 24 h, an intensity of the defect of 0.29 was recorded, probably attributable to the percentage of olives harvested late and thus overripe. In detail, the perceived defects were attributable to incorrect management of the raw material: in fact, they were 67% for the fusty defect, 29% for the vinegary and only 4% for the musty defect. During prolonged olive storage duration, the drupe tissues are damaged, resulting in the secretion of fluids favoring the growth of undesirable microorganisms [17]; increased temperature can also increase drupe respiratory activity, leading to undesirable metabolic processes accelerating fruit deterioration and characterized by the fusty sensory defect [31].

Table 5. The intensity of sensory attributes (aroma of olive fruity and green notes, flavor of bitterness, pungency and defect) of oil samples related to different olive time storage. Values are median ± standard deviation. Values followed by different letters in the same column (a, b, c) were significantly different according to Tukey's test ($p < 0.05$).

Storage Classes	Olive Fruity	Green Notes	Bitterness	Pungency	Defect
<24 h	2.25 ± 0.5 a	1.36 ± 0.65 a	2.16 ± 0.76 a	1.93 ± 0.52 a	0.29 ± 0.55 b
2–3 days	1.98 ± 0.45 b	0.92 ± 0.54 b	1.76 ± 0.62 b	1.74 ± 0.46 ab	0.34 ± 0.50 b
4–6 days	1.79 ± 0.41 b	0.64 ± 0.51 b	1.35 ± 0.53 c	1.43 ± 0.46 c	0.79 ± 0.76 a
>7 days	1.75 ± 0.38 b	0.57 ± 0.43 b	1.4 ± 0.65 bc	1.41 ± 0.51 bc	0.57 ± 0.64 b
p-value	<0.0001	<0.0001	<0.0001	<0.0001	<0.0001

The Principal Component Analysis (PCA) of the sensory data explained 91.4% of the variability and confirmed the strong influence of olive storage duration on the sensory characteristics of the oils produced (Figure 1). Most of the samples that transformed within 24 h are, in fact, positioned in the first quadrant of the PCA, showing the greatest intensities of positive attributes such as the aroma of olive fruity and green notes and bitterness. The small percentage of oil samples that, despite having been transformed within 24 h, are positioned in the fourth quadrant relative to the presence of sensory defects, is probably attributable to oils produced from overripe olives [32].

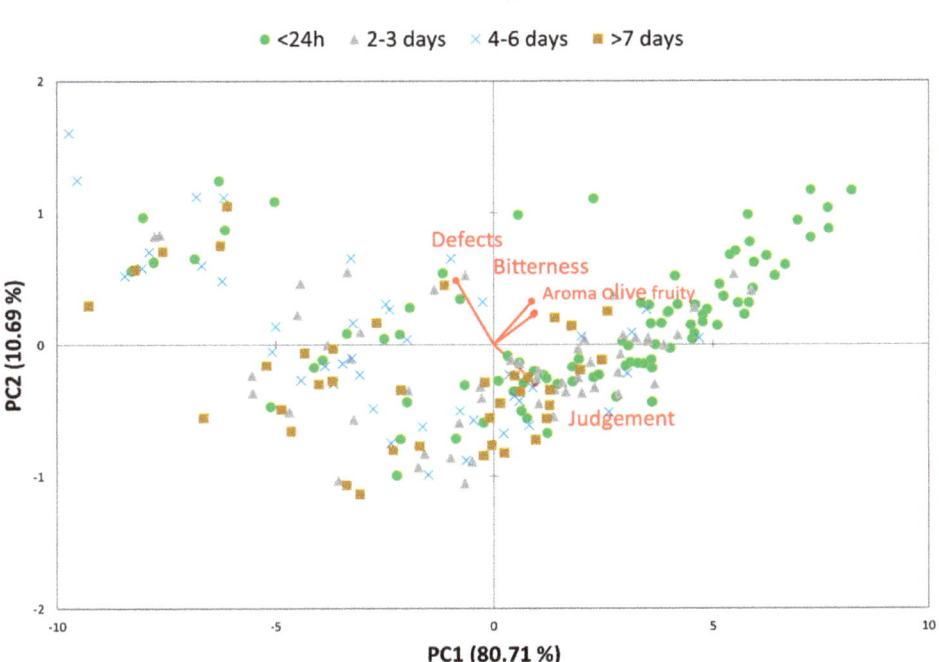

Figure 1. Principal component analysis (PCA) plot of sensory data.

4. Conclusions

The objective of this study, unlike other studies which investigated preprocessing storage as a way to modulate a positive reduction in the bitter taste of phenol-rich varieties with the aim of improving consumer acceptance [31], was to identify the criticality of the olive storage phase, highlighting its influence on the depletion of the EVOOs' chemical and sensory characteristics.

The authors are aware that this experimental design included many variables that affect the final quality of the oil: cultivar, ripeness, cultivation environment, seasonality and variation of technological parameters of extraction. However, thanks to the high rate of sampling, repeated for three consecutive production years, the single effect of the different variables was reduced. From this study, it emerged that some quality, nutritional and sensory parameters were affected by olive storage duration, independent of the varietal composition of the starting material. However, variety and ripeness degree influence the time window available to leave the olives on trees [33]. The knowledge of the effect of the harvest time window (early harvest or late harvest) on the olive oil final quality is important especially in years with late fly attacks, when it is recommended to harvest early rather than treating with chemicals, since a sustainable olive growing satisfies consumers who are increasingly attentive to the consumption of genuine and healthy products.

This study was carried out on purpose in a practical context characterized by all the limits listed above, thus the results provide an important photograph of the critical points of the olive storage phase from harvesting to pressing. By acting on the critical points, it is possible to improve the chain of olive oil production.

Free acidity, peroxide number, K232, total phenols, stability, α-tocopherol, lutein, β-carotene and organoleptic properties significantly decreased between the first and second storage interval, thus after 24 h of olive storage the final EVOO's quality was already substantially impoverished. Many specifications for PDO and PGI productions indicate

48 h as the maximum allowed storage duration; however, it is important to underline that by keeping storage within 24 h it is possible to maximize the potential of the olives, thus producing the oils with the highest nutritional and sensory properties expected by selecting cultivars known for the high quality of the final product. While PDO and IGP oils are produced according to strict production regulations, this study is aimed at blends productions that represent most of the processed olives in all of the Italian regions: for this kind of production the improving of the crucial phase of olive storage duration is important, and the results of this study are clearly significant for their applicability. The purpose of this work was to provide guidelines for obtaining a high-quality product at the time of processing, initial high quality being pivotal during the oil storage phase. The associations of olive producers guarantee their associates the supply of plastic aerated bins, together with guidelines aimed at reducing the olives' storage times from harvesting to processing. Ensuring the chemical and sensory oil quality during shelf life is, in fact, becoming the purpose of the most recent labeling regulations [34]: the community regulation states that what is indicated on the label should correspond with what is expected at the end of the product's shelf life.

Supplementary Materials: The following are available online at https://www.mdpi.com/article/10.3390/foods10102296/s1. Table S1: Definition of sensory descriptors.

Author Contributions: Conceptualization, A.R.; methodology, A.R., G.B.; formal analysis, L.M., G.B.; investigation A.R., G.B.; resources, A.R., G.B.; data curation, A.R., G.B.; writing—original draft preparation, A.R., L.M.; writing—review and editing, L.N., G.B.; supervision, A.R.; project administration, A.R. All authors have read and agreed to the published version of the manuscript.

Funding: This research received no external funding.

Acknowledgments: The authors gratefully thank Matteo Mari for his technical support, Massimiliano Magli for his support in the statistical analysis, the panel of ASSAM-Marche for sensory analysis and Simon Brown for providing language help.

Conflicts of Interest: The authors declare no conflict of interest.

References

1. Tarapoulouzi, M.; Skiada, V.; Agriopoulou, S.; Psomiadis, D.; Rébufa, C.; Roussos, S.; Theocharis, C.R.; Katsaris, P.; Varzakas, T. Chemometric Discrimination of the Geographical Origin of Three Greek Cultivars of Olive Oils by Stable Isotope Ratio Analysis. *Foods* **2021**, *10*, 336. [CrossRef]
2. EU. Council regulation (EC) No 1513/2001 of 23 July 2001 amending Regulations No 136/66/EEC and (EC) No 1638/98 as regards the extension of the period of validity of the aid scheme and the quality strategy for olive oil. *Off. J. Eur. Union* **2001**, *L201*, 4–7.
3. Fregapane, G.; Salvador, M.D. Production of superior quality extra virgin olive oil modulating the content and profile of its minor components. *Food Res. Int.* **2013**, *54*, 1907–1914. [CrossRef]
4. García, J.M.; Yousfi, K. The postharvest of mill olives. *Grasas Aceites* **2006**, *57*, 16–24. [CrossRef]
5. Clodoveo, M.L.; Delcuratolo, D.; Gomes, T.; Colelli, G. Effect of different temperatures and storage atmospheres on Coratina olive oil quality. *Food Chem.* **2007**, *102*, 571–576. [CrossRef]
6. Gutierrez, F.; Perdiguero, S.; Garcia, J.M.; Castellano, J.M. Quality of oils from olives stored under controlled atmosphere. *J. Am. Oil Chem. Soc.* **1992**, *69*, 1215–1218. [CrossRef]
7. García, J.M.; Gutiérrez, F.; Castellano, J.M.; Perdiguero, S.; Morilla, A.; Albi, M.A. Influence of storage temperature on fruit ripening and olive oil quality. *J. Agric. Food Chem.* **1996**, *44*, 264–267. [CrossRef]
8. Dourtoglou, V.G.; Mamalos, A.; Makris, D.P. Storage of olives (*Olea europaea*) under CO_2 atmosphere: Effect on anthocyanins, phenolics, sensory attributes and in vitro antioxidant properties. *Food Chem.* **2006**, *99*, 342–349. [CrossRef]
9. Koprivnjak, O.; Procida, G.; Zelinotti, T. Changes in the volatile components of virgin olive oil during fruit storage in aqueous media. *Food Chem.* **2000**, *70*, 377–384. [CrossRef]
10. Yousfi, K.; Weiland, C.M.; García, J.M. Effect of harvesting system and fruit cold storage on virgin olive oil chemical composition and quality of superintensive cultivated 'Arbequina' olives. *J. Agric. Food Chem.* **2012**, *60*, 4743–4750. [CrossRef]
11. Skiada, V.; Tsarouhas, P.; Varzakas, T. Comparison and Discrimination of Two Major Monocultivar Extra Virgin Olive Oils in the Southern Region of Peloponnese, According to Specific Compositional/Traceability Markers. *Foods* **2020**, *9*, 155. [CrossRef]
12. EEC. Commission regulation (EEC) No 2568/91 of 1July of 1991 on the characteristics of olive oil and olive-residue oil and on the relevant methods of analysis. *Off. J. Eur. Comm.* **1991**, *L248*, 1–114.

13. Cerretani, L.; Bendini, A.; Biguzzi, B.; Lercker, G.; Gallina Toschi, T. Stabilità ossidativa di oli extravergini di oliva ottenuti con diversi impianti tecnologici. *Ind. Aliment.* **2003**, *42*, 706–711.
14. Rotondi, A.; Bertazza, G.; Magli, M. Effect of olive fruits quality on the natural antioxidant compounds in extravirgin olive oil of Emilia-Romagna region. *Prog. Nutr.* **2004**, *6*, 139–145.
15. Bendini, A.; Gallina Toschi, T.; Lercker, G. Determinazione dell'attività antiossidante di estratti vegetali mediante Oxidative Stability Instrument (OSI). *Ind. Aliment.* **2001**, *403*, 525–529.
16. Skiada, V.; Tsarouhas, P.; Varzakas, T. Preliminary Study and Observation of "Kalamata PDO" Extra Virgin Olive Oil, in the Messinia Region, Southwest of Peloponnese (Greece). *Foods* **2019**, *8*, 610. [CrossRef] [PubMed]
17. Olias, J.M.; Garcia, J.M. Olive. In *Postharvest Physiology and Storage of Tropical and Subtropical Fruits*; Mitra, S.K., Ed.; CAB International: Wallingford, UK, 1997; pp. 229–243.
18. Servili, M.; Selvaggini, R.; Taticchi, A.; Esposto, S.; Montedoro, G. Volatile compounds and phenolic composition of virgin olive oil: Optimization of temperature and time of exposure of olive pastes to air contact during the mechanical extraction process. *J. Agric. Food Chem.* **2003**, *51*, 7980–7988. [CrossRef] [PubMed]
19. Vichi, S.; Romero, A.; Gallardo-Chacón, J.; Tous, J.; López-Tamames, E.; Buxaderas, S. Volatile phenols in virgin olive oils: Influence of olive variety on their formation during fruits storage. *Food Chem.* **2009**, *116*, 651–656. [CrossRef]
20. Youssef, O.; Guido, F.; Manel, I.; Youssef, N.B.; Cioni, P.L.; Mohamed, H.; Daoud, D.; Mokhtar, Z. Volatile compounds and compositional quality of virgin olive oil from Oueslati variety: Influence of geographical origin. *Food Chem.* **2011**, *124*, 1770–1776. [CrossRef]
21. Pereira, J.A.; Casal, S.; Bento, A.; Oliveira, M.B.P.P. Influence of olive storage period on oil quality of three portuguese cultivars of Olea europea, Cobrançosa, Madural, and Verdeal Transmontana. *J. Agric. Food Chem.* **2002**, *50*, 6335–6340. [CrossRef]
22. Lombardo, N.; Marone, E.; Alessandrino, M.; Godino, G.; Madeo, A.; Fiorino, P. Influence of growing season temperatures in the fatty acids (FAs) of triacilglycerols (TAGs) composition in Italian cultivars of *Olea europaea*. *Adv. Hortic. Sci.* **2008**, *22*, 49–53.
23. Youssef, N.B.; Leïla, A.; Youssef, O.; Mohamed, S.N.; Nizard, D.; Chedly, A.; Mokhtar, Z. Influence of the site of cultivation on Chétoui olive (*Olea europaea* L.) oil quality. *Plant Prod. Sci.* **2012**, *15*, 228–237. [CrossRef]
24. Gutierrez, F.; Varona, I.; Albi, M.A. Relation of acidity and sensory quality with sterol content of olive oil from stored fruit. *J. Agric. Food Chem.* **2000**, *48*, 1106–1110. [CrossRef] [PubMed]
25. Traber, M.G.; Atkinson, J. Vitamin E, antioxidant and nothing more. *Free Radic. Biol. Med.* **2007**, *43*, 4–15. [CrossRef] [PubMed]
26. Rastrelli, L.; Passi, S.; Ippolito, F.; Vacca, G.; De Simone, F. Rate of degradation of α-tocopherol, squalene, phenolics, and polyunsaturated fatty acids in olive oil during different storage conditions. *J. Agric. Food Chem.* **2002**, *50*, 5566–5570. [CrossRef] [PubMed]
27. Rotondi, A.; Lapucci, C. Nutritional properties of extra virgin olive oils from the Emilia-Romagna region: Profiles of phenols, vitamins and fatty acids. In *Olives and Olive Oil in Health and Disease Prevention*; Preedy, V.R., Watson, R.R., Eds.; Academic Press: Burlington, MA, USA, 2010; pp. 725–733.
28. Criado, M.N.; Motilva, M.J.; Goni, M.; Romero, M.P. Comparative study of the effect of the maturation process of the olive fruit on the chlorophyll and carotenoid fractions of drupes and virgin oils from Arbequina and Farga cultivars. *Food Chem.* **2007**, *100*, 748–755. [CrossRef]
29. Jabeur, H.; Zribi, A.; Abdelhedi, R.; Bouaziz, M. Effect of olive storage conditions on Chemlali olive oil quality and the effective role of fatty acids alkyl esters in checking olive oils authenticity. *Food Chem.* **2015**, *169*, 289–296. [CrossRef]
30. Psomiadou, E.; Tsimidou, M. Pigments in Greek virgin olive oils: Occurrence and levels. *J. Sci. Food Agric.* **2001**, *81*, 640–647. [CrossRef]
31. Inarejos-Garcia, A.M.; Gomez-Rico, A.; Desamparados Salvador, M.; Fregapane, G. Effect of preprocessing olive storage conditions on virgin olive oil quality and composition. *J. Agric. Food Chem.* **2010**, *58*, 4858–4865. [CrossRef]
32. Morales, M.T.; Luna, G.; Aparicio, R. Comparative study of virgin olive oil sensory defects. *Food Chem.* **2005**, *91*, 293–301. [CrossRef]
33. Morrone, L.; Rotondi, A.; Rapparini, F.; Bertazza, G. Olive Processing: Influence of Some Crucial Phases on the Final Quality of Olive Oil. In *Food Processing*; Marc, R.A., Valero Díaz, A., Posada Izquierdo, G.D., Eds.; IntechOpen: London, UK, 2019; pp. 29–46.
34. EU. Commission delegated regulation (EU) 1096/2018 of 22 May 2018 amending Implementing Regulation (EU) No 29/2012 as regards the requirements for certain indications on the labelling of olive oil. *Off. J. Eur. Union* **2018**, *L197*, 4–5.

Article

Influence of Two Innovative Packaging Materials on Quality Parameters and Aromatic Fingerprint of Extra-Virgin Olive Oils

Stefano Farris, Susanna Buratti *, Simona Benedetti, Cesare Rovera, Ernestina Casiraghi and Cristina Alamprese

Department of Food, Environmental, and Nutritional Sciences (DeFENS), Università degli Studi di Milano, Via G. Celoria 2, 20133 Milan, Italy; stefano.farris@unimi.it (S.F.); simona.benedetti@unimi.it (S.B.); cesare.rovera@unimi.it (C.R.); ernestina.casiraghi@unimi.it (E.C.); cristina.alamprese@unimi.it (C.A.)
* Correspondence: susanna.buratti@unimi.it

Abstract: The performance of two innovative packaging materials was investigated on two Sardinian extra-virgin olive oils (Nera di Gonnos and Bosana). In particular, a transparent plastic film loaded with a UV-blocker (packaging B) and a metallized material (packaging C) were compared each other and to brown-amber glass (packaging A). During accelerated shelf-life tests at 40 and 60 °C, the evolution of quality parameters (i.e., acidity, peroxide value, K_{270}, and phenolic content) was monitored, together with the aromatic fingerprint evaluated by electronic nose. Packaging B resulted in the best-performing material in protecting oil from oxidation, due to its lower oxygen transmission rate (0.1 ± 0.02 cm^3/m^2 24 h) compared to packaging C (0.23 ± 0.04 cm^3/m^2 24 h). At the end of storage, phenolic reduction was on average 25% for packaging B and 58% for packaging C, and the aromatic fingerprint was better preserved in packaging B. In addition, other factors such as the sanitary status of the olives at harvesting and the storage temperature were demonstrated to have a significant role in the shelf life of packaged extra-virgin olive oil.

Keywords: electronic nose; accelerated shelf-life tests; transparent plastic material; metallized material; brown-amber glass; oxidation; stability; packaging; olive oil quality

Citation: Farris, S.; Buratti, S.; Benedetti, S.; Rovera, C.; Casiraghi, E.; Alamprese, C. Influence of Two Innovative Packaging Materials on Quality Parameters and Aromatic Fingerprint of Extra-Virgin Olive Oils. *Foods* **2021**, *10*, 929. https://doi.org/10.3390/foods10050929

Academic Editor: Giancarlo Colelli

Received: 12 March 2021
Accepted: 21 April 2021
Published: 23 April 2021

Publisher's Note: MDPI stays neutral with regard to jurisdictional claims in published maps and institutional affiliations.

Copyright: © 2021 by the authors. Licensee MDPI, Basel, Switzerland. This article is an open access article distributed under the terms and conditions of the Creative Commons Attribution (CC BY) license (https://creativecommons.org/licenses/by/4.0/).

1. Introduction

The role of packaging throughout the food supply chain is of utmost importance since packaging contributes to preserving the food quality, maintaining the hygienic requisites thus preventing food-borne diseases, and allowing supply chain operations from the field to the consumer. Aside from these consolidated functions, modern and innovative packaging materials should be conceived to minimize their persistence into the environment, thus addressing the long-term crucial environmental issue of plastics disposal. At the same time, modern packaging is a key factor to address emerging challenges of sustainable food consumption, which involves the reduction of the environmental footprint of packed food [1]. From the environmental point of view, plastics are mistakenly perceived as the materials with the biggest environmental footprint mainly because they are almost exclusively seen under an end-of-life (EOL) perspective, with no consideration of material recyclability and impacts associated with the production and transport of the packaging materials [2,3].

However, plastic packaging materials, especially in the form of flexible configurations, provide environmental advantages and benefits over other materials, especially rigid configurations. For example, polyethylene terephthalate (PET) has demonstrated better environmental performance than traditional materials (e.g., aluminum and glass) in terms of consumption of natural resources and emissions [4]. In another work, it was demonstrated that bag-in-box and aseptic cartons had lower environmental impacts compared to single-use glass bottles for wine for all the impact categories considered by the authors, such as global warming potential, water consumption, and land use [5]. Besides, flexible packaging materials present transportation benefits since they are usually shipped either flat on a roll, thus allowing a dramatic reduction of trucks needed for transportation compared to rigid packaging [6].

Due to the high concentration of unsaturated fatty acids (oleic acid, linoleic acid, and linolenic acid), extra-virgin olive oil (EVOO) is subject to oxidation during storage even in the presence of abundant antioxidants (i.e., phenolic compounds and tocopherols). Oxidation is the main process affecting the quality of olive oil since some unstable compounds that can modify sensory and nutritional characteristics are produced. The level of EVOO oxidative degradation is strongly influenced not only by the chemical composition but also by the storage conditions. Packaging is therefore of great importance in preserving the quality of olive oil by protecting the product from oxygen and light.

Glass represents the first choice for EVOO because it is inherently impermeable to gases and vapors and it can be given light-filtering attributes, but, according to the above considerations, it may be of interest to evaluate packaging solutions for EVOO other than rigid glass. Pristouri et al. [7] compared the performance of glass bottles with plastic containers of the same volume (500 mL). They found that PET offered a moderately good performance, whereas polyethylene (PE) and polypropylene (PP) did not, mainly due to their low oxygen barrier properties. Gargouri et al. [8] evaluated the stability of Chemlali EVOO during storage with different packaging materials, i.e., clear and dark glass bottles, PE, and tin containers. They found that PE was inadequate to preserve EVOO from oxidation. Lolis et al. [9] investigated the effect of bag-in-box packaging material on the quality characteristics of EVOO using tinplate steel as the control. They demonstrated the best performance of the bag-in-box packaging materials, even when the EVOO samples were exposed to abuse temperatures (37 °C). In a more recent work, the same authors compared bag-in-box packaging material with dark-colored glass bottle [10]. They showed that samples packaged in bag-in-box material behaved in a similar way to those packaged in glass bottles.

In this work, we evaluated the performance of two innovative packaging materials on EVOO from two Sardinian olive cultivars: Nera di Gonnos and Bosana. In particular, a transparent plastic material loaded with a UV-blocker and a metallized material were compared to brown-amber glass bottles by monitoring the trends of oil quality indices (i.e., acidity, peroxide value, K_{270}, and total phenolic content) during accelerated shelf-life tests conducted at two different temperatures (40 and 60 °C).

Moreover, a commercial electronic nose (e-nose) was used to follow the evolution of the aromatic fingerprint of Nera and Bosana oils during storage. E-nose is an instrument designed to mimic the human sense of smell, widely applied in determining the quality of foods. Compared with traditional analytical techniques, including gas chromatography, high-performance liquid chromatography, and spectroscopy, e-nose is relatively inexpensive and less time-consuming; compared with sensory evaluation, e-nose provides more objective and consistent measurements [11].

2. Materials and Methods

2.1. Packaging Materials

Three different packaging materials were used in this study (Figure 1).

Brown-amber glass bottles (8 mL capacity) with a butyl/Teflon screw cap (Soffieria Vetro snc, Milano, Italy) were used as control and denoted as packaging A (Figure 1a). A transparent plastic film loaded with a UV-blocker (Cartastampa srl, Fornaci, Italy) was used as first testing material (coded as packaging B). In particular, it is a high-oxygen barrier film made of a 70 μm thick low-density polyethylene (LDPE) as the inner (in contact with oil) layer, coupled with a 12 μm high oxygen barrier-coated polyethylene terephthalate (PET) by means of a double-component polyurethane adhesive (Figure 1b). A second flexible material (coded as packaging C was made of a metallized layer (20 μm) sandwiched between an external printable layer (25 μm) and an inner sealable layer (25 μm) (Figure 1c); according to the manufacturer (TIPA, Hod Hasharon, Israel), the final material is 100% compostable and up to 65% made of bio-based materials.

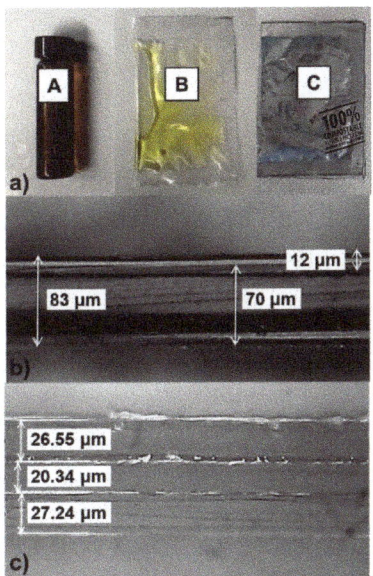

Figure 1. Packaging configurations used in the study: (**a**) glass vial (packaging A), transparent pouch (packaging B), and metallized pouch (packaging C); (**b**) optical microscope cross-sectional image (50 ×) of packaging B; (**c**) optical microscope cross-sectional image (50 ×) of packaging C.

Pouches 11.5 × 7 cm were prepared using a thermal heat sealer Polikrimper TX/08 (Alipack, Pontecurone, Italy), provided by smooth bars at 140 °C for 0.5 s and 4.5 bar pressure.

2.2. Olive Oil Samples

Two Sardinian monovarietal EVOOs (Nera di Gonnos and Bosana cultivars) that differed mainly for natural antioxidant content were subjected to accelerated shelf-life tests (ASLT): both EVOOs were divided in 6 g aliquots, stored in the three different packaging materials, and kept in the dark at 40 ± 1 °C and 60 ± 1 °C up to 96 and 32 days, respectively. During storage, at scheduled times three aliquots, for each packaging of the two EVOOs were analyzed for quality parameters and e-nose aromatic profile.

2.3. Oxygen Barrier Properties of Packaging Films

Oxygen transmission rate (OTR, mL/m^2 24 h) was measured on a 50 cm^2 surface sample using a PermeO$_2$ permeabilimeter (PermTech srl, Pieve Fosciana, Italy) equipped with an electrochemical sensor, according to ASTM 3985, with a carrier flow (N$_2$) of 10 mL/min at 23 °C and 0% relative humidity (RH) and at 1 atm pressure difference on the two sides of specimen. Three specimens were analyzed for each packaging materials.

2.4. Olive Oil Quality Parameters

The following quality parameters were considered:
- Acidity, indicative of the free fatty acid content and expressed as oleic acid (% oleic acid);
- Peroxide value (PV) corresponding to the amount of hydroperoxides (meq O$_2$kg^{-1});
- Specific extinction at 270 nm (K$_{270}$) providing a measurement of the secondary oxidation products.

All these analyses were performed in duplicate on each oil sample according to the methods reported in the European Regulation EEC no 2568/1991 and later amendments [12].

- Total phenolic content (TPC): oil samples were extracted with pure methanol as follows: 2 g oil was added to 5 mL methanol in a centrifuge tube, and the mixture was sonicated for 15 min. After sonication, the tube was centrifuged at 3500 rpm for 15 min at 15 °C, and the methanolic phase (extract) was separated; each sample was extracted in duplicate. TPC were determined by the Folin–Ciocalteu method [13], modified as follow: 0.5 mL of extract was added with 2.5 mL distilled water, 0.5 mL Folin–Ciocalteu reagent, and 2 mL Na_2CO_3 10%, and the mixture was taken to 10 mL with distilled water. After 90 min rest in the dark, the mixture was filtered with 0.2 mm Whatman filter, and the absorbance was read at 750 nm (Spectrophotometer V-650, Jasco, Japan). Results were expressed as gallic acid equivalents (mg_{GAE} kg^{-1}). Each extract was analyzed in duplicate.

2.5. Electronic Nose Analysis

Analyses were performed with the portable PEN3 e-nose (Airsense Analytics, Schwerin, Germany). The system is composed of a sampling apparatus, a sensor chamber containing the sensor array, and a pattern recognition software (Win Muster v.1.6) for data recording and processing. The sensor array consists of 10 metal oxide semiconductor (MOS) sensors: W1C (aromatic), W5S (broad range), W3C (aromatic), W6S (hydrogen), W5C (aromatic-aliphatic), W1S (broad-range), W1W (sulfur compounds), W2S (alcohols), W2W (sulfur compounds), and W3S (methane-aliphatic). The sensor response is expressed as resistivity (Ohm).

Two grams of oil samples were placed in 30 mL Pyrex® vials fitted with a pierceable silicon/teflon disk in the cap. After 10 min at 40 °C ± 1 °C for the development of the headspace, the measurement started. The volatile compounds were pumped over the sensor surfaces for 60 s (injection time) at a flow rate of 300 mL min^{-1}; the sensor signals were acquired at 50 s of sampling and statistically elaborated. After sample analysis, sensors were purged for 600 s with filtered air (purging time); then, prior to the next sample injection, the sensor baselines were re-established for 5 s. The sensor drift was estimate by using a standard solution of 0.2% ethanol included in each measurement cycle. The sensitivity of the instrument to various volatile compounds ranges from 0.1 and 5.0 ppm depending on their nature [14]. Each olive oil sample was evaluated in duplicate.

2.6. Data Analysis

Chemical data were analyzed by means of multifactor analysis of covariance (MANCOVA), considering the packaging material, EVOO cultivar, and storage temperature as main factors, and storage time as covariate. Two-way interaction effects were also evaluated. After checking the normal distribution of the responses, only TPC needed a squared transformation in order to fulfill the normality assumption. For significant factors, the Fisher's least significant difference (LSD) procedure was applied for mean comparisons ($p < 0.05$). Data were processed by Statgraphics Centurion software (v. 18.1; Statgraphics Technologies Inc., The Plains, VA, USA).

E-nose data were analyzed by principal component analysis (PCA), an unsupervised technique used to pre-process and reduce the dimensionality of high-dimensional datasets while preserving the original structure and relationships inherent to the original dataset. PCA reduces the number of the original variables into unobservable variables (principal components) that are linear combinations of the original ones. The main purpose of PCA is the explanation of the variability of the original dataset with as few principal components as possible, thus allowing one to visualize the data structure and the relationships between objects (score plot) and how strongly each variable influences a principal component (loading plot) [15].

PCA can be performed in covariance or correlation matrix: if the variables studied are measured using the same scale, it is reasonable to use the covariance matrix to obtain the PCs; on the other hand, if the variables are measured in different scales, the correlation matrix must be applied as the original variables are all standardized to unit variance [16].

In this work, PCA was performed in covariance matrix since the scale is the same for all the e-nose sensors.

E-nose data were elaborated by Minitab 17 (v. 1.0, Minitab Inc., State College, PA, USA) software package.

3. Results and Discussion

3.1. Quality Parameters of Olive Oils Stored under Different Conditions

Accelerated shelf-life tests were carried out in order to compare the ability of the proposed packaging materials to preserve the quality of EVOO by preventing oxidation, which causes loss of nutritional value and defects in the sensory properties [17,18].

Since the experimental factors were the packaging material, the oil cultivar, and the storage temperature, a multifactor analysis of variance was performed on quality parameters, using the storage time as covariate, as it was correlated to all the responses. Thus, the real effect of each experimental factor was assessed, after adjusting the storage time effect. Table 1 shows the results obtained, in terms of significance of the main and interaction effects. As expected, storage time significantly affected all the parameters, covariating with them. All the considered experimental factors were also significant, with the exception of storage temperature for acidity.

Table 1. Results of MANCOVA (F-ratio and significance level) for oil quality parameters.

Source	Acidity	PV	K270	TPC (Squared)
Covariate				
Storage time	476.41 ***	206.32 ***	93.88 ***	129.81 ***
Main effects				
Packaging	10.13 ***	16.47 ***	51.02 ***	354.02 ***
Oil cultivar	4000.51 ***	36.97 ***	15.72 ***	486.69 ***
Storage temperature	2.91 n.s.	124.73 ***	176.04 ***	60.93 ***
Interactions				
Packaging × Cultivar	8.57 ***	0.14 n.s.	1.68 n.s.	2.53 n.s.
Packaging × Temperature	10.11 ***	2.55 n.s.	5.15 **	22.34 ***
Cultivar × Temperature	4.77 *	1.79 n.s.	3.06 n.s.	0.74 n.s.

PV, peroxide value; TPC, total phenolic content; n.s., not significant; *, $p < 0.05$; **, $p < 0.01$; ***, $p < 0.001$.

Actually, a similar and progressive, though limited, acidity increase was observed at both 40 and 60 °C (Figure 2), as a consequence of the hydrolysis of triglycerides due to the action of lipases present in olives and produced by yeasts [19]. Nera oil showed acidity values significantly higher than that of Bosana oil ($p < 0.001$; Table 1), probably caused by a different sanitary status of the olives at harvesting. Indeed, excessive free fatty acids are associated with large, fully ripened, and fungus-infected drupes obtained from trees with low fruit loads. Even a small amount of such olives can spoil the oil quality [20]. However, different polyphenol content can also affected the acidity value since a higher phenolic concentration inhibits the activity of the lipase-producer yeasts [21]. At 40 °C, a similar increase of free fatty acid percentage was observed for the three packaging materials, whereas at 60 °C the highest acidity (about 0.58% and 0.32% for Nera and Bosana EVOOs, respectively) was evidenced for the oils stored in brown-amber glass bottles (packaging A) and in the transparent plastic material (packaging B). Indeed, the packaging × storage temperature effect was significant ($p < 0.001$), evidencing lower values of acidity at 60 °C for packaging C (metallized material) (Table 1). Anyhow, all the collected values did not exceed the limit of 0.8% set by the European Legislation [22].

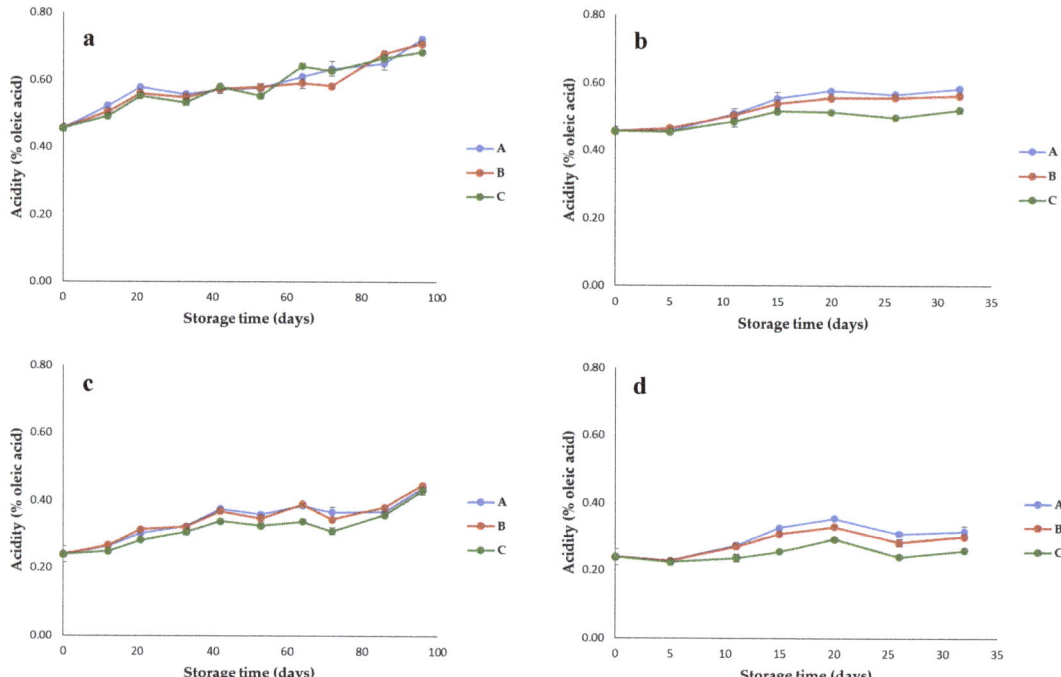

Figure 2. Trend of acidity in EVOOs stored in (A) brown-amber glass bottles, (B) transparent plastic material, and (C) metallized material; (**a**) Nera oil at 40 °C; (**b**) Nera oil at 60 °C; (**c**) Bosana oil at 40 °C; (**d**) Bosana oil at 60 °C. Error bars represent the standard deviation values (ranging from 0.001 to 0.024%).

Regarding the evolution of PV, after a first slight increase, a decrease was observed; thus, the legal limit of 20 meq $O_2 kg^{-1}$ [22] was never reached (Figure 3). This trend can be explained by considering that PV decreases with the appearance of secondary oxidation products. During oxidation, the hydroperoxides can form and at the same time decompose; when the decomposition rate prevails, PV is lowered even before exceeding the legal limit if the temperature is high and the oxygen concentration low [23].

All the considered experimental factors significantly affected PV ($p < 0.001$), whereas the two-way interactions were all not significant (Table 1). The lowest values were observed for oils stored in the metallized material (C), especially when stored at 60 °C. This means that with this packaging the oil is less protected toward oxidation since the formation of secondary oxidation products is faster. A plausible explanation for this result is the different oxygen barrier performance of the three packaging materials. Indeed, the metallized material had a permeability to gas higher than that of packaging B and A. Brown-amber glass bottles are impermeable to oxygen and packaging B had an OTR of 0.1 ± 0.02 cm^3/m^2 24 h, whereas OTR of packaging C was approximately double (0.23 ± 0.04 cm^3/m^2 24 h). Moreover, the possible contact of the oil with the metallized side of the inner layer could have catalyzed the oxidation reactions.

As expected, due to the acceleration of oxidation reactions, storage at 60 °C caused a significantly higher decrease of PV.

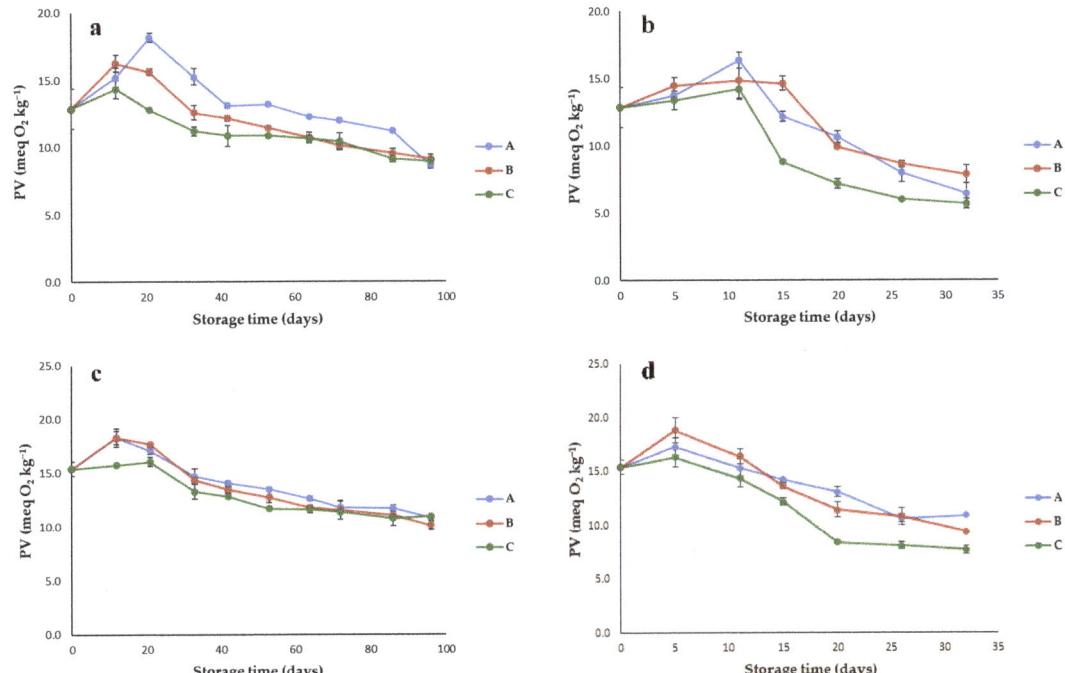

Figure 3. Trend of peroxide value (PV) in EVOOs stored in (A) brown-amber glass bottles, (B) transparent plastic material, and (C) metallized material; (**a**) Nera oil at 40 °C; (**b**) Nera oil at 60 °C; (**c**) Bosana oil at 40 °C; (**d**) Bosana oil at 60 °C. Error bars represent the standard deviation values (ranging from 0.01 to 1.51 meq $O_2 kg^{-1}$).

The two EVOO cultivars showed significantly different PV ($p < 0.001$; Table 1), with the higher values in Bosana oil. For Nera oil, the initial PV was 12.9 meq $O_2 kg^{-1}$; during storage at 40 °C, the maximum PV was reached at 21 days in the packaging A and 12 days in the packaging B and C; then, hydroperoxides readily decomposed to aldehydes, ketones, acids, esters, alcohols, and short-chain hydrocarbons [24], and PV gradually decreased to about 9.0 meq $O_2 kg^{-1}$ at the end of storage (Figure 3a). A similar trend was observed for Nera oil stored at 60 °C (Figure 3b), with lower PV at the end of storage (5.7–7.8 meq $O_2 kg^{-1}$) due to a faster degradation of peroxides to secondary oxidation products. The same PV evolution was observed for Bosana EVOO at both 40 (Figure 3c) and 60 °C (Figure 3d). Starting from an initial value of 15.4 meq $O_2 kg^{-1}$, this parameter first increased and then decreased; in particular, at 60 °C, the hydroperoxide decomposition was prevalent just after 5 days of storage, and the final PV was between 7.7 and 10.9 meq $O_2 kg^{-1}$. On average, at the end of storage Nera oil showed lower PV values than Bosana oil, probably due to a different polyphenol content affecting the protection toward oxidation phenomena.

K_{270} is known to be a good marker of oxidation secondary stage because it is related to conjugated trienes and carbonyl compounds [25]. In a recent work, Conte et al. [23] found that K_{270} is the best index for allowing one to predict EVOO shelf life when an accelerated test is applied. For both EVOO cultivars, this parameter significantly ($p < 0.001$) increased during storage, with significantly higher values in Bosana oil at 60 °C (Table 1; Figure 4), as a logical consequence of the higher initial oxidation state of Bosana samples and the acceleration of chemical reactions at higher temperatures. The packaging material had a significant effect on K_{270} ($p < 0.001$); the oils stored in the two innovative packaging materials (B and C) showed lower values than the oils stored in the brown-amber glass. In particular, the significantly lowest values were observed for the oils packaged in the metallized material stored at 60 °C; indeed, the interaction packaging × storage temperature had significant results ($p < 0.01$).

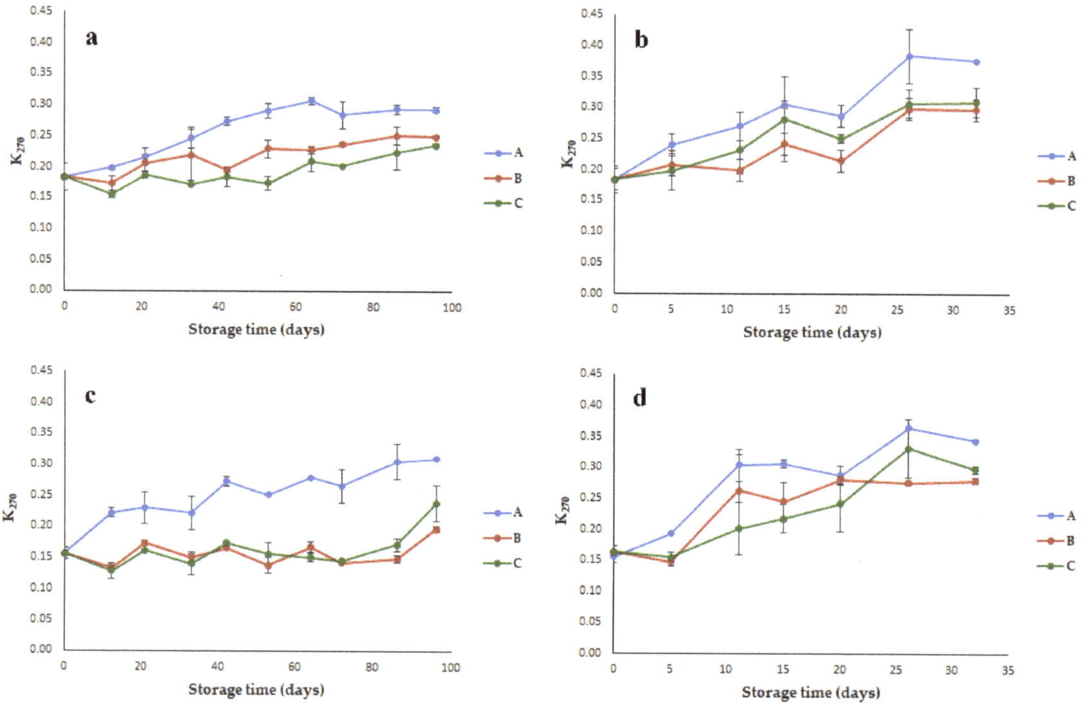

Figure 4. Trend of the spectrophotometric index K_{270} in EVOOs stored in (A) brown-amber glass bottles, (B) transparent plastic material, and (C) metallized material; (**a**) Nera oil at 40 °C; (**b**) Nera oil at 60 °C; (**c**) Bosana oil at 40 °C; (**d**) Bosana oil at 60 °C. Error bars represent the standard deviation values (ranging from 0.001 to 0.068).

For Nera oil stored in the brown-amber glass bottles (packaging A), a significant increase of K_{270} was observed, and the legal limit of 0.22 [22] was exceeded after 12 days at 40 °C and 11 days at 60 °C. For packaging B and C, the extinction at 270 nm increased considerably only at the end of storage at 40 °C, whereas at 60 °C, the legal limit was exceeded after 11 days for packaging B and 15 days for packaging C (Figure 4a,b). Similar results were obtained for Bosana oil (Figure 4c,d), with the samples packaged in brown-amber glass bottles characterized by a higher index of secondary oxidation products and exceeding the legal limit for K_{270} after 21 and 5 days of storage at 40 and 60 °C, respectively. This result could seem in contrast with the well-known protection ability of glass toward oil oxidation phenomena. However, in this case, the higher K_{270} values could be related to the higher retention capacity of glass toward low-molecular-weight compounds with respect to the tested innovative packaging materials.

Phenolic compounds are naturally present in olive oils, and they are responsible for oil stability during storage [26]. Figure 5 shows the evolution of TPC in Nera and Bosana oil as affected by packaging material and storage temperature.

Nera and Bosana oils were characterized by significantly different values of TPC, and the cultivar was indeed a significant factor ($p < 0.001$; Table 1) affecting this quality parameter during storage. Nera cultivar had a medium/low content of phenolics (300 mg$_{GAE}$ kg^{-1}), whereas Bosana was characterized by high polyphenol concentration (558 mg$_{GAE}$ kg^{-1}). During storage at both temperatures, a progressive decrease in TPC was observed (Figure 5), with a significantly ($p < 0.001$; Table 1) different trend for the tested packaging materials. In particular, the metallized material (packaging C) had a detrimental effect on phenolics, causing a higher TPC reduction, especially at 60 °C. In fact, the interaction packaging material x storage temperature was significant ($p < 0.001$). The average TPC reduction for the two EVOOs in packaging C at the end of storage was about 52% at 40 °C and 64% at 60 °C. The best-performing material in protecting oil phenolic content was the transparent plastic packaging (B), followed by the brown-amber glass (packaging A); this result can be again ascribed to the better oxygen barrier performance of the transparent pouches. During storage at 40 and 60 °C, TPC reduction was on average 21% and 30% for packaging B and 29% and 40% for packaging A. These results are in agreement with previous works showing that, during storage, phenolic compounds undergo quantitative modification due to oxidation and the temperature as well as the packaging material can have a notable influence on phenolic degradation [8,10].

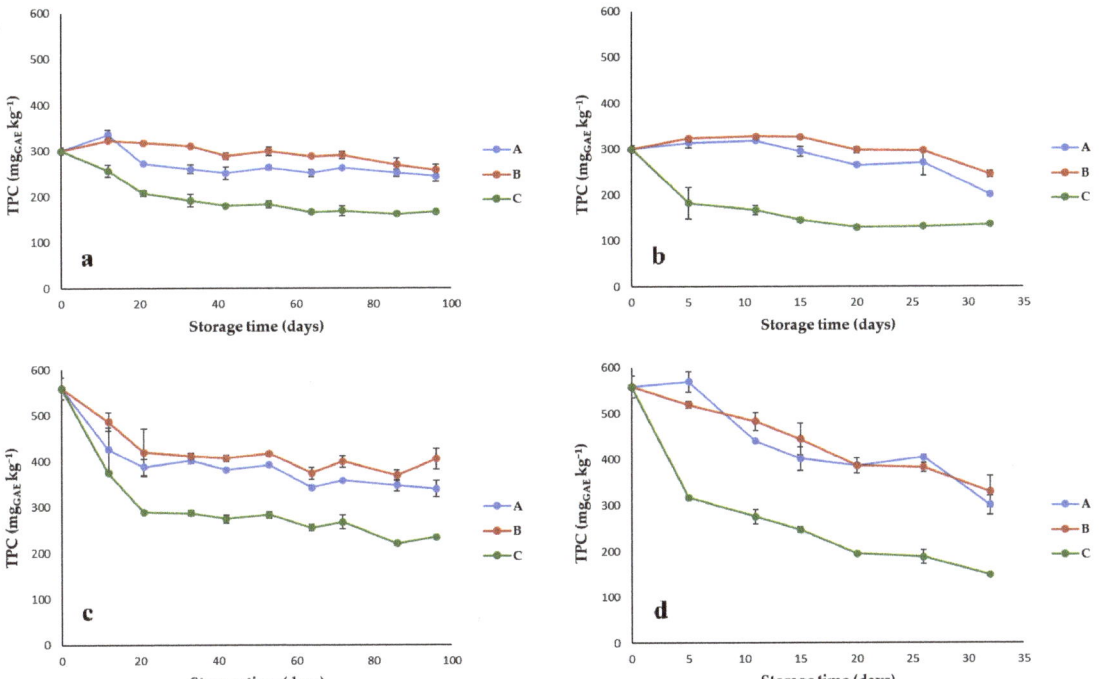

Figure 5. Trend of total phenolic content (TPC) in EVOOs stored in (A) brown-amber glass bottles, (B) transparent plastic material, and (C) metallized material; (**a**) Nera oil at 40 °C; (**b**) Nera oil at 60 °C; (**c**) Bosana oil at 40 °C; (**d**) Bosana oil at 60 °C. Error bars represent the standard deviation values (ranging from 0.1 to 47.6 mg$_{GAE}$ kg^{-1}).

3.2. E-Nose Aromatic Profile of Olive Oils Stored under Different Conditions

The e-nose is an instrument composed by non-selective or semi-selective sensors interacting with aromatic compounds to produce electronic signals. In the analysis of olive oil, e-nose has been successfully used in the determination of the geographical origin, in the detection of adulteration, and in the prediction of shelf-life [27].

In this work, a portable e-nose, with ten different MOS sensors, was applied in order to evaluate the effects of the three packaging materials on the evolution of the aromatic fingerprint of Nera and Bosana oil during storage, and the collected data were elaborated by PCA.

In order to evaluate the effects of the three packaging materials on the evolution of the aromatic fingerprint of Nera and Bosana oil during storage, a commercial e-nose was applied, and collected data were elaborated by PCA. Figures 6 and 7 show the PCA score plots and loading plots of the two oils in the plane defined by the first two principal component (PC1 and PC2) explaining almost all the variance.

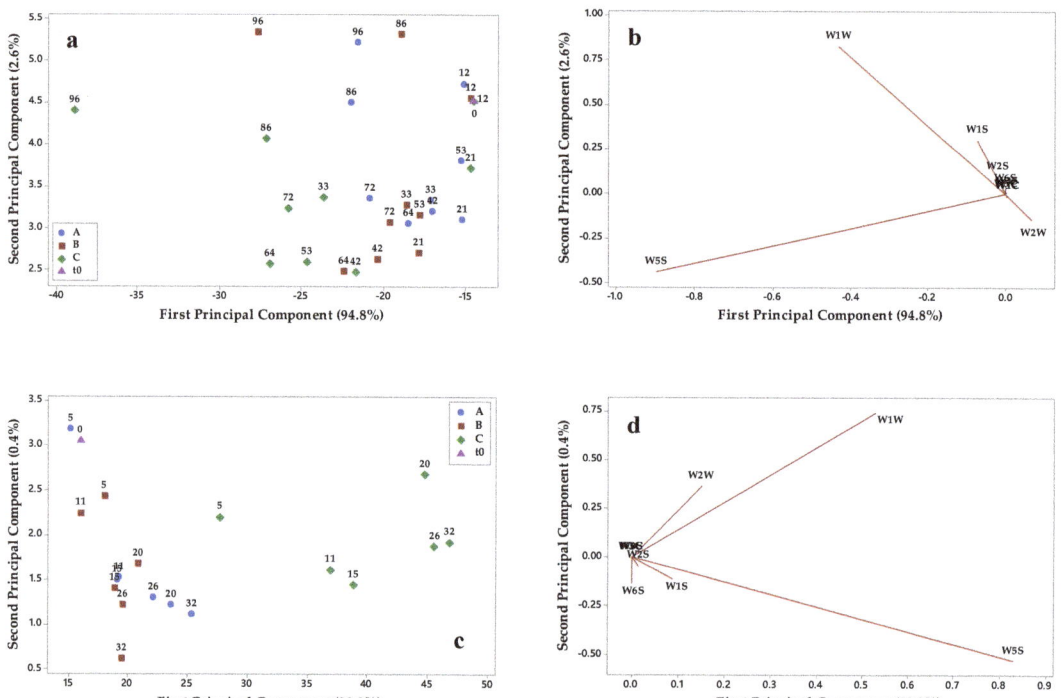

Figure 6. Results of principal component analysis of e-nose data of Nera oil stored in (A) brown-amber glass bottles, (B) transparent plastic material, and (C) metallized material: (**a**) score plot of samples stored at 40 °C; (**b**) loading plot for samples stored at 40 °C; (**c**) score plot of samples stored at 60 °C; (**d**) loading plot for samples stored at 60 °C.

By examining the PCA score plots of Nera oil (Figure 6a,c), it can be seen that, at both temperatures, the sample distribution on PC1 and PC2 followed the storage time and was affected by the packaging materials. At 40 °C, samples stored in the packaging A and B were mainly located to the right of the plot, and their aromatic profile was similar to that of fresh oil (t0) up to about 64 days of storage. For longer times (i.e., from 72 to 96 days), samples were discriminated on PC2 and located in the upper part of the score plot. The evolution trend of the oil stored in packaging C was more significant; at both temperatures, samples were distributed on PC1 according to the storage time and the aromatic profile of the oil evolved rapidly after 21 days at 40 °C and 5 days at 60 °C. Similar considerations can be drawn from Bosana score plots (Figure 7a,c); at both 40 and 60 °C, the oil samples stored in packaging C were distributed on PC1 according to the storage time, and a rapid evolution of the aromatic profile was noticeable. The oil samples stored in packaging A and B were characterized by a less modified aromatic fingerprint and by a similar evolution trend on PC1 and PC2.

The e-nose findings were quite consistent with the phenolic degradation during storage (Figure 5), in agreement with previous works reporting the influence of polyphenolic compounds on the aroma of olive oil [28,29].

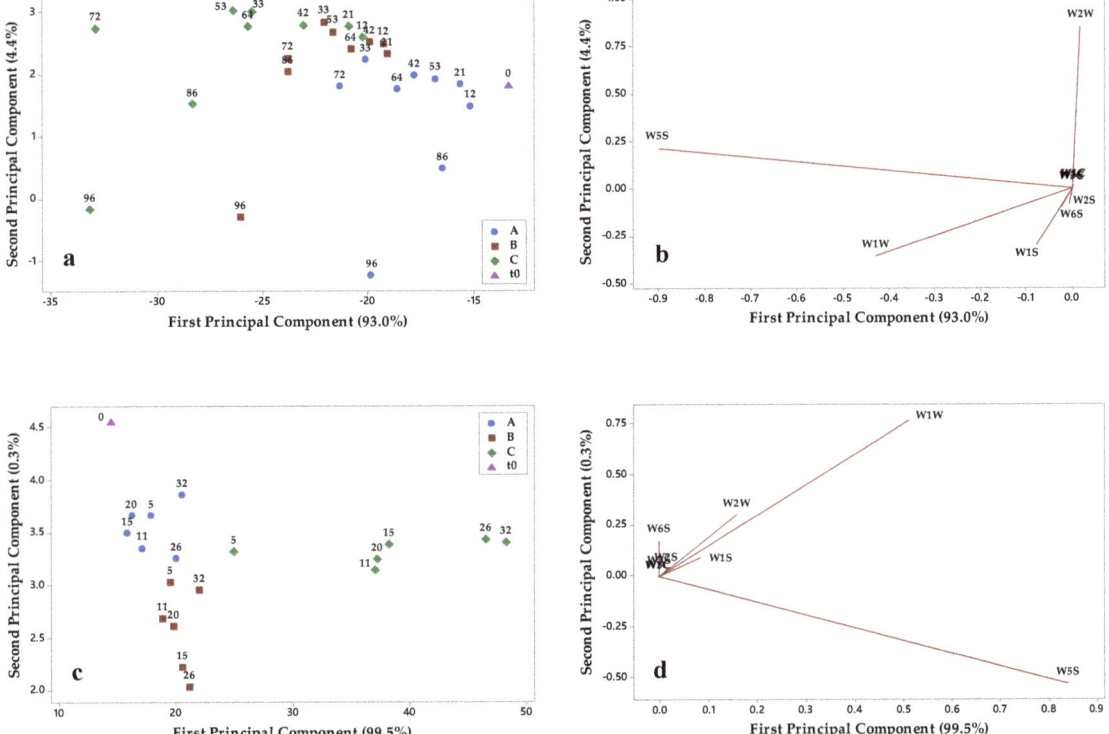

Figure 7. Results of principal component analysis of e-nose data of Bosana oil stored in (A) brown-amber glass bottles, (B) transparent plastic material, and (C) metallized material: (**a**) score plot of samples stored at 40 °C; (**b**) loading plot for samples stored at 40 °C; (**c**) score plot of samples stored at 60 °C; (**d**) loading plot for samples stored at 60 °C.

The loading plots of Nera (Figure 6b,d) and Bosana (Figure 7b,d) oils allowed one to relate e-nose sensors with the evolution trends in the score plots. W5S and W1W sensors were relevant on PC1, while W1S and W2W sensors discriminated samples on PC2. The W5S sensor was found to be the most relevant in the discriminating oil samples based on their storage conditions, and the same occurred in other works monitoring the evolution of the aromatic profile of vegetables during shelf life [30,31].

In a recent work, Xu et al. [14] evaluated the performance of PEN 3 e-nose to discriminate oils based on their oxidation rate. In agreement with our findings W5S, W1S, W1W, and W2W sensors showed different response signals to volatile compounds of oxidized oils compared to non-oxidized oils, thus demonstrating that this device provides a rapid and accurate method for characterizing the evolution aromatic profile of the oil during oxidation.

4. Conclusions

In spite of the large use of glass as main material for packaging and distribution of extra-virgin olive oil, increasing awareness of sustainability imposes a careful attention toward the selection of packaging materials with less overall environmental impact. According to this perspective, considering new materials other than glass bottles seems to be a trend that will not relent in the years ahead, especially as far as emerging markets are concerned. In this work, we have demonstrated that a flexible packaging material with outstanding oxygen barrier performance can be effectively used for extra-virgin olive oil, outperforming conventional glass as indicated by the most relevant quality parameters.

More specifically, the oxygen barrier performance seemed to play the most important role in preserving the overall quality of EVOO, especially at high temperatures (e.g., 60 °C). In this regard, the oxidation of the EVOO was less pronounced when the packaging with the lowest OTR was used (packaging B). In addition, other factors such as the sanitary status of the olives at harvesting and the storage temperature have been demonstrated to have a significant role in the shelf life of packaged EVOO. Findings arising from this work can be profitably used for the design of new packaging configurations as sustainable alternatives to glass.

Author Contributions: Conceptualization, C.A., S.B. (Susanna Buratti) and E.C.; methodology, C.A., S.F. and S.B. (Susanna Buratti); formal analysis, C.A. and S.B. (Simona Benedetti); investigation, C.R. and S.B. (Simona Benedetti); writing—original draft preparation, C.A., S.F. and S.B. (Susanna Buratti); writing—review and editing, E.C.; visualization, S.B. (Simona Benedetti); supervision, C.A. and S.B. (Susanna Buratti); project administration, C.A. and E.C.; funding acquisition, E.C. All authors have read and agreed to the published version of the manuscript.

Funding: This research was funded by AGER 2 Project, grant no. 2016-0105.

Institutional Review Board Statement: Not applicable.

Informed Consent Statement: Not applicable.

Data Availability Statement: Not applicable.

Conflicts of Interest: The authors declare no conflict of interest.

References

1. Guillard, V.; Gaucel, S.; Fornaciari, C.; Angellier-Coussy, H.; Buche, P.; Gontard, N. The next generation of sustainable food packaging to preserve our environment in a circular economy context. *Front. Nutr.* **2018**, *5*, 121–134. [CrossRef] [PubMed]
2. Boesen, S.; Bey, N.; Niero, M. Environmental sustainability of liquid food packaging: Is there a gap between Danish consumers' perception and learnings from life cycle assessment? *J. Clean. Prod.* **2019**, *210*, 1193–1206. [CrossRef]
3. Giovenzana, V.; Casson, A.; Beghi, R.; Tugnolo, A.; Grassi, S.; Alamprese, C.; Casiraghi, E.; Farris, S.; Fiorindo, I.; Guidetti, R. Environmental Benefits: Traditional vs. Innovative Packaging for Olive Oil. *Chem. Eng. Trans.* **2019**, *75*, 193–198.
4. Gomes, T.S.; Visconte, L.L.Y.; Pacheco, E. Life cycle assessment of polyethylene terephthalate packaging: An overview. *J. Polym. Environ.* **2019**, *27*, 533–548. [CrossRef]
5. Ferrara, C.; De Feo, G. Comparative life cycle assessment of alternative systems for wine packaging in Italy. *J. Clean. Prod.* **2020**, *259*, 120888–120900. [CrossRef]
6. Flexible Packaging Association. Available online: https://www.flexpack.org/sustainable-packaging (accessed on 1 January 2021).
7. Pristouri, G.; Badeka, A.; Kontominas, M.G. Effect of packaging material headspace, oxygen and light transmission, temperature and storage time on quality characteristics of extra virgin olive oil. *Food Control* **2010**, *21*, 412–418. [CrossRef]
8. Gargouri, B.; Zribi, A.; Bouaziz, M. Effect of containers on the quality of Chemlali olive oil during storage. *J. Food Sci. Technol.* **2015**, *52*, 1948–1959. [CrossRef]
9. Lolis, A.; Badeka, A.V.; Kontominas, M.G. Effect of bag-in-box packaging material on quality characteristics of extra virgin olive oil stored under household and abuse temperature conditions. *Food Packag. Shelf Life* **2019**, *21*, 100368–100377. [CrossRef]
10. Lolis, A.; Badeka, A.V.; Kontominas, M.G. Quality retention of extra virgin olive oil, Koroneiki cv. packaged in bag-inbox containers under long term storage: A comparison to packaging in dark glass bottles. *Food Packag. Shelf Life* **2020**, *26*, 100549–100559. [CrossRef]
11. Tan, J.; Xu, J. Applications of electronic nose (e-nose) and electronic tongue (e-tongue) in food quality-related properties determination: A review. *Artif. Intell. Agric.* **2020**, *4*, 104–115. [CrossRef]
12. Commission Regulation (EEC) No 2568/91 of 11 July 1991 on the characteristics of olive oil and olive-residue oil and on the relevant methods of analysis. *Off. J. Eur. Union* **1991**, *L248*, 1–128.
13. Singleton, V.L.; Rossi, J.A. Colorimetry of total phenolics with phosphomolybdic-phosphotungstic acid reagents. *Am. J. Enol. Viticult.* **1965**, *16*, 144–158.
14. Xu, L.; Yu, X.; Liu, L.; Zhang, R. A novel method for qualitative analysis of edible oil oxidation using an electronic nose. *Food Chem.* **2016**, *202*, 229–235. [CrossRef] [PubMed]
15. Abdi, H.; Williams, L.J. Principal component analysis. *Wiley Interdiscip. Rev. Comput. Stat.* **2010**, *2*, 433–459. [CrossRef]
16. Borgognone, M.G.; Bussi, J.; Hough, G. Principal component analysis in sensory analysis: Covariance or correlation matrix? *Food Qual. Prefer.* **2001**, *12*, 323–326. [CrossRef]
17. Ben-Hassine, K.; Taamalli, A.; Ferchichi, S.; Mlaouah, A.; Benincasa, C.; Romano, E.; Flamini, G.; Lazzez, A.; Grati-kamoun, N.; Perri, E.; et al. Physico-chemical and sensory characteristics of virgin olive oils in relation to cultivar, extraction system and storage conditions. *Food Res. Int.* **2013**, *54*, 1915–1925. [CrossRef]

18. Vacca, V.; Del Caro, A.; Poiana, M.; Piga, A. Effect of storage period and exposure conditions on the quality of Bosana extra-virgin olive oil. *J. Food Qual.* **2006**, *29*, 139–150. [CrossRef]
19. Ciafardini, G.; Zullo, B.A.; Iride, A. Lipase production by yeasts from extra virgin olive oil. *Food Microbiol.* **2006**, *23*, 60–67. [CrossRef] [PubMed]
20. Bustan, A.; Kerem, Z.; Yermiyahu, U.; Ben-Gal, A.; Lichter, A.; Droby, S.; Zchori-Fein, E.; Orbach, D.; Zipori, I.; Dag, A. Preharvest circumstances leading to elevated oil acidity in 'Barnea' olives. *Sci. Hortic.* **2014**, *176*, 11–21. [CrossRef]
21. Ciafardini, G.; Zullo, B.A. Effect of lipolytic activity of *Candida adriatica, Candida diddensiae* and *Yamadazyma terventina* on the acidity of extra-virgin olive oil with a different polyphenol and water content. *Food Microbiol.* **2015**, *47*, 12–20. [CrossRef] [PubMed]
22. Commission Delegated Regulation (EU) 2016/2095 of 26 September 2016 amending Regulation (EEC) No 2568/91 on the characteristics of olive oil and olive-residue oil and on the relevant methods of analysis. *Off. J. Eur. Union* **2016**, *L326*, 1–6.
23. Conte, L.; Milani, A.; Calligaris, S.; Rovellini, P.; Lucci, P.; Nicoli, M.C. Temperature dependence of oxidation kinetics of extra virgin olive oil (EVOO) and shelf-life prediction. *Foods* **2020**, *9*, 295. [CrossRef]
24. Choe, E.; Min, D.B. Mechanisms and factors for edible oil oxidation. *Compr. Rev. Food Sci. Food Saf.* **2006**, *5*, 169–186. [CrossRef]
25. Gertz, C.; Klostermann, S. A new analytical procedure to differentiate virgin or non-refined from refined vegetable fats and oils. *Eur. J. Lipid Sci. Technol.* **2000**, *102*, 329–336. [CrossRef]
26. Krichene, D.; Salvador, M.D.; Fregapane, G. Stability of virgin olive oil phenolic compounds during long-term storage (18 months) at temperatures of 5–50 °C. *J. Agric. Food Chem.* **2015**, *63*, 6779–6786. [CrossRef] [PubMed]
27. Majchrzak, T.; Wojnowski, W.; Dymerski, T.; Gębicki, J.; Namieśnik, J. Electronic noses in classification and quality control of edible oils: A review. *Food Chem.* **2018**, *246*, 192–201. [CrossRef] [PubMed]
28. Pedan, V.; Popp, M.; Rohn, S.; Nyfeler, M.; Bongartz, A. Characterization of phenolic compounds and their contribution to sensory properties of olive oil. *Molecules* **2019**, *24*, 2041. [CrossRef] [PubMed]
29. Genovese, A.; Caporaso, N.; Villani, V.; Paduano, A.; Sacchi, R. Olive oil phenolic compounds affect the release of aroma compounds. *Food Chem.* **2015**, *181*, 284–294. [CrossRef]
30. Giovenzana, V.; Beghi, R.; Buratti, S.; Civelli, R.; Guidetti, R. Monitoring of fresh-cut Valerianella locusta Laterr. shelf life by electronic nose and VIS–NIR spectroscopy. *Talanta* **2014**, *120*, 368–375. [CrossRef] [PubMed]
31. Benedetti, S.; Buratti, S.; Spinardi, A.; Mannino, S.; Mignani, I. Electronic nose as a non-destructive tool to characterise peach cultivars and to monitor their ripening stage during shelf-life. *Postharvest Biol. Technol.* **2008**, *47*, 181–188. [CrossRef]

Article

Near Infrared Spectroscopy as a Green Technology for the Quality Prediction of Intact Olives

Silvia Grassi [1], Olusola Samuel Jolayemi [1], Valentina Giovenzana [2], Alessio Tugnolo [2], Giacomo Squeo [3], Paola Conte [4], Alessandra De Bruno [5], Federica Flamminii [6], Ernestina Casiraghi [1] and Cristina Alamprese [1,*]

1. Department of Food, Environmental, and Nutritional Sciences (DeFENS), Università degli Studi di Milano, Via G. Celoria 2, 20133 Milan, Italy; silvia.grassi@unimi.it (S.G.); olusola.jolayemi@unimi.it (O.S.J.); ernestina.casiraghi@unimi.it (E.C.)
2. Department of Agricultural and Environmental Sciences (DiSAA), Università degli Studi di Milano, Via G. Celoria 2, 20133 Milan, Italy; valentina.giovenzana@unimi.it (V.G.); alessio.tugnolo@unimi.it (A.T.)
3. Department of Soil Plant and Food Sciences (DiSSPA), Università degli Studi di Bari "Aldo Moro", Via Amendola 165/A, 70126 Bari, Italy; giacomo.squeo@uniba.it
4. Department of Agricultural Sciences, Università degli Studi di Sassari, Viale Italia 39/A, 07100 Sassari, Italy; pconte@uniss.it
5. Department of Agraria, University Mediterranea of Reggio Calabria, Via dell'Università 25, 89124 Reggio Calabria, Italy; alessandra.debruno@unirc.it
6. Faculty of Bioscience and Technology for Agriculture, Food and Environment, University of Teramo, Via Balzarini 1, 64100 Teramo, Italy; fflamminii@unite.it
* Correspondence: cristina.alamprese@unimi.it; Tel.: +39-0250319187

Abstract: Poorly emphasized aspects for a sustainable olive oil system are chemical analysis replacement and quality design of the final product. In this context, near infrared spectroscopy (NIRS) can play a pivotal role. Thus, this study aims at comparing performances of different NIRS systems for the prediction of moisture, oil content, soluble solids, total phenolic content, and antioxidant activity of intact olive drupes. The results obtained by a Fourier transform (FT)-NIR spectrometer, equipped with both an integrating sphere and a fiber optic probe, and a Vis/NIR handheld device are discussed. Almost all the partial least squares regression models were encouraging in predicting the quality parameters ($0.64 < R^2_{pred} < 0.84$), with small and comparable biases ($p > 0.05$). The pair-wise comparison between the standard deviations demonstrated that the FT-NIR models were always similar except for moisture ($p < 0.05$), whereas a slightly lower performance of the Vis/NIR models was assessed. Summarizing, while on-line or in-line applications of the FT-NIR optical probe should be promoted in oil mills in order to quickly classify the drupes for a better quality design of the olive oil, the portable and cheaper Vis/NIR device could be useful for preliminary quality evaluation of olive drupes directly in the field.

Keywords: antioxidant activity; harvesting time; olive composition; olive cultivars; olive ripening; phenolic compounds; PLS regression model; portable device; quality parameters; sustainability

1. Introduction

The economic significance of olive industries to the European Union is unquestionable. Europe contributed almost 70% of the world olive oil production in the 2018–2019 harvest year campaign and the resultant revenue was to the tune of five billion euro [1]. This large and continuously expanding industry is also associated with many negative environmental problems stemmed from waste production and inappropriate disposal, soil depletion, and atmospheric emissions [2]. Every phase in the olive chain is characterized by different environmental concerns. In the agronomic phase, the use of pesticides, herbicides, and fertilizers has been identified as the principal contributor to ecological challenges [3]. In the cultivation phase, activities such as irrigation, pruning, soil management, and fertilizer applications can negatively affect the environment. The impacts of these primary phases are minor when compared to olive oil production and its unit operations. Oil extraction generates the most potentially hazardous organic compounds that accompany olive wastewater and pomace, depending on the techniques [4]. Laudable efforts have been made to adopt sustainable agricultural and industrial practices in the olive value chain to mitigate these problems. For instance, adoption of organic integrated agricultural systems in the farming and cultivation of olives is an example of sustainable agricultural practice. Industrially, practices such as the two-phase olive extraction method, which reduces water consumption, extraction of bioactive phytonutrients from by-products, and overall valorization of the olive production chain have significantly reduced the negative impacts of the industry on the environment [5,6]. However, a rather less emphasized aspect of the sustainable olive system is solvent reduction and replacement strategies during laboratory chemical analyses of olives and olive oils.

These chemical analyses are fundamental to monitor olive ripeness, estimate oil extraction efficiency, and control oil quality. Free acidity, moisture, and oil contents are examples of chemical parameters serving as quick tests on olive drupes before extraction [7]. On-field information of these chemical parameters can suggest suitable harvest time and overall orchard management [8,9]. Immediate first-hand knowledge of moisture and oil content of olive drupes prior to processing can reliably predict the economic viability of the entire production process, therefore informing producers about the raw material composition is of crucial relevance [10,11]. Similarly, prediction of minor constituents such as phenols, pigments, and antioxidants contents of olives can facilitate instant classification of the resultant oils even before production, making official standard compliance and product consistency easier. Commonly used wet methods, such as Soxhlet extraction technique, gravimetry, and chromatography have many unsustainable limitations such as excess solvent consumption, limited sample size, destructive sample preparation, slow response, and technical demand [7]. Thus, for effective processing and quality control of the olive system, application of green, sustainable eco-friendly, energy-efficient, non-destructive, non-invasive, easy-to-use, and inexpensive spectroscopic methods become inevitable.

From the technological point of view, the importance of these rapid determinations before oil extraction may lie in the possibility of modulating the extraction systems based on the drupe characteristics and type of desired product. For instance, operative conditions safeguarding the phenolic content can be adopted if phenolic substances are not so high in the drupes or, vice versa, the outstanding phenolic content of some drupes can be lowered if the final product is intended for consumers who do not like bitter/pungent oil [12,13]. Knowing how to set the equipment before starting the process instead of correcting the settings once the oil has been extracted and analyzed might be of interest.

Near infrared spectroscopy (NIRS) has gained prominence in the last decade and has contributed economically to food and feed industries by ensuring on-time processing and quality control [14,15]. The technology is a formidable green chemistry tool and environmentally sustainable analytical technique capable of handling a large sample size in solid and liquid forms and it provides quick answers to quality questions. NIRS, in conjunction with appropriate chemometrics, has become a routine analytical tool for the determination of intact olive drupes moisture and fat contents [16,17]. Using a portable Vis/NIR spectral acquisition device equipped with multiple detectors, it was possible to predict several economically important olive mill parameters such as maturity index, moisture, oil content, acidity, and dry matter [18]. Another type of NIRS system with a wavelength selection tool (acousto-optically tunable filter—AOTF) was satisfactorily applied to predict phenolic compounds and to monitor ripening of olives [19,20]. In addition to intact or crushed olive quality assessment, NIRS has been found to be handy in evaluation of olive oils and olive by-products [21,22]. However, comparative performance evaluations of NIRS using different signal acquisition devices are relatively uncommon especially for olive drupes. In this study the results obtained by a Fourier transform (FT)-NIR spectrometer (equipped with both integrating sphere and fiber optic probe) and a Vis/NIR handheld device for the prediction of quality parameters of intact olives of

13 different cultivars collected in three harvest years are discussed. In particular, the objective was to evaluate the different performance of the acquisition systems in the prediction of moisture, oil content, soluble solids, total phenol content, and antioxidant activity, in vision of suitable tools to be applied both in the field and at the mill for quick answers to quality questions in a sustainable way.

2. Materials and Methods

2.1. Olive Samples

Samples of olives belonging to 13 different cultivars from Abruzzo, Apulia, Calabria, and Sardinia regions (Italy) were used; sampling was carried out at different ripening degrees during 2016–2018 harvesting years. For each sampling and cultivar, three sample units (500 g each) were picked from different identified trees of the same grove, for a total of 267 sample units. Each unit was independently analyzed for the chemical parameters (moisture, oil content, soluble solid content, total phenolic content, and antioxidant activity). Two aliquots (100 g each) were taken from each sample unit for FT-NIR analysis with the integrating sphere. From each aliquot, 10 olives were selected as representative of the ripening stage [23] and used for analyses with both the FT-NIR and Vis/NIR fiber optical probes.

2.2. Chemical Analyses

Determination of moisture content (%) was carried out according to the AOAC 934.06 official method [24]. Oil content (% on fresh weight) was determined gravimetrically after the extraction of the oil from 10 g of dehydrated olive paste in a Soxhlet apparatus using petroleum ether as solvent [25]. Total soluble solids content (°Bx) was measured according to a previously published procedure [26]. Briefly, the sugar aqueous solution was prepared by homogenizing olive paste (20 g) in distilled water (40 mL) and stirring for 2 min. After centrifugation (11,000× g for 10 min), the supernatant solution was analyzed through a digital refractometer. Total phenol content (TPC) was determined as follows: olive pulp (1 g) was extracted using hexane (3 mL) and methanol:water (70:30 v/v; 15 mL), by stirring for 10 min at room temperature. After centrifugation (6000× g at 4 °C for 10 min), the supernatant phase was collected and further centrifuged (13,600× g, 5 min, room temperature). The obtained extracts were filtered through nylon syringe filters (pore size 0.45 µm; LLG Syringe Filter CA, Carlo Erba, Milano, Italy), properly diluted, and spectrophotometrically analyzed at 750 nm using the Folin-Ciocalteau reagent [27]. Calibration curves were made using gallic acid and the results were expressed as grams of gallic acid equivalent per kilogram olive pulp (g_{GA}/kg). Antioxidant activity (% inhibition/mg olive pulp) was determined on the same extracts used for TPC, applying the radical 2,2 diphenyl-1-picrylhydrazyl (DPPH•) method [28]. Briefly, 200 µL extract (previously diluted 1:20 in methanol) was made to react with 2.8 mL DPPH• methanol solution (6 × 10^{-5} M) for 1 h at 22 °C, measuring the discoloration at 515 nm. All reagents were from Sigma-Aldrich (St. Louis, MO, USA).

2.3. Spectra Collection

Spectra were collected by using a benchtop FT-NIR spectrometer (MPA, Bruker Optics, Milan, Italy), equipped with both an integrating sphere and a fiber-optic probe, and a handheld portable Vis/NIR device (Jaz, OceanOptics Inc., Dunedin, FL, USA). The FT-NIR spectra of the two aliquots (100 g each) of each olive sample unit were collected in duplicate in diffuse reflectance by means of the integrating sphere system. The optical fiber was used to acquire, in duplicate, the FT-NIR spectra of the 10 single olives selected from each aliquot based on ripening degree [23]. For both FT-NIR sampling systems, spectra were collected within a 12,500–3600 cm^{-1} spectral range, at 8 cm^{-1} resolution and with 32 scans. The background for the integrating sphere was performed by closing the internal reference wheel of the module, while for the fiber-optic probe a Spectralon standard was used. A dedicated software (OPUS v. 6.5, Bruker Optics, Ettlingen, Germany) was used to manage the instrument. The same single olives were analyzed in duplicate also by using the Vis/NIR portable device (500–1000 nm, i.e., 20,000–10,000 cm^{-1}; 0.3 nm resolution; 5 scans) equipped with a bifurcated optical fiber provided with a cap that standardizes the distance between the head of the probe and the sample (about 2 mm) and reduces the environmental light interference. A white reference (99% reflection) was used to set the maximum reflection. Spectrum acquisition lasted 18 s for both the integrating sphere and the probe of the benchtop FT-NIR spectrometer, and 1 s for the portable Vis/NIR device. Measurements were conducted with both instruments on the same day, thus making sample storage between analyses unnecessary.

2.4. Data Analysis

Data elaborations were performed using the Unscrambler X software (v. 10.4, CAMO ASA, Oslo, Norway). The replicated spectra were averaged in order to have one spectrum for each sample unit. For FT-NIR probe and sphere, spectral ranges were reduced to eliminate non-informative and noisy regions (i.e., 3600–4000 and 10,500–12,500 cm^{-1}), whereas in the case of the portable Vis/NIR device, the whole spectral range was used. The spectral data were independently pre-processed by standard normal variate (SNV), which removes possible interferences due to light scattering [29]. Chemical variables and all spectral data were merged in a single matrix (267 sample units × 5024 variables) and used to perform principal component analysis (PCA), autoscaling all the variables to overcome the heteroscedasticity nature of the data. The coordinate transformation of the merged spectral–chemical data matrix allowed for the selection of a calibration and a prediction data set, using the Kennard–Stone (KS) algorithm [30]. The algorithm partitioned the data in order to have 70% of samples (187 sample units) in the calibration set and 30% (80 sample units) in the prediction set.

Prediction of olive chemical characteristics based on spectral data was performed applying the partial least squares (PLS) regression to the calibration set of each spectral matrix (187 sample units × 1686 variables for the FT-NIR systems; 187 sample units × 1647 variables for the Vis/NIR equipment) using nonlinear iterative partial least squares (NIPALS) algorithm. Different pre-treatments of spectral data were tested: SNV, first derivative (d1; Savitzky–Golay algorithm, second order polynomial, 11-window size), which allows removal of baseline offset [31], and their combination. After calibration, the models were validated internally, through cross-validation (Venetian blind, 10 cancellation segments). The number of components to be considered for each model was determined based on the plot of calibration and cross-validation errors as a function of the number of latent variables (LVs). The optimal number of LVs was chosen as the number of LV allowing to minimize the cross-validation error. Afterwards, the models were externally validated by independently using the prediction set previously created with KS. Model performance was evaluated in terms of determination coefficients for calibration (R^2_{cal}), cross-validation (R^2_{cv}), and prediction (R^2_{pred}), as well as by root mean square error of calibration (RMSEC), cross-validation (RMSECV), and prediction (RMSEP), and standard error of prediction (SEP).

Prediction performances of the models obtained by the three spectral acquisition systems were compared by different approaches: (i) comparison of intermediate precisions expressed as standard error of laboratory (SEL); (ii) comparison of SEP with SEL of reference analyses; (iii) statistical tests proposed in the scientific literature [32,33]. SEL of the reference analyses and NIRS acquisition systems was calculated as follows [34]:

$$SEL = \sqrt{\frac{\sum_1^m (x_1 - x_2)^2}{m}}$$

where m is the number of olive samples and $x_1 - x_2$ is the absolute value of the difference between replicate results. In the third approach (i.e., statistical tests), first, the model biases, i.e., differences between the reference method results and those of the models predicting the chemical parameters, were compared by a *t* confidence interval for paired samples with a 95% confidence interval. The null hypothesis (H_0) states that model biases are not different. If the calculated Fisher value is higher than the F critical value, the H_0 is rejected and the hypothesis H_1 is true (i.e., differences between models are significant) [32]. Furthermore, a pairwise comparison of the model standard deviations was performed by the calculation of the correlation coefficient between each two sets of prediction errors (r). Then, K index is calculated by the following equation:

$$K = 1 + \{[2(1-r^2)t^2_{n-2, 0.025}]/(n-2)\} \quad (1)$$

where $t_{n-2, 0.025}$ is the upper 2.5% point of the *t* distribution on $n - 2$ degrees of freedom. Subsequentially, L index is calculated as follows [33]:

$$L = \sqrt{[K + \sqrt{((K^2-1))}]} \quad (2)$$

Then, the 95% confidence interval for the ratio of the standard deviations (L-lower and L-upper limits) was calculated. If the L interval includes 1, the standard deviations are not significantly different ($p > 0.05$). The model comparison was performed in MATLAB environment (v. R2017b, The MathWorks, Inc., Natick, MA, USA).

3. Results and Discussion

3.1. Chemical Parameters

Descriptive statistics of the chemical variables are presented in Figure 1 as box and whisker plots. The box lines represent the first and third quartiles and the median. The mean value is indicated by a cross sign. Whiskers correspond to the minimum and maximum measured values. Genetic, environmental, and cultivation factors affect olive composition, which changes during growth together with the drupe weight [5]. Actually, the tested cultivars and the different ripening stages and crop seasons accounted for a high range of variability of all the chemical parameters. This is an important point for the development of prediction models useful for different production sites. Variation ranges of the chemical parameters for the different olive cultivars are reported in Table S1.

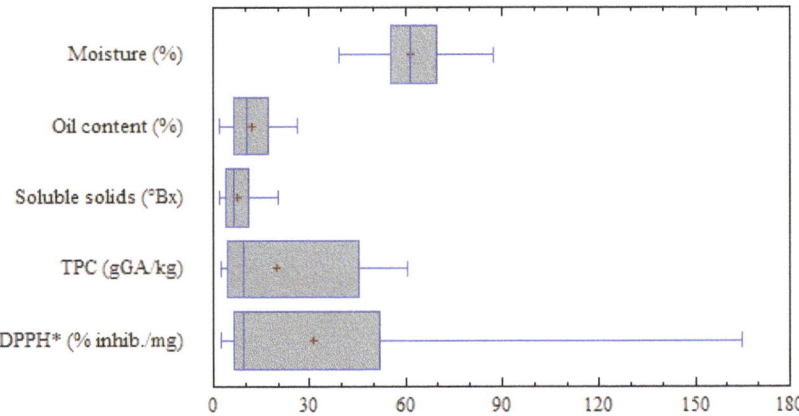

Figure 1. Box and whisker plots showing the descriptive statistics for the chemical variables tested on olive drupes. TPC: total phenol content; GA: gallic acid equivalent; DPPH•: radical 2,2 diphenyl-1-picrylhydrazyl; inhib.: inhibition.

Moisture represents the main constituent alongside oil. In the considered drupes, moisture content ranged from 39.3 to 87.2%. The obtained results agree with previously published data [18,35], considering that the moisture mean value was 63.3%, while the highest values (>80%) were obtained only in three out of thirteen cultivars, all from Calabria region. Excluding those three cultivars, the maximum value for moisture was 73.7%.

Commonly, olives intended for oil production have approximately 20% oil [36]. The samples here considered had a wide range of oil content (1.9–26.0%), suggesting the high influence of cultivar and ripening degree on this parameter. A general increase in oil content ranging from 2 to 12% was observed over ripening, depending on the considered cultivar.

TPC is an approximate estimation of total phenolic acids, phenolic alcohols, flavonoids, and secoiridoids in olive drupes. These compounds confer the bitter taste and pungent sensation on olive oils and are responsible for the well-known antioxidant properties. TPC values of the samples had a wide range of variation (2.5–60.6 g_{GA}/kg), with the highest levels (>35 g_{GA}/kg) found in three cultivars from Sardinia region. The antioxidant activity too was very different in the various samples, ranging from 2.4 to 165.0% inhibition/mg. Unexpectedly, the highest values (>70% inhibition/mg) were not found in the olives with the highest TPC, but in two cultivars from the Apulia region.

3.2. Spectral Features

Figure 2 shows the spectra of the olives obtained from the three acquisition systems. Visual features and patterns of the spectra conform with those previously reported for intact olive drupes [37,38].

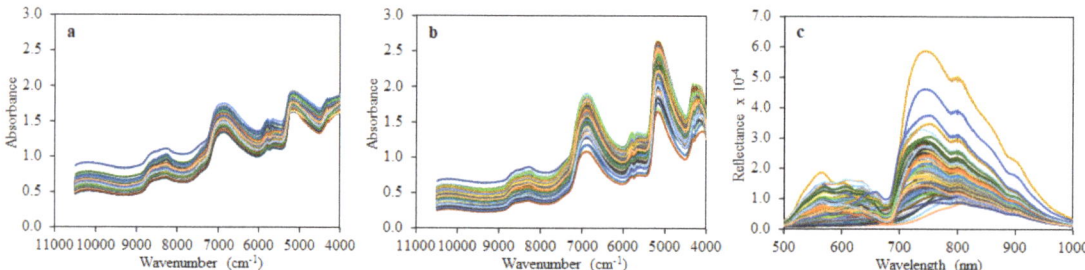

Figure 2. Spectra of olive drupes acquired with: (**a**) FT-NIR integrating sphere; (**b**) FT-NIR fiber-optic probe; (**c**) portable Vis/NIR device.

Aside from the visual differences in band intensities among samples, FT-NIR spectra from the integrating sphere and the fiber-optic probe (Figure 2a,b) are quite similar, with the latter exhibiting higher absorbances in most of the observable peaks.

The low absorbance band around 8600 cm^{-1} represents a combined symmetric and asymmetric OH stretching and bending vibrations. This is followed by the second overtone of CH stretching vibrations at 8300 cm^{-1} that corresponds to methyl (-CH$_3$), methylene (-CH$_2$), and olefin (-CH=CH-) bonds [37]. The high water content of the olive drupes (39–87%) explains the two absorption bands at 7500–6100 and 5400–4500 cm^{-1}. These bands are designated as the combination of first overtone of symmetric and asymmetric OH-bending and OH-stretching bands (6900 cm^{-1}) and combined OH-bending and OH-stretching bands (5200 cm^{-1}), respectively [39]. Similarly to the second overtone of CH stretching vibrations at 8300 cm^{-1}, the two bands at 5800 and 5650 cm^{-1} represent the first overtone of CH-stretching vibrations present in the same CH$_3$, CH$_2$, and CH=CH functional groups. At the far end of the FT-NIR spectral range, two peaks at 4335 and 4262 cm^{-1} represent CH and CH$_2$ s overtones, respectively [35]. However, the intermediate bands between the overtones (i.e., 8600, 5800–5650, and 4350–4250 cm^{-1}) have been attributed to the oil content of the drupes [40]. Regarding olive fruit phenols, there are no reported NIR correlated bands in the literature. However, a previous study suggested that some regions (i.e., 8700–8300 and 5800–5650 cm^{-1}) are correlated with TPC of olives [19].

In the case of Vis/NIR spectra (Figure 2c), more peak variations among samples were observed, especially within the visible (550–680 nm) and near-infrared (700–790 nm) regions. The changes around 550–680 nm correspond to some varying pigment indices. Specifically, the peak around 540 nm has been associated with anthocyanin, while that at 680 nm has been linked to chlorophyll [41]. Thus, changes in reflectance along these peaks may be due to maturation differences among the drupes. Other parameters, such as soluble solids, pH, and firmness, have been implicated within these regions in pears, especially around 340–740 nm [42]. Changes in the two absorption peaks around 750 and 850 nm could be assigned to the third overtone of H_2O and C-H functional group, respectively [43].

3.3. Principal Component Analysis

Figure 3 shows the score and loading plots of the PCA model built on the merged chemical and spectral database. The first two principal components (PCs) represent 59% of total data variance. The application of KS algorithm after PCA allowed to select evenly distributed samples for the calibration and prediction sets, highlighted with different colors in the score plot of Figure 3a. Few samples were seemingly outliers, but they were not removed in order to avoid presumptive assumption that they might adversely affect the model. Anyway, KS data splitting algorithm retained to a large extent as much variability as possible within the calibration and validation sets and this is a prerequisite for model robustness and validity in prediction. The loading plot (Figure 3b) shows a balanced contribution of both the chemical parameters and the three spectral ranges to sample distribution and consequently to the dataset partitioning.

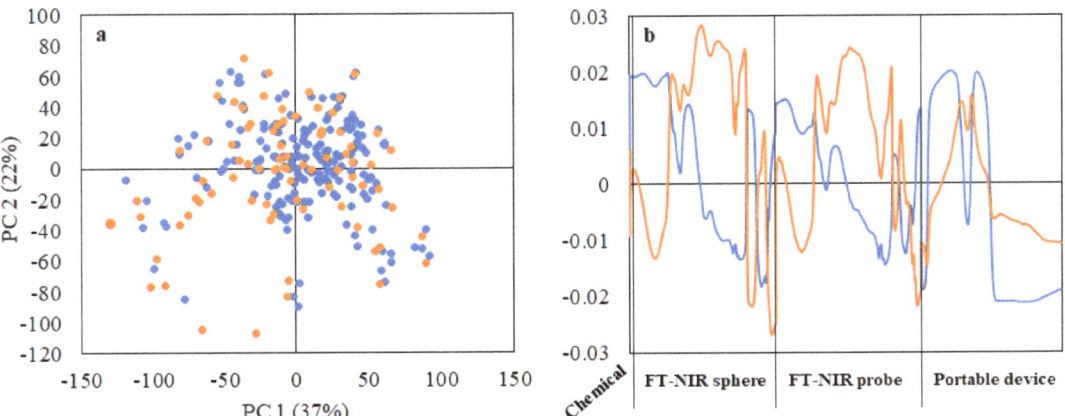

Figure 3. PCA results: (**a**) score plot showing the distribution of calibration (blue) and prediction (orange) set samples selected by Kennard-Stone algorithm applied on the merged chemical and spectral dataset of olive drupes; (**b**) loading plot of PC1 (blue) and PC2 (orange).

3.4. Regression Models

PLS regression models were built with FT-NIR and Vis/NIR spectra to quantify moisture, oil content, soluble solids, TPC, and antioxidant activity of olive drupes. In order to make data more evenly distributed, TPC and DPPH• results were transformed in the inverse and the logarithmic values, respectively. The best models based on determination coefficients and errors are reported in Table 1 for each spectral acquisition system. Predicted vs. measured plots of the models are reported in the Supplementary Figure S1. In general, performances of the three acquisition systems were similar in calibration and cross-validation, while in prediction FT-NIR spectra gave better results, maybe due to the wider NIR range and the low complexity of the models resulting in a higher stability.

Table 1. Figures of merit of the best PLS regression models for olive chemical parameter prediction based on spectroscopic data.

Parameter	NIR System	Pre-treatment	LVs	Calibration		Cross-Validation		Prediction	
				R^2_{cal}	RMSEC	R^2_{cv}	RMSECV	R^2_{pred}	RMSEP
Moisture content (%)	Sphere	SNV+d1	8	0.92	2.67	0.85	3.66	0.77	4.59
	Probe	SNV+d1	7	0.88	3.56	0.85	3.87	0.84	3.97
	Portable	d1	16	0.87	3.68	0.77	4.77	0.77	4.75
Oil content (%)	Sphere	SNV+d1	9	0.93	1.62	0.82	2.62	0.77	2.92
	Probe	SNV+d1	5	0.79	2.87	0.77	2.99	0.78	2.86
	Portable	SNV+d1	16	0.81	2.72	0.67	3.58	0.64	3.74
Soluble solids (°Bx)	Sphere	SNV+d1	9	0.90	1.45	0.75	2.36	0.70	2.39
	Probe	SNV+d1	11	0.87	1.66	0.80	2.06	0.74	2.23
	Portable	SNV	13	0.79	2.11	0.75	2.34	0.58	3.02
1/TPC (kg/g_{GA})	Sphere	SNV	13	0.89	0.04	0.81	0.04	0.77	0.04
	Probe	SNV+d1	13	0.87	0.04	0.76	0.05	0.76	0.04
	Portable	SNV+d1	9	0.83	0.05	0.79	0.05	0.69	0.05
logDPPH• (log % inhib./mg)	Sphere	SNV	15	0.84	0.20	0.68	0.29	0.68	0.29
	Probe	SNV+d1	16	0.93	0.14	0.79	0.24	0.73	0.27
	Portable	d1	13	0.79	0.23	0.72	0.27	0.41	0.39

TPC: total phenolic content; GA: gallic acid equivalent; inhib.: inhibition; DPPH•: radical 2,2 diphenyl-1-picrylhydrazyl; LVs: latent variables; R^2_{cal}: calibration coefficient of determination; R^2_{cv}: cross-validation coefficient of determination; R^2_{pred}: prediction coefficient of determination; RMSEC, RMSECV, and RMSEP: root mean square errors of calibration, cross-validation, and prediction, respectively; SNV: standard normal variate; d1: first derivative.

With respect to moisture content prediction, the three acquisition systems exhibited promising and similar prediction outcomes. The determination coefficients ranged from 0.77 to 0.92, with reasonably low values of errors (from 2.67 to 4.75%). However, the model calculated with the Vis/NIR spectra transformed in d1 showed a higher number of LVs (16 vs. 8 and 7 for the sphere and the probe, respectively), maybe due to the higher resolution of the spectra and the limited NIR range considered.

Oil content was better predicted by FT-NIR spectra, pre-treated with SNV and d1. Prediction coefficients of determination were higher than those of the portable acquisition system (0.77 and 0.78 vs. 0.64), with lower RMSEP values (2.92 and 2.86% vs. 3.74%) and LVs (9 and 5 vs. 16). The outcomes of calibration and cross-validation coefficients of determination for the FT-NIR sphere and probe (0.77–0.93) were comparable to those reported in the literature (0.78–0.84) for a smaller number of samples (183) [18].

Considering soluble solids, the regression model reliability appeared even more promising for FT-NIR spectrometer than for the portable device. Both the FT-NIR sphere and fiber-optic spectra pre-treated with a combination of SNV and d1 resulted in satisfactory determination coefficients in prediction (0.70 and 0.74, respectively) and low RMSEP (2.39 and 2.23 °Bx, respectively). The precision of the models was comparable to those observed for other fruits as, to the best of our knowledge, there is no study on NIR prediction of soluble solids in intact olives. For instance, quantitative determination of soluble solid content for quality prediction of intact strawberries using a handheld micro-electromechanical NIR system, resulted in R^2_{pred} of 0.37–0.47 and RMSEP of 1.02–0.87% [44]. With the spectra in the Vis/NIR range, the coefficient of determination in prediction decreased to 0.58, with a RMSEP of 3.02 °Bx.

Similar model performances in calibration and cross-validation were obtained for 1/TPC for all the spectral acquisition systems (R^2 range, 0.76–0.89), whereas in prediction FT-NIR spectra, gave better results (R^2_{pred} = 0.77–0.76) than Vis/NIR spectra (R^2_{pred} = 0.69). FT-NIR models are better than those reported in the literature for a filter-based NIR spectrometer [35]. The authors attributed the unsatisfactory output of their model (R^2_{cal}, 0.72; SEP, 13.35 g oleuropein/kg_{dm}) to the exclusion of 8600–6900 cm^{-1} range from the spectral bands, which was instead here considered. Our models were more promising also when compared to grape TPC prediction models developed using a portable NIR-

AOTF [45]; the authors observed determination coefficient values of 0.77 and 0.62 in calibration and cross-validation, respectively.

To the best of our knowledge, there is no other published paper in which the antioxidant activity of olive drupes is tentatively determined using rapid spectroscopic techniques. Therefore, our models seem fair especially when the FT-NIR probe was used, which generated comparatively highest R^2_{pred} and lowest RMSEP among the three spectral acquisition systems. The dynamic nature of this in vitro antioxidant activity makes its adaptation to spectroscopic techniques somewhat difficult. A more accurate NIR prediction of DPPH• radical scavenging activity was recorded in bean flours (R^2_{cal}, 0.94–0.99; R^2_{val}, 0.85–0.97) [46]. On the contrary, for a more bioactive horticultural product like *Hibiscus sabdariffa*, calibration and prediction determination coefficients are reported in the literature in the ranges 0.82–0.87 and 0.75–0.86, respectively, depending on spectra pretreatments [47].

From the inspection of the weighted regression coefficients of PLS models, for both the FT-NIR sphere and the probe the relevance of 7500–6100 and 5400–4500 cm^{-1} regions for moisture and soluble solid prediction was confirmed. Moreover, the PLS model developed for oil content prediction were highly influenced by the 5800–5650 and 4350–4250 cm^{-1} regions, attributed to the oil content of the drupes [40]. The same regions showed high weighted regression coefficients for TPC and DPPH• models, which were also characterized by high relevance of the 8700–8300 cm^{-1} region, previously related to TPC of olives [19].

As for the model developed with the Vis/NIR spectra, the inspection of the weighted regression coefficients revealed that both visible and NIR range influenced the prediction of moisture, oil content, and soluble solids. In particular, the range 880–970 nm showed the highest influence in the models for moisture and oil content prediction, whereas the maximum recorded weight for soluble solids corresponded to 970 nm. Moving to TPC and DPPH• prediction, it has been noticed that the highest values of the weighted regression coefficients were related to the visible range (550–700 nm), maybe linked to the olive color modification occurring during ripening, due to compounds like chlorophylls, carotenoids, anthocyanins, and polyphenols. Actually, other authors demonstrated that during olive ripening a rise in some bands of the visible range occurs (i.e., 600–650 and 550–625 nm), due to the presence of anthocyanin and other pigments related to reddish as well as green and yellow color [18].

3.5. Regression Model Comparison

The effectiveness of the prediction ability was at first established comparing the intermediate precisions (SEL) of the regression models with those of the reference methods (Table 2). The SEL values for the different NIR systems were generally higher than those obtained for the reference analyses, except for the SEL of the oil content predicted by the FT-NIR probe measurements. Indeed, the SEL values of NIR systems are more affected by the drupe heterogeneity, since spectra are collected on entire olives without the sample preparation phase of the chemical analyses, which is carried out by grinding and homogenizing the olive pulp.

The SEL_{ref} values were also compared with the prediction performances of the models in terms of SEP. As expected, SEP values were always higher than those of SEL_{ref}, because they include not only the sampling and analysis errors, but also the spectroscopy and model errors. The SEP obtained for the FT-NIR probe models were the lowest and the closest to the corresponding SEL_{ref} values. If the SEP is <2SEL_{ref}, the prediction performance of the model should be considered as good [48]. This was the case of models developed from FT-NIR probe spectra for moisture, oil content, and 1/TPC prediction.

Furthermore, the *t*-test for paired samples demonstrated that the biases for the models developed with the three spectral acquisition systems were comparable, i.e., the null hypothesis was confirmed (*p* values between 0.1 and 0.8; data not shown). On the other hand, the comparison between the standard deviations of the models [33] returned some differences as reported in the last three columns of Table 2. For moisture, the FT-NIR probe model resulted significantly different from those based on sphere and portable device spectra, due to a better performance resulting in a lower RMSEP (Table 1). All the other comparisons resulted in similar performance of the FT-NIR sphere and probe models, whereas the portable device models resulted significantly different because of the worse performance in terms of R^2_{pred}, RMSEP, and SEP.

Table 2. Comparison of regression models calculated for olive chemical parameter prediction based on three different FT-NIR and Vis-NIR acquisition systems.

Parameter	SEL$_{ref}$	NIR System	SEL$_{NIR}$	SEP	NIR System		
					Sphere	Probe	Portable Device
Moisture content (%)	2.00	Sphere	4.41	4.56	-	*	n.s.
		Probe	3.21	3.99	*	-	*
		Portable	4.49	4.72	n.s.	*	-
Oil content (%)	2.29	Sphere	3.13	2.94	-	n.s.	*
		Probe	2.18	2.88	n.s.	-	*
		Portable	2.95	3.77	*	*	-
Soluble solids (°Bx)	1.02	Sphere	2.21	2.41	-	n.s.	*
		Probe	2.31	2.24	n.s.	-	*
		Portable	1.88	3.03	*	*	-
1/TPC (kg/gGAE)	0.023	Sphere	0.045	0.044	-	n.s.	*
		Probe	0.044	0.043	n.s.	-	*
		Portable	0.036	0.052	*	*	-
logDPPH• (log % inhib./mg)	0.106	Sphere	0.257	0.287	-	n.s.	*
		Probe	0.282	0.267	n.s.	-	*
		Portable	0.223	0.390	*	*	-

TPC: total phenolic content; GA: gallic acid equivalent; DPPH•: radical 2,2 diphenyl-1-picrylhydrazyl; inhib.: inhibition; SEL$_{ref}$: standard error of laboratory for reference analyses; SEL$_{NIR}$: standard error of laboratory for NIR systems; SEP: standard error of prediction; n.s.: not significantly different standard deviation values ($p > 0.05$); *: statistically different standard deviation values ($p \leq 0.05$).

4. Conclusions

The benefits of different NIRS acquisition systems as green technology for quality characterization of intact olive drupes were explored. Generally, the calculated PLS models were remarkably encouraging in terms of determination coefficients and errors, both in internal validation and prediction. The model comparison highlighted a general better performance of both the FT-NIR sphere and probe acquisition systems with respect to the handheld device. However, the Vis/NIR device, being portable and relatively cheaper, is worthy of further investigations, because its use could be in any case very useful for preliminary quick quality assessment of olive drupes directly in the field. On the contrary, an on-line or in-line application of the FT-NIR optical probe in the olive mill should be promoted in order to quickly classify the drupes for a better quality design of the olive oil and a more sustainable management of the production chain.

Supplementary Materials: The following are available online at https://www.mdpi.com/article/10.3390/foods10051042/s1, Figure S1: Regression lines obtained for the prediction of entire olive chemical parameters with models developed by FT-NIR integrating sphere, FT-NIR fiber-optic probe, and portable Vis/NIR device. Table S1: Variation ranges of the chemical parameters for the different olive cultivars.

Author Contributions: Conceptualization, S.G., V.G., E.C. and C.A.; methodology, S.G., V.G., A.T., G.S., P.C., A.D.B., C.A.; formal analysis, S.G., O.S.J.; investigation, S.G., A.T., G.S., P.C., A.D.B.; resources, G.S., P.C., A.D.B., F.F.; data curation, S.G., O.S.J., C.A.; writing—original draft preparation, S.G., O.S.J., C.A.; writing—review & editing: V.G., A.T., G.S., P.C., A.D.B., F.F.; supervision, E.C., C.A.; funding acquisition, E.C. All authors have read and agreed to the published version of the manuscript.

Funding: This research was funded by AGER 2 Project, grant no. 2016-0105.

Acknowledgments: The authors wish to thank prof. Francesco Caponio for the supervision of the whole research project.

Conflicts of Interest: The authors declare no conflict of interest. The funder had no role in the design of the study; in the collection, analyses, or interpretation of data; in the writing of the manuscript, or in the decision to publish the results.

References

1. Casson, A.; Beghi, R.; Giovenzana, V.; Fiorindo, I.; Tugnolo, A.; Guidetti, R. Visible near infrared spectroscopy as a green technology: An environmental impact comparative study on olive oil analyses. *Sustainability* **2019**, *11*, 2611. [CrossRef]
2. Arvanitoyannis, I.S.; Kassaveti, A. Current and potential uses of composted olive oil waste. *Int. J. Food Sci. Technol.* **2017**, *42*, 281–295. [CrossRef]
3. Banias, G.; Achillas, C.; Vlachokostas, C. Environmental impacts in the life cycle of olive oil: A literature review. *J. Sci. Food Agric.* **2017**, *97*, 1686–1697. [CrossRef] [PubMed]
4. Souilem, S.; El-Abbassi, A.; Kiai, H.; Hafidi, A.; Sayadi, S.; Galanakis, C.M. Olive oil production sector: Environmental effects and sustainability challenges. In *Olive Mill Waste. Recent Advances for Sustainable Management*; Galanakis, C.M., Ed.; Academic Press: Oxford, UK, 2017; pp. 1–28. [CrossRef]
5. Pantziaros, A.G.; Trachili, X.A.; Zentelis, A.D.; Sygouni, V.; Paraskeva, C.A. A new olive oil production scheme with almost zero wastes. *Biomass Conv. Bioref.* **2020**. [CrossRef]
6. Rosello-Soto, E.; Koubaa, M.; Moubarik, A.; Lopes, R.P.; Saraiva, J.A.; Boussetta, N.; Grimi, N.; Barba, F.J. Emerging opportunities for the effective valorization of wastes and by-products generated during olive oil production process: Non-conventional methods for the recovery of high-added value compounds. *Trends Food Sci. Technol.* **2015**, *45*, 296–310. [CrossRef]
7. Salguero-Chaparro, L.; Palagos, B.; Peña-Rodríguez, F.; Roger, J.M. Calibration transfer of intact olive NIR spectra between a pre-dispersive instrument and a portable spectrometer. *Comp. Electron. Agric.* **2013**, *96*, 202–208. [CrossRef]
8. Hernández-Sánchez, N.; Gómez-Del-Campo, M. From NIR spectra to singular wavelengths for the estimation of the oil and water contents in olive fruits. *Grasas Aceites* **2018**, *69*, 1–13. [CrossRef]
9. Morrone, L.; Neri, L.; Cantini, C.; Alfei, B.; Rotondi, A. Study of the combined effects of ripeness and production area on Bosana oil's quality. *Food Chem.* **2018**, *245*, 1098–1104. [CrossRef]
10. Correa, E.C.; Roger, J.M.; Lléo, L.; Hernández-Sánchez, N.; Barreiro, P.; Diezma, B. Optimal management of oil content variability in olive mill batches by NIR spectroscopy. *Sci. Rep.* **2019**, *9*, 1–11. [CrossRef]
11. Giovenzana, V.; Beghi, R.; Romaniello, R.; Tamborrino, A.; Guidetti, R.; Leone, A. Use of visible and near infrared spectroscopy with a view to on-line evaluation of oil content during olive processing. *Biosyst. Eng.* **2018**, *172*, 102–109. [CrossRef]
12. Bianchi, B.; Tamborrino, A.; Giametta, F.; Squeo, G.; Difonzo, G.; Catalano, P. Modified rotating reel for malaxer machines: Assessment of rheological characteristics, energy consumption, temperature profile, and virgin olive oil quality. *Foods* **2020**, *9*, 813. [CrossRef] [PubMed]
13. Caponio, F.; Squeo, G.; Brunetti, L.; Pasqualone, A.; Summo, C.; Paradiso, V.M.; Catalano, P.; Bianchi, B. Influence of the feed pipe position of an industrial scale two-phase decanter on extraction efficiency and chemical-sensory characteristics of virgin olive oil. *J. Sci. Food Agric.* **2018**, *98*, 4279–4286. [CrossRef]
14. Porep, J.U.; Kammerer, D.R.; Carle, R. On-line application of near infrared (NIR) spectroscopy in food production. *Trends Food Sci. Technol.* **2015**, *46*, 211–230. [CrossRef]
15. Qu, J.H.; Liu, D.; Cheng, J.H.; Sun, D.W.; Ma, J.; Pu, H.; Zeng, X.A. Applications of near-infrared spectroscopy in food safety evaluation and control: A review of recent research advances. *Crit. Rev. Food Sci. Nutr.* **2015**, *55*, 1939–1954. [CrossRef]
16. Leon, L.; Garrido, A.; Downey, G. Parent and harvest year effects on near-infrared reflectance spectroscopic analysis of olive (*Olea europaea* L.) fruit fruits. *J. Agric. Food Chem.* **2004**, *52*, 4957–4962. [CrossRef] [PubMed]
17. Saha, U.; Jackson, D. Analysis of moisture, oil, and fatty acid composition of olives by near-infrared spectroscopy: Development and validation calibration models. *J. Sci. Food Agric.* **2018**, *98*, 1821–1831. [CrossRef] [PubMed]
18. Cayuela, A.J.; Camino, M.D.P. Prediction of quality of intact olives by near infrared spectroscopy. *Eur. J. Lipid Sci. Technol.* **2010**, *112*, 1209–1217. [CrossRef]
19. Bellincontro, A.; Taticchi, A.; Servili, M.; Esposto, S.; Farinelli, D.; Mencarelli, F. Feasible application of a portable NIR-AOTF tool for on-field prediction of phenolic compounds during the ripening of olives for oil production. *J. Agric. Food Chem.* **2012**, *60*, 2665–2673. [CrossRef]
20. Cirilli, M.; Bellincontro, A.; Urbani, S.; Servili, M.; Esposto, S.; Mencarelli, F.; Muleo, R. On-field monitoring of fruit ripening evolution and quality parameters in olive mutants using a portable NIR-AOTF device. *Food Chem.* **2016**, *199*, 96–104. [CrossRef] [PubMed]
21. Azizian, H.; Mossoba, M.M.; Fardin-Kia, A.R.; Karunathilaka, S.R.; Kramer, J.K.G. Developing FT-NIR and PLS1 methodology for predicting adulteration in representative varieties/blends of extra virgin olive oils. *Lipids* **2016**, *51*, 1309–1321. [CrossRef]
22. Comino, F.; Ayora-Cañada, M.J.; Aranda, V.; Díaz, A.; Domínguez-Vidal, A. Near-infrared spectroscopy and X-ray fluorescence data fusion for olive leaf analysis and crop nutritional status determination. *Talanta* **2018**, *188*, 676–684. [CrossRef]

23. Tugnolo, A.; Giovenzana, V.; Beghi, R.; Grassi, S.; Alamprese, C.; Casson, A.; Casiraghi, E.; Guidetti, R. A diagnostic visible/near infrared tool for a fully automated olive ripeness evaluation in a view of a simplified optical system. *Comput. Electron. Agr.* **2020**, 105887.
24. AOAC. *Official Methods of Analysis of the Association of Official Analytical Chemists International*, 17th ed.; Official Method 934.06. Moisture in Dried Fruits; Journal of AOAC International: Gaithersburg, MD, USA, 2000.
25. Thiex, N.J.; Anderson, S.; Gildemeister, B. Crude fat, diethyl ether extraction, in feed, cereal grain, and forage (Randall/Soxtec/submersion method): Collaborative study. *J. AOAC Int.* **2003**, *86*, 888–898. [CrossRef] [PubMed]
26. Migliorini, M.; Cherubini, C.; Mugelli, M.; Gianni, G.; Trapani, S.; Zanoni, B. Relationship between the oil and sugar content in olive oil fruits from Moraiolo and Leccino cultivars during ripening. *Sci. Hort.* **2011**, *129*, 919–921. [CrossRef]
27. Singleton, V.L.; Rossi, J. Colorimetry of total phenolics with phosphomolybdic-phosphotungstic acid reagents. *Am. J. Enol. Vitic.* **1965**, *16*, 144–158.
28. Conte, P.; Squeo, G.; Difonzo, G.; Caponio, F.; Fadda, C.; Del Caro, A.; Urgeghe, P.P.; Montanari, L.; Montinaro, A.; Piga, A. Change in quality during ripening of olive fruits and related oils extracted from three minor autochthonous Sardinian cultivars. *Emir. J. Food Agric.* **2019**, *31*, 196–205. [CrossRef]
29. Rabatel, G.; Marini, F.; Walczak, B.; Roger, J.M. VSN: Variable sorting for normalization. *J. Chemom.* **2020**, *34*, 1–16. [CrossRef]
30. Kennard, R.W.; Stone, L.A. Computer aided design of experiments. *Technometrics* **1969**, *11*, 137–148. [CrossRef]
31. Rinnan, Å.; Van Den Berg, F.; Engelsen, S.B. Review of the most common pre-processing techniques for near-infrared spectra. *TrAC* **2009**, *28*, 1201–1222. [CrossRef]
32. Roggo, Y.; Duponchel, L.; Ruckebus ch, C.; Huvenne, J.P. Statistical tests for comparison of quantitative and qualitative models developed with near infrared spectral data. *J. Mol. Struct.* **2003**, *654*, 253–262. [CrossRef]
33. Fearn, T. Comparing standard deviations. *NIR News* **1996**, *7*, 5–6. [CrossRef]
34. The European Agency for the Evaluation of Medicinal Products. Note for Guidance on the Use of Near Infrared Spectroscopy by the Pharmaceutical Industry and the Data Requirements for New Submissions and Variations. Available online: https://www.ema.europa.eu/en/documents/scientific-guideline/note-guidance-use-near-infrared-spectroscopy-pharmaceutical-industry-data-requirements-new_en.pdf (accessed on 15 April 2021).
35. Trapani, S.; Migliorini, M.; Cecchi, L.; Giovenzana, V.; Beghi, R.; Canuti, V.; Fia, G.; Zanoni, B. Feasibility of filter-based NIR spectroscopy for the routine measurement of olive oil fruit ripening indices. *Eur. J. Lipid Sci. Technol.* **2017**, *119*, 1600239. [CrossRef]
36. de la Casa, J.A.; Castro, E. Fuel savings and carbon dioxide emission reduction in a fired clay bricks production plant using olive oil wastes: A simulation study. *J. Clean Prod.* **2018**, *185*, 230–238. [CrossRef]
37. Fernández-Espinosa, A.J. Combining PLS regression with portable NIR spectroscopy to on-line monitor quality parameters in intact olives for determining optimal harvesting time. *Talanta* **2016**, *148*, 216–228. [CrossRef] [PubMed]
38. Giovenzana, V.; Beghi, R.; Civelli, R.; Trapani, S.; Migliorini, M.; Cini, E.; Zanoni, A.; Guidetti, R. Rapid determination of crucial parameters for the optimization of milling process by using visible/near infrared spectroscopy on intact olives and olive paste. *Ital. J. Food Sci.* **2017**, *29*, 357–369. [CrossRef]
39. Dupuy, N.; Galtier, O.; Le Dréau, Y.; Pinatel, C.; Kister, J.; Artaud, J. Chemometric analysis of combined NIR and MIR spectra to characterize French olives. *Eur. J. Lipid Sci. Technol.* **2010**, *112*, 463–475. [CrossRef]
40. Salguero-Chaparro, L.; Peña-Rodríguez, F. On-line versus off-line NIRS analysis of intact olives. *LWT Food Sci. Technol.* **2014**, *56*, 363–369. [CrossRef]
41. Beghi, R.; Giovenzana, V.; Marai, S.; Guidetti, R. Rapid monitoring of grape withering using visible near-infrared spectroscopy. *J. Sci. Food Agric.* **2015**, *95*, 3144–3149. [CrossRef]
42. Li, J.; Huang, W.; Zhao, C.; Zhang, B. A comparative study for the quantitative determination of soluble solids content, pH and firmness of pears by Vis/NIR spectroscopy. *J. Food Eng.* **2013**, *116*, 324–332. [CrossRef]
43. Xia, Y.; Huang, W.; Fan, S.; Li, J.; Chen, L. Effect of spectral measurement orientation on online prediction of soluble solids content of apple using Vis/NIR diffuse reflectance. *Infrared Phys. Technol.* **2019**, *97*, 467–477. [CrossRef]
44. Sánchez, M.T.; De La Haba, M.J.; Benítez-López, M.; Fernández-Novales, J.; Garrido-Varo, A.; Pérez-Marín, D. Non-destructive characterization and quality control of intact strawberries based on NIR spectral data. *J. Food Eng.* **2012**, *110*, 102–108. [CrossRef]
45. Barnaba, F.E.; Bellincontro, A.; Mencarelli, F. Portable NIR-AOTF spectroscopy combined with winery FTIR spectroscopy for an easy, rapid, in-field monitoring of Sangiovese grape quality. *J. Sci. Food Agric.* **2014**, *94*, 1071–1077. [CrossRef] [PubMed]
46. Carbas, B.; Machado, N.; Oppolzer, D.; Queiroz, M.; Brites, C.; Rosa, E.A.S.; Barros, A.I.R.N.A. Prediction of phytochemical composition, in vitro antioxidant activity and individual phenolic compounds of common beans using MIR and NIR spectroscopy. *Food Bioproc. Technol.* **2020**, *13*, 962–977. [CrossRef]
47. Tahir, H.E.; Xiaobo, Z.; Jiyong, S. Rapid determination of antioxidant compounds and antioxidant activity of Sudanese karkade (*Hibiscus sabdariffa* L.) using near infrared spectroscopy. *Food Anal. Methods* **2016**, *9*, 1228–1236. [CrossRef]
48. Shenk, J.S.; Westerhaus, M.O. Calibration the ISI Way. In *Near Infrared Spectroscopy: The Future Waves*; Davies, P.C., Williams, A.M.C., Eds.; NIR Publications: Chichester, UK, 1996; pp. 198–202.

Article

Application of ^{1}H and ^{13}C NMR Fingerprinting as a Tool for the Authentication of Maltese Extra Virgin Olive Oil

Frederick Lia [1,*], Benjamin Vella [1], Marion Zammit Mangion [2] and Claude Farrugia [1]

1 Department of Chemistry, University of Malta, 2080 MSD Msida, Malta; ben.vella.16@um.edu.mt (B.V.); claude.farrugia@um.edu.mt (C.F.)
2 Department of Physiology and Biochemistry, University of Malta, 2080 MSD Msida, Malta; marion.zammit-mangion@um.edu.mt
* Correspondence: fredericklia@gmail.com

Received: 24 April 2020; Accepted: 22 May 2020; Published: 26 May 2020

Abstract: The application of ^{1}H and ^{13}C nuclear magnetic resonance (NMR) in conjunction with chemometric methods was applied for the discrimination and authentication of Maltese extra virgin olive oils (EVOOs). A total of 65 extra virgin olive oil samples, consisting of 30 Maltese and 35 foreign samples, were collected and analysed over four harvest seasons between 2013 and 2016. A preliminary examination of ^{1}H NMR spectra using unsupervised principle component analysis (PCA) models revealed no significant clustering reflecting the geographical origin. In comparison, PCA carried out on ^{13}C NMR spectra revealed clustering approximating the geographical origin. The application of supervised methods, namely partial least squares discriminate analysis (PLS-DA) and artificial neural network (ANN), on ^{1}H and ^{13}C NMR spectra proved to be effective in discriminating Maltese and non-Maltese EVOO samples. The application of variable selection methods significantly increased the effectiveness of the different classification models. The application of ^{13}C NMR was found to be more effective in the discrimination of Maltese EVOOs when compared to ^{1}H NMR. Furthermore, results showed that different ^{1}H NMR pulse methods can greatly affect the discrimination of EVOOs. In the case of ^{1}H NMR, the Nuclear Overhauser Effect (NOESY) pulse sequence was more informative when compared to the zg30 pulse sequence, since the latter required extensive spectral manipulation for the models to reach a satisfactory level of discrimination.

Keywords: extra virgin olive oil; authentication; chemometrics; proton NMR; carbon NMR; machine learning; artificial neural networks; PLS-DA

1. Introduction

Several international organisations, including the European Union through its directives (EC No. 2568/1991 and its amendments) [1] and the International Olive Oil Council (COI/T.15/NC No. 3/Rev 6) [2], have been at the vanguard in the development of methods and establishing limits for physicochemical parameters of extra virgin olive oil (EVOO) to protect against frauds. The typical approach relies on comparison of the chemical composition with official limits, as it is expected that the presence of adulterants will modify the concentration of these constituents. Nonetheless, this procedure may be inadequate, especially for oils which are classified as 'virgin' but do not conform to official limits of certain constituents due to local climatic or soil peculiarities [3]. Furthermore, these methodologies do not address the problem of geographical traceability and tend to be rather time-consuming with a very low throughput.

During the last decade, nuclear magnetic resonance (NMR) spectroscopy has been shown to be highly effective in the study of oils of vegetable origin [4–8]. In 1999, Vlahov [9] proposed NMR spectroscopy as a new analytical tool to compete with the existing methods for studying olive oil chemistry. Among the vast applications of NMR spectroscopy to the study of EVOO, target analysis of triacylglycerides, fatty acids, unsaturated fatty chains for quantification, seed oil adulteration, and degradation of EVOOs encompass some of the techniques that could employ the use of NMR. Furthermore, NMR spectroscopy could also be extended to the study of minor constituents including phenolic compounds, sterols, and phospholipids for both detection and quantification of markers for geographical origin and cultivar information. The main methods used in NMR include ^{1}H and ^{13}C NMR spectroscopy as reviewed by a number of authors [9–14] together with ^{31}P NMR as employed by Spyros and Dais [15]. Apart from target-based analytical approaches, NMR metabolic fingerprinting [16–18] employs the use of whole NMR spectral data to classify a relevant number of samples according to their origin, harvest, and age. In most cases, fingerprinting analysis is used in conjunction with sophisticated statistical and mathematical procedures.

^{1}H NMR has been much more widely used in the field of olive oil chemistry than ^{13}C NMR. While requiring more concentrated samples than ^{1}H NMR, ^{13}C NMR spectra have a much wider radiofrequency range. Coupled with proton decoupling techniques, this leads to sharp spectra which rarely have overlapping carbon peaks, allowing easy detection of impurities and making the peaks readily interpretable. The main disadvantage in ^{13}C NMR is the long acquisition times which reduces the sample throughput, unlike ^{1}H NMR which takes around 10 min for the entire run to be completed. Preedy and Watson [19] suggest that each type of NMR spectroscopy could be used for a different type of analysis into the composition of olive oil—the ^{13}C technique is useful in characterisation of the genotype of the oil, while the ^{1}H NMR technique is more suited to geographical characterisation of the oils.

The combination of ^{1}H and ^{13}C NMR fingerprinting with multivariate analysis provides a promising approach to studying the profile of olive oils in relation to their geographical origin. The Maltese olive oil industry makes an interesting case, as the industry has only recently been regenerated using an indigenous olive stock. Considering the small state of the market, mislabeled EVOO originating from other countries sold as Maltese EVOO could severely impede the growth of the industry, with severe negative economic repercussions. Recent studies have shown that Maltese EVOOs have a significantly different phenolic composition and mineral composition [20–22]. In this study, a variety of olive oils selected from different areas around the Maltese islands and countries around the Mediterranean were studied. No data is present in the literature regarding the use of ^{13}C and ^{1}H NMR for the authentication of Maltese EVOOs. The aim of this study was to explore the use of ^{13}C NMR and ^{1}H NMR (specifically ^{1}H zg30 and ^{1}H NOESY), in conjunction with chemometrics in order to differentiate the Maltese EVOOs from other EVOOs derived from other countries within the Mediterranean region, thus developing an easy and cost-saving verification method for the origin of EVOOs from the Maltese islands ensuring olive oil chain sustainability.

2. Materials and Methods

2.1. Sample Preparation

For this preliminary study, a total of 65 extra virgin olive oil samples were collected from the Maltese islands over four harvest seasons from 2013–2016 and from other neighboring Mediterranean countries. The cultivars used in this study and their country of origin can be seen in Table S1. The samples were all taken from different oil producers to cover a representative sample of the Maltese islands in terms of pedological and microclimatic conditions, whilst also accounting for manufacturing techniques and the different presses employed. Foreign olive oils obtained were bought with a protected designation of origin in order to ensure traceability of the product. All the samples were stored at 4 °C in the absence of light prior to the analysis. The samples were preheated to 35 °C in a water bath for 1 h and mixed to ensure homogeneity. For ^{1}H NMR, 20 µL of the EVOO were placed in 5 mm NMR tubes and dissolved in 700 µL of deuterated chloroform, followed by the addition of 20 µL of deuterated DMSO and vortex mixing for 20 s. For ^{13}C NMR, 440 µL of sample was dissolved in 420 µL of deuterated chloroform without the addition of DMSO [23].

2.2. ^{1}H and ^{13}C NMR Spectra Acquisition

The analysis was performed on a model AVANCE III 500 MHz NMR spectrometer equipped with a 5 mm ^{1}H/D-BB probehead with z-gradient, automated tuning and matching accessory, and a BTO-2000 accessory for temperature control (Bruker BioSpin GmbH, Rheinstetten, Germany). Samples were measured at 300.0 K after a 5 min resting period for temperature equilibration. NMR spectra were acquired using Topspin 3.5 (Bruker). Automated tuning and matching, locking and shimming using the standard Bruker routines, ATMA (automatic tuning and matching in automatic mode), LOCK (frequency-field lock to offset the effect of the natural drift of the NMR's magnetic field B0) and TopShim, were used to optimise the NMR conditions. Samples were analysed using the zgpg30 pulse method for ^{13}C NMR, while the zg30 and NOESY 1D noesypr1d NMR pulse sequence using a standard presaturation were used for ^{1}H NMR. Every extract sample was run twice with a ^{1}H NMR standard single pulse experiment zg30 for 100 scans. The samples were run twice automatically under the control of ICON-NMR. Each run had two prior dummy scans, resulting in 65,536 data points with a resolution of 0.305 Hz acquired with an acquisition time and a relaxation delay time of 3.27 and 4 s, respectively. The 90° flip angle for free induction decay was adjusted to 10 µs. In the case of one-dimensional Nuclear Overhauser Effect spectroscopy, 100 scans were acquired, each run having two dummy scans, which resulted in 32,768 data points with a resolution of 0.489 Hz, acquired with an acquisition time and relaxation time of ~2.04 and 4 s, respectively. In the case of ^{13}C NMR, 250 scans were recorded for each sample, with an acquisition time of 21 s to allow sufficient time for complete relaxation of ^{13}C nuclei between scans. The acquisition delay was set at 2 s. The receiver gain was set at 203, and the temperature was locked at 298.0 K by means of a BTO-2000 accessory. Broadband 1 H decoupling techniques were employed. The above parameters and settings could run samples with a turnover time of 1 h and 40 min each, excluding an initial 5-min temperature equilibration period.

Prior to Fourier transformation, the free induction decays (FIDs) were zero-filled to 64 k and a 0.3 Hz line-broadening factor was applied. The chemical shifts are expressed in d scale (ppm), referenced to the residual signal of chloroform. For ^{1}H NMR, this was found at 7.24 ppm [21] whilst for ^{13}C NMR, this was found as a triplet centerd around 77.01 ppm [22]. The corrected spectra were exported as ASCII files from Topspin 3.5 (TopSpin™ version 5, Bruker, Billerica, MA, USA) and imported directly into The Unscrambler X 10.3 (CAMO Software, Oslo, Norway) for all subsequent mathematical data processing. Each spectrum was automatically binned by the software into 32,768 buckets, each bucket being 0.0072223 ppm wide. The signal-to-noise ratio was calculated using the peak at 172.8 ppm for ^{13}C NMR corresponding to C1 of the glycerol chain, which resulted in a signal-to-noise ratio of 520:1. For ^{1}H NMR, the signal-to-noise ratio was calculated using the peak at 9.70 ppm, corresponding to the aldehyde proton in hexanal, and a signal-to-noise ratio of 1.26:1 and 1.46:1 was obtained for zg30 and NOESY pulse sequences, respectively.

The spectrum obtained was subjected to different spectroscopic signal processing techniques, which were evaluated and compared. The spectra were normalised, a transformation that put all spectra on the same scale, thus eliminating the fluctuations in intensities between spectra arising from slightly different sample concentrations. Both peak normalisation and area normalisation were carried out separately on the baseline corrected spectrum. Normalisation was followed by detrending and deresolving procedures. The detrend transformation removes the effects of nonlinear trends, showing only the absolute changes in values across spectra by removing the least-squares line of best fit from the data, thus focusing only on fluctuations between data. Deresolve is a noise-reducing transformation that operates by artificially lowering the resolution of the spectra. Other treatments applied to the baseline corrected spectrum include multiplicative and orthogonal scatter corrections (MSC and OSC), and standard normal variate (SNV). MSC was corrected for scaling effects by performing a regression of a spectrum against a reference spectrum, thereby correcting the spectrum using the slope of the fit was obtained from the regression. OSC removes variance from the factors that is not related to the response, by finding directions in X that describe large variances while being orthogonal to Y and subtracting them from the data. The SNV transformation works similarly to MSC, however, it standardises each spectrum using data from the spectrum itself rather than data averaged from all the spectra. A number of derivatising procedures (1st and 2nd derivatives, Savitzky-Golay) were also carried out. The 1st derivative removes baseline effects while the 2nd derivative also removes the slope of the spectrum by measuring the change in slope, thereby sharpening spectral features. The Savitzky-Golay derivative fits a low-degree polynomial to adjacent points in a spectrum, thereby smoothing the spectrum while minimally affecting the signal-to-noise ratio.

2.3. Data Analysis

A principle component analysis (PCA) was carried out using Unscrambler X 10.3 in order to identify any gross outliers and determine any preliminary clustering reflecting the geographical origin. An inspection of the PCA loadings was carried out in order to determine whether the loadings had a spectral shape indicating that observed variation was due to the NMR spectra and not due to noise. PCA was carried out on all treated spectra to reduce all the spectral information down to seven principal components (PCs), which retained the information of the original dataset. The first PC accounted for most of the variation in the dataset, with successive principal components accounting for decreasing amounts of the variation. The resulting PC-1 vs. PC-2 plots could be examined for any clustering that might arise from each spectral pretreatment. Similarly, to PLS, PCA generates loading plots which indicate those x-values which are most responsible for the variability between the different spectra. The loading plots for the first two principal components (which explain most of the variability in the dataset) were used to determine which ppm values had the largest influence on the separation of PC-1 and PC-2. Following a PCA, supervised chemometric methods were carried out using JMP®, Version 10 (SAS Institute Inc., Cary, NC, USA), including the partial least squares discriminate analysis (PLS-DA). The whole dataset was split into two sets, termed the training and test sets (the former to build the model, the latter to validate it). In order to preserve the diversity in the training and test sets and to account for the fact that different pretreatments had to be tested, a unique sample splitting scheme was used.

In order to determine the suitability of the whole NMR spectra for discrimination of EVOOs of Maltese origin, an artificial neural network (ANN) analysis was carried out. The main advantage of a neural network model is that it can efficiently model different response surfaces due to its nonlinearity, allowing a better fit to the data given enough hidden nodes and layers, providing an accurate prediction for many kinds of data. Unlike other modeling and discriminate methods (PLS) the main disadvantage of a neural network model is that the results are not easily interpretable, due to the presence of several intermediate hidden layers. In this experiment, 25 iterations were carried out using a TanH activation function as the standard neuron activation function in JMP software. In the case of ANN, three different cross-validation techniques were employed in order to prevent model overfitting; the k-fold (CV-10), hold back (33.3%), and excluded rows (Venetian blinds). Thirty-three percent of the samples were held back from the model during holdback validation, which operates by randomly splitting the dataset into training and validation sets. Thirty-three percent of the data was thus 'held back' to form the validation set. Excluded rows holdback uses those rows that were excluded by the Venetian blinds method as the validation set. K-fold validation divides the dataset into 'k' number of subsets where each subset contains a fraction '1/k' of the data. Each of these sets is used to validate the model thereby fitting 'k' number of models. The best fitting model is presented as the final output. In this study, K-fold validation was carried out using 5 k-folds.

3. Results and Discussion

3.1. Geographical Classification of EVOO Using NMR Spectroscopy

Figure 1 and Table 1 show ^{1}H NMR signals of the major and some minor compounds together with their chemical shifts and their assignments to protons of the different functional groups [10,24–28]. Figure 2 and Table 2 show the major peaks obtained using ^{13}C NMR and identified using the literature [10,13,14,17,27–33].

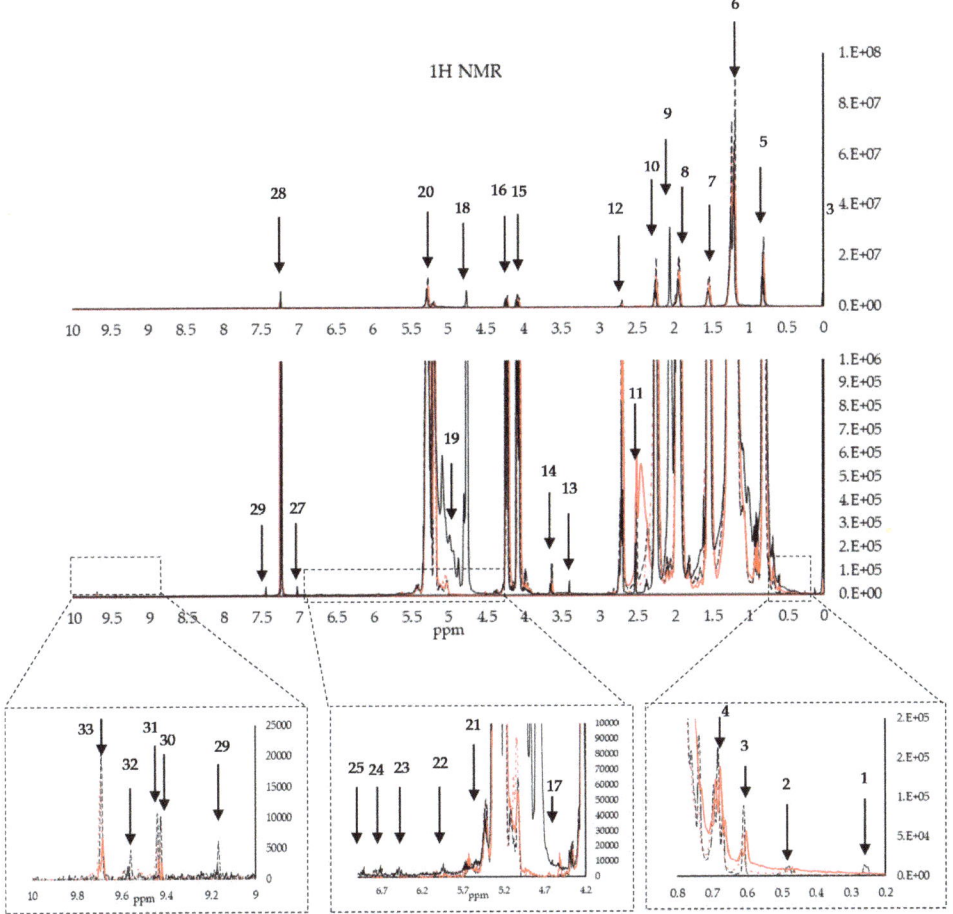

Figure 1. The major peaks of interest obtained from the nuclear magnetic resonance (NMR) of extra virgin olive oils (EVOOs) using the zg30 pulse sequence (black) and NOESY pulse sequence (red).

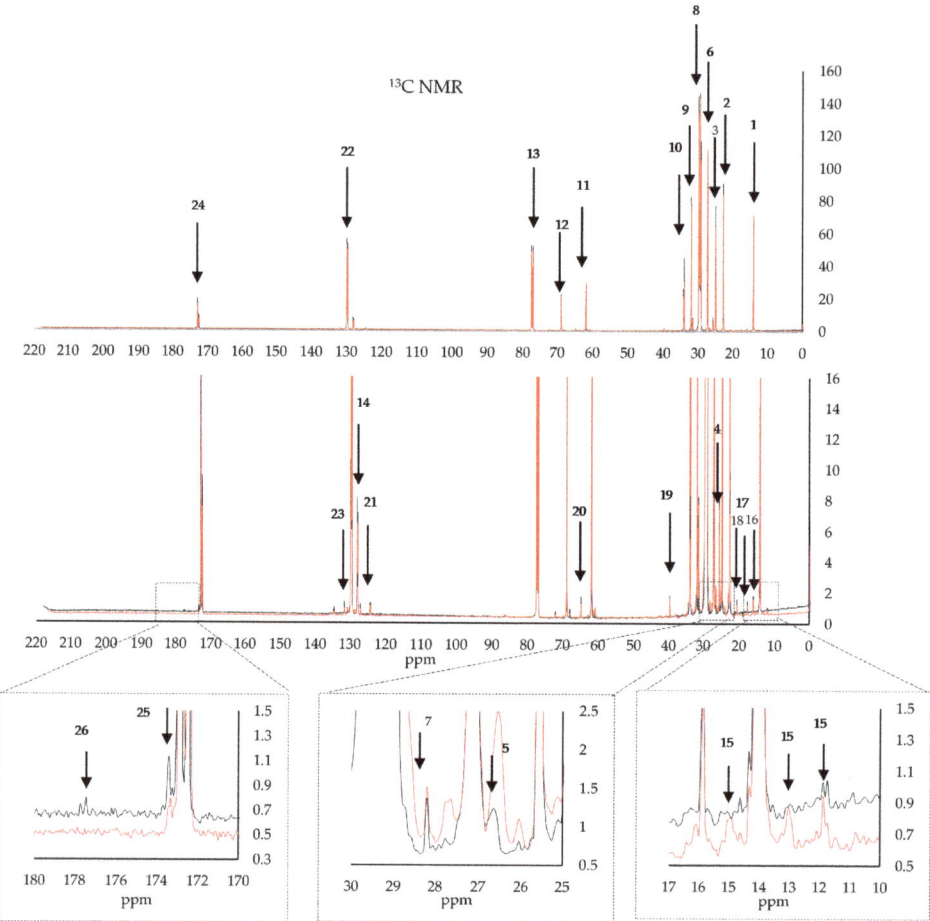

Figure 2. The major peaks of interest obtained using the ^{13}C NMR of EVOOs (black line Maltese EVOOs, red line non-Maltese EVOOs).

Whilst the chemical shifts of the major constituents are well known and easily identified, the ^{1}H and ^{13}C signals of the minor oil components are only observed when their signals do not overlap with those of the main components, and when their concentrations are high enough to be detected [11]. Minor constituents which are expected to yield NMR signals include mono- and diglycerides, sterols, tocopherols, aliphatic alcohols, hydrocarbons, fatty acids, pigments, and phenolic compounds [32]. Figure 1 shows the most common ^{1}H NMR signals of the major and some minor compounds together with their chemical shifts and their assignments to protons of the different functional groups. The main identified compounds include; cycloartenol at 0.29 and 0.54 ppm, β-sitosterol at 0.62, 0.67 ppm, stigmasterol at 0.69 ppm, wax at 0.98 ppm, squalene at 1.66 ppm, sn-1,2 diglyceryl group protons at 3.71 and 5.28 ppm, and two unknown terpenes at 4.53, 4.65, and 4.95 ppm, hexanal at 9.7 ppm, and phenolic protons at, 6.95, and 6.72 ppm. These compounds have already been observed and identified by other authors [10,11,14,17,30]. In the case of ^{13}C NMR, the minor constituents observed were restricted to chemical shifts corresponding to squalene, with a shouldering peak at 26.6 ppm and another minor peak at 28.2 ppm attributed to the allylic methylene group [26].

Table 1. Chemical shifts and the corresponding chemical functional group observed for ^{1}H NMR.

	Chemical Shift	Compound Functional Group		Chemical Shift	Compound Functional Group
1	0.29	-CH$_2$-(cyclopropanic ring) cycloartenol	17	4.53	Terpene
2	0.54	-CH$_2$-(cyclopropanic ring) cycloartenol	18	4.65	Terpene
3	0.62	-CH$_3$(C18-steroid group) β-sitosterol	19	4.95	Terpene
4	0.69	-CH$_3$(C18-steroid group) β-sitosterol	20	5.28	>CHOCOR (glyceryl group)
5	0.81	-CH$_3$(acyl group)	21	5.55	Unk 2-Tocopherols
6	1.19	-(CH$_2$)n-(acyl group)	22	5.91	-CH=CH-CH=CH-(cis, trans conjugated dienediene system)
7	1.54	-OCO-CH$_2$-CH$_2$-(acyl group)	23	6.56	-CH=CH-CH=CH-(cis, trans conjugated dienediene system)
8	1.95	-CH$_2$-CH=CH-(acyl group)	24	6.72	-Ph-H (phenolic ring)
9	2.08	-CH$_2$-CH=CH-(acyl group)	25	6.95	-Ph-H (phenolic ring)
10	2.26	CH-CH$_2$-CH=(acyl group)	26	7.02	Chloroform ^{13}C satellite
11	2.54	CH-CH$_2$-CH=(acyl group) satellite	27	7.24	Chloroform
12	2.71	CH-CH$_2$-CH=(acyl group)	28	7.44	Chloroform ^{13}C satellite
13	3.39	Unk 1-alcohol	29	9.17	Unk 4-hydrocarbon
14	3.71	-CH$_2$OCOR (glyceryl group)	30	9.46	Unk 5-hydrocarbon
15	4.10	-CH$_2$OCOR (glyceryl group)	31	9.47	Unk 5-hydrocarbon
16	4.22	-CH$_2$OCOR (glyceryl group)	32	9.58	Unk 6-hydrocarbon
			33	9.70	Hexanal

Table 2. Chemical shifts and the corresponding chemical functional group observed for ^{13}C NMR.

	Chemical Shift	Compound Functional Group		Chemical Shift	Compound Functional Group
1	14	C18(ω1) terminal carbon of fatty acyl chain	14	128.01 129.82	C9, C10 oleoyl unsaturated carbons between 2- and 1(3) of glycerol
2	22.64	C17(ω2) penultimate carbons from the fatty acyl chains	15	11.97,13.12,14.95	Unk 1 possibly being attributed to waxes
3	24.74	C3 methylenic group in β position with respect to the carbonylic group	16	15.95	Unk 2 possibly C8′a and C4′a of tocopherols
4	25.53	C11 Linoleyl Linolenyl	17	17.51	Unk 3 possibly C12′a of tocopherols
5	26.61	C8 allylic methylenes of sqaulene	18	20.47	C17(ω2) all acyl chains
6	27.12	C8 allylic carbons of oleoyl chains	19	39.68	Unk 4-C1 of tocopherol
7	28.6	C12 allylic methylenes of sqaulene	20	64.89	Unk 5 possibly C2 of elenolic acid derivative of tyrosol or hydroxytyrosol
8	29.28	C4–C7, C12–C15, C8–C15, C8–C13 methylenic groups in fatty acid central chain	21	124.40	C3′, C7′, C11′ of tocopherols
9	31.88	C16 methylenic acylic chains ω	22	131.70	C9 Linoleyl and linolenyl, C13 Linoleyl
10	33.91	C2, sn-2 acyl chains	23	134.80	C4′, C8′ of tocopherols
11	61.93	CH$_2$O-1(3) glycerol carbons of triglycerides	24	172.8	C1, sn-2 2-glycerol chain
12	68.86	CH$_2$O-2-glycerol carbon of triglycerides resonates	25	173.2	C1, sn-1,3 1(3) glycerol chain positions
13	77.39	CDCl$_3$ Solvent	26	177.92	Unk 6-COOCH$_3$ of elenolic acid

The discriminatory models for the traceability of EVOOs from the Maltese islands coupled ^{1}H and ^{13}C NMR spectroscopy with chemometrics. In order to overcome the instrumental limitation and to account for scattering and other minor variations which would hinder the performance of the classification model, different kinds of spectral pretreatments were tested and compared. A total of

10 spectral pretreatment methods were used. In each case, after pretreatment, a PCA was carried in order to dimensionally reduce the number of variables into a small set of principal components whilst retaining all the information of the larger set. PCA enabled the preliminary identification of which pretreatment offered the highest variability and possible sample grouping based on the geographical origin but also enabled the identification of outliers and noise modeling.

Figure 3 shows some of the different forms of spectral pretreatments employed and the corresponding PCA plot for the first two principal components. In the case of ^{1}H NMR, although clustering was observed in most of the spectral pretreatments, it did not fully discriminate the EVOOs of Maltese origin from those obtained from other Mediterranean countries. Only a weak clustering resembling the geographical origin was observed by using PCA. For ^{1}H NMR, the raw data was presented in Figure 3 as these were seen as the most representative data for highlighting clustering in PCA. Other spectral transformations can be viewed in the Supplementary Materials Figures S1–S3. In the case of ^{13}C NMR, the clustering obtained using OSC and SNV spectral transformations highly resembled the geographical origin of EVOO.

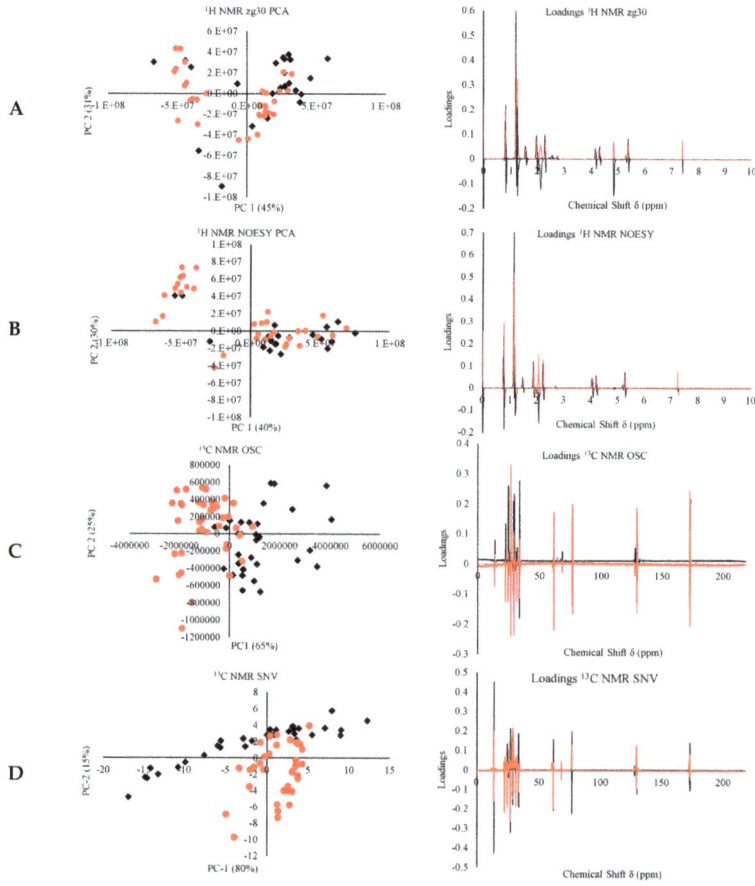

Figure 3. The principle component analysis (PCA) biplots (black boxes = Maltese red dots = non-Maltese) and loading plots for PC1 (black line) and PC2 (red line) for the untreated raw data for the zg30 (**A**) NOESY (**B**), ^{13}C NMR orthogonal scatter corrections (OSC) (**C**), and ^{13}C NMR standard normal variate (SNV) (**D**) spectra.

Inspection of the PC loadings revealed a spectral form, which suggests that the variation observed was due to the actual NMR spectra and not due to noise. In the case of zg30, it was observed that the chemical shifts observed at 0.8 and 1.2–1.25 ppm and 0.5–1.25 ppm for the NOESY experiment seem to have a larger influence on the first and second principal component separation. These observations suggest that the phytosterol content, namely β-sitosterol, campesterol, cycloartenol together with 1-eicosanol and α-tocopherol, which show chemical shifts between 0.5–1.25 ppm, have a greater influence on the variation observed along the first two principal components. In the case of zg30, other peaks observed in the 4.7–4.9 ppm range also seem to be influential, especially in the 1st PC, these peaks correspond to terpenic compounds present in EVOOs. Alonso-Salces et al., [17,30] identified three peaks at 4.57, 4.65, and 4.70 ppm, which were attributed to unknown terpenes during their study on the unsaponifiable fraction of EVOOs. For ^{13}C NMR, inspection of the PC loading plots corresponding to the previously identified chemical shifts were found to offer the most variation, with the peak at 14 ppm assigned to the terminal –CH$_3$ of all acyl chains explaining most of the variation in the SNV spectra.

3.2. Application of PLS-DA for the Discrimination of Maltese EVOOs

The Maltese and the non-Maltese samples were grouped in ascending order so that the first 30 samples would represent Maltese EVOOs whilst the rest corresponded to non-Maltese EVOOs. A Venetian blinds cross-validation method was then employed, which selected every sth sample from the data by making data splits such that all samples are left out exactly once (s = 5). This sampling method excluded 20% of the dataset so that they would be retained as the testing set. The remaining 80% of the dataset was used to build the training set. In the case of PLS-DA, an inspection of the variable importance plot (VIP) scores was carried out. Variables having a smaller VIP than 0.8 were removed, and an adjusted PLS model was built after the removal of these variables. The goodness of fit of the adjusted model was evaluated and compared to the original model. Table 3 shows the accuracy (% correct classification during training) and the precision (% correct classification during testing) obtained on using different spectral pretreatments for the two NMR methods. For the zg30 NMR spectra obtained after deresolve, SNV and quantile normalisation showed the best model performance with a % accuracy ranging from 93.1–87.9% and % predictability ranging from 72.7–81.8%, whilst for the NOESY experiment, spectra treated using normalisation and Savitzky-Golay showed the best performance with an accuracy of 94.8% and predictability of 90.9%. In the case of the zg30 experiment, all the spectral pretreatments showed an improvement in the % predictability when compared to the raw data, whilst in the NOESY experiment, spectra treated using SNV and detrending functions showed a lower % predictability and % accuracy when compared to actual nonpretreated raw data. This observation suggests that, in the case of NOESY, the signal suppression of the major peaks improves the signal to noise ratio, and the resulting spectra can be used without the need of extensive pretreatments. Results obtained by Longobardi et al., [18] showed that the presaturation of the dominating lipid signals resulted in an increased receiver gain which in turn resulted in a signal-to-noise gain close to 10 compared to the zg30 spectra. In the case of ^{13}C NMR, higher rates of accuracy and predictability were observed when compared to ^{1}H NMR methods with a % predictability ranging from 66.7–100%, with OSC reaching 100% correct classification in both the training and validation sets. The higher rates of predictability of ^{13}C NMR spectra were attributed to a higher signal-to-noise ratio, less coupling interactions resulting in a cleaner signal, proof of this is the % predictability of the raw untreated ^{13}C spectra with respect to ^{1}H spectra.

The next step was to build another PLS model, this time using only variables which had a VIP score > 0.8. Table 3 also shows the results obtained by using the adjusted PLS model for ^{13}C and ^{1}H NMR. An improvement in the overall % accuracy and predictability of the model. Furthermore, the models obtained using only VIP > 0.8 variables showed an increase in both %X and %Y explained, and a higher % accuracy and % precision indicating enhanced model performance. In the case of the zg30 experiment, it was found that normalised spectra and Savitzky-Golay derived spectra had the optimal performance, whilst detrended and SNV spectra had optimal performance when the whole data set was used. In the case of the NOESY experiment, the models obtained using VIP > 0.8 showed an increase in the performance when compared to those obtained with whole data.

Table 3. The PLS-DA analysis on both the entire (**a**) ^{1}H NMR and (**b**) ^{13}C NMR spectra and selected variables having a VIP > 0.8 for the different spectral pretreatments. The results obtained on the training dataset are given in terms of % accuracy of correct classification whilst for the testing data set these are given in terms of % predictability of correct classification.

(a)

	zg30 ^{1}H NMR			
Pretreatment	Whole Spectrum		VIP > 0.8	
	% Accuracy	% Predictability	% Accuracy	% Predictability
Raw	77.59	27.27	82.76	45.45
Normalised	91.38	63.64	94.83	90.91
Q Norm	93.10	72.73	94.83	72.73
Detrend	70.69	36.36	68.97	36.36
Deresolve	87.93	63.64	82.76	45.45
SNV	93.10	81.82	60.34	36.36
MSC	91.38	81.82	67.24	63.64
OSC	72.41	45.45	94.83	72.73
Savitzky-Golay	74.14	54.55	98.28	90.91
1st Derivative	68.97	45.45	94.83	72.73
2nd Derivative	77.59	63.64	93.10	63.64

	NOESY ^{1}H NMR			
Pretreatment	Whole Spectrum		VIP > 0.8	
	% Accuracy	% Predictability	% Accuracy	% Predictability
Raw	82.76	45.45	93.10	75.00
Normalised	94.83	90.91	94.83	83.33
Q Norm	94.83	72.73	96.55	91.67
Detrend	68.97	36.36	74.14	66.67
Deresolve	82.76	45.45	94.83	83.33
SNV	60.34	36.36	70.69	75.00
MSC	67.24	63.64	68.97	75.00
OSC	94.83	72.73	96.55	83.33
Savitzky-Golay	98.28	90.91	89.66	75.00
1st Derivative	94.83	72.73	93.10	91.67
2nd Derivative	93.10	63.64	93.10	91.67

(b)

	^{13}C NMR			
Pretreatment	Whole Spectrum		VIP > 0.8	
	% Accuracy	% Predictability	% Accuracy	% Predictability
Raw	100.00	73.33	100.00	80.00
Normalised	100.00	86.68	94.64	100.00
Q Norm	100.00	80.00	100.00	100.00
Detrend	100.00	66.67	100.00	100.00
Deresolve	78.57	73.33	91.07	100.00
SNV	100.00	86.67	100.00	100.00
MSC	100.00	86.67	100.00	100.00
OSC	100.00	100.00	100.00	100.00
Savitzky-Golay	100.00	73.33	100.00	100.00
1st Derivative	100.00	80.00	100.00	100.00
2nd Derivative	100.00	86.67	100.00	100.00

These observations indicate that different spectral pretreatments are affected differently to variable selection techniques since each one of them attempts to maximise spectral variations and corrections, therefore, removal of a small number of predictors can have a devastating effect on the model performance. In the case of ^{13}C NMR, variable selection greatly improved the discrimination with most of the pretreated spectra reaching 100% accuracy and predictability. The noticeable increase in the model performance has been attributed to the removal of redundant variables which correct for overfitting by excluding noise variables from the data, therefore, preventing them from affecting the model. Reducing the number of variables around which the model is built also increases the model's reliability.

3.3. Whole ^{1}H and ^{13}C-NMR Modeling Using Feed-Forward Predictive Artificial Neural Networks

The use of feed-forward predictive neural networks on the NMR data as a method for classification was assessed using three different forms of validation, namely 33.3% data holdback, CV-10 k-fold, and excluded row validation. Since ANNs are more powerful than any other classification method in terms of their flexibility and noise insensitivity, the algorithm was fitted on the training set using the entire NMR spectrum without any form of variable selection. Table 4a,b shows % accuracy and % predictability for the different forms of cross-validation carried out on different spectral pretreatments of ^{13}C NMR and ^{1}H NMR under the zg30 and NOESY NMR spectra. In general, contrary to what was observed in PLS-DA, it was observed that models obtained under ^{1}H NMR models had higher rates of classification when compared to ^{13}C NMR. Similarly, to what was observed in PLS-DA, raw data derived from the NOESY experiment had a higher model performance throughout the three different validation methods used when compared to the zg30 experiment. With respect to validation it was observed that, irrelevant to the spectrum used, the 33% holdback cross-validation resulted in overfitted models which were identified as spectral transformations that had very good training models but failed to predict new samples with the exception of MSC. In general, the best performing cross-validation method was the excluded row validation followed by k-fold validation. This could be attributed to the fact that these cross-validation methods are not completely random as the 33% holdback. In the case of the excluded row validation the samples were selected in such a way that, the groups contained approximately equal amounts of local and foreign EVOOs in the training stage. Thus, the models obtained where equally capable of recognising and predicting local and foreign EVOOs.

Table 4. Application of artificial neural network (ANN) on the (a) ^{1}H NMR and (b) ^{13}C NMR data using three forms of cross-validation.

Pretreatment	ANN					
	Holdback		CV-10		Excluded Row	
	Training	Validation	Training	Validation	Training	Validation
(a)						
zg30 ^{1}H NMR						
Raw	81.03	81.82	96.55	81.82	86.21	90.91
Normalised	94.83	81.82	94.83	100	81.03	63.64
Q Norm	98.28	90.91	98.28	90.91	82.76	81.82
Detrend	77.59	54.55	91.38	90.91	75.86	45.45
Deresolve	91.38	90.91	93.1	90.91	79.31	81.82
SNV	96.55	90.91	98.28	100.00	74.14	72.73
MSC	98.28	100.00	96.55	90.91	93.10	90.91
OSC	77.59	36.36	89.66	63.64	84.48	81.82
2nd Derivative	96.55	90.91	98.28	100.00	86.21	45.45
1st Derivative	84.48	90.91	98.28	90.91	81.03	63.64
Savitzky-Golay	91.38	81.82	98.28	90.91	85.00	90.91
NOESY ^{1}H NMR						
Raw	93.33	91.67	93.33	91.67	93.33	91.67
Normalised	95.00	91.67	95.00	100.00	93.33	100.00
Q Norm	98.33	100.00	98.33	100.00	93.33	100.00
Detrend	70.00	83.33	96.67	91.67	95.00	83.33
Deresolve	96.67	100.00	96.67	100.00	91.67	83.33
SNV	93.33	91.67	98.33	100.00	98.33	100.00
MSC	93.33	100.00	93.33	83.33	95.00	91.67
OSC	81.67	75.00	91.67	75.00	90.00	91.67
2nd Derivative	96.67	100	98.33	100.00	96.67	91.67
1st Derivative	93.33	91.67	98.33	100.00	93.33	100.00
Savitzky-Golay	88.33	100	98.33	100.00	90.00	91.67

Table 4. *Cont.*

Pretreatment	ANN					
	Holdback		CV-10		Excluded Row	
	Training	Validation	Training	Validation	Training	Validation
(b)						
^{13}C NMR						
Raw	83.93	60.00	100.00	80.00	92.62	80.00
Normalised	91.07	46.67	100.00	100.00	100.00	100.00
Q Norm	78.57	80.00	98.21	86.67	98.21	100.00
Detrend	91.07	80.00	100.00	80.00	98.21	86.67
Deresolve	78.57	73.33	100.00	40.00	87.50	86.67
SNV	94.64	86.67	98.21	93.33	92.86	100.00
MSC	80.36	53.33	100.00	73.33	100.00	86.67
OSC	75.00	66.67	75.00	66.67	100.00	73.33
2nd Derivative	85.71	40.00	100.00	66.67	96.43	86.67
1st Derivative	94.64	73.33	100.00	46.67	100.00	73.33
Savitzky-Golay	83.93	40.00	100.00	73.33	83.93	93.33

4. Conclusions

It was shown that different NMR methods in conjunction with chemometric methods provided a new insight in the identification of Maltese EVOOs. From the preliminary assessment using only unsupervised PCA models, no significant clustering was observed, and this was attributed to the high levels of similarity between the two classes of EVOOs studied, therefore, this method was deemed to be unsatisfactory when it comes to discrimination of geographical origin. The application of supervised methods of classification, namely PLS-DA and ANN, were shown to be highly effective in discriminating local and nonlocal EVOO samples. The use of the variable selection methods significantly increased the effectiveness of PLS-DA models in discriminating Maltese EVOOs. ANN models were also shown to offer similar classification rates to PLS-DA models and thus they corroborate the results obtained. Results showed that different NMR pulse methods can greatly affect the discrimination of EVOOs. The most informative method was ^{13}C NMR, which resulted in a cleaner spectrum which was void of coupling, followed by the ^{1}H NOESY pulse sequence, in which suppression of strong signals greatly improved the signal-to-noise ratio when compared to the zg30 ^{1}H NMR spectra. NMR data acquired using the zg30 pulse sequence required an extensive spectral elaboration in order to obtain a comparable model performance to that of ^{1}H NOESY and ^{13}C NMR. It was concluded that apart from the initial and running costs of the instrumentation, NMR proved to be a cheap and reliable technique for the discrimination of Maltese EVOOs from non-Maltese EVOOs. Whilst ^{13}C NMR was very successful in the discrimination of Maltese EVOOs, the long acquisition time proved to be unsatisfactory for a high throughput analysis and thus it is proposed to be used as a confirmatory method for the identification of origin.

Supplementary Materials: The following are available online at http://www.mdpi.com/2304-8158/9/6/689/s1, Table S1: The cultivars used in this study and their country of origin. Figure S1. The principle component analysis and loading plots for different ^{13}C NMR spectra. Figure S2. The principle component analysis biplots and loading plots for different ^{1}H zg30 NMR spectra. Figure S3. The principle component analysis biplots and loading plots for different ^{1}H NOESY NMR spectra.

Author Contributions: F.L., data acquisition, research paper conceptualisation, methodology, software, validation, formal analysis, investigation, data curation, writing—original draft preparation, writing—review and editing, and funding acquisition; B.V., data acquisition, methodology, software, validation, writing—review and editing; M.Z.M., conceptualisation, writing—original draft preparation, writing—review and editing, and supervision; C.F., conceptualisation, software, writing—original draft preparation, formal analysis supervision, and project administration. All authors have read and agreed to the published version of the manuscript.

Funding: This research was funded by the Malta Government Scholarships Post-Graduate Scheme for 2014 (MGSS-PG 2014).

Acknowledgments: Robert Borg for his constant support and training on the NMR spectrometer present at the University of Malta.

Conflicts of Interest: The authors declare no conflict of interest.

References

1. European Community Commission Regulation (EEC) no. 2568/1991 on the characteristics of olive and olive pomace oils and their analytical methods. *Off. J. Eur. Communities* **1991**, *L248*, 1–83.
2. International Olive Council (IOC). Trade Standard Applying to Olive Oils and Olive-Pomace Oils, COI/T.15/NC No. 3/Rev 6. 2011. Available online: http://www.internationaloliveoil.org (accessed on 11 February 2020).
3. Ceci, L.N.; Carelli, A.A. Relation between oxidative stability and composition in Argentinian olive oils. *J. AOCS* **2010**, *87*, 1189–1197. [CrossRef]
4. Frankel, E.N. Chemistry of extra virgin olive oil: Adulteration, oxidative stability, and antioxidants. *J. Agric. Food Chem.* **2010**, *58*, 5991–6006. [CrossRef] [PubMed]
5. Canabate-Diaz, B.; Segura Carretero, A.; Fernandez-Gutierrez, A.; Belmonte Vega, A.; Garrido Frenich, A.; Martinez Vidal, J.L. Separation and determination of sterols in olive oil by HPLC-MS. *Food Chem.* **2007**, *102*, 593–598. [CrossRef]
6. Murkovic, M.; Lechmer, S.; Pietzka, A.; Bratacos, M.; Katzoiamnos, E. Analysis of minor components in olive oil. *J. Biochem. Biophys. Methods* **2004**, *61*, 155–160. [CrossRef]
7. Suárez, M.; Macià, A.; Romero, M.P.; Motilva, M.J. Improved liquid chromatography tandem mass spectrometry method for the determination of phenolic compounds in virgin olive oil. *J. Chromatogr. A* **2008**, *1214*, 90–99. [CrossRef]
8. Morales, M.T.; Luna, G.; Aparicio, R. Comparative study of virgin olive oil sensory defects. *Food Chem.* **2005**, *91*, 293–301. [CrossRef]
9. Vlahov, G. Application of NMR to the study of olive oils. *Prog. Nucl. Magn. Reson. Spectrosc.* **1999**, *35*, 341–357. [CrossRef]
10. Sacchi, R.; Patumi, M.; Fontanazza, G.; Barone, P.; Fiordiponti, P.; Mannina, L.; Segre, A.L. A high-field ^{1}H nuclear magnetic resonance study of the minor components in virgin olive oils. *J. AOCS* **1996**, *23*, 747–758.
11. Guillen, M.D.; Ruiz, A. High resolution ^{1}H nuclear magnetic resonance in the study of edible oils and fats. *Trends Food Sci. Technol.* **2001**, *12*, 328–338. [CrossRef]
12. Hidalgo, F.J.; Gómez, G.; Navarro, J.L.; Zamora, R. Oil stability prediction by high-resolution ^{13}C nuclear magnetic resonance spectroscopy. *J. Agric. Food Chem.* **2002**, *50*, 5825–5831. [CrossRef] [PubMed]
13. Mannina, L.; Marini, F.; Gobbino, M.; Sobolev, A.P.; Capitani, D. NMR and chemometrics in tracing European olive oils: The case study of Ligurian samples. *Talanta* **2010**, *80*, 2141–2148. [CrossRef] [PubMed]
14. Mannina, L.; Patumi, M.; Proietti, N.; Bassi, D.; Segre, A.L. Geographical characterization of Italian extra virgin olive oils using high field ^{1}H-NMR spectroscopy. *J. Agric. Food Chem.* **2001**, *49*, 2687–2696. [CrossRef] [PubMed]
15. Spyros, A.; Dais, P. Application of ^{31}P NMR spectroscopy in food Analysis, Quantitative determination of the mono- and di-glyceride composition of olive oils. *J. Agric. Food Chem.* **2000**, *48*, 802–805. [CrossRef] [PubMed]
16. Rezzi, S.; Axelson, D.E.; Heberger, K.; Reniero, F.; Mariani, C.; Guillou, C. Classification of olive oils using high throughput flow ^{1}H-NMR fingerprinting with principal component analysis, linear discriminant analysis and probabilistic neural networks. *Anal. Chim. Acta* **2005**, *55*, 13–24. [CrossRef]
17. Alonso-Salces, R.M.; Héberger, K.; Holland, M.V.; Moreno-Rojas, J.M.; Mariani, C.; Bellan, G.; Guillou, C. Multivariate analysis of NMR fingerprint of the unsaponifiable fraction of virgin olive oils for authentication purposes. *Food Chem.* **2010**, *118*, 956–965. [CrossRef]
18. Longobardi, F.; Ventrella, A.; Napoli, C.; Humpfer, E.; Schütz, B.; Schäfer, H.; Sacco, A. Classification of olive oils according to geographical origin by using ^{1}H NMR fingerprinting combined with multivariate analysis. *Food Chem.* **2012**, *130*, 177–183. [CrossRef]
19. Preedy, V.R.; Watson, R.R. *Olives and Olive Oil in Health and Disease Prevention*; Elsevier Academic Press: Amsterdam, The Netherlans, 2010.
20. Lia, F.; Farrugia, C.; Zammit-Mangion, M. Application of Elemental Analysis via Energy Dispersive X-Ray Fluorescence (ED-XRF) for the Authentication of Maltese Extra Virgin Olive Oil. *Agriculture* **2020**, *10*, 71. [CrossRef]

21. Lia, F.; Farrugia, C.; Zammit-Mangion, M. A First Description of the Phenolic Profile of EVOOs from the Maltese Islands Using SPE and HPLC: Pedo-Climatic Conditions Modulate Genetic Factors. *Agriculture* **2019**, *9*, 107. [CrossRef]
22. Lia, F.; Formosa, J.P.; Zammit-Mangion, M.; Farrugia, C. The First Identification of the Uniqueness and Authentication of Maltese Extra Virgin Olive Oil Using 3D-Fluorescence Spectroscopy Coupled with Multi-Way Data Analysis. *Foods* **2020**, *9*, 498. [CrossRef]
23. Merchak, N.; Bacha, E.L.; Bou Khouzam, R.; Rizk, T.; Akoka, S.; Bejjani, J. Geoclimatic, morphological, and temporal effects on Lebanese olive oils composition and classification: A ^{1}H NMR metabolomic study. *Food Chem.* **2017**, *217*, 379–388. [CrossRef] [PubMed]
24. Hoffman, R.E. Standardization of chemical shifts of TMS and solvent signals in NMR solvents. *Magn. Reson. Chem.* **2006**, *44*, 606–616. [CrossRef] [PubMed]
25. Sacchi, R.; Addeo, F.; Paolillo, L. ^{1}H and ^{13}C NMR of virgin olive oil. An overview. *Magn. Reson. Chem.* **1997**, *35*, 133–145. [CrossRef]
26. Nam, A.; Bighelli, A.; Tomi, F. Quantification of Squalene in Olive Oil Using ^{13}C Nuclear Magnetic Resonance Spectroscopy. *Magnetochemistry* **2017**, *3*, 34. [CrossRef]
27. Sacco, A.; Brescia, M.A.; Liuzzi, V.; Reniero, F.; Guillou, C.; Ghelli, S.; van der Meer, P. Characterization of Italian olive oils based on analytical and nuclear magnetic resonance determinations. *JAOCS* **2000**, *77*, 619–625. [CrossRef]
28. Vlahov, G.; Del Re, P.; Simone, N. Determination of geographical origin of olive oils using ^{13}C nuclear magnetic resonance spectroscopy. I—Classification of olive oils of the Puglia region with denomination of protected origin. *J. Agric. Food Chem.* **2003**, *51*, 5612–5615. [CrossRef]
29. Shaw, A.D.; di Camillo, A.; Vlahov, G.; Jones, A. Discrimination of the variety and region of origin of extra virgin olive oil using ^{13}C NMR and multivariate calibration with variable reduction. *Anal. Chim. Acta* **1997**, *348*, 357–374. [CrossRef]
30. Alonso-Salces, R.M.; Moreno-Rojas, J.M.; Holland, M.V.; Reniero, F.; Guillou, C.; Heberger, K. Virgin Olive Oil Authentication by Multivariate Analyses of ^{1}H NMR Fingerprints and ^{13}C and ^{2}H Data. *J. Agric. Food Chem.* **2010**, *58*, 5586–5596. [CrossRef]
31. McKenzie, J.M.; Koch, K.R. Rapid analysis of major components and potential authentication of South African olive oils by quantitative ^{13}C nuclear magnetic resonance. *S. Afr. J. Sci.* **2004**, *100*, 349–354.
32. Harwood, J.L.; Aparicio, R. *Handbook of Olive Oil: Analysis and Properties*; Aspen. Henna: Gaithersburg, MD, USA, 2000.
33. D'Imperio, M.; Dugo, G.; Alfa, M.; Mannina, L.; Segre, A.L. Statistical analysis on Sicilian olive oils. *Food Chem.* **2007**, *102*, 956–965. [CrossRef]

© 2020 by the authors. Licensee MDPI, Basel, Switzerland. This article is an open access article distributed under the terms and conditions of the Creative Commons Attribution (CC BY) license (http://creativecommons.org/licenses/by/4.0/).

Article

Physical and Thermal Evaluation of Olive Oils from Minor Italian Cultivars

Maria Paciulli [1,*], Graziana Difonzo [2], Paola Conte [3], Federica Flamminii [4], Amalia Piscopo [5] and Emma Chiavaro [1]

1. Department of Food and Drug, University of Parma, Parco Area delle Scienze 27/A, 43124 Parma, Italy; emma.chiavaro@unipr.it
2. Department of Soil Plant and Food Sciences, University of Bari "Aldo Moro", Via Amendola 165/A, 70126 Bari, Italy; graziana.difonzo@uniba.it
3. Department of Agriculture, University of Sassari, Viale Italia 39/A, 07100 Sassari, Italy; pconte@uniss.it
4. Faculty of Bioscience and Technology for Agriculture, Food and Environment, University of Teramo, 64100 Teramo, Italy; fflamminii@unite.it
5. Department of AGRARIA, University Mediterranea of Reggio Calabria, 89124 Reggio Calabria, Italy; amalia.piscopo@unirc.it
* Correspondence: maria.paciulli@unipr.it; Tel.: +39-0521-905891

Abstract: Authentication of extra virgin olive oils is a key strategy for their valorization and a way to preserve olive biodiversity. Physical and thermal analysis have been proposed in this study as fast and green techniques to reach this goal. Thirteen extra virgin olive oils (EVOOs) obtained from minor olive cultivars, harvested at three different ripening stages, in four Italian regions (Abruzzo, Apulia, Sardinia, and Calabria) have been studied. Thermal properties, viscosity and color, as influenced by fatty acid composition and chlorophyll content, have been investigated. The thermal curves of EVOOs, obtained by differential scanning calorimetry, were mostly influenced by the oleic acid content: a direct correlation with the cooling and heating enthalpy and an indirect correlation with the cooling transition range were observed. The minor fatty acids, and particularly arachidic acid, showed an influence, mostly on the heating thermograms. Viscosity and color showed respectively a correlation with fatty acids composition and chlorophyll content, however they didn't result able to discriminate between the samples. Thanks to the principal component analysis, the most influencing thermal parameters and fatty acids were used to cluster the samples, based on their botanical and geographical origin, resulting instead the harvesting time a less influential variable.

Keywords: extra virgin olive oil; authenticity; biodiversity; differential scanning calorimetry; color; chlorophyll; harvesting time; geographical origin; botanical origin; principal component analysis

1. Introduction

Extra virgin olive oil (EVOO)—which is considered an essential component of the Mediterranean diet, as well as its main source of fats—is appreciated all over the world for its nutritional value and associated health benefits [1,2]. When talking about EVOO, however, it should be considered that there is a wide variety of oils on the market that are often characterized by different quality standards and sensory profiles [3]. In recent years, the increasing demand for olive oil has led to the rapid spread of high-density and super-high-density olive plantations that, although only possible using a limited number of cultivars, maximize productivity and efficiency, providing a more standardized product with an affordable selling price [3–5]. At the same time, the main producing countries have vigorously implemented a policy of using as many local cultivars as possible, aiming to preserve olive tree biodiversity and diversify and promote sensory specificity and high-quality local olive oil production [3,5]. This trend has also received the endorsement of the European Union (EU) that, as far back as 1992, introduced the quality trademark Protected Designation of Origin (PDO) to protect and promote typical foods with strong roots in a specific geographic region [6]. Conservation of biological diversity is, in fact, the best tool to ensure species survival, through their adaptability to new environmental conditions and climate change and, in turn, to guarantee long-term sustainability of the entire supply chain [7]. In this scenario, Italy and its very rich olive germplasm—estimated to include about 800 cultivars—play a dominant role, not only in the preservation of olive biodiversity, but also in the production of high-quality olive oils with strong sensory specificity [8,9]. In recent years, several studies have focused on the rediscovery and valorization of minor local Italian cultivars in an attempt to provide valuable genetic resources to be used as strategic elements to increase the sustainability of the future of oil production, pursuing, at the same time, enrichment and diversification of EVOOs to be placed on the market [5,7,9–13].

The authentication of extra virgin olive oils represents a key strategy for their valorization and diversification. Traditionally the traceability of extra virgin olive oils involves their chemical characterization, which is influenced by genotype and different agronomic, environmental, and technological factors [14].

Closely related to chemical composition, but less debated in the literature, is the physical and thermal characterization of olive oils, which could be considered of large interest for consumers and industries.

Differential scanning calorimetry (DSC) has been proposed as an alternative and reproducible method for olive oils identification, through the study of their thermal behavior upon cooling and heating [15]. This technique has been successfully applied in the field of olive oil with the aim to discriminate between commercial categories [16], oxidative status [17], agronomic practices [18], or to detect fraudulent mixtures with other vegetable oils [19]. Some studies have also applied DSC to study the authentication and traceability of extra virgin olive oils by applying chemometric data processing. Chatziantoniou and co-workers [20] successfully determined the botanical origin and geographical origin of six monovarietal extra virgin olive oils originating from four geographical regions of Greece, by applying linear discriminant analysis (LDA) on the data obtained from DSC heating and cooling profiles. DSC in combination with principal component analysis (PCA) was applied to identify EVOO from different Mediterranean countries, revealing how the thermogram obtained upon heating contains important information for sample characterization [21]. An approach based on HPLC-DSC in combination with partial least-square (PLS) regression was used to clarify the influence of triacylglycerol composition on the shape of the cooling curves of EVOOs, to a subsequent authentication of the olive oils [22]. DSC exhibits some advantages compared to the classical analytical methods as it is rapid, does not require sample preparation or solvent utilization, and has a reduced environmental impact.

The measurement of oil viscosity is essential at an industrial level for the selection of proper equipment, such as settling and centrifugation devices, including pumps, pipes, filtration systems, etc. Moreover, from the sensorial point of view, the viscosity can be associated with the term 'fluidity', where oil with low viscosity means a higher fluidity. Although this subject is not included in the official method [23], the differences perceived between samples can be linked with the oil fatty acid composition [24].

Color is a basic criterion affecting consumer preferences, although the European Union does not require its measurement for the assessment of the virgin olive oil quality. [23] The green shades of olive oil are strictly related to the olive fruit pigments, especially chlorophylls, that are transferred to the oil during the extraction process [25]. Their composition changes during the olive ripening time, influencing both drupe and oils' color [26]. Olive oil pigments have also been proposed as markers of olive oil's genetic and environmental make-up [27].

The aim of this work was to analyze thermal profiles and color of thirteen monovarietal extra virgin olive oils, obtained from minor autochthonous Italian olive cultivars, at three different ripening stages, to evaluate the potentiality of these two fast and green methods to differentiate samples based on cultivar-environment-agronomic practice interaction, in relation to the FA and chlorophyll composition.

This approach can be strategic to create a unique and recognizable hallmark for authentication and traceability of extra virgin olive oils from minor Italian cultivars, with the final goal to pursue their valorization and preserve their biodiversity.

2. Materials and Methods

2.1. Plant Material

Drupes of 13 minor olive Italian cultivars, from 4 Italian regions, were harvested in the 2017 harvest seasons.

The selected cultivars were: Tortiglione (TOR), Dritta (DR) and Gentile dell'Aquila (GEN) from Abruzzo; Sivigliana da olio (SIV), Semidana (SEM) and Corsicana da olio (COR) from Sardinia; Cima di Melfi (CM), Oliva Rossa (OR) and Bambina (BAM) from Apulia; the two clones Ottobratica Cannavà (OTT) and Ottobratica Calipa (OTTC), Tonda di Filogaso (TDF), Ciciarello (CIC) from Calabria.

Temperature data from 2017, in each olive production area, are reported in Table 1.

Table 1. Temperatures (°C) recorded in 2017 in the selected Italian provinces [1].

Province	Cultivar	Maximum	Minimum	Average
L'Aquila	TOR, DR	18.0	5.2	11.6
Teramo	GEN	19.9	8.4	14.2
Sassari	SIV, SEM, COR	22.2	11.5	16.9
Bari	CM, OR, BAM	21.4	10.5	16.0
Reggio Calabria	OTT, OTTC, TDF, CIC	23.1	15.9	19.5

[1] Source: Italian Ministry of Agriculture, Food and Forestry [28]. Tortiglione (TOR), Dritta (DR), Gentile dell'Aquila (GEN), Sivigliana da olio (SIV), Semidana (SEM), Corsicana da olio (COR), Cima di Melfi (CM), Oliva Rossa (OR), Bambina (BAM), Ottobratica Cannavà (OTT), Ottobratica Calipa (OTTC), Tonda di Filogaso (TDF), Ciciarello (CIC).

All the olive trees were located in commercial orchards and grown traditionally. Ten kilograms of drupes were sampled from ten different olive trees, every 15 days (Sampling 1 (t1); Sampling 2 (t2); Sampling 3 (t3)), starting around the middle of October (±one week) at the physiological maturity stage, defined at about 50–70% véraison of the fruits, according to the growers harvesting experience, and as confirmed by the maturity index assessment, as reported by Alamprese et al. [29], on the same olive samples. For each sampling, the drupes were divided into 3 aliquots (around 3 kg each), representing the biological replicates. For each harvesting time, the olive drupes were collected from the same trees and stored at refrigerated temperature overnight, before extraction. VOOs were extracted starting from each cultivar at each harvesting time.

For olive oil extraction, the drupes were milled with a hammer mill. The obtained paste was malaxated at a temperature below 20–25 °C for 30 min and pressed using a hydraulic press (pressure up to 200 bar) in a small olive oil press mill Mini 30 system (Agrimec Valpesana, Firenze, Italy). After centrifugation and filtration through paper, olive oils were then stored in dark glass bottles at room temperature [30].

2.2. Fatty Acids Composition

The fatty acid composition was determined after sample transesterification with KOH 2N in methanol [23,31] using a gas-chromatograph system composed of an Agilent Technologies 7890 (Agilent Technologies Inc., Santa Clara, CA, USA), equipped with an FID detector (set at 220 °C) and an SP™ 2340 fused silica capillary column (Supelco, Bellefonte, PA, USA), 60 m length × 0.25 mm i.d. and 0.20 µm film thickness. The temperature of the split injector was 210 °C, with a splitting ratio of 1:100; the detector temperature was 220 °C. The oven temperature was programmed as follows: at the very beginning, the temperature was set at 160 °C then gradually raised to 240 °C. Helium was used as the carrier gas at a flow of 1 mL min^{-1}. The identification of each fatty acid was carried out by comparing the retention time with that of the corresponding standard methyl ester (Sigma-Aldrich,

St. Louis, MO, USA). The amount of single fatty acids was expressed as area % with respect to the total area [23,31].

2.3. Thermal Analysis

EVOO samples (8–10 mg) were weighed in non-hermetic aluminum pans and analyzed by differential scanning calorimetry with a DSC Q100 (TA Instruments, New Castle, DE), following the method of Cerretani et al. [32]. Indium (melting temperature 156.6 °C, ΔHf = 28.45 J/g) and n-dodecane (melting temperature −9.65 °C, ΔHf = 216.73 J/g) were used to calibrate the instrument and an empty pan was used as reference. Oil samples were equilibrated at 30 °C for 8 min and then cooled at −80 °C at the rate of 2 °C/min, equilibrated at −80 °C for 8 min and then heated from −80 to 30 °C at 2 °C/min. Dry nitrogen was purged in the DSC cell at 50 cm^3/min. DSC curves were analyzed with Universal Analysis Software (Version 3.9A, TA Instruments) to obtain the enthalpy change for transition (ΔH, J/g), onset temperature of transition (Ton,°C), offset temperature of transition (Toff,°C), and peak temperature at the maximum (Tp) for the main events of cooling and heating transitions (p1c and p2c, p1h, p2h, and p3h, °C). The range of transition was calculated as the temperature difference between Ton and Toff for both the cooling and heating transitions.

2.4. Viscosity Measurement

Measurements were made using an Advanced Rheometric Expansion System (ARES, Rheometrics (Co)). The viscosity value, in mPas, was calculated on the basis of the speed (100 s^{-1}) and the geometry of the probe (Couette cell geometry). Temperature (25 °C) was controlled with a water bath connected to the rheometer. The experiment was carried out using 15 mL of sample. Shear stress was plotted as a function of shear rate using the Orchestrator™® software package and the viscosity (μ) value was obtained from Newton's law (Equation (1)).

$$\sigma = \mu \dot{\gamma} \quad (1)$$

where σ is shear stress (mPa), ẏ is the shear rate (1/s) and p is viscosity (mPa s).

2.5. Chlorophyll Content

Chlorophylls were determined according to Zago et al. [33]. The chlorophyll content was evaluated by the absorption spectrum according to the American Oil Chemists' Society [34] and expressed as mg of pheophytin a per kg of oil.

2.6. Color

The olive oil color was measured using the software package ImageJ, v.1.38x, fitted with the plugin Color Inspector 3D v. 2.3 [18]. Each time 20 mL of samples were put into a glass Petri dish. The images of each Petri dish were acquired with a scanner (Hewlett Packard, Palo Alto, CA, USA) at 600 dots per inch (dpi). Based on the CIELAB colorimetric system, the measured colorimetric parameters were L^* (lightness); $-a^*$ (green shade); b^* (yellowness).

2.7. Statistical Analysis

Means and standard deviations were calculated with the SPSS (version 27.0 SPSS Inc., Chicago, IL, USA) statistical software package. SPSS was used to perform a one-way analysis of variance (ANOVA) and Tukey's honest significant difference test (HSD) at a 95% confidence level ($p < 0.05$) to identify differences between samples. Pearson correlation coefficients were calculated between the variables at 95% and 99% confidence levels ($p < 0.05$ and $p < 0.01$). Principal component analysis (PCA) was also performed, on normalized data, by means of the Statistica software package (version 8.0, Stat-Soft, Tulsa, OK, USA). PCA was used as a descriptive statistical technique by plotting the normalized independent variables (analytical parameters) versus all cases (samples) with the aim to identify the variables able to discriminate between the cases.

3. Results

3.1. Fatty Acid Composition

The fatty acid composition is a quality parameter and authenticity indicator of virgin olive oils. The fatty acid composition of the thirteen olive oil samples analyzed in this study is reported in Table 2. Based on these data, all the samples may be classified in the category extra virgin olive oil, according to the European Regulation 2568/91 [23].

The three most abundant fatty acids were oleic (C18:1), palmitic (C16:0), and linoleic (C18:1) acid, as expected. They showed significantly different values between the cultivars. Comparing the samples at t1, C18:1 ranged from 75.5% of TOR and CIC, to 65% of COR; C16:0 ranged from 16.5% to 11% for COR and OR, respectively. C18:2 ranged from 13.5% of SIV and COR to 6.5 of TOR. In general, an opposite trend between C18:1 and C16:0, C18:2 was observed. A clear justification of the observed differences is not that immediate; genetic, environmental, and agronomic factors, alone or in combination, have been reported to influence the composition of olive oils [35]. In particular, the differences in oleic, palmitic, and linoleic acid content seem to be mostly related to the weather: it was reported that lower temperatures could be correlated with a higher content of oleic acid and higher temperatures with a lower content of palmitic and/or linoleic acids [35]. This assumption is partially confirmed by the results of this study (Tables 1 and 2). The content of some minor fatty acids such as linolenic (C18:3) and arachidic (C20:0) deserves special attention, as their levels are determining factors for the olive oil merceological classification [23]. While C18:3 did not show many differences between the samples, C20:0 showed more variability. In particular, all the cultivars from Abruzzo (TOR, DR, GEN) showed the lowest values (~0.20%), while percentages ranging from 0.45 of OTT to 0.31 of CM were observed for the oils from the other Italian regions.

Considering the differences between the harvesting time, different behaviors were observed between the cultivars. All the Abruzzi cultivars (TOR, DR, GEN), other than CM, OTT, and OTTC, showed an increase in C18:1, passing from t1 to t3. Decreasing values of C18:1 were instead observed for SIV, SEM, TDF, and CIC, from t1 to t3. For the c16:0 content, the opposite was observed in the same cultivars. COR, OR, and BAM did not have differences in oleic acid content, and, moreover, BAM was the most stable cultivar, showing very poor variations of all the fatty acids over time. These trends are also dependent on varietal characters, such as a response to environmental factors, as evidenced by [5,36]. It is reported that higher temperatures during the phases of oil accumulation involve a decrease in oleic acid content [37].

Table 2. Fatty acids (%) of 13 olive oils belonging from minor Italian cultivars harvested at three different maturation stages [1].

		$C_{16:0}$	$C_{16:1}$	$C_{18:0}$	$C_{18:1}$	$C_{18:2}$	$C_{18:3}$	$C_{20:0}$
TOR	t1	14.05 ± 0.83 efA	0.56 ± 0.03 efA	2.33 ± 0.04 bcdeA	75.75 ± 0.76 aB	6.50 ± 0.08 cB	0.57 ± 0.01 abB	0.24 ± 0.02 dB
	t2	12.41 ± 0.20 cB	0.54 ± 0.00 efA	2.16 ± 0.04 deB	77.13 ± 0.13 aA	6.87 ± 0.04 efA	0.63 ± 0.00 bcA	0.27 ± 0.01 eA
	t3	11.70 ± 0.41 efgB	0.52 ± 0.04 fA	2.23 ± 0.02 defB	77.83 ± 0.43 abA	6.88 ± 0.02 fA	0.57 ± 0.01 abB	0.27 ± 0.00 dA
DR	t1	15.05 ± 0.43 bcdA	1.59 ± 0.03 abA	1.77 ± 0.05 efA	72.40 ± 0.34 bB	8.25 ± 0.13 cB	0.74 ± 0.08 aA	0.20 ± 0.01 dAB
	t2	13.64 ± 0.20 abcB	1.75 ± 0.09 bcA	1.72 ± 0.10 fA	73.08 ± 0.34 bcdB	8.83 ± 0.19 cdA	0.74 ± 0.02 aA	0.23 ± 0.00 fA
	t3	13.19 ± 0.38 cdefB	1.27 ± 0.07 cdB	1.87 ± 0.02 fA	74.84 ± 0.32 cdA	8.07 ± 0.13 eB	0.58 ± 0.22 abA	0.19 ± 0.02 eB
GEN	t1	16.12 ± 0.73 abA	1.96 ± 0.27 aA	1.54 ± 0.02 fB	69.06 ± 0.63 cB	10.46 ± 0.37 bAB	0.63 ± 0.04 abA	0.22 ± 0.01 dA
	t2	16.24 ± 0.30 aA	2.25 ± 0.12 aA	1.55 ± 0.03 fB	67.82 ± 0.94 efB	11.32 ± 0.69 bA	0.61 ± 0.01 bcA	0.20 ± 0.01 fB
	t3	14.66 ± 0.41 efgB	1.99 ± 0.09 aA	2.13 ± 0.10 defA	71.09 ± 0.49 fA	9.34 ± 0.16 dB	0.59 ± 0.02 abA	0.21 ± 0.00 eAB
SIV	t1	15.55 ± 0.07 bcdB	1.37 ± 0.01 bcdB	2.49 ± 0.01 abcdB	66.01 ± 0.07 deA	13.74 ± 0.01 aC	0.47 ± 0.01 bA	0.38 ± 0.00 abcA
	t2	15.30 ± 0.01 abB	1.36 ± 0.01 cdB	2.77 ± 0.00 bA	64.50 ± 0.02 gB	15.20 ± 0.03 aA	0.49 ± 0.01 efA	0.38 ± 0.00 bcdA
	t3	16.34 ± 0.28 aA	1.48 ± 0.02 bA	2.40 ± 0.12 cdeB	64.73 ± 0.17 hB	14.24 ± 0.01 aB	0.48 ± 0.02 bA	0.34 ± 0.01 cB
SEM	t1	14.80 ± 0.15 bcdA	0.96 ± 0.01 cdeA	2.57 ± 0.01 abcB	72.12 ± 0.13 bA	8.50 ± 0.02 bcC	0.65 ± 0.01 abB	0.39 ± 0.01 abB
	t2	14.02 ± 0.07 abcB	0.97 ± 0.01 deA	2.61 ± 0.02 bcB	71.98 ± 0.08 cdA	9.31 ± 0.00 cB	0.68 ± 0.00 abA	0.43 ± 0.01 aA
	t3	13.58 ± 0.13 cdeC	0.88 ± 0.01 eB	2.74 ± 0.01 abcA	71.19 ± 0.12 fB	10.54 ± 0.01 cA	0.63 ± 0.01 abA	0.43 ± 0.01 bA
COR	t1	16.69 ± 0.17 aA	1.41 ± 0.01 bcA	2.42 ± 0.02 abcdA	65.17 ± 0.15 eA	13.35 ± 0.02 aC	0.61 ± 0.01 abA	0.35 ± 0.01 bcA
	t2	16.21 ± 0.04 aB	1.30 ± 0.00 dB	2.37 ± 0.00 cdeB	65.14 ± 0.03 fgA	14.03 ± 0.00 aB	0.60 ± 0.01 cdA	0.36 ± 0.00 cdA
	t3	15.92 ± 0.18 abB	1.20 ± 0.02 cdC	2.42 ± 0.02 bcdeA	65.40 ± 0.16 hA	14.16 ± 0.02 aA	0.54 ± 0.00 abB	0.36 ± 0.01 cA
CM	t1	15.40 ± 0.97 abcdA	0.90 ± 0.40 defA	1.88 ± 0.34 defA	72.44 ± 0.69 bB	8.48 ± 0.51 bcA	0.58 ± 0.02 abA	0.31 ± 0.01 cA
	t2	13.72 ± 0.71 abcAB	0.47 ± 0.01 fA	2.37 ± 0.03 cdeAB	74.72 ± 0.55 abcA	7.89 ± 0.14 deA	0.48 ± 0.02 efB	0.35 ± 0.01 cdA
	t3	11.13 ± 1.64 gB	0.48 ± 0.02 fA	2.46 ± 0.02 abcdA	76.24 ± 1.27 bcA	8.73 ± 0.35 deA	0.59 ± 0.02 abA	0.38 ± 0.01 cA
OR	t1	11.06 ± 0.28 gA	0.49 ± 0.01 fB	2.78 ± 0.02 abA	74.19 ± 0.40 abA	10.46 ± 0.09 bA	0.62 ± 0.26 abA	0.41 ± 0.00 abA
	t2	12.10 ± 3.12 cA	0.47 ± 0.03 fB	2.57 ± 0.01 bcB	75.23 ± 2.40 abA	8.50 ± 0.61 cdB	0.75 ± 0.06 aA	0.38 ± 0.03 bcA
	t3	12.13 ± 1.41 efgA	0.56 ± 0.02 fA	2.48 ± 0.03 abcdC	75.32 ± 1.05 cdA	8.44 ± 0.30 deB	0.68 ± 0.05 aA	0.37 ± 0.01 cA
BAM	t1	14.12 ± 0.07 defA	1.46 ± 0.00 bA	1.98 ± 0.01 cdefB	74.01 ± 0.04 abA	7.49 ± 0.03 cB	0.60 ± 0.02 abA	0.35 ± 0.01 bcA
	t2	13.39 ± 0.83 bcA	1.08 ± 0.51 dA	2.32 ± 0.06 cdeA	73.51 ± 1.89 bcdA	8.82 ± 0.58 cdA	0.53 ± 0.04 deA	0.36 ± 0.01 cdA
	t3	13.50 ± 0.11 cdeA	1.33 ± 0.01 bcA	2.27 ± 0.00 defA	72.72 ± 0.13 efA	9.23 ± 0.02 dA	0.59 ± 0.01 abA	0.36 ± 0.01 cA
	t1	16.04 ± 0.22 abA	1.39 ± 0.09 bcB	2.94 ± 0.15 aA	68.22 ± 0.38 cdC	10.44 ± 0.18 bA	0.53 ± 0.01 abB	0.45 ± 0.01 aA

Table 2. *Cont.*

		$C_{16:0}$	$C_{16:1}$	$C_{18:0}$	$C_{18:1}$	$C_{18:2}$	$C_{18:3}$	$C_{20:0}$
OTT	t2	14.40 ± 0.13 abcB	1.91 ± 0.04 abA	2.07 ± 0.17 eB	71.75 ± 0.71 dB	8.87 ± 0.52 cdB	0.66 ± 0.03 bcA	0.35 ± 0.01 dB
	t3	12.45 ± 0.10 defgC	1.17 ± 0.06 dC	2.79 ± 0.14 abcA	74.31 ± 0.23 deA	8.38 ± 0.18 deB	0.52 ± 0.01 abB	0.36 ± 0.01 cB
OTTC	t1	15.04 ± 0.01 bcdA	2.01 ± 0.01 aA	1.97 ± 0.15 cdefC	72.17 ± 0.12 bC	7.77 ± 0.11 cA	0.69 ± 0.01 abA	0.35 ± 0.00 bcB
	t2	12.69 ± 0.17 bcB	1.12 ± 0.05 dB	2.43 ± 0.11 cdB	76.80 ± 0.39 aB	6.10 ± 0.28 fB	0.45 ± 0.02 fB	0.41 ± 0.01 abA
	t3	11.46 ± 0.13 fgC	0.80 ± 0.11 eC	2.91 ± 0.06 aA	78.46 ± 0.47 aA	5.45 ± 0.30 gC	0.47 ± 0.05 fB	0.46 ± 0.03 abA
TDF	t1	14.44 ± 0.16 cdefB	1.44 ± 0.06 bB	1.62 ± 0.56 fB	73.98 ± 1.04 abA	7.45 ± 0.37 cC	0.68 ± 0.01 abA	0.39 ± 0.01 abA
	t2	15.23 ± 0.15 abA	1.40 ± 0.02 cdB	3.17 ± 0.11 aA	67.69 ± 0.76 efC	11.62 ± 0.51 bA	0.47 ± 0.01 efB	0.41 ± 0.02 abA
	t3	14.35 ± 0.14 bcdB	1.89 ± 0.05 aA	2.00 ± 0.36 efB	71.81 ± 0.69 fB	8.95 ± 0.48 deB	0.65 ± 0.02 abA	0.34 ± 0.00 cB
CIC	t1	13.25 ± 0.26 fB	1.16 ± 0.28 bcdA	2.23 ± 0.27 bcdeA	75.46 ± 2.65 aA	6.96 ± 2.33 cB	0.53 ± 0.01 abA	0.42 ± 0.07 abA
	t2	14.79 ± 0.30 abcA	1.31 ± 0.05 cdA	2.80 ± 0.25 bA	68.47 ± 0.68 eB	11.73 ± 0.59 bA	0.46 ± 0.01 efB	0.44 ± 0.01 aA
	t3	14.83 ± 0.28 abcA	1.28 ± 0.02 cdA	2.86 ± 0.35 abA	68.06 ± 0.45 gB	12.02 ± 0.90 bA	0.46 ± 0.01 bB	0.48 ± 0.03 aA

[1]. Data are expressed as mean ± standard deviation of three replicates. C18:3 is the sum of alpha and gamma-linolenic acid. A, B, C in the same column, between the three harvesting times for the same cultivar, indicate significant differences between the means ($p < 0.05$). a, b, c, d, e, f, g, h in the same column, at the same harvesting time for the different cultivars, indicate significant differences between the means ($p < 0.05$).

3.2. Thermal Analysis

The phase transitions of olive oils measured by DSC are affected by molecular composition changes [38,39]. Figure 1 shows the cooling (A, B, C, D) and heating (E, F, G, H) thermograms of the thirteen studied monovarietal extra virgin olive oils at t1, divided by region of origin. All the curves show common traits: two main transitions upon cooling (p1c, p2c), three main transitions upon heating (p1h, p2h, p3h); analogous thermograms have been already observed for extra virgin olive oils [38,39]. The thermal phenomena observed during cooling are basically influenced by the chemical composition of the samples [38]. In particular, the main exothermic event, peaked at lower temperatures (p1c, Figure 1) has been related to the crystallization of TAG rich in oleic acid. The shape of this transition always appeared as a symmetrical Gaussian curve; suggesting an ordered and cooperative event involving homogenous molecules. The second major exothermic event peak occurred at higher temperatures upon cooling, p2c, and had an asymmetrical shape, indicating the involvement of more heterogeneous molecules, previously identified as saturated triglycerides (TAG) fractions [39].

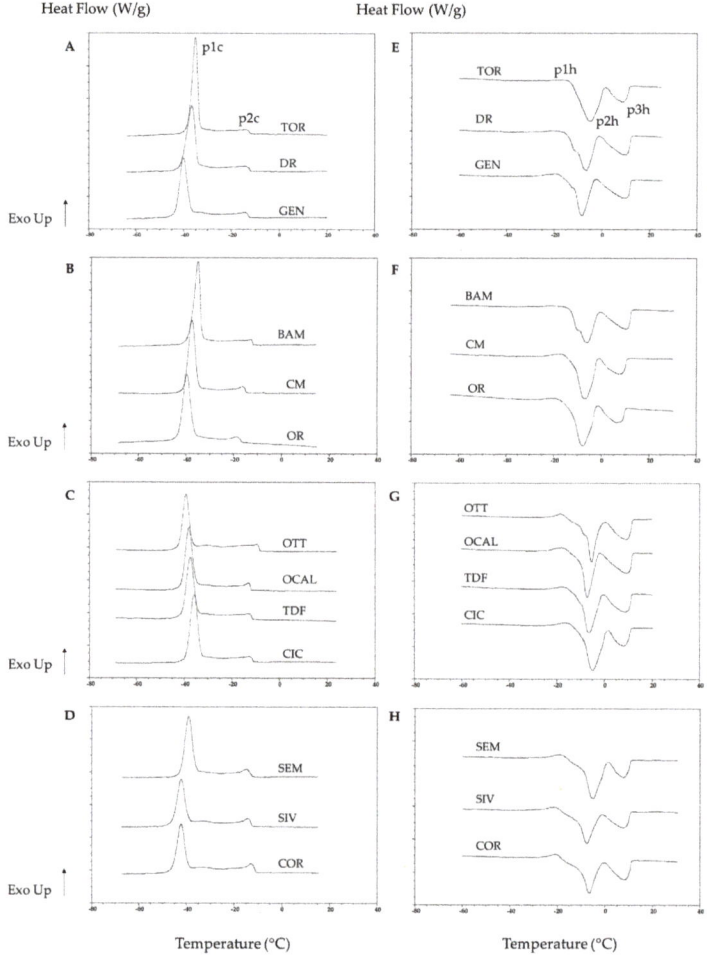

Figure 1. DSC thermograms of the thirteen olive oils (t1) divided by region of origin. **A–D**: cooling thermograms; **E–H**: heating thermograms. p1c, p2c: main thermal events on cooling; p1h, p2h, p3h: main thermal events on heating.

Intuitively, it might be presumed that the thermograms found upon heating would mirror the ones observed during cooling, in which the formed crystals melt. However, the heating thermograms are more complex. In detail, the first thermal event, p1h, is an exothermic transition, related to a solid-state transformation of the TAG crystals towards more stable forms [40]. p2h and p3h are two endothermic events related to the melting of other TAG polymorphic forms [40]. Bayés-García et al. [40] described the nature of these phenomena very well, by explaining how three main groups of TAGs: triunsaturated OOO and OOL, saturated-unsaturated-unsaturated POO, POL, and SOO, and saturated-saturated-unsaturated PPO, were responsible for the polymorphic behavior observed.

Besides these similarities, each sample showed specific transition temperatures and peak amplitudes and shapes. In some cases, additional thermal events, often visible as shoulders of the main thermal events, both upon cooling and heating, were observed. These minor transitions have not been examined in this study.

Tables 3 and 4 report the thermal parameters extrapolated from the cooling and heating thermograms, respectively.

Looking at the cooling parameters (Table 3), significant differences between the samples were observed, also showing high correlations with the fatty acid composition. The range of cooling (Range_C), calculated as the difference between Ton_C and Toff_C, at t1 varied from 32.5 °C of TOR, CM, and OR to 42.5 °C of COR. The larger is the cooling transition, the more heterogeneous are the molecules involved in the crystallization [41]. In general, the Apulian and Abruzzi samples had narrower ranges of transition, with lower Ton_C and higher Toff_C, than the Sardinian and Calabrian ones. Narrow cooling transition ranges and low Ton_C have been previously associated with olive oils rich in oleic acid [39], able to perform cooperative crystallization phenomena at lower temperatures. In support of this hypothesis, negative correlations have been found between oleic acid content, Ton_C, and Range_C ($p < 0.01$; R = -0.326; -0.633). The cooling enthalpy (ΔH_C) was also influenced by the oleic acid content. In particular, it was positively correlated with C18:1 ($p < 0.01$; R= 0.397) and negatively with C16:0, C18:2 and C:20 ($p < 0.01$, R= -0.270; -0.392; -0.238). The cooling enthalpy, calculated as the area under the cooling curve, is influenced by the number of molecules involved in the exothermic phenomenon [32]. In this study, at t1, it ranges from 67.5 J/g of TOR and DR, to 62 J/g of SIV. The temperature of the major crystallization peak (Tp1_C) ranged from -42 °C of SIV and COR to -35 °C of BAM. This thermal event, previously associated with the crystallization of oleic rich TAGs [40], in this study, showed positive correlations with C18:1 ($p < 0.01$; R = 0.686) and negative with C16:0, C16:1, C18:2 ($p < 0.01$; R = -0.538; -347; 0.665). The minor exothermal event (p2_c) peaked in a range of temperatures from -9.7 °C of OTT, to -18 °C of OR. Tp2_C correlated positively with C16:0, C16:1, and C20:0 ($p < 0.01$; R = 0.287; 0.401; 0.238) and negatively with C18:1 ($p < 0.05$; R = -0.182) and C18:3 ($p < 0.01$; R = -0.270). These results confirm that p2_c occurs at higher temperatures for oils richer in saturated fatty acids [40].

Table 3. DSC cooling parameters of 13 Italian minor olive oils from harvested at three different maturation stages [1].

		Ton_C	Toff_C	Range_C	ΔH_C	Tp1_C	Tp2_C
TOR	t1	−11.08 ± 0.51 bcdA	−43.59 ± 0.49 aA	32.51 ± 0.09 gA	67.73 ± 1.64 aA	−35.78 ± 0.62 abA	−13.75 ± 0.17 bcA
	t2	−12.03 ± 0.33 eAB	−43.77 ± 0.67 aA	31.74 ± 1.00 fA	64.45 ± 3.10 abA	−35.09 ± 0.86 abA	−14.99 ± 0.57 cdeB
	t3	−12.82 ± 0.34 efB	−43.47 ± 0.59 aA	30.65 ± 0.82 gA	67.43 ± 2.63 abA	−34.60 ± 0.10 abA	−15.55 ± 0.61 efB
DR	t1	−12.04 ± 0.37 deA	−46.79 ± 0.82 bA	34.75 ± 0.44 fgA	67.12 ± 1.96 aAB	−36.46 ± 0.21 bcA	−14.96 ± 0.67 dcA
	t2	−12.84 ± 0.28 fA	−47.46 ± 0.45 bcdeA	34.62 ± 0.73 deA	65.71 ± 1.02 abB	−36.87 ± 0.37 cdeA	−16.27 ± 0.09 efB
	t3	−12.60 ± 0.41 deA	−46.34 ± 0.99 bA	33.74 ± 1.26 fA	70.45 ± 2.40 aA	−35.56 ± 0.93 bA	−16.90 ± 0.19 gB
GEN	t1	−11.39 ± 0.43 bcdA	−47.48 ± 1.98 bcA	36.09 ± 1.92 efA	64.05 ± 3.20 abcA	−40.19 ± 0.20 fB	−14.00 ± 0.48 bA
	t2	−12.12 ± 0.33 efA	−49.36 ± 0.34 defA	37.24 ± 0.60 cdA	63.16 ± 1.53 abA	−40.62 ± 0.55 iB	−14.70 ± 0.88 cdeA
	t3	−11.44 ± 0.56 cA	−48.49 ± 0.88 cdeA	37.05 ± 0.56 cdeA	67.04 ± 2.54 abcA	−38.98 ± 0.35 cdA	−13.83 ± 0.37 bcA
SIV	t1	−11.25 ± 0.05 bcdAB	−51.97 ± 0.20 efA	40.72 ± 0.22 abC	61.80 ± 1.85 cA	−42.15 ± 0.13 gA	−14.13 ± 0.44 bcA
	t2	−11.00 ± 0.18 bcdA	−52.70 ± 0.20 ghB	41.70 ± 0.16 aB	61.09 ± 2.91 bA	−43.11 ± 0.24 jB	−14.49 ± 0.53 deA
	t3	−11.33 ± 0.12 cB	−53.53 ± 0.03 fC	42.21 ± 0.14 aA	61.49 ± 2.82 cA	−43.27 ± 0.10 eB	−14.93 ± 0.36 deA
SEM	t1	−10.87 ± 0.08 bcdA	−50.29 ± 0.18 deA	39.43 ± 0.26 bcdA	65.02 ± 0.13 abcA	−39.23 ± 0.13 efA	−15.09 ± 0.04 cdA
	t2	−11.42 ± 0.38 deA	−49.82 ± 0.60 efA	38.40 ± 0.49 bcA	63.48 ± 1.38 abA	−39.61 ± 0.16 hiB	−15.18 ± 0.32 deA
	t3	−11.71 ± 0.49 cA	−49.62 ± 0.51 eA	37.91 ± 0.92 cdA	63.73 ± 2.89 bcA	−40.32 ± 0.11 dC	−15.40 ± 0.18 eA
COR	t1	−9.72 ± 0.06 abA	−52.43 ± 0.39 fA	42.71 ± 0.45 aA	65.00 ± 1.47 abcA	−42.44 ± 0.15 gA	−13.25 ± 0.28 bA
	t2	−10.26 ± 0.34 bB	−52.84 ± 0.89 ghA	42.58 ± 1.23 aA	62.96 ± 0.69 abB	−42.59 ± 0.02 jAB	−13.32 ± 0.38 bcA
	t3	−10.41 ± 0.32 bB	−52.58 ± 0.90 fA	42.20 ± 0.99 aA	63.80 ± 0.71 bcAB	−42.70 ± 0.19 eB	−14.25 ± 0.31 bcdB
CM	t1	−13.59 ± 0.11 efB	−47.33 ± 0.21 bcA	33.74 ± 0.31 fgA	64.71 ± 1.02 abcA	−37.58 ± 0.03 cdB	−16.28 ± 0.14 deB
	t2	−12.85 ± 0.04 fA	−47.37 ± 0.59 bcdA	34.53 ± 0.63 deA	65.83 ± 1.20 abA	−35.79 ± 0.18 abcA	−15.12 ± 0.23 deA
	t3	−13.69 ± 0.00 fB	−47.40 ± 0.39 bcA	33.71 ± 0.39 fA	64.51 ± 1.25 bcA	−35.51 ± 0.67 bA	−16.47 ± 0.43 fgB
OR	t1	−14.22 ± 1.90 fA	−47.11 ± 0.10 bcA	32.89 ± 1.80 gB	63.41 ± 0.05 abcB	−38.77 ± 0.95 deB	−17.23 ± 1.61 eA
	t2	−13.70 ± 0.15 gA	−47.20 ± 0.01 bcdA	33.50 ± 0.14 efAB	67.17 ± 0.41 aA	−37.55 ± 0.55 defB	−17.25 ± 0.38 fA
	t3	−11.88 ± 0.22 cdA	−47.63 ± 0.05 bcdB	35.75 ± 0.26 defA	67.31 ± 1.02 abA	−34.21 ± 0.66 abA	−15.04 ± 0.27 deA
BAM	t1	−11.57 ± 0.15 cdA	−44.41 ± 0.53 aA	32.84 ± 0.45 gB	66.79 ± 1.66 abA	−34.83 ± 0.23 aA	−13.30 ± 0.24 bA
	t2	−11.05 ± 0.20 cdA	−45.61 ± 1.26 abAB	34.56 ± 1.27 deAB	65.47 ± 2.25 abA	−36.17 ± 0.66 bcdB	−13.42 ± 0.45 bcdA
	t3	−11.36 ± 0.27 cA	−46.93 ± 0.47 bcB	35.57 ± 0.73 efA	66.02 ± 2.04 abcA	−37.64 ± 0.01 cC	−13.90 ± 0.27 bcA

Table 3. Cont.

		Ton_C	Toff_C	Range_C	ΔH_C	Tp1_C	Tp2_C
OTT	t1	−8.31 ± 0.28 aA	−48.94 ± 0.55 dA	40.63 ± 0.83 abA	62.655 ± 0.63 bcA	−38.66 ± 0.61 deA	−9.69 ± 0.05 aA
	t2	−8.81 ± 0.10 aAB	−49.04 ± 0.27 cdefA	40.22 ± 0.33 abA	63.64 ± 1.00 abA	−39.30 ± 0.87 ghiA	−10.33 ± 0.26 aAB
	t3	−9.10 ± 0.34 aA	−49.33 ± 0.37 deA	40.23 ± 0.71 abA	63.72 ± 1.66 bcA	−38.62 ± 0.25 cA	−10.63 ± 0.40 aB
OTTC	t1	−10.51 ± 0.34 bcdA	−48.52 ± 0.43 bcdA	38.01 ± 0.33 cdeA	64.40 ± 1.20 abcA	−37.97 ± 0.10 deA	−12.99 ± 0.13 bA
	t2	−10.57 ± 0.51 bcA	−50.86 ± 1.87 fgA	40.29 ± 2.03 abA	62.44 ± 2.86 abA	−38.47 ± 0.29 fghA	−12.67 ± 0.61 bA
	t3	−11.30 ± 0.09 bcA	−50.20 ± 0.04 eA	38.90 ± 0.05 bcA	65.29 ± 1.68 abcA	−37.62 ± 0.87 cA	−13.65 ± 0.27 bB
TDF	t1	−9.79 ± 0.47 abcA	−50.15 ± 0.49 deA	40.36 ± 0.42 abcB	64.11 ± 1.18 abcA	−37.89 ± 0.58 dA	−12.81 ± 0.30 bA
	t2	−11.09 ± 0.17 cdB	−53.52 ± 1.63 hB	42.43 ± 1.80 aA	63.85 ± 1.23 abA	−37.83 ± 0.23 efgA	−13.28 ± 0.37 bcA
	t3	−11.78 ± 0.17 cdB	−49.81 ± 0.24 eA	38.03 ± 0.07 cC	64.26 ± 0.16 bcA	−37.89 ± 0.25 cA	−14.20 ± 0.18 bcdB
CIC	t1	−10.12 ± 0.26 bcA	−47.34 ± 0.34 bcA	37.23 ± 0.60 deA	66.81 ± 1.33 abA	−35.23 ± 0.86 abA	−13.03 ± 0.21 bA
	t2	−11.13 ± 0.08 cdB	−46.90 ± 0.80 bcA	35.77 ± 0.88 cdeA	67.26 ± 1.73 aA	−34.55 ± 1.34 aA	−13.83 ± 0.12 bcdB
	t3	−11.75 ± 0.33 cdC	−47.12 ± 1.76 bcA	35.37 ± 2.09 efA	67.63 ± 1.57 abA	−33.28 ± 1.38 aA	−14.63 ± 0.04 cdeC

[1] Data are expressed as mean ± standard deviation of three replicates. Different capital letters in the same column, between the three harvesting times for the same cultivar, indicate significant differences between the means ($p < 0.05$). Different small letters in the same column, at the same harvesting time for the different cultivars, indicate significant differences between the means ($p < 0.05$).

Table 4. DSC heating parameters of 13 minor Italian olive oils harvested at three different maturation stages [1].

		Ton_H	Toff_H	Range_H	ΔH_H	Tp1_H	Tp2_H	Tp3_H
TOR	t1	−37.29 ± 0.31 gA	13.36 ± 0.05 bcdA	50.65 ± 0.36 aA	69.91 ± 0.53 abcdA	−16.06 ± 0.04 aA	−4.85 ± 0.03 abB	8.59 ± 0.12 cdefA
	t2	−36.96 ± 0.15 fA	12.94 ± 0.34 bcdAB	49.90 ± 0.39 aAB	66.29 ± 2.47 deA	−17.90 ± 1.55 bcAB	−4.71 ± 0.14 aB	8.15 ± 0.05 cdB
	t3	−37.02 ± 0.24 fA	12.47 ± 0.05 cdB	49.49 ± 0.28 aB	70.79 ± 2.58 abcA	−19.27 ± 0.78 cdeB	−4.46 ± 0.10 aA	7.69 ± 0.02 deC
DR	t1	−36.98 ± 0.45 gA	13.59 ± 0.37 abcdA	50.57 ± 0.39 aA	68.83 ± 1.86 abcdA	−20.82 ± 0.32 fgA	−6.80 ± 0.50 efgA	9.29 ± 0.44 bA
	t2	−36.63 ± 0.34 efA	12.77 ± 0.63 cdeA	49.40 ± 0.97 aA	66.70 ± 0.87 deA	−20.49 ± 1.44 fgA	−7.31 ± 0.64 deA	8.21 ± 0.92 cdA
	t3	−36.30 ± 1.09 fA	13.00 ± 0.25 abcA	49.30 ± 1.22 aA	71.47 ± 3.26 abA	−20.45 ± 0.16 defA	−5.75 ± 0.87 bcdA	8.40 ± 0.28 bcA
GEN	t1	−31.47 ± 0.32 eB	14.10 ± 0.26 abA	45.56 ± 0.27 bA	66.95 ± 2.05 deB	−19.19 ± 0.12 deA	−8.07 ± 0.26 iB	9.79 ± 0.22 abA
	t2	−30.58 ± 0.09 dA	13.53 ± 0.28 abAB	44.11 ± 0.36 bB	68.28 ± 0.68 bcdeAB	−19.89 ± 0.35 efA	−8.28 ± 0.60 eB	9.46 ± 0.12 abA
	t3	−31.03 ± 0.28 dAB	13.14 ± 0.26 abB	44.17 ± 0.54 bB	72.03 ± 2.38 abA	−19.28 ± 1.15 cdeA	−7.09 ± 0.10 eA	8.67 ± 0.07 bB

Table 4. Cont.

		Ton_H	Toff_H	Range_H	ΔH_H	Tp1_H	Tp2_H	Tp3_H
SIV	t1	−29.86 ± 0.11 dA	11.89 ± 0.06 fA	41.75 ± 0.16 cA	64.64 ± 1.96 eA	−21.35 ± 0.05 gA	−7.70 ± 0.33 hiA	7.65 ± 0.03 gA
	t2	−29.27 ± 0.19 cA	11.17 ± 0.27 gB	40.44 ± 0.43 gB	64.76 ± 2.09 eA	−21.77 ± 0.04 gB	−7.57 ± 0.66 eA	7.28 ± 0.14 eB
	t3	−29.57 ± 0.43 cdA	11.67 ± 0.19 eA	41.24 ± 0.53 cAB	65.52 ± 2.53 cA	−21.84 ± 0.06 fB	−7.03 ± 0.21 eA	7.24 ± 0.08 eB
SEM	t1	−28.45 ± 0.57 cA	12.45 ± 0.09 eA	40.90 ± 0.47 bcdeA	67.84 ± 0.88 bcdeA	−19.35 ± 0.08 deA	−5.02 ± 0.03 abA	7.83 ± 0.01 efgA
	t2	−29.26 ± 0.33 cA	12.36 ± 0.11 deA	41.62 ± 0.41 cA	66.63 ± 1.50 deA	−19.61 ± 0.02 defB	−5.50 ± 0.34 abAB	7.65 ± 0.09 deB
	t3	−29.00 ± 0.24 bcA	11.54 ± 0.17 fB	40.54 ± 0.08 cB	67.41 ± 3.18 bcA	−19.80 ± 0.04 cdefC	−5.65 ± 0.15 bcB	7.23 ± 0.08 eC
COR	t1	−28.66 ± 0.18 cA	12.34 ± 0.30 efA	41.00 ± 0.48 cdA	66.42 ± 0.38 deA	−20.85 ± 0.08 fgA	−6.42 ± 0.10 defA	8.36 ± 0.17 defgA
	t2	−29.11 ± 0.36 cA	12.02 ± 0.28 efA	41.12 ± 0.64 cA	66.08 ± 0.42 deA	−21.10 ± 0.01 fgB	−7.30 ± 0.08 deA	7.93 ± 0.08 deB
	t3	−28.90 ± 0.42 bcA	12.23 ± 0.34 deA	41.13 ± 0.56 cA	65.10 ± 0.45 cB	−21.13 ± 0.10 efB	−6.61 ± 0.12 deA	7.68 ± 0.15 deB
CM	t1	−26.46 ± 0.59 aA	12.45 ± 0.18 efA	38.91 ± 0.77 fA	70.83 ± 0.60 abA	−17.46 ± 0.54 bA	−6.76 ± 0.14 efgB	7.74 ± 0.04 fgA
	t2	−26.48 ± 0.56 aA	11.47 ± 0.25 fgB	37.95 ± 0.30 dB	71.78 ± 1.67 abcA	−16.18 ± 0.10 aB	−5.55 ± 0.13 abA	6.61 ± 0.24 fB
	t3	−27.08 ± 0.45 aA	10.74 ± 0.04 gC	37.82 ± 0.40 eB	71.47 ± 1.69 abA	−16.38 ± 0.28 abA	−5.43 ± 0.04 bA	5.99 ± 0.01 fC
OR	t1	−26.40 ± 0.06 aAB	10.62 ± 0.94 gB	37.01 ± 1.00 gAB	70.74 ± 0.01 abcA	−19.03 ± 1.32 cdeA	−7.20 ± 0.65 ghB	6.03 ± 1.34 hA
	t2	−26.08 ± 0.16 aA	10.01 ± 0.06 hB	36.09 ± 0.10 eB	75.57 ± 0.62 aA	−17.48 ± 0.38 abcA	−6.57 ± 0.33 cdB	5.23 ± 0.11 gA
	t3	−26.91 ± 0.62 aB	11.18 ± 0.28 fgA	38.09 ± 0.91 deA	73.43 ± 1.85 aB	−18.01 ± 2.68 bcdA	−4.92 ± 0.57 abA	6.34 ± 0.06 fA
BAM	t1	−35.25 ± 0.34 fA	14.25 ± 0.24 aA	49.50 ± 0.57 aA	69.91 ± 0.78 abcdA	−19.57 ± 0.76 efA	−5.93 ± 0.11 cdA	10.61 ± 0.05 aA
	t2	−35.52 ± 0.98 eA	13.74 ± 0.00 aAB	49.26 ± 0.98 aA	67.15 ± 1.85 cdeA	−20.72 ± 0.11 fgA	−6.19 ± 0.14 bcA	9.72 ± 0.11 aB
	t3	−34.26 ± 1.78 eA	13.29 ± 0.61 abB	47.55 ± 2.39 aA	67.87 ± 0.42 abcdeA	−18.80 ± 1.80 bcdeA	−6.50 ± 0.52 cdeA	8.73 ± 0.66 bC
OTT	t1	−27.82 ± 0.04 bcA	13.51 ± 0.33 abcdA	41.33 ± 0.37 cdA	67.23 ± 0.07 cdeA	−18.23 ± 0.37 bcdA	−5.56 ± 0.33 bcA	8.98 ± 0.01 bcdA
	t2	−27.37 ± 0.22 abA	12.89 ± 0.28 bcdB	40.26 ± 0.23 cAB	69.14 ± 1.76 bcdeA	−18.66 ± 0.38 cdeA	−6.28 ± 0.26 bcB	8.21 ± 0.22 cdB
	t3	−27.09 ± 0.79 aA	12.69 ± 0.29 bcdB	39.78 ± 1.08 cdeB	67.25 ± 1.51 bcA	−18.38 ± 0.01 bcdA	−6.54 ± 0.22 cdeB	7.98 ± 0.25 cdB
OTTC	t1	−27.85 ± 0.10 bcAB	13.67 ± 0.10 abcA	41.52 ± 0.20 cdA	68.39 ± 0.99 abcdA	−17.84 ± 0.24 bcA	−7.03 ± 0.08 fghA	9.48 ± 0.10 bcA
	t2	−28.05 ± 0.09 bcB	13.40 ± 0.34 abcA	41.45 ± 0.43 cA	66.52 ± 3.10 cdeA	−17.90 ± 0.44 bcA	−7.71 ± 0.37 eB	8.79 ± 0.30 bcB
	t3	−27.71 ± 0.21 abA	13.35 ± 0.28 aA	41.06 ± 0.49 cA	71.42 ± 2.34 abA	−19.56 ± 1.75 cdefA	−8.16 ± 0.35 fB	9.32 ± 0.33 aAB
TDF	t1	−27.32 ± 0.43 abA	12.99 ± 0.26 cdeA	40.30 ± 0.64 cdeA	68.17 ± 0.62 bcdeA	−18.28 ± 0.61 bcdA	−6.16 ± 0.41 cdeA	8.72 ± 0.10 cdeA
	t2	−28.02 ± 0.64 bcA	12.61 ± 0.29 deAB	40.62 ± 0.35 cA	69.69 ± 1.46 bcdA	−18.17 ± 0.13 bcdA	−7.36 ± 0.01 deB	7.57 ± 0.25 deB
	t3	−27.83 ± 0.74 abcA	12.26 ± 0.05 deB	40.08 ± 0.79 cdA	69.45 ± 0.69 abcA	−17.78 ± 0.15 abcdA	−6.60 ± 0.12 cdeA	7.27 ± 0.13 eB
CIC	t1	−26.54 ± 0.67 aA	12.89 ± 0.15 deA	39.42 ± 0.52 efA	71.80 ± 2.06 aA	−15.70 ± 0.74 cdeA	−4.75 ± 0.08 aA	8.10 ± 0.00 defgA
	t2	−28.35 ± 1.27 bcB	12.52 ± 0.39 deAB	40.87 ± 1.66 cA	72.70 ± 2.15 abA	−16.47 ± 0.78 abA	−5.31 ± 0.04 abB	7.93 ± 0.20 deA
	t3	−27.32 ± 0.83 abAB	12.38 ± 0.19 cdB	39.69 ± 0.64 cdeA	73.35 ± 1.83 aA	−15.54 ± 0.53 aA	−4.38 ± 0.46 aA	7.37 ± 0.01 eB

1. Data are expressed as the mean ± standard deviation of three replicates. Different capital letters in the same column, between the three harvesting times for the same cultivar, indicate significant differences between the means ($p < 0.05$). Different small letters in the same column, at the same harvesting time for the different cultivars, indicate significant differences between the means ($p < 0.05$).

Fewer differences have been observed comparing the different harvesting times. The parameters that were more affected by olive ripening were Tp1_C and Tp2_C. In detail, Tp1_C increased over time for GEN, CM, OR and decreased for SIV, SEM, and BAM; these trends may be related to the changes of C18:1 observed during ripening (Table 1). Interestingly, for samples not affected by the change in Tp1_C over time, a change in Tp2_C was instead observed. In particular, a decrease in this temperature was observed for TOR, DR, COR, OTT, OTTC, TDF, CIC; it can be related to a decrease in the saturated fatty acids and an increase in the unsaturated ones, with the exception of TDF, CIC for which the opposite trend was observed. Chiavaro et al. [39] observed a significant shift of Ton towards higher temperatures and enlargement of the temperature range as a consequence of ripening on the cooling curves of three monovarietal extra virgin Italian olive oils. These authors suggested an increase in the complexity of oil composition, due to TAG lysis and lipid oxidation. In this study only SIV, OR, and BAM demonstrated a broadening of the crystallization range; however, TDF even showed a narrowing of the transition.

Looking at the heating parameters (Table 4), at t1 significant differences have been observed between the samples. Ton_H ranged from −26.5 °C of CM, OR and CIC, to −37 °C of TOR and DR. Toff_H ranged from 14.5 °C of BAM to 10.5 °C of OR. From these results, it is visible that OR had the narrowest heating transition (Range_H: 37 °C), while TOR, DR, and BAM showed a broader transition (Range_H: 50 °C). Range_H was negatively correlated with C18:0 and C20:0 ($p < 0.05$, R = −465, −0.651); it suggests that the presence of heterogeneous TAGs containing saturated fatty acids formed different polymorphic crystals during the cooling phase, which melt over a wider range of temperatures. The enthalpy of the heating transition ranged from 71.8 J/g of CIC to 64.64 of SIV. This parameter was positively correlated with C18:1 ($p < 0.01$, R = 0.484) and negatively with C16:0 and C18:2 ($p < 0.01$, R = 0.474, 0.431), as already reported in previous studies [18].

Tp1_H ranged from −21.35 °C of SIV to −16 °C of TOR and CIC. Tp2_H ranged from −8 °C of GEN to −4.74 °C of CIC, while Tp3_H ranged from 6 °C of OR to 10.5 °C of BAM. Obtaining a clear correlation of this phenomena with the fatty acid composition is complicated by the kinetic nature of peak p1h and the polymorphisms that characterize p2h and p3h. However, the presence of additional characteristic melting phenomena in the region between peak 1 and peak 2, makes the DSC heating curves of extra virgin olive oil a unique fingerprint for this kind of sample [21].

Comparing the different ripening times, only the olive oil from the cultivar DR did not show any modification. Moreover, Ton_H and ΔH_H were almost stable for all the studied samples. The temperature ranges of the heating transition (Range_H) showed a narrowing tendency for TOR, GEN, SIV, CM, and OTT, as a consequence of the Toff_H shifting towards lower temperatures. Among the three thermal events observed during heating, the first one (p1_h) was exothermic and shifted towards lower temperatures during ripening only for TOR and the three Sardinian cultivars (SIV, SEM, COR). It is not easy to find an explanation of this trend, as it is more related to kinetic phenomena. Tp2_H shifted towards higher temperatures during ripening for TOR, GEN, CM, OR, while it moved to lower temperatures for OTT and OTTC. For these last two samples, it was more clear that this phenomenon could be due to a decrease in the saturated and polyunsaturated fatty acids and an increase in the monounsaturated ones. Tp3_H was the most affected during ripening time; except DR and OR, this thermal event in all other samples shifted towards lower temperatures. These events may be related to the melting of the most saturated TAG polymorphic forms, which tend to decrease over time.

3.3. Viscosity

In this study, all the tested olive oils exhibited a linear relationship between shear stress and shear rate, as expected [18,42], allowing olive oil to be classified as a Newtonian fluid. The viscosity of the samples (Figure 2), calculated by Newton's law (Equation (1)), at t1 ranged from 65.97 mPa*s of SIV to 69.83 mPa*s of TOR, without significant differences between them. These values were in the same order of magnitude as that reported by other authors on virgin olive oils at 25 °C [18,42]. Comparing the viscosity values of the oils obtained from the same cultivar at different ripening times, few differences were measured: COR and OTTC showed, respectively, a decrease and increase in viscosity passing from t1 to t3.

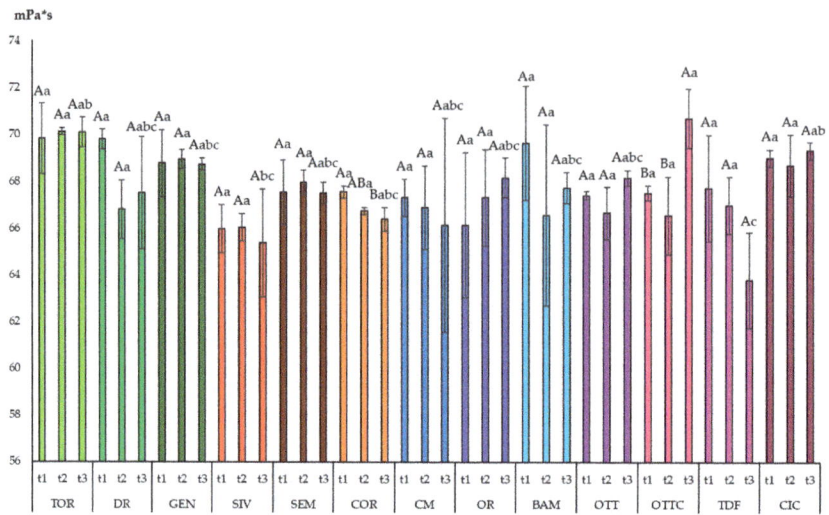

Figure 2. Viscosity of 13 olive oils from minor Italian cultivars harvested at three different maturation stages. Data are expressed as the mean of three replicates ± standard deviations. Different capital letters, between the three harvesting times for the same cultivar, indicate significant differences ($p < 0.05$). Different small letters, at the same harvesting time for the different cultivars, indicate significant differences ($p < 0.05$).

Exploring possible correlations with fatty acids composition, a positive Pearson correlation was found between viscosity and oleic acid ($p < 0.01$; R = 0.276), while an inverse correlation was found with linoleic acid ($p < 0.01$; R = -0.333,). These findings have been already reported by other authors [42,43], as fatty acids with more double bonds, being loosely packed, and exhibiting a more fluid-like behavior.

3.4. Chlorophyll Content

Even though the color is not considered a quality attribute in olive oil quality assessment by panel experts [23], consumers use color as a parameter to evaluate olive oil quality and authenticity [44].

The green color of an extra virgin olive oil is due to the presence of chlorophyll; a photosynthetic pigment extracted from olives during milling. During olive ripening, due to catabolic enzymes, chlorophyll undergoes chemical modifications, involving a shift in color from brilliant green to black while going through several shades of purple/pink [25]. This phenomenon, called véraison, literally means a change of color, and is used as an indicator of the ripening stage. Olive farmers start harvesting the olives when they are in the middle of véraison, before full ripeness [13]. The change of color is due to chlorophyll loss and a concomitant increase in anthocyanin pigmentation [45].

It is assumed that the degree of ripening of the olive fruit, and consequently its chlorophyll content, will determine the amount of chlorophyll in the final oil [26].

Figure 3 shows the levels of chlorophyll found in the thirteen olive oils obtained from minor olive Italian cultivars, harvested in three different periods, shifted about two weeks from each other. At time 1 (t1), which represents the optimal olive ripening period, according to the farmers' experience, the chlorophyll levels ranged from 58.5 mg/kg of BAM, to 5.6 mg/kg of TDF. In general, at t1, the Apulian cultivars (CM, OR, BAM) and SEM, which is a Sardinian cultivar, showed the highest levels of chlorophyll. On the other hand, the Calabrian cultivars (OTT, OTTC, TDF, CIC) had the lowest level of chlorophyll. The amount of chlorophylls in olive oil depends on the olive cultivar, pedoclimatic conditions, and agronomic practices [46].

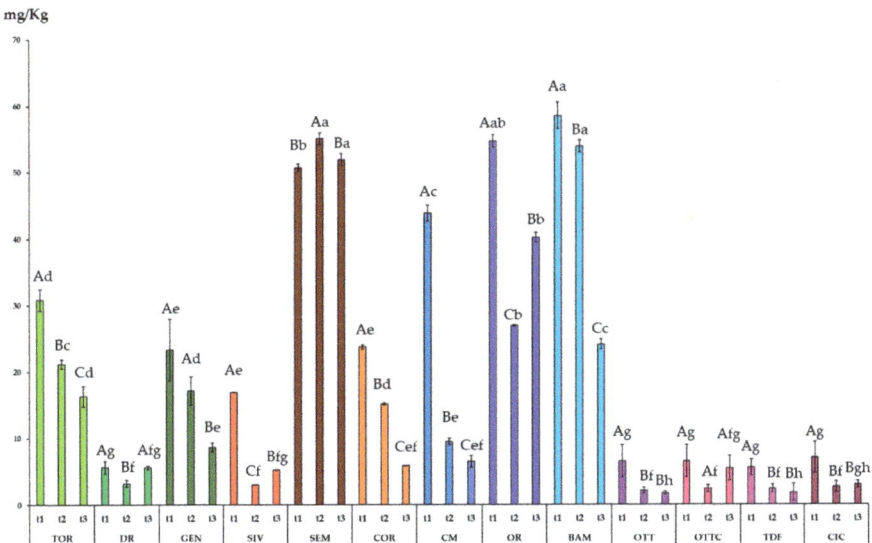

Figure 3. Chlorophyll content of 13 olive oils from minor Italian cultivars harvested at three different maturation stages. Data are expressed as the mean of three replicates ± standard deviations. Different capital letters, between the three harvesting times for the same cultivar, indicate significant differences ($p < 0.05$). Different small letters, at the same harvesting time for the different cultivars, indicate significant differences ($p < 0.05$).

Comparing the different olives' harvesting times, in most of the cases the amount of chlorophyll decreased over time. In particular, passing from t1 to t3, the highest chlorophyll loss was registered for CM, which undergoes an 85% loss. Similarly, Criado et al. [47] studied the pigment content in fruit from different olive varieties in six consecutive stages of ripeness. They found that the concentrations of chlorophyll decreased continuously in all the varieties during ripening.

In a few cases, the amount of chlorophyll remained rather constant between t1 and t3 (DR, SEM, OTTC).

3.5. Color

As previously reported, the color of olive oil is strictly connected with its pigment content. Confirming this hypothesis, significant correlations between chlorophyll content and colorimetric parameters (Table 5) have been found. In particular, negative Pearson correlations between chlorophylls and the chromatic parameters L ($p < 0.01$ R = -0.799) and a^* ($p < 0.01$ R = -0.637) were observed. The higher the chlorophyll content, the darker and greener the olive oil. Surprisingly, the correlation between chlorophylls and the chromatic parameters b^* was positive ($p < 0.01$ R = 0.668). The b^* parameter represents the yellow tones; it was previously related to the carotenoid content in olive oil [47]. We assume that, in this study, the method used for chlorophyll detection measured both chlorophylls a and b, known to generate intense blue-green and yellow-green shades, respectively [25]. Possibly, the amount of chlorophyll b may have influenced this result.

Table 5. Color parameters of 13 minor Italian olive oils harvested at three maturation stages [1].

		L	a^*	b^*
TOR	t1	53.67 ± 0.58 Bc	−7.33 ± 0.58 Acd	53.00 ± 3.46 Aab
	t2	55.00 ± 0.00 Acd	−7.00 ± 0.00 Ade	46.00 ± 2.65 ABb
	t3	55.33 ± 0.58 Ac	−7.00 ± 0.00 Ad	42.33 ± 3.51 Bb
DR	t1	56.33 ± 0.58 Aa	−6.00 ± 0.00 Aa	25.33 ± 3.21 Bd
	t2	56.67 ± 0.58 Aab	−6.00 ± 0.00 Acd	25.00 ± 0.00 Bc
	t3	56.33 ± 0.58 Abc	−6.67 ± 0.58 Acd	36.00 ± 5.20 Abc
GEN	t1	55.67 ± 0.58 Aab	−7.00 ± 0.00 Abc	43.33 ± 8.14 Aabc
	t2	55.00 ± 1.00 Acd	−7.67 ± 0.58 Ae	43.67 ± 4.16 Ab
	t3	56.00 ± 0.00 Abc	−7.33 ± 0.58 Ad	44.33 ± 5.51 Ab
SIV	t1	55.33 ± 0.58 Bab	−8.00 ± 0.00 Cd	46.00 ± 0.00 Aabc
	t2	58.00 ± 0.00 Aa	−6.00 ± 0.58 Acd	26.67 ± 0.58 Cc
	t3	57.67 ± 0.58 Aa	−7.00 ± 0.58 Bd	30.33 ± 0.58 Bc
SEM	t1	46.33 ± 0.58 Be	−8.00 ± 0.58 Bd	50.67 ± 0.58 Cabc
	t2	50.67 ± 0.58 Ag	−7.67 ± 0.58 ABe	53.67 ± 0.58 Ba
	t3	51.00 ± 0.00 Ae	−7.00 ± 0.00 Ad	55.00 ± 0.00 Aa
COR	t1	55.00 ± 0.00 Babc	−8.00 ± 0.00 Bd	56.00 ± 0.00 Aa
	t2	56.00 ± 0.00 Abc	−8.00 ± 0.00 Be	54.00 ± 0.00 Ba
	t3	56.50 ± 0.50 Aabc	−7.00 ± 0.00 Ad	44.00 ± 1.00 Cb
CM	t1	51.67 ± 0.58 Bd	−7.33 ± 0.58 Acd	55.33 ± 1.15 Ba
	t2	56.00 ± 0.00 Abc	−7.00 ± 0.00 Ade	58.67 ± 0.58 Aa
	t3	56.00 ± 0.00 Abc	−7.00 ± 0.00 Ad	59.00 ± 0.00 Aa
OR	t1	47.00 ± 0.00 Ce	−9.00 ± 0.00 Be	51.00 ± 0.00 Cabc
	t2	52.67 ± 0.58 Af	−7.67 ± 0.58 Ae	56.00 ± 0.00 Aa
	t3	49.67 ± 0.58 Bf	−7.00 ± 0.00 Ad	53.67 ± 0.58 Ba
BAM	t1	47.33 ± 0.58 Ce	−9.00 ± 0.00 Be	51.33 ± 0.58 Cabc
	t2	56.00 ± 0.00 Abc	−7.00 ± 0.00 Ade	58.67 ± 0.58 Aa
	t3	54.00 ± 0.00 Bd	−7.33 ± 0.58 Ad	56.33 ± 0.58 Ba
OTT	t1	55.00 ± 0.00 Babc	−6.33 ± 0.58 Bab	39.67 ± 2.52 Abc
	t2	54.33 ± 0.58 Bde	−4.33 ± 0.58 Aab	20.67 ± 2.52 Bcd
	t3	56.50 ± 0.50 Aabc	−4.00 ± 0.00 Aab	18.50 ± 2.50 Bd
OTTC	t1	55.00 ± 0.00 Babc	−6.33 ± 0.58 Bab	38.67 ± 10.79 Acd
	t2	55.00 ± 0.00 Bcd	−5.00 ± 0.00 Abc	26.33 ± 3.21 ABc
	t3	56.00 ± 0.00 Abc	−4.50 ± 0.50 Ab	17.50 ± 0.50 Bd
TDF	t1	54.67 ± 0.58 ABbc	−7.00 ± 0.00 Bbc	43.00 ± 3.00 Aabc
	t2	53.33 ± 0.58 Bef	−3.33 ± 0.58 Aa	15.67 ± 1.15 Bd
	t3	55.67 ± 0.58 Abc	−3.33 ± 0.58 Aa	15.67 ± 0.58 Bd
CIC	t1	54.33 ± 1.15 Bbc	−7.00 ± 0.00 Bbc	48.67 ± 7.77 Aabc
	t2	55.33 ± 0.58 ABbcd	−5.67 ± 0.58 Ac	25.33 ± 4.62 Bc
	t3	56.67 ± 0.58 Aab	−5.67 ± 0.58 Ac	30.3 ± 35.13 Bc

[1.] Data are expressed as the mean ± standard deviation of three replicates. Different capital letters in the same column, between the three harvesting times for the same cultivar, indicate significant differences between the means ($p < 0.05$). Different small letters in the same column, at the same harvesting time for the different cultivars, indicate significant differences between the means ($p < 0.05$).

Looking at the differences between the cultivars at t1, the L values ranged from 46–47 of SEM and OR to 56 of DR, resulting in, respectively, the darkest and the lightest samples. Negative a^* values indicate the green color. At t1, OR, and BAM were the greenest samples, with values of −9. On the other hand, DR was the least green sample with values of −6. The color parameter b^* indicates yellow tones; a large variability of this parameter was observed between the samples at t1. The values of b^* ranged from 25 of DR to 56 of COR, resulting in, respectively, olive oils that were less or more yellow.

Focusing on the differences between the harvesting times, for L the general tendency was to increase over time, in relation to the chlorophyll decrease. This phenomenon was especially visible passing from t1 to t2 for most of the cultivars. BAM underwent the highest lightening from t1 to t3 (14%), while DR and GEN did not show any significant change of L. The parameter a^* underwent a general increase from t1 to t3, indicating a progressive loss of greenness. TOR, DR, GEN, and the Apulian cultivar CM did not show significant differences of a^*. On the other hand, the highest loss of greenness was observed for the Calabrian cultivars already at t1. In particular, TDF suffered around a 52% loss of this value prolonging the harvesting time. A large variability between the cultivars was observed in the b^* value trend. A general decrease in the b^* value was observed for the Calabrian cultivars (OTT, OTTC, TDF, CIC), particularly for TDF with a 63.5% loss. On the other hand, the b^* value of the Apulian cultivars (CM, OR, BAM) increased during this time, and DR showed the highest increase in b^* from t1 to t3 (42%). Criado and co-workers [47] observed a decrease in L, an increase in a^*, and a decrease in b^* in two olive oil samples, in relation to the ripening stage of the olive fruit.

3.6. PCA

The use of principal component analysis to discriminate between olive oil samples, based on their chemical, physical, and thermal properties, has already been applied successfully [18,41]. In this study, the use of this multivariate statistical technique has been applied to tentatively discriminate between the 13 examined Italian olive oils based on their geographical origin, botanical origin, and olive harvesting time. Starting from twenty-three variables, only nine of them were significant after factor extraction, using an eigenvalue higher than 0.7 as selection criteria. The first two principal components in the PCA accounted for 74.64% of the total variance. Figure 4 shows the projection of the variables on a factor plane.

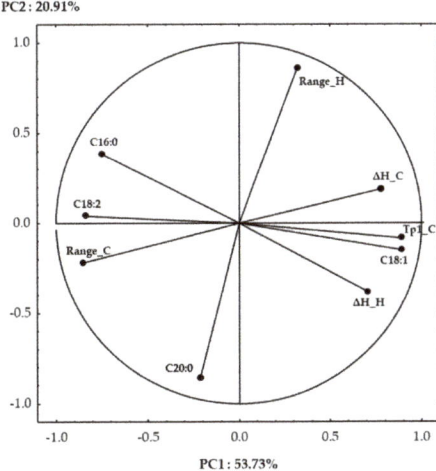

Figure 4. Projections of the variables on a factor plane.

Most of the selected variables were better described in PC1, which was the most influencing component, able to describe 53.73% of the total variance. Among the fatty acids, C18:1 showed positive factor loadings on PC1, while C16:0 and C18:2 showed negative ones. Considering the thermal properties, heating and cooling enthalpies (ΔH_H, ΔH_C) and the temperature of peak 1 (Tp1_C) showed positive loadings, while the range of cooling (Range_C) showed negative ones. The shift of Tp1_C towards lower temperature and the intensification of ΔH_H and ΔH_C with an increase in C18:1 and an opposite trend with the amount of C16:0 and C18:2 has been already documented [32]. Only two variables were described on PC2, which had a lower influence on the total variance (20.91%). Range_H and C20:0 showed positive and negative values on PC2, respectively. Although Bayés-García and co-workers [40] stated that minor fatty acids do not have an influence on the olive oil thermal transition, in this study, arachidic acid was an influencing parameter for olive oil sample discrimination.

From the factor analysis, the viscosity, chlorophyll content, and all the color parameters were not able to discriminate between the studied olive oils, and, thus, they were excluded from the test.

Figure 5 shows the projection of the cases on a factor plan. It is evident that the main variables influencing clustering were cultivar and geographical of origin, and the harvesting time had less influence. In detail, three main clusters may be distinguished. In the first cluster, the two Sardinian cultivars SIV and COR were well described by the higher concentration of C16:0 and C18:2, with consequent lower enthalpies and values of Tp1_C. The second cluster grouped all the Calabrian cultivars (OTT, OTTC, TDF, CIC), the Apulian cultivars CDM and OLR, together with the Sardinian cultivar SEM; it was characterized by higher levels of C:20, and lower Range H. Within this cluster, the samples CDM, OLR, and OTTC at the latest harvesting stages (t2–t3), shifted towards more positive PC1, indicating the influence of the C18:1 increase during ripening. This phenomenon was even more evident for the three cultivars harvested in Abruzzo (GEN, DR, TOR) for which the t3 largely shifted to the right, indicating an increase in the oleic acid at the last stage of ripening. GEN, DR, and TOR formed together with the Apulian cultivar BAM the third cluster. This cluster, in which a larger distance between the samples was observed, was characterized by high values of both Tp1_C and oleic acid.

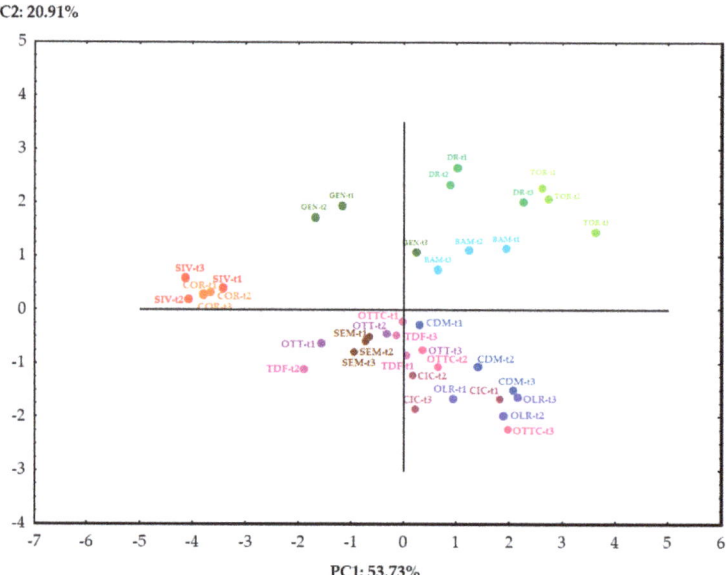

Figure 5. Projections of the cases on a factor plane.

Interestingly, also a distribution of the samples according to their region of origin can be observed. In particular, all the Sardinian cultivars showed negative loadings on PC1, while all the Apulian cultivars showed positive loadings on PC1; the Abruzzi and Calabrian cultivars showed, on the other hand, positive and negative loadings on PC2, respectively.

4. Conclusions

Thermal analysis is a useful tool to discriminate the EVOOs according to their botanical and geographical origin, which influences the chemical composition of the oil. The Abruzzi cultivars (TOR, DR, GEN), together with BAM, were well differentiated from the others, especially for their higher arachidic acid content, which negatively influenced the range of the heating transition. However, while the Abruzzi cultivars showed an increase in $Tp1\_C$ during ripening, BAM showed an opposite trend, in correlation to the oleic acid content. The Sardinian cultivars SIV and COR were mostly characterized by lower values of oleic acid and, consequently, lower transition enthalpies both upon cooling and heating. The Apulian CM, OR, the Calabrian TDF, OTT, OTTC, CIC, and the Sardinian SEM showed similar, intermediate behaviors among others. The olive ripening stage did not particularly influence the olive thermal behavior. The EVOOs' viscosity and color parameters, despite the correlation with the fatty acid composition and chlorophyll content, respectively, were not selected as a discriminating variables.

Author Contributions: Conceptualization, M.P. and E.C.; methodology, M.P., G.D., P.C., F.F., A.P; formal analysis, M.P., G.D., P.C., F.F., A.P.; data curation, M.P., G.D.; writing—original draft preparation, M.P., G.D., P.C.; writing—review and editing, M.P., P.C., E.C.; supervision, E.C.; funding acquisition, E.C. All authors have read and agreed to the published version of the manuscript.

Funding: This research was funded by AGER 2 PROJECT, grant number 2016-0105.

Conflicts of Interest: The authors declare no conflict of interest.

References

1. Willett, W.C.; Sacks, F.; Trichopoulou, A.; Drescher, G.; Ferro-Luzzi, A.; Helsing, E.; Trichopoulos, D. Mediterranean diet pyramid: A cultural model for healthy eating. *Am. J. Clin. Nutr.* **1995**, *61*, 1402S–1406S. [CrossRef]
2. Escrich, E.; Moral, R.; Solanas, M. Olive oil, an essential component of the Mediterranean diet, and breast cancer. *Public Health Nutr.* **2011**, *14*, 2323–2332. [CrossRef]
3. Ilarioni, L.; Proietti, P. Olive tree cultivars. In *The Extra-Virgin Olive Oil Handbook*; John Wiley & Sons, Ltd.: Hoboken, NJ, USA, 2014; pp. 59–67. ISBN 9781118460412.
4. Allalout, A.; Krichène, D.; Methenni, K.; Taamalli, A.; Daoud, D.; Zarrouk, M. Behavior of super-intensive spanish and greek olive cultivars grown in northern tunisia. *J. Food Biochem.* **2011**, *35*, 27–43. [CrossRef]
5. Conte, P.; Squeo, G.; Difonzo, G.; Caponio, F.; Fadda, C.; Del Caro, A.; Urgeghe, P.P.; Montanari, L.; Montinaro, A.; Piga, A. Change in quality during ripening of olive fruits and related oils extracted from three minor autochthonous Sardinian cultivars. *Emir. J. Food Agric.* **2019**, *31*, 196–205. [CrossRef]
6. The Council of the European Communities. Council regulation (EEC) N. 2081/92 of 14 July 1992 on the protection of geographical indications and designations of origin for agricultural products and foodstuffs. *Off. J. Eur. Communities* **1992**, *208*, 1–8.
7. Miazzi, M.M.; di Rienzo, V.; Mascio, I.; Montemurro, C.; Sion, S.; Sabetta, W.; Vivaldi, G.A.; Camposeo, S.; Caponio, F.; Squeo, G.; et al. Re.Ger.O.P.: An Integrated Project for the Recovery of Ancient and Rare Olive Germplasm. *Front. Plant Sci.* **2020**, *11*, 73. [CrossRef]
8. Muzzalupo, I. Olive Germplasm—Italian Catalogue of Olive Varieties. In *Olive Germplasm—Italian Catalogue of Olive Varieties*; InTech: Rijeka, Croatia, 2012; pp. 1–5.
9. Rotondi, A.; Alfei, B.; Magli, M.; Pannelli, G. Influence of genetic matrix and crop year on chemical and sensory profiles of Italian monovarietal extra-virgin olive oils. *J. Sci. Food Agric.* **2010**, *90*, 2641–2648. [CrossRef]
10. Rotondi, A.; Ganino, T.; Beghè, D.; Di Virgilio, N.; Morrone, L.; Fabbri, A.; Neri, L. Genetic and landscape characterization of ancient autochthonous olive trees in northern Italy. *Plant Biosyst.* **2018**, *152*, 1067–1074. [CrossRef]
11. Piscopo, A.; De Bruno, A.; Zappia, A.; Ventre, C.; Poiana, M. Characterization of monovarietal olive oils obtained from mills of Calabria region (Southern Italy). *Food Chem.* **2016**, *213*, 313–318. [CrossRef] [PubMed]
12. Di Vaio, C.; Nocerino, S.; Paduano, A.; Sacchi, R. Characterization and evaluation of olive germplasm in southern Italy. *J. Sci. Food Agric.* **2013**, *93*, 2458–2462. [CrossRef]
13. Squeo, G.; Silletti, R.; Mangini, G.; Summo, C.; Caponio, F. The Potential of Apulian Olive Biodiversity: The Case of Oliva Rossa Virgin Olive Oil. *Foods* **2021**, *10*, 369. [CrossRef]
14. Skiada, V.; Tsarouhas, P.; Varzakas, T. Comparison and discrimination of two major monocultivar extra virgin olive oils in the southern region of Peloponnese, according to specific compositional/traceability markers. *Foods* **2020**, *9*, 155. [CrossRef] [PubMed]
15. Ou, G.; Hu, R.; Zhang, L.; Li, P.; Luo, X.; Zhang, Z. Advanced detection methods for traceability of origin and authenticity of olive oils. *Anal. Methods* **2015**, *7*, 5731–5739. [CrossRef]

16. Chiavaro, E.; Rodriguez-Estrada, M.T.; Barnaba, C.; Vittadini, E.; Cerretani, L.; Bendini, A. Differential scanning calorimetry: A potential tool for discrimination of olive oil commercial categories. *Anal. Chim. Acta.* **2008**, *625*, 215–226. [CrossRef] [PubMed]
17. Vecchio Ciprioti, S.; Paciulli, M.; Chiavaro, E. Application of different thermal analysis techniques to characterize oxidized olive oils. *Eur. J. Lipid Sci. Technol.* **2017**, *119*, 1600074. [CrossRef]
18. Ben Rached, M.; Paciulli, M.; Pugliese, A.; Abdallah, M.; Boujnah, D.; Zarrouk, M.; Guerfel, M.; Chiavaro, E. Effect of the soil nature on selected chemico-physical and thermal parameters of extra virgin olive oils from cv Chemlali. *Ital. J. Food Sci.* **2017**, *29*, 74–89.
19. Jafari, M.; Kadivar, M.; Keramat, J. Detection of adulteration in Iranian olive oils using instrumental (GC, NMR, DSC) methods. *J. Am. Oil Chem. Soc.* **2009**, *86*, 103–110. [CrossRef]
20. Chatziantoniou, S.E.; Triantafillou, D.J.; Karayannakidis, P.D.; Diamantopoulos, E. Traceability monitoring of Greek extra virgin olive oil by differential scanning calorimetry. *Thermochim. Acta* **2014**, *576*, 9–17. [CrossRef]
21. Mallamace, D.; Vasi, S.; Corsaro, C.; Naccari, C.; Clodoveo, M.L.; Dugo, G.; Cicero, N. Calorimetric analysis points out the physical-chemistry of organic olive oils and reveals the geographical origin. *Physica A* **2017**, *486*, 925–932. [CrossRef]
22. Maggio, R.M.; Barnaba, C.; Cerretani, L.; Paciulli, M.; Chiavaro, E. Study of the influence of triacylglycerol composition on DSC cooling curves of extra virgin olive oil by chemometric data processing. *J. Therm. Anal. Calorim.* **2014**, *115*, 2037–2044. [CrossRef]
23. Commission of the European Communities. Commission regulation (EEC) No 2568/91 of 11 July of 1991 and subsequent integrations and amendments. *Off. J. Eur. Communities* **1991**, *248*, 1–83.
24. Koriyama, T.; Wongso, S.; Watanabe, K.; Abe, H.J. Fatty acid compositions of oil species affect the 5 basic taste perceptions. *J. Food Sci.* **2002**, *67*, 868–873. [CrossRef]
25. Paciulli, M.; Palermo, M.; Chiavaro, E.; Pellegrini, N. Chlorophylls and Colour Changes in Cooked Vegetables. In *Fruit and Vegetable Phytochemicals: Chemistry and Human Health*, 2nd ed.; Elhadi, M.Y., Ed.; John Wiley & Sons: Hoboken, NJ, USA, 2017; Volume 2, pp. 703–719.
26. Giuliani, A.; Cerretani, L.; Cichelli, A. Chlorophylls in olive and in olive oil: Chemistry and occurrences. *Crit. Rev. Food Sci. Nutr.* **2011**, *51*, 678–690. [CrossRef]
27. Gandul-Rojas, B.; Cepero, M.R.L.; Mínguez-Mosquera, M.I. Use of chlorophyll and carotenoid pigment composition to determine authenticity of virgin olive oil. *J. Am. Oil Chem. Soc.* **2000**, *77*, 853–858. [CrossRef]
28. Italian Ministry of Agricultural, Food and Forestry Policies. Available online: https://www.politicheagricole.it/ (accessed on 18 April 2021).
29. Alamprese, C.; Grassi, S.; Tugnolo, A.; Casiraghi, E. Prediction of olive ripening degree combining image analysis and FT-NIR spectroscopy for virgin olive oil optimisation. *Food Control.* **2021**, *123*, 107755. [CrossRef]
30. Piscopo, A.; Zappia, A.; De Bruno, A.; Poiana, M. Effect of the harvesting time on the quality of olive oils produced in Calabria. *Eur. J. Lipid Sci. Technol.* **2018**, *120*, 1700304. [CrossRef]
31. Difonzo, G.; Pasqualone, A.; Silletti, R.; Cosmai, L.; Summo, C.; Paradiso, V.M.; Caponio, F. Use of olive leaf extract to reduce lipid oxidation of baked snacks. *Food Res. Int.* **2018**, *108*, 48–56. [CrossRef]
32. Cerretani, L.; Bendini, A.; Rinaldi, M.; Paciulli, M.; Vecchio, S.; Chiavaro, E. DSC evaluation of extra virgin olive oil stability under accelerated oxidative test: Effect of fatty acid composition and phenol contents. *J. Oleo Sci.* **2012**, *61*, 303–309. [CrossRef] [PubMed]
33. Zago, L.; Squeo, G.; Bertoncini, E.I.; Difonzo, G.; Caponio, F. Chemical and sensory characterization of Brazilian virgin olive oils. *Food Res. Int.* **2019**, *126*, 108588. [PubMed]
34. AOCS official method Cc 13i-96. Determination of Chlorophyll Pigments in Crude Vegetable Oils. In *Official Methods and Recommended Practices of the AOCS*, 7th ed.; AOCS Press: Washington, DC, USA, 2017.
35. Inglese, P.; Famiani, F.; Galvano, F.; Servili, M.; Esposto, S.; Urbani, S. Factors Affecting Extra-Virgin Olive Oil Composition. In *Horticultural Reviews*; Janik, J., Ed.; Johnwiley & Sons, Ltsd.: West Sussex, UK, 2011; Volume 38, pp. 83–148.
36. Rondanini, D.P.; Castro, D.N.; Searles, P.S.; Rousseaux, M.C. Contrasting patterns of fatty acid composition and oil accumulation during fruit growth in several olive varieties and locations in a non-Mediterranean region. *Eur. J. Agron.* **2014**, *52*, 237–246. [CrossRef]
37. Rondanini, D.P.; Castro, D.N.; Searles, P.S.; Rousseaux, M.C. Fatty acid profiles of varietal virgin olive oils (*Olea europaea* L.) From mature orchards in warm arid valleys of Northwestern Argentina (La Rioja). *Grasas Aceites.* **2011**, *62*, 399–409.
38. Chiavaro, E.; Vittadini, E.; Rodriguez-Estrada, M.T.; Cerretani, L.; Bendini, A. Monovarietal extra virgin olive oils. Correlation between thermal properties and chemical composition: Heating thermograms. *J. Agric. Food Chem.* **2008**, *56*, 496–501. [CrossRef]
39. Chiavaro, E.; Vittadini, E.; Rodriguez-Estrada, M.T.; Cerretani, L.; Bonoli, M.; Bendini, A.; Lercker, G. Monovarietal extra virgin olive oils: Correlation between thermal properties and chemical composition. *J. Agric. Food Chem.* **2007**, *55*, 10779–10786. [CrossRef]
40. Bayés-García, L.; Calvet, T.; Cuevas-Diarte, M.A.; Ueno, S. From trioleoyl glycerol to extra virgin olive oil through multicomponent triacylglycerol mixtures: Crystallization and polymorphic transformation examined with differential scanning calorimetry and X-ray diffration techniques. *Food Res. Int.* **2017**, *99*, 476–484.
41. Caponio, F.; Chiavaro, E.; Paradiso, V.M.; Paciulli, M.; Summo, C.; Cerretani, L.; Gomes, T. Chemical and thermal evaluation of olive oil refining at different oxidative levels. *Eur. J. Lipid Sci. Technol.* **2013**, *115*, 1146–1154. [CrossRef]

42. Gila, A.; Jiménez, A.; Beltrán, G.; Romero, A. Correlation of fatty acid composition of virgin olive oil with thermal and physical properties. *Eur. J. Lipid Sci. Technol.* **2015**, *117*, 366–376. [CrossRef]
43. Kim, J.; Kim, D.N.; Lee, S.H.; Yoo, S.-H.; Lee, S. Correlation of fatty acid composition of vegetable oils with rheological behaviour and oil uptake. *Food Chem.* **2010**, *118*, 398–402.
44. Perito, M.A.; Sacchetti, G.; Di Mattia, C.D.; Chiodo, E.; Pittia, P.; Saguy, I.S.; Cohen, E. Buy local! Familiarity and preferences for extra virgin olive oil of Italian consumers. *J. Food Prod. Mark.* **2019**, *25*, 462–477.
45. Roca, M.; Mínguez-Mosquera, M.I. Changes in chloroplast pigments of olive varieties during fruit ripening. *J. Agric. Food Chem.* **2001**, *49*, 832–839. [CrossRef]
46. Piscopo, A.; Mafrica, R.; De Bruno, A.; Romeo, R.; Santacaterina, S.; Poiana, M. Characterization of Olive Oils Obtained from Minor Accessions in Calabria (Southern Italy). *Foods* **2021**, *10*, 305. [CrossRef]
47. Criado, M.N.; Motilva, M.J.; Goni, M.; Romero, M.P. Comparative study of the effect of the maturation process of the olive fruit on the chlorophyll and carotenoid fractions of drupes and virgin oils from Arbequina and Farga cultivars. *Food Chem.* **2007**, *100*, 748–755. [CrossRef]

Article

Paving the Way to Food Grade Analytical Chemistry: Use of a Natural Deep Eutectic Solvent to Determine Total Hydroxytyrosol and Tyrosol in Extra Virgin Olive Oils

Vito Michele Paradiso [1,2,*], Francesco Longobardi [3,*], Stefania Fortunato [2], Pasqua Rotondi [3], Maria Bellumori [4], Lorenzo Cecchi [4], Pinalysa Cosma [3], Nadia Mulinacci [4] and Francesco Caponio [2]

1. Department of Biological and Environmental Sciences and Technologies, University of Salento, Centro Ecotekne, S.P. 6 Lecce-Monteroni, I-73100 Lecce, Italy
2. Department of Soil, Plant and Food Sciences, University of Bari, Via Amendola 165/a, I-70126 Bari, Italy; stefania.fortunato@uniba.it (S.F.); francesco.caponio@uniba.it (F.C.)
3. Department of Chemistry, University of Bari, Via Orabona 4, I-70126 Bari, Italy; pattyrotondi83@alice.it (P.R.); pinalysa.cosma@uniba.it (P.C.)
4. Department of NEUROFARBA, and Multidisciplinary Centre of Research on Food Sciences (M.C.R.F.S.-Ce.R.A.), University of Firenze, Via Ugo Schiff 6, I-50019 Sesto F.no (Firenze), Italy; maria.bellumori@unifi.it (M.B.); lo.cecchi@unifi.it (L.C.); nadia.mulinacci@unifi.it (N.M.)

* Correspondence: vito.paradiso@unisalento.it (V.M.P.); francesco.longobardi@uniba.it (F.L.); Tel.: +39-334-3427275 (V.M.P.); +39-328-2851916 (F.L.)

Citation: Paradiso, V.M.; Longobardi, F.; Fortunato, S.; Rotondi, P.; Bellumori, M.; Cecchi, L.; Cosma, P.; Mulinacci, N.; Caponio, F. Paving the Way to Food Grade Analytical Chemistry: Use of a Natural Deep Eutectic Solvent to Determine Total Hydroxytyrosol and Tyrosol in Extra Virgin Olive Oils. *Foods* **2021**, *10*, 677. https://doi.org/10.3390/foods10030677

Academic Editor: Francesca Venturi

Received: 4 March 2021
Accepted: 16 March 2021
Published: 22 March 2021

Publisher's Note: MDPI stays neutral with regard to jurisdictional claims in published maps and institutional affiliations.

Copyright: © 2021 by the authors. Licensee MDPI, Basel, Switzerland. This article is an open access article distributed under the terms and conditions of the Creative Commons Attribution (CC BY) license (https://creativecommons.org/licenses/by/4.0/).

Abstract: Extra virgin olive oil (EVOO) is well known for containing relevant amounts of healthy phenolic compounds. The European Food Safety Authority (EFSA) allowed a health claim for labelling olive oils containing a minimum amount of hydroxytyrosol (OHTyr) and its derivatives, including tyrosol (Tyr). Therefore, harmonized and standardized analytical protocols are required in support of an effective application of the health claim. Acid hydrolysis performed after extraction and before chromatographic analysis has been shown to be a feasible approach. Nevertheless, other fast, green, and easy methods could be useful for on-site screening and monitoring applications. In the present research, a natural deep eutectic solvent (NADES) composed of lactic acid and glucose was used to perform a liquid/liquid extraction on EVOO samples, followed by UV-spectrophotometric analysis. The spectral features of the extracts were related with the content of total OHTyr and Tyr, determined by the acid hydrolysis method. The second derivative of spectra allowed focusing on three single wavelengths (i.e., 299 nm, 290 nm, and 282 nm) significantly related with total OHTyr, total Tyr, and their sum, respectively. In particular, the sum of OHTyr and Tyr could be determined with a root mean square error of prediction of 29.5 mg kg^{-1}, while the limits of quantitation and detection were respectively 11.8 and 4.9 mg kg^{-1}. The proposed method, therefore, represents an easy screening tool, with the use of a green, food-derived solvent, and could be considered as an attempt to pave the way for food grade analytical chemistry.

Keywords: phenolic compounds; acidic hydrolysis; derivative UV spectroscopy; green chemistry; screening methods; health claim

1. Introduction

The health properties of phenolic compounds contained in extra virgin olive oil (EVOO) have been clearly established by many scientific papers [1–3]. On this basis, according to the European Food Safety Authority (EFSA), since 2012 European regulation has allowed the use of a health claim in olive oil labeling [4]. The health claim states as follows: "Olive oil polyphenols contribute to the protection of blood lipids from oxidative stress" and can be applied to olive oils containing at least 5 mg of hydroxytyrosol (OHTyr) and its derivatives (e.g., oleuropein complex and tyrosol) per 20 g of product.

Based on this regulation, scholars focused their attention on analytical issues related to the assessment of the suitability of oils to be labelled with the health claim. Several analytical methods have been proposed, usually comprising an extraction step followed by separation by means of liquid chromatography, gas chromatography, or capillary electrophoresis, and a suitable detection method.

Irrespective of the separation and detection methods, the quantification step of the target compounds remains a critical issue. In fact, OHTyr and tyrosol (Tyr) are present in EVOO in their free form, as well as esterified in several derivatives [5]. Therefore, the quantitation of each derivative is difficult to obtain. A widely adopted approach to overcome this difficulty is to carry out a hydrolysis step prior to separation and analysis, in order to quantify, as free forms, the total of free and linked OHTyr and Tyr. This approach has been adopted either directly on the oil or on the phenolic extract, with quite satisfactory performances [6–9].

Nevertheless, the adoption of separation methods for the analytical determination, though being affordable and sensitive, requires expensive equipment, toxic/pollutant reagents, and trained operators. The availability of easy, less expensive, and operator- and environment-friendly methods, rather than being an alternative, could be a valid analytical complement, useful for screening purposes, even in oil mills or bottling plants as a quality monitoring or a decision supporting tool [10]. Some efforts in this direction have been made. The Folin–Ciocalteu assay, a widely used spectrophotometric method, has provided good preliminary results [11].

In addition, natural deep eutectic solvents (NADES) have been proposed as solvents for easy screening methods. NADES are green solvents consisting of mixtures of one or more hydrogen bond acceptor–donor pair that, in appropriate molar ratios, generate strong intermolecular interactions [12]. Compared to deep eutectic solvents (DES), NADES are obtained from molecules naturally present in living organisms as metabolites [13]. They are being increasingly used for analytical purposes [12,14]. NADES have been successfully applied in the analysis of EVOO phenolic compounds as extraction media prior to liquid chromatography [15], electrochemical analysis [16], and spectrophotometric analysis [10,17]. In one case, NADES extraction and direct spectrophotometric analysis allowed assessing the amount of OHTyr and Tyr derivatives determined as the sum of the free and linked forms determined by high-performance liquid chromatography (HPLC), thus providing a useful tool to label EVOO according to the EU health claim [10].

In view of a better harmonization of the green screening methods with the candidate official methods related to the health claim, there is still the need for easy and green methods allowing assessing total OHTyr and Tyr as free forms, as determined by emerging hydrolysis methods. The present research was therefore aimed at setting up the spectrophotometric determination of total OHTyr and Tyr, free and linked, after extraction with a NADES composed of lactic acid, glucose, and water.

2. Materials and Methods

2.1. Reagents and Oil Samples

Glucose ($\geq$99.5%), lactic acid (90%), formic acid, sulfuric acid (95.0–98.0%), methanol and acetonitrile of HPLC grade, methanol and ethanol of analytical grade, hydroxytyrosol, and tyrosol were purchased from Sigma-Aldrich (Sigma-Aldrich Co. LLC, St. Louis, MO, USA). Ultrapure water was obtained from an Elga Purelab Option R system (Veolia Environnement S.A., Paris, France). Extra virgin olive oil (EVOO) samples ($n = 26$) were obtained from producers and research laboratories. They differed by geographical origin (different Italian regions), cultivar, olive maturity, and extraction technology.

2.2. NADES Preparation

The NADES was obtained according to a previous work, with slight modifications [17]. Lactic acid, glucose, and water (5:1:3 molar ratio) were mixed by means of a magnetic stirrer at 50 °C for about 90 min, until obtaining a clear solution. Further dilution of the components in such molar ratio was carried out with 20% (v/v) water to reduce solvent viscosity. Previous studies reported the possibility of using water to tailor the solvent properties, mainly viscosity [18,19]. In particular, Pisano et al. proved that the same NADES used in this study held its supramolecular structure throughout dilutions [19].

2.3. Extraction with NADES and Spectrophotometric Analysis of the Extract (NADES-UV Method)

The EVOO sample (0.5 g) was submitted to extraction with 5 mL of NADES. After intense agitation with a vortex (5 min), centrifugation was performed for 10 min at 6000 rpm. The lower layer (NADES plus phenolics) was recovered, centrifuged again at 9000 rpm for 5 min and, after being recovered, finally filtered at 0.45 µm using nylon filters (VWR International, Center Valley, PA, USA).

The NADES extracts were analyzed by UV-Vis spectrophotometry in the wavelength range 250–400 nm by means of an Agilent Cary 60 spectrophotometer (Agilent Technologies, Santa Clara, CA, USA). The acquisition parameters were the following: 1 cm optical path, 2 nm slit, and 60 nm min^{-1} scan rate. Pure NADES was used for baseline correction. For quantitation purposes, calibration curves of OHTyr and Tyr in NADES were built in the ranges 3–30 mg L^{-1} and 5–40 mg L^{-1}, and at wavelengths 299 nm and 290 nm, respectively. A total phenolic calibration curve was also built at 282 nm in the range of total 8–70 mg L^{-1} by using mixtures of OHTyr and Tyr, and employing for each compound the same concentration range reported above.

2.4. HPLC Analysis of Total Phenolic Compounds Free OHTyr and Tyr, and Total OHTyr and Tyr

The analysis of free OHTyr and Tyr and of total phenolic compounds was performed according to the IOC official method [20]. Phenolic compounds were extracted from the EVOO sample (2.0 g) with 5 mL MeOH:H_2O 80:20 v/v by shaking for 1 min, extracting in an ultrasonic bath for 15 min, and centrifuging at 5000 rpm for 25 min. The supernatant was filtered on PVDC filters (0.45 µm) and analyzed by HPLC. An HP 1100 system coupled to a diode array detector (Agilent Technologies, Santa Clara, CA, USA) was used. The column was a SphereClone ODS-2, 5 µm, 250 × 4.6 mm id kept at room temperature during chromatographic separation. The eluents were formic acid solution (pH 3.2), acetonitrile, and methanol. The analyses were carried out at a flow rate of 1 mL min^{-1}, with an injection volume of 20 µL and a total analysis time of 82 min, applying the gradient reported in the IOC method. The areas were registered at 280 nm, with syringic acid as the internal standard. The content of phenolic compounds was expressed as mg of Tyr per kg of oil.

Total OHTyr and Tyr were determined by HPLC after acid hydrolysis (with sulfuric acid at 80 °C for 2 h) of the hydroalcoholic extract obtained according to the IOC method [21]. For HPLC analysis of the hydrolyzed extracts, formic acid solution at pH 3.2 (solvent A) and acetonitrile (solvent B) were used as eluents. An HP1200 liquid chromatograph coupled to a diode array detector (Agilent Technologies, Santa Clara, CA, USA) was used with a 150 × 3 mm (5 µm) Gemini RP18 column (Phenomenex, Torrance, CA, USA). The flow rate and the injection volume were 0.4 mL min^{-1} and 20 µL, respectively. A linear gradient was applied starting from 95% A to 70% A in 5 min, to 50% A in 5 min, to 2% A in 5 min with a final plateau of 5 min. Total analysis time was 22 min. For UV detection, the wavelength of 280 nm was used. The results were expressed as mg of Tyr per kg of oil for tyrosol, and as mg of OHTyr per kg of oil for hydroxytyrosol after application of the formula reported by Bellumori et al. [22], for keeping into account the 35% overestimation of OHTyr when the calibration curve of tyrosol is used.

2.5. NADES-UV Method Validation

The NADES-UV method was validated by measuring the linearity, the limits of detection (LOD) and quantitation (LOQ), as well as the precision (repeatability), and accuracy (recovery). Linearity of the response in the analytical range was assessed by regression analysis and R^2 values. LOD was expressed as 3.3 × S.D. of the blank ($n = 3$), while LOQ was expressed as 10 × S.D. of the blank ($n = 3$). Sunflower oil was used as blank sample. Repeatability was assessed with repeated intra-day ($n = 3 \times 5$ samples) and inter-day ($n = 3 \times 1$ sample $\times$ 3 days) trials. Recovery was determined as apparent recovery compared to the reference method [21,22] and expressed in percent.

2.6. Statistical Analysis

Spectra, after subtraction of the pure solvent spectrum, were pre-processed using Solo 8.6.2 (Eigenvector Research, Inc., Manson, WA, USA), by applying the second derivative Savitzky–Golay algorithm (polynomial order: 2, window: 11 pt). Analysis of correlation, least squares regression, and analysis of variance (ANOVA) were performed with Origin 2021 (OriginLab Corporation, Northampton, MA, USA).

3. Results and Discussion

3.1. Spectral Characterization of NADES Standard Solutions

The spectra of OHTyr and Tyr in NADES were acquired and analyzed. Figure 1 reports the mean spectra (a), and the second derivatives (b) of both compounds at the concentration of 20 mg L^{-1}. The application of a second derivative treatment allows highlighting minor or subtle spectral features, and resolves spectral overlapping [23–25]. Valleys in the second derivative spectra correspond to peaks in direct absorption spectra, and vice versa.

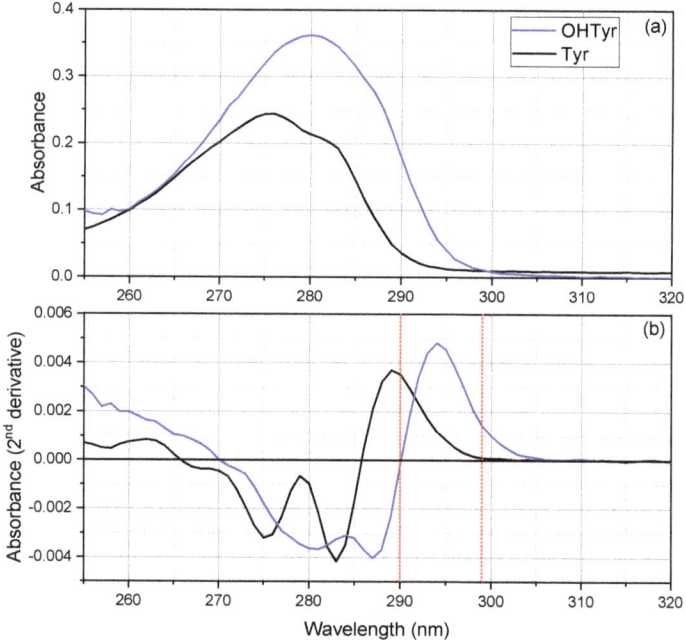

Figure 1. Mean spectra (**a**), and second derivatives (**b**) of OHTyr and Tyr solutions in natural deep eutectic solvent (NADES) (20 mg L^{-1}, n = 3). The red dotted reference lines correspond to the wavelengths selected for the calibration curves.

As regards OHTyr, the spectrum showed a large peak at 280 nm (λ_{max} = 280 nm, $\varepsilon_{\lambda max}$ = 2793 M^{-1} cm^{-1}). The molar extinction (e) determined for OHTyr in NADES was lower than that reported for OHTyr in water [26]. The second derivative showed more spectral features, with two negative peaks at 281 nm and 287 nm, and a positive peak at 294 nm, with a slight shoulder on the right. Spectra of NADES solutions at different concentrations of OHTyr (3, 5, 10, 20, 30 mg L^{-1}) were acquired. Then, a rapid screening via analysis of correlation (data not shown) indicated that the wavelengths that showed the best correlation of the second derivative of the spectrum with the concentrations of

OHTyr were 294, 299, 286, and 280 nm ($r = 0.99878$, $r = 0.9986$, $r = 0.99853$, and $r = 0.99839$, respectively, $n = 15$).

As regards Tyr, the spectrum showed a peak at 276 nm and a shoulder at about 283 nm ($\lambda_{max} = 276$ nm, $\varepsilon_{\lambda max} = 1676$ M^{-1} cm^{-1}). Therefore, hypsochromic and hypochromic shifts of the absorbance were observed for the monophenolic group of Tyr compared to the o-diphenolic structure of OHTyr. Spectra of NADES solutions at different concentrations of Tyr (5, 10, 20, 30, 40 mg L^{-1}) were acquired. The analysis of correlation (data not shown) indicated that the wavelengths that showed the best correlation of the second derivative of the spectrum with the concentrations of Tyr were 290, 274, and 282 nm ($r = 0.99911$, $r = 0.99897$, $r = 0.99883$ respectively, $n = 15$).

3.2. NADES Extraction and Extracts Characterization

A sample set of EVOOs ($n = 26$) was submitted to NADES extraction. Figure 2 reports the spectra of the NADES extracts (A), as well as the second derivatives (B) of the spectra. As a blank sample, a sunflower oil was submitted to NADES extraction in order to assess possible matrix effects due to components other than phenolic compounds, of which sunflower oil is void. The red dotted line in the panel of the second derivative spectra (B) represents the second derivative spectrum of the NADES extract obtained from the sample of sunflower oil (mean of three replicates). As can be seen, the curve appears flat and very near to zero. The second derivative spectra of the NADES extracts clearly reflect the spectral features of the mono- and o-diphenolic structures observed in the spectra of the standard compounds. In particular, negative peaks at 276–277 nm and 283 nm could be observed in all samples, as well as a positive peak at 292–294 nm. These wavelengths proved relevant in the analysis of EVOO phenolic compounds via NADES extraction [10,17].

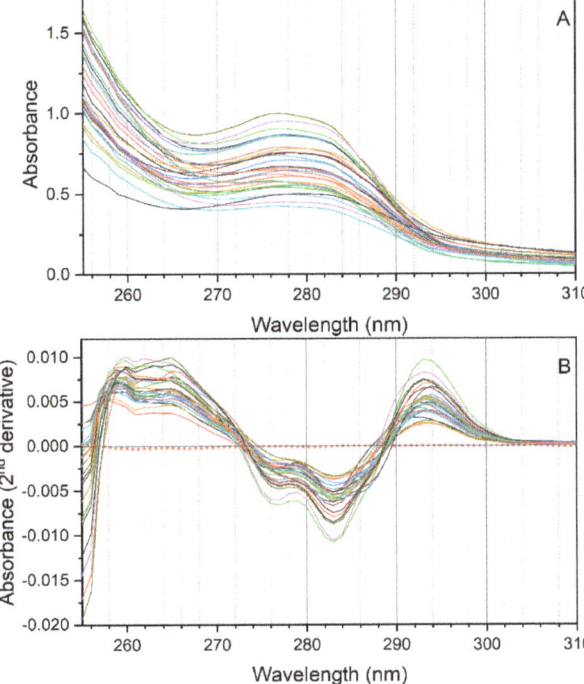

Figure 2. Spectra (**A**), and second derivatives of spectra (**B**) of the NADES extracts of the EVOO samples ($n = 26$). The red dotted line in (**B**) reports the second derivative of the NADES extract of a sample of sunflower oil.

3.3. Analysis of Correlation

An analysis of correlation was carried out on the second derivative of absorption of the NADES extracts of the 26 EVOO samples with their contents of free and total OHTyr and Tyr (reported in Supplementary Table S1). The correlation analysis confirmed the relationships between the spectral features of the extracts and the content of phenolic compounds in the oils (Figure 3). OHTyr contents were associated with a peak of negative correlation with the second derivative of absorption at 280 and 285 nm. On the other hand, a positive correlation was observed with the second derivative of absorption in the range 293–302 nm (maximum at 299 nm, where a r value of 0.9563 was reached). Hypsochromic shifts were observed for the peaks of correlation with the contents of Tyr (negative correlation with the second derivatives at 276 and 283 nm, and a maximum positive correlation at 290 nm). These results confirm that the spectral information contained in the NADES extracts of the oils was strictly related to their content of OHTyr and Tyr.

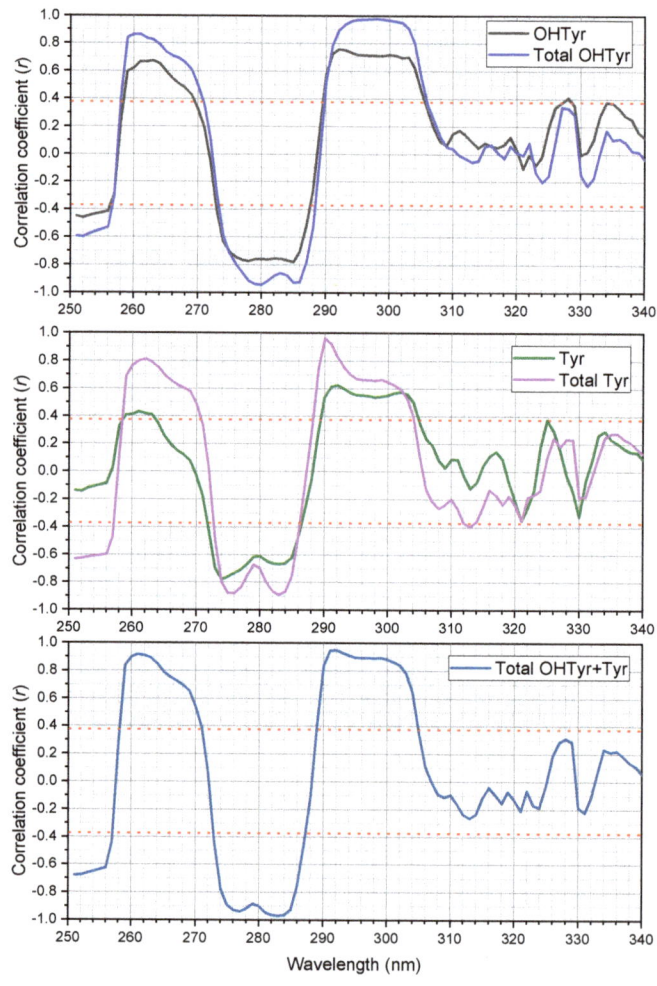

Figure 3. Correlation plots of the second derivatives of absorbance of the NADES extracts ($n = 26$) versus the content of free (before hydrolysis) and total (after hydrolysis) OHTyr and Tyr, and their sum. The red dotted lines and the light blue areas indicate the thresholds of significance ($p = 0.05$).

Notably, correlations were clearly higher when relating total OHTyr and Tyr rather than their free forms. This means that absorbance of the NADES extracts in the UV range can be attributed, rather than to specific molecules (i.e., free OHTyr and free Tyr), to specific moieties of a wide range of phenolic antioxidants (i.e., both free and esterified OHTyr and Tyr moieties). This is confirmed by the data in Figure 4, were the ratio between the values of the second derivative of absorbance at 299 nm and 290 nm (normalized by the extinction coefficients respectively of OHTyr and Tyr) of the NADES extracts is plotted against the corresponding ratio of either free OHTyr to free Tyr, or total OHTyr to total Tyr.

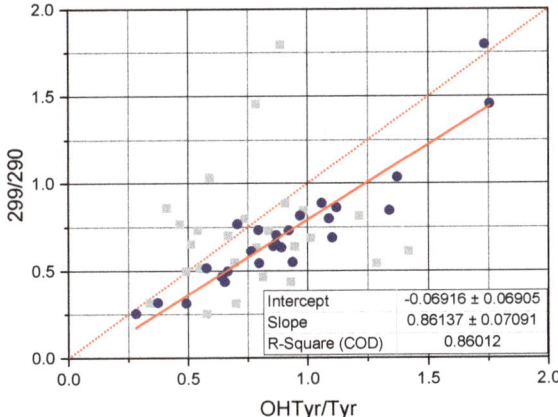

Figure 4. The scatter plot reports the ratio, 299/290, between the second derivative of the absorbance of NADES extracts at 299 nm and the second derivative of the absorbance of NADES extracts at 290 nm versus the ratio OHTyr/Tyr between the content of OHTyr and the content of Tyr, both free (grey scatter) and total (blue scatter). The second derivatives at 299 and 290 have been normalized by the molar extinction coefficient of respectively OHTyr and Tyr. The box reports the parameters of the linear regression of the 299/290 ratio versus the total OHTyr/total Tyr ratio.

3.4. Determination of Total OHTyr and Tyr

On the basis of the UV spectral information of the NADES extracts, a quantitation of total OHTyr and Tyr could be approached. According to the analysis of OHTyr and Tyr spectra, as well as to the correlation analysis of the NADES extracts of the EVOO sample set, two different wavelengths were selected for the determination of total OHTyr and Tyr. As regards OHTyr, considering the overlaps with the spectra of Tyr, the wavelength selected for the calibration curve was 299 nm. For Tyr, the wavelength 290 nm was considered optimal, since it corresponded with a zero-point of the derivative of OHTyr spectrum.

The selected wavelengths allowed obtaining calibration curves of both OHTyr and Tyr in NADES, in the ranges 3–30 mg L^{-1} for OHTyr, and 5–40 mg L^{-1} for Tyr, with R^2 values of 0.997 and 0.998, respectively.

Table 1 reports the parameters of the calibration curves:

$$A_\lambda = a + b \cdot C \text{ (mg L}^{-1}\text{)}, \tag{1}$$

where A_λ is the second derivative of the absorbance at the selected wavelength l, C is the concentration of the analyte in NADES, a is the intercept, and b is the slope.

Table 1. Parameters of the calibration curves of OHTyr and Tyr in NADES.

	Tyr	OHTyr
Wavelength of second derivative (λ)	290 nm	299 nm
Intercept (a)	-1.62972×10^{-4}	-2.87704×10^{-5}
Slope (b)	1.82318×10^{-4}	7.11259×10^{-5}
R^2	0.998	0.997
Range	5–40 mg L^{-1}	3–30 mg L^{-1}

The determination of Tyr was carried out on the basis of the second derivative of absorption at 290 nm of the extracts and of the parameters of the calibration curve of Tyr in NADES. Figure 5 reports the contents of total Tyr determined by HPLC after hydrolysis of the extracts plotted versus the contents of total Tyr determined by NADES extraction and UV analysis.

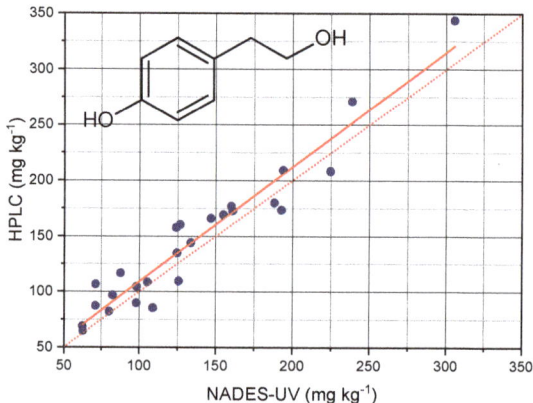

Figure 5. Contents of total Tyr determined by HPLC after hydrolysis of the extracts plotted versus the contents of total Tyr determined by NADES extraction and UV analysis. The red dotted line has the function y = x, the red solid line is the fitted regression.

The analysis of regression showed that the contents of Tyr determined by the NADES-UV method allowed predicting the contents determined by acid hydrolysis and HPLC with an accuracy corresponding to a R^2 value of 0.925. The parameters of the regression are reported in Table 2. The prediction performance of the method was clearly better compared to those observed in our previous works [10,17]. The root mean square error (RMSE) was 18 mg kg^{-1}, corresponding to a RMSE% of 5.8%. The improved performance of this method could be due to the use of derivative processing of the spectra. Moreover, the comparison with the method of determination of total Tyr after hydrolysis could highlight the sensitivity of the NADES-UV method towards specific structural features in the phenolic pattern of EVOO. The slope of the curve was very near to 1; the amounts determined nu NADES- UV were very close to those determined by acid hydrolysis and HPLC analysis.

The quantitation of OHTyr provided the results reported in Figure 6 and Table 3. The analysis of regression showed the possibility of predicting the content of total OHTyr in EVOO with an accuracy corresponding to a R^2 value of 0.942, a RMSE of 14 mg kg^{-1}, and a RMSE% of 5.0%. It should be noted that the NADES-UV method provided an overestimation of total OHTyr, as showed by the slope value of about 0.73. However, the good fitting performance allowed satisfactorily correcting the obtained value and predicting the result of the HPLC analysis.

Table 2. Parameters of the regression analysis of total Tyr determined by NADES-UV versus total Tyr determined by acid hydrolysis and HPLC.

	Value	Significance
Intercept	6.26613	0.49
Slope	1.02911	<0.001
R^2	0.925	
RMSE	18.0	
RMSE%	5.8	

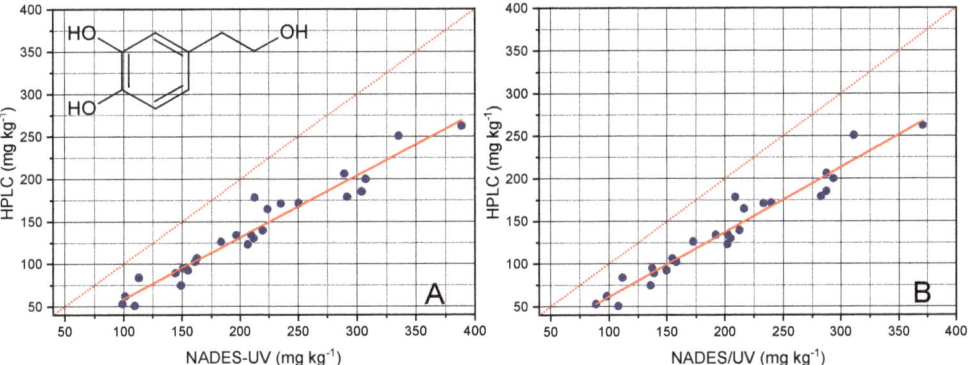

Figure 6. Contents of total OHTyr determined by HPLC after hydrolysis of the extracts plotted versus the contents of total OHTyr determined by NADES extraction and UV analysis. The red dotted line has the function y = x, the red solid line is the fitted regression. The panel on the left (**A**) plots the values of total OHTyr determined prior the correction of the second derivative of absorbance at 299 nm, while the panel on the right (**B**) plots the values obtained after the correction (please see text for details on the correction procedure).

Table 3. Parameters of the regression analysis of total OHTyr determined by NADES-UV, versus total OHTyr determined by acid hydrolysis and HPLC.

	Value	Significance
	Before correction	
Intercept	−13.84855	0.10
Slope	0.72584	<0.001
R^2	0.942	
RMSE	14.0	
RMSE%	5.0	
	After correction	
Intercept	−13.48954	0.10
Slope	0.75556	<0.001
R^2	0.945	
RMSE	13.7	
RMSE%	4.9	

A possible explanation for such overestimation could be related to the matrix effects attributable to the phenolic pattern of EVOO; the NADES-UV method implies the direct analysis of the extract, without chromatographic separation. Therefore, a possible interference of Tyr, as well as possible differences in absorptivity of the esterified forms with respect to free forms could have occurred. This hypothesis seems to be confirmed by the fact that total Tyr was also correlated with the 299 nm wavelength ($r = 0.95$); this would indicate a contribution of Tyr to the absorbance at that wavelength.

Therefore, in order to reduce the total Tyr contribution, the values of the second derivative at 299 nm were corrected as follows, prior to applying the parameters of the calibration curve of OHTyr: A calibration curve for Tyr was built at 299 nm; then, the contents of Tyr determined using the calibration curve built at 290 nm were applied to calculate the corresponding value of the second derivative at 299 nm; the value calculated was then subtracted to the value measured; the difference was finally used to determine the content of OHTyr.

As shown in Figure 6B, even if a slight decrease of the overestimation and a contemporary slight improvement of the fitting was obtained after correction, a not negligible overestimation of OHTyr was still observed, suggesting the hypothesis of a matrix effect as the basis of the overestimation of OHTyr.

3.5. Sum of OHTyr and Tyr

An attempt to determine the sum of total OHTyr and Tyr was made by building a calibration curve with mixtures of OHTyr and Tyr. OHTyr and Tyr were solubilized in te NADES in different concentrations, with a total range of 8 to 70 mg L^{-1}. Analysis of correlation (data not shown) indicated 282 nm as the wavelength with the highest correlation with the content of total OHTyr + Tyr. This agrees with the results of the correlation analysis of the spectra of the NADES extracts with the contents of OHTyr and Tyr in EVOOs. The calibration curve built with the second derivative of the absorption at 282 nm as a function of total OHTyr + Tyr showed the parameters reported in Table 4.

Table 4. Parameters of the calibration curve of OHTyr + Tyr in NADES.

	Tyr + OHTyr
Wavelength of second derivative (λ)	282 nm
Intercept (a)	1.30785×10^{-4}
Slope (b)	-1.80048×10^{-4}
R^2	0.995
Ranges	5–40 mg L^{-1} (Tyr); 3–30 mg L^{-1} (OHTyr); 8–70 mg L^{-1} (OHTyr + Tyr)

The total content of OHTyr + Tyr in the EVOO samples was measured based on the calibration curve and on the second derivative of the absorption at 282 of the NADES extracts. The values determined were plotted against the values obtained by HPLC after acid hydrolysis of the methanolic extracts, and a regression analysis was carried out. The results are reported in Figure 7 and Table 5. The contents of OHTyr + Tyr were, also in this case, overestimated with respect to the method by HPLC, with a mean residual of 37 mg L^{-1}. The mean recovery was 117 ± 15%. The R^2 value of the regression was 0.931, confirming a satisfactory predictive capacity of the NADES-UV method with respect to the method by HPLC. The overestimation was confirmed by the regression analysis with the significance of the intercept. On the other hand, the slope of the regression was very close to 1.

Table 5. Parameters of the regression analysis of total OHTyr + Tyr determined by NADES-UV versus total OHTyr + Tyr determined by acid hydrolysis and HPLC.

	Value	Significance
Intercept	−55.61922	0.00679
Slope	1.05626	<0.001
R^2	0.931	
RMSE	29.5	
RMSE%	5.6	

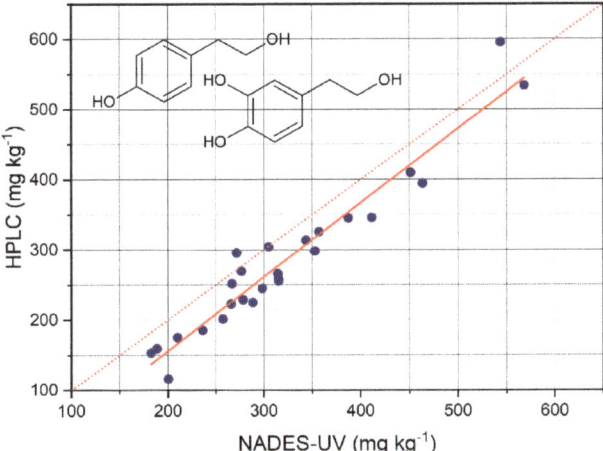

Figure 7. Contents of total OHTyr + Tyr determined by HPLC after hydrolysis of the extracts plotted versus the contents of total OHTyr + Tyr determined by NADES extraction and UV analysis. The red dotted line has the function y = x, the red solid line is the fitted regression.

Therefore, the wavelength 282 nm resulted in a common wavelength of absorption of both OHTyr and Tyr, unlike 299 nm and 290 nm that were specific spectral features. The values of RMSE and RMSE% were 29.5 mg kg^{-1} and 5.6%, respectively. Compared to the performance of SVM regression applied to the whole spectra of NADES extract to determine the sum of OHTyr and Tyr derivatives assessed by HPLC, as reported in our previous work [10], the RMSE of prediction resulted lower (29.5 vs. 35.5 mg kg^{-1}), while it reached even higher levels when considering the single wavelength of 283 nm (58.7 mg kg^{-1}). As regards the R^2 value, it increased to 0.931 from the 0.84 and 0.64 of SVM and linear regression, respectively, of our previous work.

Table 6 reports the figures of merit of the method. The LOD and LOQ were respectively 3.9 and 11.8 mg kg^{-1}. As regards repeatability, the CV% of the method remained around or below 5%.

Table 6. Figures of merit of the method of analysis of total OHTyr + Tyr in EVOOs by NADES extraction and UV analysis of the second derivative spectra.

	OHTyr + Tyr
LOQ (limit of quantification)	11.8 mg kg^{-1}
LOD (limit of detection)	3.9 mg kg^{-1}
Apparent recovery (%) [1]	116.8 ± 15.2
Intra-day repeatability (CV%, n = 3 × 5 samples)	2.7%
Inter-day repeatability (CV% n = 3 × 1 sample × 3 days)	5.1%

[1] Compared to the amounts determined by the reference method [21].

The proposed analytical approach can be considered as a screening method applicable to assess the levels of total OHTyr and Tyr, as determined by the emerging hydrolysis methods [6]. Some key aspects could support their use as a tool with complementary functions to the more accurate and precise chromatographic methods:

- Use of environment- and operator-friendly solvents;
- Use of low-cost equipment;
- Easy analytical procedure;
- Short analysis time.

Due to these features, this analytical method could be applied to the field (e.g., oil mills, bottling plants, storage plants) and allow the monitoring of the levels of phenolic compounds related to the health claim. A continuous monitoring of this parameter and its trends during storage before bottling, as well as after bottling, could make operators aware of the evolving quality of individual EVOO batches, and therefore more confident in the use of the health claim on labels [27,28]. The appropriate adoption of the health claim on labels could also support focused communication and be considered for possible its effects on consumer perception and choices and the segment the trade category of EVOO, with eventual benefits for the EVOO value chain [29,30].

To the best of our knowledge, there are few alternative screening methods with the same aim. Reboredo-Rodríguez et al. [11] previously showed that the traditional Folin–Ciocalteu assay provided, for a sample set of 12 EVOOs, results comparable to the amounts of total OHTyr and Tyr determined by acid hydrolysis and HPLC. Nevertheless, the performance of this approach, compared to the reference method, was lower ($r = 0.94$). Moreover, the analytical reagents are not environment friendly.

Shabani et al. [16] recently proposed the electrochemical detection of EVOO phenolic compounds extracted with the same NADES used in our previous work [17]. The method proposed by the authors was easy and environment- and user-friendly, though the analytical target includes all the phenolic compounds contained in EVOO and not only those accounting for the health claim.

The method presented in this paper, could provide, therefore, a feasible opportunity for operators of the EVOO chain to access relevant chemical information. The NADES adopted, composed of lactic acid, glucose, and water, could pave the way for a new category of green analytical chemistry: food grade analytical chemistry. It is not simply a paradox: after sustainable reagents, can we move towards edible reagents?

4. Conclusions

A green and simple method, using food grade reagents, was set up as an on-site, environment- and operator-friendly tool for screening of olive oils according to the content of total (i.e., free and linked) OHTyr and Tyr. The reagents (glucose, lactic acid, and water) and the equipment (a spectrophotometer) required make this method affordable and feasible in contexts other than analytical labs. The limits of detection and quantitation (3.9 and 11.8 mg kg^{-1}) and the repeatability (CV% ≤ 5%) can be considered satisfactory for screening purposes. The results obtained, although leading to a slight overestimation, fit with those obtained by the reference HPLC method, and showed good predictive capacity. The NADES-UV method could be a suitable complement to more complex methods (involving extraction, acid hydrolysis, and HPLC analysis) that seem to be the best approach to address the requirements of the EFSA health claim. The availability of a complete analytical platform comprising both on-site screening methods and accurate laboratory determination could support the value chain operators in EVOO differentiation and segmentation, as well as consumers in gaining increasing awareness.

Supplementary Materials: The following are available online at https://www.mdpi.com/2304-8158/10/3/677/s1. Table S1: Contents of OHTyr an Tyr in the EVOO samples set, determined by the reference method before and after acid hydrolysis.

Author Contributions: Conceptualization, V.M.P. and F.C.; methodology, V.M.P., F.L. and N.M.; formal analysis, V.M.P. and F.L.; investigation, S.F., L.C., M.B., P.R.; resources, F.C., P.C. and N.M.; writing—original draft preparation, V.M.P. and F.L.; writing—review and editing, M.B., L.C., P.C. and F.C.; visualization, V.M.P.; supervision, F.C. and N.M.; project administration, V.M.P. and F.L.; funding acquisition, V.M.P. and F.C. All authors have read and agreed to the published version of the manuscript.

Funding: This research was funded by AGER 2 Project (grant no. 2016-0105) and by Regione Puglia (POR Puglia FESR-FSE 2014–2020—Innonetwork call—"Multifunctional microsystems for the monitoring of olive oils oxidation"—project code XMPYXR1).

Institutional Review Board Statement: Not applicable.

Informed Consent Statement: Not applicable.

Data Availability Statement: The data presented in this study are available on request from the corresponding author.

Conflicts of Interest: The authors declare no conflict of interest. The funders had no role in the design of the study; in the collection, analyses, or interpretation of data; in the writing of the manuscript, or in the decision to publish the results.

References

1. Gavahian, M.; Khaneghah, A.M.; Lorenzo, J.M.; Munekata, P.E.; Garcia-Mantrana, I.; Collado, M.C.; Meléndez-Martínez, A.J.; Barba, F.J. Health benefits of olive oil and its components: Impacts on gut microbiota antioxidant activities, and prevention of noncommunicable diseases. *Trends Food Sci. Technol.* **2019**, *88*, 220–227. [CrossRef]
2. Romani, A.; Ieri, F.; Urciuoli, S.; Noce, A.; Marrone, G.; Nediani, C.; Bernini, R. Health Effects of Phenolic Compounds Found in Extra-Virgin Olive Oil, By-Products, and Leaf of Olea europaea L. *Nutrients* **2019**, *11*, 1776. [CrossRef] [PubMed]
3. Francisco, V.; Ruiz-Fernández, C.; Lahera, V.; Lago, F.; Pino, J.; Skaltsounis, A.-L.; González-Gay, M.A.; Mobasheri, A.; Gómez, R.; Scotece, M.; et al. Natural Molecules for Healthy Lifestyles: Oleocanthal from Extra Virgin Olive Oil. *J. Agric. Food Chem.* **2019**, *67*, 3845–3853. [CrossRef] [PubMed]
4. European Commission. Regulation EC No. 432/2012 establishing a list of permitted health claims made on foods, other than those referring to the reduction of disease risk and to children's development and health. *Off. J. Eur. Union* **2012**, *L136*, 1–40.
5. Tsimidou, M.Z.; Nenadis, N.; Servili, M.; García-González, D.L.; Toschi, T.G.; Gonzáles, D.L.G. Why Tyrosol Derivatives Have to Be Quantified in the Calculation of "Olive Oil Polyphenols" Content to Support the Health Claim Provisioned in the EC Reg. 432/2012. *Eur. J. Lipid Sci. Technol.* **2018**, *120*, 1800098. [CrossRef]
6. Mulinacci, N.; Giaccherini, C.; Ieri, F.; Innocenti, M.; Romani, A.; Vincieri, F.F. Evaluation of lignans and free and linked hydroxy-tyrosol and tyrosol in extra virgin olive oil after hydrolysis processes. *J. Sci. Food Agric.* **2006**, *86*, 757–764. [CrossRef]
7. Romero, C.; Brenes, M. Analysis of Total Contents of Hydroxytyrosol and Tyrosol in Olive Oils. *J. Agric. Food Chem.* **2012**, *60*, 9017–9022. [CrossRef]
8. Mastralexi, A.; Nenadis, N.; Tsimidou, M.Z. Addressing Analytical Requirements to Support Health Claims on "Olive Oil Polyphenols" (EC Regulation 432/2012). *J. Agric. Food Chem.* **2014**, *62*, 2459–2461. [CrossRef]
9. Purcaro, G.; Codony, R.; Pizzale, L.; Mariani, C.; Conte, L. Evaluation of total hydroxytyrosol and tyrosol in extra virgin olive oils. *Eur. J. Lipid Sci. Technol.* **2014**, *116*, 805–811. [CrossRef]
10. Paradiso, V.M.; Squeo, G.; Pasqualone, A.; Caponio, F.; Summo, C. An easy and green tool for olive oils labelling according to the contents of hydroxytyrosol and tyrosol derivatives: Extraction with a natural deep eutectic solvent and direct spectrophotometric analysis. *Food Chem.* **2019**, *291*, 1–6. [CrossRef]
11. Reboredo-Rodríguez, P.; Valli, E.; Bendini, A.; Di Lecce, G.; Simal-Gándara, J.; Toschi, T.G. A widely used spectrophotometric assay to quantify olive oil biophenols according to the health claim (EU Reg. 432/2012). *Eur. J. Lipid Sci. Technol.* **2016**, *118*, 1593–1599. [CrossRef]
12. de los Ángeles Fernández, M.; Boiteux, J.; Espino, M.; Gomez, F.J.; Silva, M.F. Natural deep eutectic solvents-mediated extractions: The way forward for sustainable analytical developments. *Anal. Chim. Acta* **2018**, *1038*, 1–10. [CrossRef]
13. Choi, Y.H.; Van Spronsen, J.; Dai, Y.; Verberne, M.; Hollmann, F.; Arends, I.W.; Witkamp, G.-J.; Verpoorte, R. Are Natural Deep Eutectic Solvents the Missing Link in Understanding Cellular Metabolism and Physiology? *Plant. Physiol.* **2011**, *156*, 1701–1705. [CrossRef]
14. Chen, J.; Li, Y.; Wang, X.; Liu, W. Application of Deep Eutectic Solvents in Food Analysis: A Review. *Molecules* **2019**, *24*, 4594. [CrossRef]
15. Fanali, C.; Della Posta, S.; Dugo, L.; Russo, M.; Gentili, A.; Mondello, L.; De Gara, L. Application of deep eutectic solvents for the extraction of phenolic compounds from extra-virgin olive oil. *Electrophoresis* **2020**, *41*, 1752–1759. [CrossRef]
16. Shabani, E.; Zappi, D.; Berisha, L.; Dini, D.; Antonelli, M.L.; Sadun, C. Deep eutectic solvents (DES) as green extraction media for antioxidants electrochemical quantification in extra-virgin olive oils. *Talanta* **2020**, *215*, 120880. [CrossRef]
17. Paradiso, V.M.; Clemente, A.; Summo, C.; Pasqualone, A.; Caponio, F. Towards green analysis of virgin olive oil phenolic compounds: Extraction by a natural deep eutectic solvent and direct spectrophotometric detection. *Food Chem.* **2016**, *212*, 43–47. [CrossRef]
18. Dai, Y.; Witkamp, G.-J.; Verpoorte, R.; Choi, Y.H. Tailoring properties of natural deep eutectic solvents with water to facilitate their applications. *Food Chem.* **2015**, *187*, 14–19. [CrossRef] [PubMed]
19. Pisano, P.L.; Espino, M.; de los Ángeles Fernández, M.; Silva, M.F.; Olivieri, A.C. Structural analysis of natural deep eutectic solvents. Theoretical and experimental study. *Microchem. J.* **2018**, *143*, 252–258. [CrossRef]
20. International Olive Council. *COI/T.20/Doc. No 29—Determination of Biophenols in Olive Oils by HPLC*; International Olive Council: Madrid, Spain, 2009.
21. Bellumori, M.; Cecchi, L.; Innocenti, M.; Clodoveo, M.L.; Corbo, F.; Mulinacci, N. The EFSA Health Claim on Olive Oil Polyphenols: Acid Hydrolysis Validation and Total Hydroxytyrosol and Tyrosol Determination in Italian Virgin Olive Oils. *Molecules* **2019**, *24*, 2179. [CrossRef] [PubMed]

22. Bellumori, M.; Cecchi, L.; Romani, A.; Mulinacci, N.; Innocenti, M. Recovery and stability over time of phenolic fractions by an industrial filtration system of olive mill wastewaters: A three-year study. *J. Sci. Food Agric.* **2018**, *98*, 2761–2769. [CrossRef]
23. Govindaraj, N.; Gangadoo, S.; Truong, V.K.; Chapman, J.; Gill, H.; Cozzolino, D. The use of derivatives and chemometrics to interrogate the UV–Visible spectra of gin samples to monitor changes related to storage. *Spectrochim. Acta Part. A Mol. Biomol. Spectrosc.* **2020**, *227*, 117548. [CrossRef] [PubMed]
24. Ojeda, C.B.; Rojas, F.S. Recent applications in derivative ultraviolet/visible absorption spectrophotometry: 2009–2011. *Microchem. J.* **2013**, *106*, 1–16. [CrossRef]
25. Parmar, A.; Sharma, S. Derivative UV-vis absorption spectra as an invigorated spectrophotometric method for spectral resolution and quantitative analysis: Theoretical aspects and analytical applications: A review. *TrAC Trends Anal. Chem.* **2016**, *77*, 44–53. [CrossRef]
26. Bouzid, O.; Navarro, D.; Roche, M.; Asther, M.; Haon, M.; Delattre, M.; Lorquin, J.; Labat, M.; Asther, M.; Lesage-Meessen, L. Fungal enzymes as a powerful tool to release simple phenolic compounds from olive oil by-product. *Process. Biochem.* **2005**, *40*, 1855–1862. [CrossRef]
27. López-Huertas, E.; Lozano-Sánchez, J.; Segura-Carretero, A. Olive oil varieties and ripening stages containing the antioxidants hydroxytyrosol and derivatives in compliance with EFSA health claim. *Food Chem.* **2021**, *342*, 128291. [CrossRef]
28. Criado-Navarro, I.; López-Bascón, M.A.; Priego-Capote, F. Evaluating the Variability in the Phenolic Concentration of Extra Virgin Olive Oil According to the Commission Regulation (EU) 432/2012 Health Claim. *J. Agric. Food Chem.* **2020**, *68*, 9070–9080. [CrossRef]
29. Roselli, L.; Clodoveo, M.L.; Corbo, F.; De Gennaro, B. Are health claims a useful tool to segment the category of extra-virgin olive oil? Threats and opportunities for the Italian olive oil supply chain. *Trends Food Sci. Technol.* **2017**, *68*, 176–181. [CrossRef]
30. Pichierri, M.; Pino, G.; Peluso, A.M.; Guido, G. The interplay between health claim type and individual regulatory focus in determining consumers' intentions toward extra-virgin olive oil. *Food Res. Int.* **2020**, *136*, 109467. [CrossRef] [PubMed]

Article

Physical and Sensory Properties of Mayonnaise Enriched with Encapsulated Olive Leaf Phenolic Extracts

Federica Flamminii, Carla Daniela Di Mattia *, Giampiero Sacchetti, Lilia Neri, Dino Mastrocola and Paola Pittia

Faculty of Bioscience and Technology for Agriculture, Food and Environment, University of Teramo, Via Balzarini 1, 64100 Teramo, Italy; fflamminii@unite.it (F.F.); gsacchetti@unite.it (G.S.); lneri@unite.it (L.N.); dmastrocola@unite.it (D.M.); ppittia@unite.it (P.P.)
* Correspondence: cdimattia@unite.it; Tel.: +39-0861-266947

Received: 1 July 2020; Accepted: 22 July 2020; Published: 24 July 2020

Abstract: This work aimed to study the physical, structural, and sensory properties of a traditional full-fat mayonnaise ($\approx$ 80% oil) enriched with an olive leaf phenolic extract, added as either free extract or encapsulated in alginate/pectin microparticles. Physical characterization of the mayonnaise samples was investigated by particle size, viscosity, lubricant properties, and color; a sensory profile was also developed by a quantitative descriptive analysis. The addition of the extract improved the dispersion degree of samples, especially when the olive leaf extract-loaded alginate/pectin microparticles were used. The encapsulated extract affected, in turn, the viscosity and lubricant properties. In particular, both of the enriched samples showed a lower spreadability and a higher salty and bitter perception, leading to a reduced overall acceptability. The results of this study could contribute to understanding the effects of the enrichment of emulsified food systems with olive by-product phenolic extracts, both as free and encapsulated forms, in order to enhance real applications of research outcomes for the design and development of healthy and functional formulated foods.

Keywords: olive leaf polyphenols; encapsulation; functional food; mayonnaise; alginate/pectin beads; phenolic extract; food enrichment

1. Introduction

Nowadays, the increasing amount of attention being paid to health and wellbeing has modified consumer choices toward foods that, besides meeting nutritional requirements and providing hedonistic gratification, may also guarantee benefits by exerting desired functional properties. Mayonnaise is a traditional sauce prepared by the gentle mixing of oil with egg yolk, vinegar, salt, and spices to form an oil-in-water emulsion with a dispersed lipid phase ranging between 60 and 80%. It represents one of the most consumed sauces worldwide and is highly appreciated for its special flavor and creamy mouthfeel.

The words "functional" and "mayonnaise" do not seem to get along; however, if such a combination of words is considered from another point of view, it could lead to an innovative and challenging functional food [1]. Several strategies have been proposed to make mayonnaise a healthier product: along with the reduction of fat, which is the most widespread approach used up to now, a more recent approach is based on the enrichment of emulsified sauces with beneficial ingredients that can respond to the health-related needs of people (antioxidant, prebiotics, and probiotics). Moreover, from a food science point of view, functional ingredients, such as natural antioxidants, are known to improve the oxidative stability of the product and allow the replacement of synthetic and debated antioxidants. The interest in the use of natural ingredients is often associated with the opportunity of recovering functional and bioactive compounds from food waste and by-products, such as those derived from the olive oil chain. Recently, several studies have proposed the enrichment of different food products with an olive leaf phenolic extract (OLE), as in the case of vegetable oils, due to their antioxidant activity (i.e., sunflower, soybean, maize, and frying oils) [2], or meat products against oxidative [3] and microbial spoilage [4]. Moreover, the use of OLE was also tested in cereal [5] and dairy products [6]. Apart from having healthy, antioxidant, and antimicrobial properties, olive polyphenols have been proven to exert important technological functionality, such as a water/oil holding capacity and emulsifying activity [7], and can represent a useful multifunctional ingredient that can help in the production and stabilization of complex food products such as emulsions. However, consumers' willingness to accept foods produced with olive by-product ingredients depends on the perception of different factors, mainly related to the general attitude of the consumer, rather than a product-specific choice. Indeed, information about the characteristics of olive by-products and the perception of the benefits from sustainable consumption can possibly offset the consumers' choice with a positive association between the use of vegetable by-products and sustainable production and environmental responsibility [8].

One main barrier to a large usage in food manufacturing is that phenolic-rich extracts are characterized by chemical and physical instability under conditions commonly encountered in processing and storage (temperature, oxygen, and light) [9], as well as by an unpleasant bitter and pungent taste [10].

Therefore, encapsulation could represent a valid strategy for overcoming all of these drawbacks and various investigations have been carried out in this field by using different technologies and approaches [11–14]. Nonetheless, up to now, few works have reported the use of encapsulated OLE in foodstuffs: In one study, an OLE encapsulated by nano-emulsions was added in soybean oil, increasing the solubility and controlling the release of olive leaf phenolic compounds, as well as enhancing its antioxidant activity when compared to non-encapsulated OLE and synthetic antioxidants [15]. In another study, OLE encapsulated in water-in-oil-in-water double emulsions ($W_1/O/W_2$) was used in meat systems, improving the oxidative stability of the product, while providing a healthier lipid profile with respect to the free extract [16]. Furthermore, the enrichment of cereal, dairy, beverages, and spreadable products with OLE encapsulated with different techniques was assessed, in order to improve their shelf-lives [17–20], but additional studies are needed, in order to widen the knowledge on the qualitative properties of OLE-enriched formulated products.

Therefore, the aim of this research was to study the physical, structural, and sensory properties of a full-fat mayonnaise (≈ 80% oil) enriched with an olive leaf phenolic extract (OLE), encapsulated in alginate/pectin microparticles. The physical characterization of the mayonnaise samples after preparation was investigated by measuring the particle size, viscosity, tribology, and color; a sensory profile was also developed by a quantitative descriptive analysis. For comparison purposes, samples enriched with pure OLE and without OLE (control) were prepared. The results will provide a new formulation and technological insights on the effect of the enrichment of a full-fat mayonnaise with encapsulated OLE on its physical and structural properties and, at the same time, evaluate the effect of encapsulation on OLE performances.

2. Materials and Methods

2.1. Materials

A dried olive leaf extract (OLE) with a standardized concentration of oleuropein (40%) was kindly donated by Oleafit srl (Isola del Gran Sasso, Teramo, Italy) Dried OLE-loaded alginate/pectin microcapsules (Alg/Pec) with an average volume weighted mean diameter ($D_{4,3}$) of 62.6 ± 0.2 μm were obtained by applying an emulsion/internal gelation technique [11]. Sunflower oil, eggs, salt, vinegar, and lemon juice were purchased in a local supermarket. All of the reagents used were of analytical grade.

2.2. Mayonnaise Preparation

Mayonnaise samples were prepared according to a reference recipe with the following formulation: Oil (500 g ≈ 78%), five egg yolks (27 g ≈ 12%), vinegar (10 g ≈ 2%), salt (5 g ≈ 1%), lemon juice (37.5 g ≈ 7%), and sodium azide 0.05% (w/w). The samples were prepared using a lab-scale mixer (Bimby TM31, Vorwerk, Wuppertal, Germany) in a two-step standardized process: Eggs, vinegar, salt, and lemon juice were preliminary mixed (100 g·min, 1 min), and oil was then slowly added under vigorous mixing (from 500 up to 1000 g·min) for 20 min, allowing complete oil incorporation. In order to evaluate the exploitability of pure OLE and Alg/Pec + OLE beads as functional ingredients, a final concentration of 200 mg of total phenolic compounds per kg of product was added to mayonnaise. In particular, ≈ 1 g/kg (≈ 0.09%) of OLE and ≈ 4 g/kg (≈ 0.32%) of Alg/Pec + OLE, both in a dried form, were added to the water phase during the preliminary mixing (first step). Three following series of samples were thus prepared: A control (without phenolic extract added), Mayo + OLE (enriched with pure OLE), and Mayo + Alg/Pec (enriched with Alg-Pec + OLE microspheres). Just after preparation, the following evaluations were carried out on the samples: Particle size distribution, microstructure, color, viscosity, tribology, and sensory profile evaluations.

2.3. Particle Size

Particle size distribution was measured by laser diffraction analysis and the use of a particle size analyzer (Mastersizer Hydro 3000, Malvern Instruments Ltd., Worcestershire, UK). Mayonnaise samples (0.2 g) were diluted with 20 mL of a 1% (w/v) sodium dodecyl sulphate (SDS) solution, gently stirred with a magnetic stirrer until complete dispersion [21], and thereafter added to the Hydro 2000S dispersion unit containing distilled water at 2500 rpm until an obscuration of 5–6% was reached. Each sample was analysed in triplicate and five records for each measurement were collected. The optical properties of the sample were defined as follows: Refractive index of 1.46 and absorption of 0.00. Droplet size measurements are reported as particle size distribution curves (PSD), the surface mean diameter ($D_{3,2}$), and the volume mean diameter ($D_{4,3}$).

2.4. Microstructure

An Olympus BX53F optical microscope (Olympus, Tokyo, Japan) was used to evaluate the mayonnaise's microstructure. An amount of each sample was deposited until a very thin and homogeneous layer developed on the slide and images were obtained with 100× magnification and acquired with a QCAM fast 1934 (QImaging, Surrey, BC, Canada), equipped with a 55 mm objective. Images were elaborated through the Software "Image Pro Plus 7.0" (Media cybernetic, Inc., Rockville, MD, USA).

2.5. Color Evaluation

The color of the mayonnaise was evaluated with a CHROMA METER CR5 instrument (Konica Minolta, Osaka, Japan), illuminant D65; each sample was homogeneously distributed into a glass vessel and the color was recorded at five different points. The average of the five measurements was assessed for all the colorimetric parameters (L*, a*, and b*). In the CIE Lab colour space, L* represents the lightness within the range of 0 (black) to 100 (white); a* is the redness, from green (−a*) to red (+a*); and b* is the yellowness, from blue (−b*) to yellow (+b*). The a* and b* parameters were used to calculate the tonality angle (Hue angle, h°) according to Equation (1):

$$h° = tan^{-1}\left(\frac{b^*}{a^*}\right) \quad (1)$$

2.6. Flow Behavior and Tribological Measurements

Mayonnaise rheological measurements were performed at 25 °C by a controlled stress–strain rheometer (MCR 302, Anton Paar, Graz, Austria) connected to a circulating water bath for temperature control. The flow behavior of the samples was evaluated by using a parallel plate configuration (50 mm) and a gap distance of 1 mm, and excess sample protruding from the edge of the sensor was carefully trimmed off with a thin blade. Flow curves were measured by recording viscosity values shearing the samples at logarithmic increasing shear rates from 3 to 300 s^{-1}. Flow curves were built for all of the tested samples and fitted to the Herschel–Bulkley model by using Excel's solver software to obtain the yield stress (τ_0), consistency index (k), and flow index (n). The friction and lubricant properties of model-like mayonnaise samples were evaluated using a tribology apparatus (Anton Paar, Graz, Austria) fitted to the rheometer. The measuring cell consisted of a ball-on-three-plates geometry consisting of a rotating glass ball and three stationary cylindrical polydimethylsiloxane (PDMS) pins fixed to the sample holder. The ball measured 12.7 mm in diameter and the pins measured 6 mm in diameter and height. The friction and lubricating properties of mayonnaise samples in between two surfaces were recorded at 30 °C with a normal force (FN) of 1 N. The Stribeck curves for each sample were shaped by plotting the friction factor (–) versus the sliding velocity (Vs).

2.7. Sensory Evaluation

Sensory evaluation was carried out through quantitative descriptive analysis (QDA) by a panel consisting of 15 assessors. The panelists were students and technical staff of the Faculty of Bioscience and Technology for Agriculture, Food and Environment (University of Teramo, Teramo, Italy). Before analysis, brief training was carried out, in order to verify and discuss the vocabulary and to explain the scales being used. The following attributes were evaluated: Spreadability, consistency, and color uniformity for the appearance and structure; saltiness, bitterness, sourness, and astringency for flavor and taste; creaminess and grittiness for texture and mouthfeel; and then the overall acceptability. The ranking was defined as follows: 1 = lowest intensity and 10 = highest intensity. The samples (about 10 g) were served at room temperature in white plastic dishes with teaspoons. Judges were asked to first observe all of the samples, in order to evaluate the appearance and structure, and then, to put the mayonnaise into their mouth and evaluate the other attributes (flavor, taste, texture, and overall acceptability). Water was used for cleaning the mouth between tasting different samples. Data were normalized to the average score given by each panelist, for each attribute. Then, data were recalculated using the calibration curve obtained by plotting each attribute's raw data (*y*-axis) versus normalized data (*x*-axis). Through this approach, differences in the scale of values of each non-calibrated panel member were avoided. The average scores for each attribute were determined and reported as spider plots.

2.8. Statistical Analysis

All of the data are the average of at least three measurements and reported as the mean and corresponding standard deviation; for the viscosity parameters, the median was considered. One-way ANOVA analysis was applied to experimental data. Linear regression analyses were carried out on data and the goodness of fit of the models was checked by the determination coefficient R^2. Tukey's test was used to establish the significance of differences among the mean values at the 0.05 significance level. Data were processed with XLSTAT software (Addinsoft SARL, New York, NY, USA).

3. Results and Discussion

The effect of the enrichment with Alg/Pec + OLE microparticles was studied by evaluating several physical properties of the mayonnaise samples just after preparation. In particular, the particle size, viscosity, tribology, and color were determined for their effect on the structural properties and stability of the emulsified structure and the sensory attributes.

3.1. Particle Size and Microstructure

In Table 1, the particle size of the mayonnaise samples, expressed as both $D_{4,3}$ and $D_{3,2}$, which are related to the size of the bulk of the droplets constituting the mayonnaise and the size of small droplets [22], respectively, are reported together with the specific surface area (SSA).

Table 1. Particle size parameters of mayonnaise samples after preparation, as affected by olive leaf phenolic extract (OLE) and alginate/pectin (Alg/Pec) + OLE enrichment.

	D [4;3] μm	D [3;2] μm	SSA m²/kg
Control	2.7 ± 0.7 [a]	1.9 ± 0.2 [a]	3270 ± 408 [c]
Mayo + OLE	2.1 ± 0.0 [b]	1.7 ± 0.0 [b]	3663 ± 59 [b]
Mayo + Alg/Pec	1.8 ± 0.0 [b]	1.5 ± 0.0 [c]	4272 ± 9 [a]

Values are means ± SD. Different superscript letters in the same column indicate significant differences ($p < 0.05$).

The results showed a reduction of the volume mean diameter ($D_{4,3}$) and the surface mean diameter ($D_{3,2}$) values for both of the enriched samples (Mayo + Alg/Pec and Mayo + OLE), with respect to the control, and a corresponding increase of the specific surface area with values of 4272, 3663, and 3270 for Mayo + Alg/Pec, Mayo + OLE, and the control, respectively. In general, the particle size distribution curves of the samples showed a polymodal distribution with two main populations and a shift towards a lower size in both of the enriched systems, which was more evident in the Mayo + Alg/Pec sample (Figure 1).

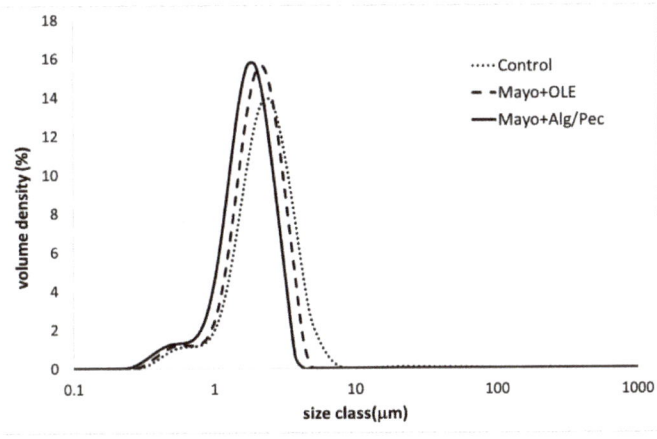

Figure 1. Particle size distribution curves of the differently enriched mayonnaise samples after preparation.

The decrease of the particle size in both of the enriched systems can be associated with the presence of olive leaf amphiphilic phenolic compounds, which have been proven to lower the oil/water interfacial tension and exert emulsifying properties, especially when tested in acidic conditions (acetate buffer, pH 4.5) [7]. In the case of Mayo + Alg/Pec, the presence of microparticles could have further improved the formation and stabilization of the emulsified structure, likely due to a Pickering stabilizing mechanism. Indeed, particles are able to stabilize oil/water interfaces, making them mechanically stronger, and are able to provide a sufficient steric repulsion force to inhibit droplet coalescence during emulsification [23]. The positive effect of OLE on the dispersion degree of the oil droplets is in accordance with other authors who prepared mayonnaise samples enriched with increasing amounts of olive phenolic extract [24]. In particular, the mayonnaise prepared with sunflower oil (SO) with an amount of oil (≈ 80%) comparable to the systems under investigation in this study, showed quite similar values of $D_{4,3}$ with respect to the control samples. However, when the enrichment was carried out at a similar concentration (200 ppm) to the one used in this work, a less broad particle distribution with a lower degree of polydispersity was obtained. Moreover, in the same study, olive oil mayonnaise-like emulsions formulated with different naturally phenolic-rich extra virgin olive oils displayed similar results when a low phenolic olive oil (270 mg/kg) was used as the source of phenolic compounds. Therefore, such results strengthen the argument for the use, for the formulation of high oil content emulsions, of a phenolic-rich extract recovered by olive by-products, with respect to a more expensive ingredient, such as extra virgin olive oil.

Optical microphotographs were captured to characterize the microstructural features of the samples (Figure 2a–c). In general, a well-dispersed oil phase could be observed in all of the systems, in agreement with the particle size results. While the presence of larger droplets was evident in samples a and b, defining a higher degree of polydispersity, a more finely dispersed structure and a closely packed distribution of oil droplets were visible in the mayonnaise enriched with OLE-loaded alginate/pectin microparticles.

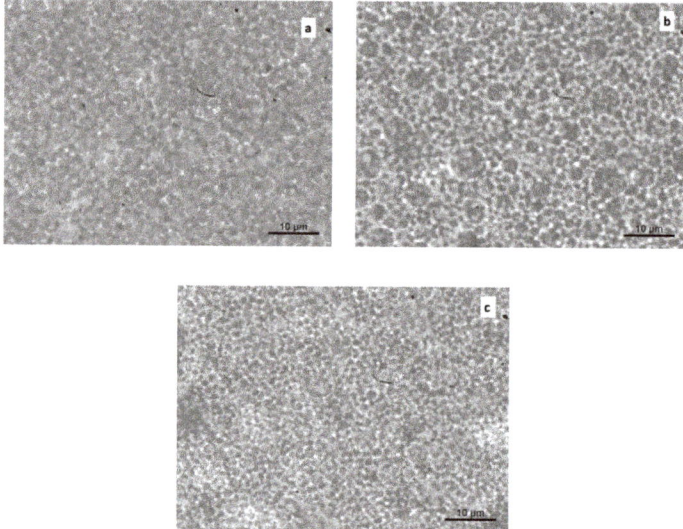

Figure 2. Optical micrograph images of mayonnaise samples just after preparation. Control (**a**), Mayo + OLE (**b**), and Mayo + Alg/Pec (**c**).

3.2. Color Properties

Color has a main impact on consumers' choice, as it is one of the most important sensory features that affect the willingness to purchase or taste a food. In general, the typical pale yellow color of mayonnaise originates from the egg yolk and oil and may be further influenced by the addition of mustard, additives, or some other spices with coloring effects (i.e., Annatto E160 and Turmeric). The enrichment of mayonnaise with unconventional ingredients, which are different with respect to those used in a standard recipe, could lead to physical and chemical changes that may affect, in turn, the color of the final products. In Figure 3, the chromatic indices of lightness and hue angle (L^* and $h°$, respectively) of mayonnaise samples, prepared with the addition of free OLE or OLE encapsulated in alginate/pectin microparticles, are reported, along with the control sample.

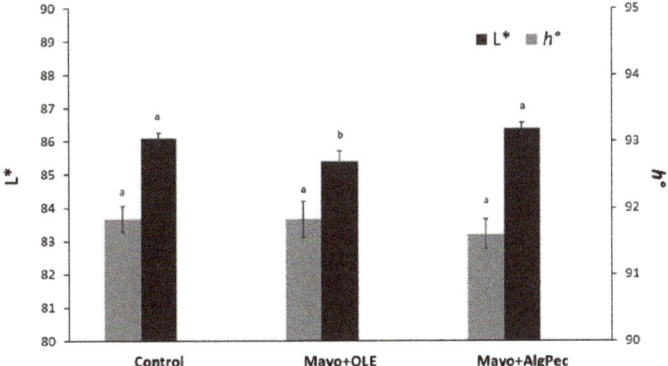

Figure 3. Color properties in terms of the lightness (L*) and hue angle ($h°$) of the mayonnaise samples under investigation. Different letters for parameters are significantly different by Tukey's HSD test ($p < 0.05$).

The lightness value (L*), which is the colorimetric parameter referring to the ability of the system to reflect and scatter the light, was, in general, similar to data reported in the literature for full-fat mayonnaise [25–27]. Significant differences ($p < 0.05$) between the L* value of the control and Mayo + Alg/Pec sample with respect to that of the Mayo + OLE sample were observed and this result may reflect the different dispersion degrees of the systems. In particular, the Mayo + OLE product displayed a significantly lower lightness due to the color of the added OLE extract, which, in the case of Mayo + Alg/Pec samples, was mitigated by the encapsulation.

On the contrary, the results of the hue angle were close to 90°, confirming the yellow color, without significant differences ($p > 0.05$) among the samples. Overall, these results highlight that the addition of the olive leaf extract, either as a pure or encapsulated form, did not alter the visual appearance of the emulsion.

3.3. Effect of OLE Enrichment on Flow Behavior and Lubricant Properties

The viscosity properties of concentrated emulsions, such as mayonnaise, are strictly related to the close packing of the dispersed oil droplets as they interact with one another in the matrix: The closer the droplets, the higher the viscosity, as a consequence of the higher droplet–droplet interaction [24].

The flow curves of mayonnaise samples are reported in Figure 4. In general, all of the mayonnaise samples exhibited a non-Newtonian shear-thinning behavior and a yield point related to the initial resistance of the systems to flow. Control and Mayo + OLE samples are both characterized by a similar trend, with a tendency to break-down at a higher shear rate (300 s^{-1}). At a similar shear rate, OLE-enriched mayonnaise displayed lower shear stress values with respect to the control, while the Alg/Pec mayonnaise samples showed higher yield stress and shear stress values. This latter behavior may be related to several factors, including the higher packing degree of the lower size oil droplets (Figure 2c) and the thickening effect of beads swelling due to the presence of alginate and pectin in the aqueous continuous phase. Recent studies observed an increase in viscosity for mayonnaise enriched with carbohydrate-based capsules due to the thickening effect of the polymers (i.e., dextran, pullulan, glucose syrup, and zein) used as shell material upon their disintegration [28,29]. Moreover, it must be pointed out that the pH values of the samples were 3.87 ± 0.06, 3.92 ± 0.01, and 3.89 ± 0.01 for the control, Mayo + OLE, and Mayo + Alg/Pec, respectively, which are values close to the average isoelectric point of egg yolk proteins, when the protein charge is minimized and both the viscoelasticity and stability of mayonnaise are generally enhanced [30].

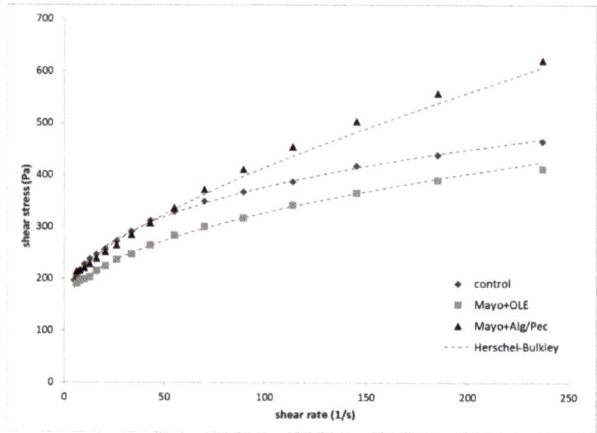

Figure 4. Flow curves of the mayonnaise after preparation.

Mayonnaise flow curves were fitted with the Herschel–Bulkley model, which is commonly used to describe the flow properties of concentrated emulsified systems such as mayonnaise [31]; the flow parameters yield stress τ_0 (Pa), consistency index K (Pa s^n), and flow index n of the samples after preparation are listed in Table 2.

Table 2. Herschel–Bulkley viscosity parameters of the mayonnaise samples.

	τ_0 (Pa)	n	K (Pa s^n)
Control	78.8 [b]	0.3 [b]	71.3 [a]
Mayo + OLE	136.1 [a,b]	0.5 [a,b]	21 [b]
Mayo + Alg/Pec	171.4 [a]	0.7 [a]	11.4 [c]

Values are the median ± SD. The flow parameters are the yield stress τ_0 (Pa), consistency index K (Pa s^n), and flow index n. Different superscript letters in the same column indicate significant differences ($p < 0.05$).

When compared to the control, both of the enriched samples were statistically different ($p < 0.05$) in terms of K, τ_0, and n, with Mayo + Alg/Pec showing the highest τ_0 and n values. As reported in the literature, yield stress values are an index of the strength of the attractive forces between the oil droplets [24,32] and this is in agreement with the results related to the Mayo + Alg/Pec sample, which exhibited the finest oil dispersion (Table 1).

Moreover, all of the mayonnaises presented a flow index (n)—the parameter that describes a pseudoplastic shear thinning behavior—lower than 1. It is interesting to note that the flow index tended to increase from the control to the enriched samples, with Mayo + Alg/Pec mayonnaise showing the highest value, corresponding to a lower pseudoplastic behavior. A shear-thinning profile in an oil-in-water emulsion is an indication that the system flows more readily as the dispersed particles align with the direction of flow; the enrichment with alginate/pectin OLE-loaded beads, which undergo swelling upon hydration, caused an increase of solids in the dispersing phase that likely hindered the alignment of the oil droplets.

It is known that flow behavior can affect the mouthfeel of food emulsions, while other sensory attributes, such as creaminess, can be correlated with the lubrication properties of the emulsified structure. Overall, the perception of a food emulsion in the mouth is initially dominated by the bulk viscosity at the front of the oral cavity, where the gap between the tongue and palate is rather large (>10 μm) [33]; upon chewing, the product mixed with the saliva is squeezed between moving surfaces, such as the tongue and palate (or food substrate and palate), and the oral perception depends on its thin film rheological behavior. To better analyse the rheological properties of the mayonnaise samples in the form of thin films, tribological investigations were carried out. The lubrication properties of the control and differently enriched mayonnaise samples were represented in the form of a Stribeck curve (Figure 5), where the friction coefficient (−) was plotted against the sliding velocity (mm s^{-1}) in the regions from 1 mm s^{-1}, generally associated with mouth-like conditions during food consumption, up to 200 mm s^{-1}, which is the maximum speed that the human tongue can move [34].

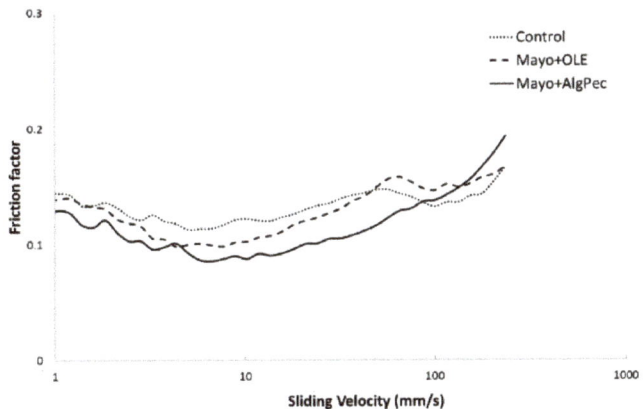

Figure 5. Stribeck curves of mayonnaise samples.

For all samples, it is possible to observe the occurrence of a typical Stribeck curve where the different regimes can be distinguished: In particular, the boundary regime, at low sliding speeds, with surface–surface interactions dominating the friction response with high friction factors; the mixed regime, at mid-speeds, when the sample starts to separate from the contact surfaces, decreasing the friction; and, finally, the hydrodynamic regime, where the surfaces are completely separated by the fluid with an increase of the friction values. For all of the mayonnaise samples, a similar behavior against the entertainment speed was noticed, with some differences in the friction factor values with the following ranking: Mayo + Alg/Pec < Mayo + OLE < control, and a clear discrimination among samples. With the exception of the highest sliding speeds, where all of the samples present similar trends, the mayonnaise enriched with Alg/Pec microparticles showed the lowest friction values, probably associated with a higher content of solids in the formulation and the finest oil droplet dispersion, with a corresponding better lubricant behavior. The highest lubricant properties of Mayo + Alg/Pec could also be attributed to the swelling of the hydrogel particles, which made them more deformable, allowing a decrease in the friction [35]. Another interesting result is related to the fact that Mayo + Alg/Pec sample exhibited a shift of the transition from a mixed to hydrodynamic regime toward a higher speed with respect to control and Mayo + OLE samples.

Despite the general effect of the addition of OLE, in both the free form and encapsulated in Alg/Pec beads, on the lubricant properties, which decreased the friction factors obtained by tribological measurements, additional investigations are needed to describe the behavior of such systems during oral processing, where other parameters, including the presence of saliva and mouth conditions (e.g., temperature), may play a key role.

3.4. Effect of OLE Enrichment on the Sensory Profile

Mayonnaises enriched with pure OLE or OLE encapsulated in alginate/pectin microparticles were assessed by a group of 15 panelists using a quantitative descriptive analysis. Before each session, the assessors were trained on each attribute, the scale used, and how to score the samples. The average scores of each selected attribute are reported in the spider plot shown in Figure 6. As expected, polyphenols addition, either in a pure or encapsulated form, significantly affected the bitterness perception with respect to the control sample, while no marked differences were reported for the two enriched mayonnaise samples, which were both perceived with a medium bitterness intensity (score ≈ 5.8) not appreciated by the assessors. The enrichment of mayonnaise did not influence the color, astringency, consistency, creaminess, and graininess parameters, which could be related to the presence of OLE or microcapsule powder. The saltiness, sourness, and spreadability were partially influenced by the addition of alginate/pectin microcapsules. It can be hypothesized that the saltiness increase may have been caused by the presence of calcium alginate of the microgels structure; moreover, in the literature, it is reported that bitter compounds such as caffein can enhance the effect of salt [36]. However, to the best of the authors' knowledge, to date, no literature data has been reported on the interaction between the saltiness and bitterness of olive leaf phenolic compounds. The enriched samples scored low in spreadability with respect to the control, indicating some resistance to spread, which is a result that is in accordance with the higher values of yield stress observed in the flow curves. Eventually, the sensory evaluation highlighted that both of the enriched mayonnaise samples, but in particular the system fortified with OLE-loaded microparticles, displayed the lowest overall acceptability from the panel. The results reflect the low ability of alginate/pectin microparticles to mask the bitterness of an OLE extract and present new challenges for further investigations, especially in terms of alternative encapsulation methods, in order to overcome such issues.

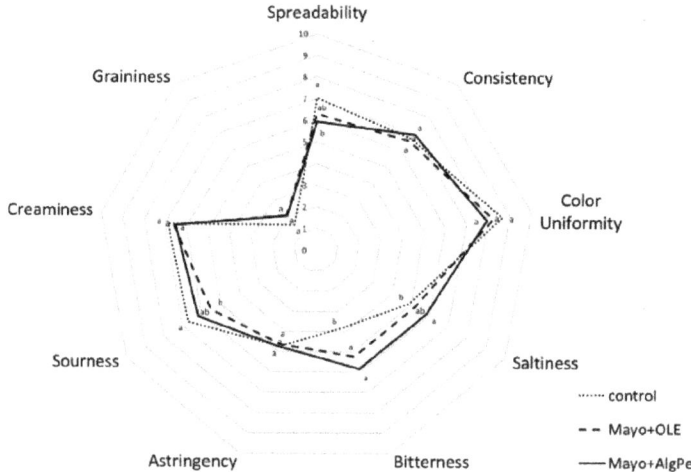

Figure 6. Spider plot of specific sensory attributes of the mayonnaise formulations. Lowercase letters for each attribute indicate significant differences ($p < 0.05$).

4. Conclusions

The enrichment of mayonnaise with OLE-loaded microparticles (Mayo + Alg/Pec) was effective in improving its physical properties in terms of both the dispersion degree of the oil droplets and lubricant properties of the emulsions; however, the highest viscosity, which in turn was reflected by the lowest spreadability, led to a different sensory perception with respect to both the control and Mayo + OLE samples. The addition of OLE, either as pure extract or encapsulated in alginate/pectin microparticles, slightly affected the chromatic parameters with respect to the control sample. The sensory evaluation depicted a lower overall acceptability for both of the enriched samples, likely due to the increased salty and bitter perception. The encapsulation technique adopted was thus not able to mask or control the bitterness perception, leading to the conclusion that further investigations are needed in terms of encapsulation methods, bead compositions and structures, and mayonnaise formulations to overcome such issues. The proposed research could represent a valid contribution in the wide panorama of functional foods and opens new opportunities for formulating healthier foodstuffs while valorizing olive by-products.

Author Contributions: Conceptualization, F.F. and C.D.D.M.; formal analysis, F.F., G.S., and C.D.D.M.; investigation, F.F.; resources, C.D.D.M. and L.N.; data curation, F.F., L.N., G.S., and C.D.D.M.; writing—original draft preparation, F.F.; writing—review and editing, C.D.D.M., P.P., and D.M.; visualization, C.D.D.M.; supervision, C.D.D.M., D.M., and P.P.; funding acquisition, C.D.D.M.; project administration, C.D.D.M. All authors have read and agreed to the published version of the manuscript.

Funding: This research was supported by AGER 2 Project–"S.O.S—Sustainability of the Olive-Oil System"—grant no. 2016-0105.

Acknowledgments: The authors are grateful to Domenico Cifà from Oleafit srl (Isola del Gran Sasso, Teramo, Italy) for providing the olive leaf extract and Chiara Di Giacinto for her contribution to this work.

Conflicts of Interest: The authors declare no conflicts of interest.

References

1. Mirzanajafi-Zanjani, M.; Yousefi, M.; Ehsani, A. Challenges and approaches for production of a healthy and functional mayonnaise sauce. *Food Sci. Nutr.* **2019**, *7*, 2471–2484. [CrossRef] [PubMed]
2. Sánchez De Medina, V.; Priego-Capote, F.; Jiménez-Ot, C.; Luque De Castro, M.D. Quality and stability of edible oils enriched with hydrophilic antioxidants from the olive tree: The role of enrichment extracts and lipid composition. *J. Agric. Food Chem.* **2011**, *59*, 11432–11441. [CrossRef] [PubMed]
3. Aouidi, F.; Okba, A.; Hamdi, M. Valorization of functional properties of extract and powder of olive leaves in raw and cooked minced beef meat. *J. Sci Food Agric.* **2017**, *97*, 3195–3203. [CrossRef] [PubMed]
4. Djenane, D.; Gómez, D.; Yangüela, J.; Roncalés, P.; Ariño, A. Olive leaves extract from algerian oleaster (Olea europaea var. Sylvestris) on microbiological safety and shelf-life stability of raw halal minced beef during display. *Foods* **2019**, *8*, 10. [CrossRef]
5. Difonzo, G.; Pasqualone, A.; Silletti, R.; Cosmai, L.; Summo, C.; Paradiso, V.M.; Caponio, F. Use of olive leaf extract to reduce lipid oxidation of baked snacks. *Food Res. Int.* **2018**, *108*, 48–56. [CrossRef]
6. Palmeri, R.; Parafati, L.; Trippa, D.; Siracusa, L.; Arena, E.; Restuccia, C.; Fallico, B. Addition of Olive Leaf Extract (OLE) for Producing Fortified Fresh Pasteurized Milk with An Extended Shelf Life. *Antioxidants* **2019**, *8*, 255. [CrossRef]
7. Flamminii, F.; Di Mattia, C.D.; Difonzo, G.; Neri, L.; Faieta, M.; Caponio, F.; Pittia, P. From by-product to food ingredient: Evaluation of compositional and technological properties of olive-leaf phenolic extracts. *J. Sci. Food Agric.* **2019**, *99*, 6620–6627. [CrossRef]
8. Perito, M.A.; Di Fonzo, A.; Sansone, M.; Russo, C. Consumer acceptance of food obtained from olive by-products: A survey of Italian consumers. *Br. Food J.* **2019**. [CrossRef]
9. Fang, Z.; Bhandari, B. Encapsulation of polyphenols—A review. *Trends Food Sci. Technol.* **2010**, *21*, 510–523. [CrossRef]
10. Vitaglione, P.; Savarese, M.; Paduano, A.; Scalfi, L.; Fogliano, V.; Sacchi, R. Healthy Virgin Olive Oil: A Matter of Bitterness. *Crit Rev. Food Sci. Nutr.* **2015**, *55*, 1808–1818. [CrossRef]
11. Flamminii, F.; Di Mattia, C.D.; Nardella, M.; Chiarini, M.; Valbonetti, L.; Neri, L.; Difonzo, G.; Pittia, P. Structuring alginate beads with different biopolymers for the development of functional ingredients loaded with olive leaves phenolic extract. *Food Hydrocoll.* **2020**, 105849. [CrossRef]
12. González-Ortega, R.; Faieta, M.; Di Mattia, C.D.; Valbonetti, L.; Pittia, P. Microencapsulation of olive leaf extract by freeze-drying: Effect of carrier composition on process efficiency and technological properties of the powders. *J. Food Eng.* **2020**, *285*, 110089. [CrossRef]
13. González, E.; Gómez-Caravaca, A.M.; Giménez, B.; Cebrián, R.; Maqueda, M.; Martínez-Férez, A.; Segura-Carretero, A.; Robert, P. Evolution of the phenolic compounds profile of olive leaf extract encapsulated by spray-drying during in vitro gastrointestinal digestion. *Food Chem.* **2019**, *279*, 40–48. [CrossRef] [PubMed]
14. Mohammadi, A.; Jafari, S.M.; Assadpour, E.; Faridi Esfanjani, A. Nano-encapsulation of olive leaf phenolic compounds through WPC-pectin complexes and evaluating their release rate. *Int. J. Biol. Macromol.* **2016**, *82*, 816–822. [CrossRef]
15. Mohammadi, A.; Jafari, S.M.; Esfanjani, A.F.; Akhavan, S. Application of nano-encapsulated olive leaf extract in controlling the oxidative stability of soybean oil. *Food Chem.* **2016**, *190*, 513–519. [CrossRef]
16. Robert, P.; Zamorano, M.; González, E.; Silva-Weiss, A.; Cofrades, S.; Giménez, B. Double emulsions with olive leaves extract as fat replacers in meat systems with high oxidative stability. *Food Res. Int.* **2019**, *120*, 904–912. [CrossRef]

17. Ganje, M.; Jafari, S.M.; Dusti, A.; Dehnad, D.; Amanjani, M.; Ghanbari, V. Modeling quality changes in tomato paste containing microencapsulated olive leaf extract by accelerated shelf life testing. *Food Bioprod. Process.* **2016**, *97*, 12–19. [CrossRef]
18. Kranz, P.; Braun, N.; Schulze, N.; Kunz, B. Sensory quality of functional beverages: Bitterness perception and bitter masking of olive leaf extract fortified fruit smoothies. *J. Food Sci.* **2010**, *75*, S308–S311. [CrossRef]
19. Tavakoli, H.; Hosseini, O.; Jafari, S.M.; Katouzian, I. Evaluation of Physicochemical and Antioxidant Properties of Yogurt Enriched by Olive Leaf Phenolics within Nanoliposomes. *J. Agric. Food Chem.* **2018**, *66*, 9231–9240. [CrossRef]
20. Urzúa, C.; González, E.; Dueik, V.; Bouchon, P.; Giménez, B.; Robert, P. Olive leaves extract encapsulated by spray-drying in vacuum fried starch–gluten doughs. *Food Bioprod. Process.* **2017**, *106*, 171–180. [CrossRef]
21. Di Mattia, C.; Balestra, F.; Sacchetti, G.; Neri, L.; Mastrocola, D.; Pittia, P. Physical and structural properties of extra-virgin olive oil based mayonnaise. *LWT Food Sci. Technol.* **2015**, *62*, 764–770. [CrossRef]
22. McClements, D. *Food Emulsions: Principles, Practices, and Techniques*; CRC Press: Boca Raton, FL, USA, 2015.
23. Mao, L.; Miao, S. Structuring Food Emulsions to Improve Nutrient Delivery During Digestion. *Food Eng. Rev.* **2015**, *7*, 439–451. [CrossRef]
24. Giacintucci, V.; Di Mattia, C.; Sacchetti, G.; Neri, L.; Pittia, P. Role of olive oil phenolics in physical properties and stability of mayonnaise-like emulsions. *Food Chem.* **2016**, *213*, 369–377. [CrossRef] [PubMed]
25. Cerezal Mezquita, P.; Morales, J.; Palma, J.; Ruiz, M.D.C.; Jáuregui, M. Stability of Lutein Obtained from *Muriellopsis sp* biomass and used as a natural colorant and antioxidant in a mayonnaise-like dressing sauce. *CyTA J. Food.* **2019**, *17*, 517–526. [CrossRef]
26. Sikimić, V.M.; Popov-Raljić, J.V.; Zlatković, B.P.; Lakić, N. Colour Determination and Change of Sensory Properties of Mayonnaise with Different Contents of Oil Depending on Length of Storage 1. *Sens. Transducers J.* **2010**, *112*, 138–165.
27. Worrasinchai, S.; Suphantharika, M.; Pinjai, S.; Jamnong, P. β-Glucan prepared from spent brewer's yeast as a fat replacer in mayonnaise. *Food Hydrocoll.* **2006**, *20*, 68–78. [CrossRef]
28. Hermund, D.; Jacobsen, C.; Chronakis, I.S.; Pelayo, A.; Yu, S.; Busolo, M.; Lagaron, J.M.; Jónsdóttir, R.; Kristinsson, H.G.; Akoh, C.C.; et al. Stabilization of Fish Oil-Loaded Electrosprayed Capsules with Seaweed and Commercial Natural Antioxidants: Effect on the Oxidative Stability of Capsule-Enriched Mayonnaise. *Eur. J. Lipid Sci. Technol.* **2019**, *121*, 1800396. [CrossRef]
29. Miguel, G.A.; Jacobsen, C.; Prieto, C.; Kempen, P.J.; Lagaron, J.M.; Chronakis, I.S.; García-Moreno, P.J. Oxidative stability and physical properties of mayonnaise fortified with zein electrosprayed capsules loaded with fish oil. *J. Food Eng.* **2019**, *263*, 348–358. [CrossRef]
30. Kiosseoglou, V.D.; Sherman, P. Influence of egg yolk lipoproteins on the rheology and stability of o/w emulsion and mayonnaise 1. Viscoelasticity of groundnut oil-in-water emulsions and mayonnaise. *J. Texture Stud.* **1983**, *14*, 397–417. [CrossRef]
31. Liu, H.; Xu, X.M.; Guo, S.D. Rheological, texture and sensory properties of low-fat mayonnaise with different fat mimetics. *LWT Food Sci. Technol.* **2007**, *40*, 946–954. [CrossRef]
32. Mun, S.; Kim, Y.L.; Kang, C.G.; Park, K.H.; Shim, J.Y.; Kim, Y.R. Development of reduced-fat mayonnaise using 4αGTase-modified rice starch and xanthan gum. *Int. J. Biol. Macromol.* **2009**, *44*, 400–407. [CrossRef]
33. Ruan, Q.; Yang, X.; Zeng, L.; Qi, J. Physical and tribological properties of high internal phase emulsions based on citrus fibers and corn peptides. *Food Hydrocoll.* **2019**, *95*, 53–61. [CrossRef]
34. Hiiemae, K.M.; Palmer, J.B. Tongue movements in feeding and speech. *Crit. Rev. Oral Biol. Med.* **2003**, *14*, 413–429. [CrossRef] [PubMed]
35. Chojnicka-Paszun, A.; Doussinault, S.; De Jongh, H.H.J. Sensorial analysis of polysaccharide-gelled protein particle dispersions in relation to lubrication and viscosity properties. *Food Res. Int.* **2014**, *56*, 199–210. [CrossRef]
36. Amerine, M.A.; Pangborn, R.M.; Roessler, E.B. *Principles of Sensory Evaluation of Food*; Departments of Viticulture and Enology, Food Science and Technology, and Mathematics, University of California: Davis, CA, USA, 2013.

© 2020 by the authors. Licensee MDPI, Basel, Switzerland. This article is an open access article distributed under the terms and conditions of the Creative Commons Attribution (CC BY) license (http://creativecommons.org/licenses/by/4.0/).

Article

Gluten-Free Breadsticks Fortified with Phenolic-Rich Extracts from Olive Leaves and Olive Mill Wastewater

Paola Conte [1,*], Simone Pulina [1], Alessandra Del Caro [1], Costantino Fadda [1], Pietro Paolo Urgeghe [1], Alessandra De Bruno [2], Graziana Difonzo [3], Francesco Caponio [3], Rosa Romeo [2] and Antonio Piga [1]

1. Department of Agricultural Sciences, Università degli Studi di Sassari, Viale Italia 39/A, 07100 Sassari, Italy; spulina1@uniss.it (S.P.); delcaro@uniss.it (A.D.C.); cfadda@uniss.it (C.F.); paolou@uniss.it (P.P.U.); pigaa@uniss.it (A.P.)
2. Department of Agraria, University Mediterranea of Reggio Calabria, 89124 Reggio Calabria, Italy; alessandra.debruno@unirc.it (A.D.B.); rosa.romeo@unirc.it (R.R.)
3. Department of Soil Plant and Food Sciences, University of Bari Aldo Moro, Via Amendola 165/A, 70126 Bari, Italy; graziana.difonzo@uniba.it (G.D.); francesco.caponio@uniba.it (F.C.)
* Correspondence: pconte@uniss.it; Tel.: +39-079-229277

Abstract: Nowadays, food processing by-products, which have long raised serious environmental concerns, are recognized to be a cheap source of valuable compounds. In the present study, incorporation of phenolic-rich extracts (500 and 1000 mg kg^{-1}) from olive leaves (OL) and olive mill wastewater (OMW) into conventional gluten-free formulations has been exploited as a potential strategy for developing nutritious and healthy breadsticks with extended shelf-life. To this end, moisture, water activity (a_w), visual and textural properties, the composition of biologically active compounds (soluble, insoluble, and bio-accessible polyphenols), antioxidant activity, oxidation stability, and consumer preference of the resulting breadsticks were investigated. Fortified breadsticks had higher moisture and a_w, lower hardness, and similar color in comparison to the control, especially in the case of OL extract supplementation. All enriched formulations significantly affected the phenolic composition, as evidenced by the decrease in insoluble/soluble polyphenols ratio (from 7 in the control up to 3.1 and 4.5 in OL and OMW, respectively), and a concomitant increase in polyphenol bio-accessibility (OL: 14.5–23% and OMW: 10.4–15% rise) and antioxidant activity (OL: 20–36% and OMW: 11–16% rise). Moreover, a significant shelf-life extension was observed in all fortified breadsticks (especially in case of OMW supplementation). Sensory evaluation evidenced that 61% of the assessors showed a marked, but not significant, tendency to consider the sample supplemented with high levels of OL as a more palatable choice.

Keywords: breadsticks; gluten-free; olive oil by-products; antioxidants; oxidation stability

1. Introduction

For a long time, food processing waste and by-products have been of great concern for both agro-food industries and society due to the environmental and economic impacts induced by their safe disposal [1]. In the past few year, however, this view has changed in order to recognize such industry by-products as a potential and cheap source of both nutraceuticals and functional compounds (mainly bio-phenols) to be re-used in the development of highly-nutritious and healthy foods with extended shelf-life [2]. In addition, re-using products that are wasted during industrial processing and transformation could reduce food waste level, contribute to environmental preservation, and provide long-term sustainability to the food production systems [3].

The olive oil industry, apart from its main valuable product, usually produces–in a short period of time (from November to February)–huge amounts of other waste, such as olive leaves (OL), thin branches, woods, olive mill wastewater (OMW) and olive pomace, whose management, treatment and disposal are often the cause of major problems for olive oil producers [4,5]. The first residue generated during the virgin olive oil extraction process is represented by the OL, which are accumulated in large quantities (about 10% of the total weight of the harvested olives) during the early stages of olive cleaning or during pruning of olive trees [6]. These lignocellulosic residues are a rich source of phenolic compounds (at about 2.5%), mainly oleuropein, whose content can exceed 15% in dry material, but also verbascoside, hydroxy-tyrosol, tyrosol, caffeic acid, vanillic acid, luteolin, rutin, and apigenin, which are well known for their recognized antioxidant and antimicrobial activity [7–9]. OL also contain water (51%), carbohydrates (27%), protein (7%), crude fiber (7%), oil (3%), and ashes (2.5%) [10].

However, the most polluting by-product produced by the olive oil extraction industry is represented by the liquid effluent OMW, which is generated during the separation phase of the oil from the malaxed paste in amounts that vary depending on the type of system used for oil extraction. In fact, with the exclusion of the traditional discontinuous press system, which is now disused due to both economic and quality disadvantages, the largest amount of OMW is generated by the widely used three-phase centrifugal extraction technology that requires the addition of up to 50 L of water per 100 kg of olive paste [11]. Despite its variable composition–mainly due to factors such as olive variety, fruit ripeness, climatic conditions, and processing method–OMW contains high amounts of organic substances (from 4 to 18 g 100 g^{-1}), including sugars, pectins, lignin, tannins, and lipids, a high content of phenolic compounds (0.5–24 mg L^{-1}), as well as smaller amounts of inorganic substances (0.4–2.5 g 100 g^{-1}), mainly potassium and phosphatic salts [12]. In particular, phenolic compounds, which are mainly responsible for the high polluting load and consequent phytotoxic activity of this discharging effluent, are also well known for their antioxidant properties, health-benefits and good bioavailability [13]. In particular, the most abundant bio-phenol found in OMW, as well as the most interesting from a nutritional point of view, is hydroxy-tyrosol, which exerts several biological activities, including free radical scavenging action, protection against the oxidation of human low-density lipoproteins (LDL), anti-inflammatory, antithrombotic, and in vitro anti-microbial activity [4,12,14,15]. Therefore, through the recovery and reutilization of such valuable compounds, the polluting waste from the olive oil industry could be turned into alternative and low-cost sources of bioactive phenols to be used as natural antioxidants in foods [16].

Besides improving nutritional value and providing health benefits, enrichment of food products with natural antioxidants might also be an effective tool to delay the lipid oxidation process and slow down the formation of off-flavors, thus extending the shelf life of the product [17–20]. In addition, natural antioxidants can be used as substitutes for other synthetic preservatives, such as butylated hydroxytoluene (BHT), butylated hydroxy anisole (BHA), *tert*-butyl hydroquinone (TBHQ), and propyl gallate (PG), that, despite their high efficiency in improving the shelf life of food products, have been shown to have harmful effects on human health [19,20].

Regularly eaten and affordable baked snacks like breadsticks–typical and widespread Italian bakery products appreciated in many other European countries–due to the way they intercept consumer and cultural food trends, could represent an ideal vehicle for the addition of natural bioactive compounds to the diet [21,22]. Breadsticks are crispy products with low moisture content and long shelf-life. The major causes of their quality deterioration are associated with loss of crunchiness and lipid oxidation [23]. In fact, despite being foods with low a_w, breadsticks are usually prepared by adding high amounts of lipids–in some cases olive oil, but mostly vegetable oils and animal fat–to the basic formulation [22,24,25]. These phenomena are even more pronounced in the production of gluten-free (GF) baked products that, besides an unattractive appearance, poor mouthfeel and flavor, and a low nutritional value (lack of protein, iron, calcium, and vitamins), are more prone to staleness than their gluten-containing counterparts [26]. Such behavior may be due mainly to the higher amounts of water required to obtain workable doughs with acceptable consistency that, in the absence of gluten and in the presence of high amounts of starchy ingredients, foster moisture redistribution and starch retrogradation during storage, thus shortening the shelf life of the final product [27].

Recently, the authors of [25] reported the application of OL extracts as a useful tool to improve the oxidative stability and, in turn, the shelf-life of salty wheat-based baked snacks. However, after an extensive review of the available literature, no research on the potential use of olive oil by-products (neither OL nor OMW) in the development of both gluten containing and GF fortified baked snacks has been found.

In this context, the aim of this study is to evaluate the effects of the addition of two different amounts (500 and 1000 mg kg^{-1}) of phenolic-rich extracts obtained from both OL and OMW on the textural and nutritional characteristics, as well as on the oxidative stability and consumer preferences, of GF breadsticks. Special emphasis will be placed on the assessment of both polyphenol fractions and antioxidant activity of the enriched baked products.

2. Materials and Methods

2.1. Raw Materials

Commercial rice flour, corn starch, guar gum and *Psyllium* fiber were from Chimab Campodarsego (Padova, PD, Italy). Fresh compressed yeast, salt and sugar were purchased from a local supermarket. Sunflower oil was from Carapelli Firenze (Florence, Italy).

2.2. Preparation of By-Product Extracts

2.2.1. Olive Leaf Extract

The OL were collected in the crop season 2020/2021 from Coratina olive cultivar, stored at 4 °C, and processed in less than 24 h. After washing with tap water at room temperature, the olive leaves were dried at 120 °C for 8 min in a ventilated oven (Argolab, Carpi, Italy) to reach a moisture content <1%, and then milled with a blender (Waring-Commercial, Torrington, CT, USA). The extraction process was ultrasound-assisted (CEIA, Viciomaggio, Italy) and water was added in a ratio 1/20 (w/v). After three washings, each performed for 30 min at a temperature of 35 ± 5 °C, the extracts were filtered through Whatman filter paper (GE Healthcare, Milan, Italy), freeze-dried, and stored at −20 °C. Before the analysis, the extract was further filtered by using nylon filters (0.45 µm pore-size).

2.2.2. Olive Mill Wastewater Extract

The OMW was collected in the crop season 2020/2021 from Ottobratica olive cultivar and produced according to a three-phase centrifugation process. The phenolic extract was obtained following the method previously reported by [28], with some modifications. Briefly, two liters of OMW were acidified to pH 2 with hydrochloric acid and washed three times with hexane (1:1, v/v) to remove the lipid fraction. The obtained mixture was vigorously shaken and centrifuged at 1550× g rpm for 3 min at 10 °C. The extraction procedure was carried out in separate funnels and repeated three times using ethyl acetate (1:4, v/v) as a solvent. The organic phase was then separated and filtered through a Buchner apparatus. The ethyl acetate was evaporated under vacuum in a rotary evaporator (Laborota400, Heidolph Instrument, Schwabach, Germany) at 25 °C. Finally, the obtained residue was dissolved in water to a final volume of 100 mL, further filtered by using a PTFE syringe filter (0.45 µm pore-size) and stored at 4 °C before analysis. To ensure that both hexane and ethyl acetate were removed (or were present in trace amounts) from the obtained extract, a headspace analysis using GC Thermo Trace 1310 apparatus (Waltham, MA, USA) equipped with a Single Quadrupole Mass Spectrometer ISQ LT system and a fused-silica capillary column (Thermo Scientific, Waltham, MA, USA) was carried out. It was found that the concentration of both the solvents used was below the limit of detection of the instrument.

2.2.3. Total Polyphenol Content and Antioxidant Activity of OL and OMW Extracts

The total polyphenol content of both OL and OMW extracts (hereinafter referred to as OL_E and OMW_E) was measured using the Folin-Ciocalteau reagent in a spectrophotometer (mod. Cary 3500, Agilent, Cernusco, Milan, Italy) set at 750 nm [29]. Briefly, after a proper dilution, an aliquot of 0.5 mL of extract were mixed with 0.5 mL of Folin-Ciocalteau reagent, 10 mL of sodium carbonate (7.5%) and adjusted to 25 mL with distilled water. The mixture was incubated in the dark for 1 h at room temperature before readings. Calibration curves were made using gallic acid and the results (mean of three replicates) were expressed as mg of gallic acid equivalents (GAE) per g of extract.

The free radical scavenging activity of both extracts was determined using a discoloration curve of the stable radical 2,2-diphenyl-1-picrylhydrazyl (DPPH). After proper dilution, an aliquot of 0.3 mL of extract was added to 2.7 mL of 0.0634 $\mu mol\ mL^{-1}$ DPPH methanol solution for 1 h at 515 nm and 22 °C. Results (mean of three replicates) are expressed as decrease in absorbance (%) per mg of extract when 0.17 µmol of DPPH are available to react.

The determination of the phenolic compounds and antioxidant activity was carried out simultaneously in both the extracts and the breadstick samples only a few days after the extraction process (specifically, no more than 15 days later).

2.3. Preparation of GF Breadsticks

GF control breadsticks were prepared using a conventional GF formulation consisting of 50% rice flour and 50% corn starch as the basic recipe. The other ingredients used, which were added as % on flour basis, were: 55% water (26 °C), 10% sunflower oil, 4% compressed yeast, 3% sugar, 1.8% salt, 1.5% guar gum, and 1.5% *Psyllium* fiber. For the preparation of the enriched GF samples, OL_E and OMW_E were singly added to the basic formulation (on flour basis) a two different level of supplementation: low (500 mg kg^{-1}) and high (1000 mg kg^{-1}). Breadstick sample codes were defined according to the type of extract and the level of supplementation used, as follows: Control (no extract addition), Leaf50 (low addition of OL_E), Leaf100 (high addition of OL_E), WasteW50 (low addition of OMW_E) and WasteW100 (high addition of OMW_E). For each formulation, GF breadsticks were prepared by firstly suspending extracts, yeast, salt and sugar in aliquots of warm water. Then, these dissolved ingredients and the sunflower oil were slowly added to the pre-mixed dry ingredients and kneaded using a mixer (KitchenAid Professional, Model 5KSM7990, St. Joseph, MI, USA) equipped with a dough hook at speed 1 for 5 min followed by other 8 min at speed 2. After mixing, the obtained doughs (three for each sample) were proofed in a climate chamber for 30 min (33 °C–90% RH), manually shaped into 28 g weighted and 30 cm length sticks, placed on rectangular baked pans, and proofed once again for 30 min (33 °C–90% RH). Finally, the GF breadsticks were baked in an electric oven (Europa, Malo, VI, Italy) following a two-step baking process: they were firstly baked at 180 °C for 13 min, allowed to cool at room temperature for 30 min and then baked once again for other 22 min at 160 °C. After baking, breadstick samples were cooled for 1 h, before the analysis.

2.4. Breadsticks Measurements

2.4.1. Moisture Content and Water Activity

The moisture content of the GF breadstick samples was determined using a moisture analyzer (Model Kern-DAB 100-3, KERN & SOHN GmbH, Balingen, Germany) equipped with a halogen quartz glass heater (400 W) and set with a standard heating profile at 105 °C. The results were expressed in % as the average of five repetitions.

Water activity was measured on ground and homogenized samples with an electronic hygrometer (model Aw-Win, Rotronic, Bassersdorf, Switzerland) equipped with a Karl-Fast probe previously calibrated in the range of 0.1–0.95 with solutions of lithium chloride (LiCl) of known activity. A total of five repetitions for each sample were made.

2.4.2. Color Determination

Color measurements were carried out on the day of baking by using a tristimulus colorimeter (Minolta CR-300, Konica Minolta Sensing, Osaka, Japan) equipped with a measuring head CR-300 and previously calibrated against a white tile. To avoid inaccurate measurements due to the limited width of the samples, 60 breadsticks per batch were finely ground and placed into the granular material attachment (CR-A50, Konica Minolta Sensing, Osaka, Japan) of the colorimeter. The results were expressed in accordance with the Hunter Lab color space and the parameters acquired were lightness L^*, redness (a^*), and yellowness (b^*). Total color difference (ΔE) was also calculated by using the following equation:

$$\Delta E = ((\Delta L^2) + (\Delta a)^2 + (\Delta b)^2)^{1/2} \tag{1}$$

For each sample, a total of ten repetitions were made.

2.4.3. Textural Properties

Evaluation of breadsticks' textural properties was carried out on twenty freshly prepared samples 1 h after baking by means of a three-point bending test. A texture analyzer (TA-XT2 Texture Analyzer, Stable Microsystems, Surrey, UK) equipped with a 30 kg load cell and a three-point bending rig (HDP/3PB) was used. After placing and blocking the two adjustable supports of the rig base at a span distance of 60 mm, a half breadstick–about 15 cm long–was placed centrally over the supports and broken by a blade probe moving downwards at a pre-test speed of 1 mm s^{-1} and a test speed of 3 mm s^{-1}. The maximum peak force (N) required to break the sample and the distance to break (mm) were determined from the obtained force–distance curves and further referred to as hardness and brittleness. The software Texture Exponent TEE32 (v. 6.1.10.0 Stable Micro System, Surrey, UK) was used for data processing.

2.4.4. Determination of Polyphenol Fractions and Antioxidant Activity

To determine the soluble, insoluble phenolic fractions of breadstick samples, the procedures previously described in GF breads by [26] were applied. Briefly, 2 g of finely ground breadstick samples were extracted twice with a solution (4 mL) of 37% hydrochloric acid/methanol/water (1/80/10, $v/v/v$). The supernatants were collected, filtered, and used for the determination of the soluble polyphenols. To obtain the insoluble phenolic fraction, sample residues from the soluble polyphenols' extraction were digested with 5 mL of a methanol/concentrated sulfuric acid solution (10:1, v/v) for 20 h by shacking in a thermostatic water bath set at 85 °C. The obtained extracts were analyzed using the Folin-Ciocalteau reagent in a spectrophotometer (mod. Cary 3500, Agilent, Cernusco, Milan, Italy) set at 750 nm [29].

To estimate polyphenols' bio-accessibility, the experimental samples were digested in vitro according to the procedure described by [26]. Briefly, 1 g of each ground sample was firstly digested with 0.5 mL of pepsin in a shaking water bath for 1 h at 37 °C to accurately simulate the gastric digestion, and then with 2.5 mL of a solution containing bile salts and the pancreatin enzyme and 2.5 mL of sodium chloride/potassium chloride solution (at room temperature for 2 h) to simulate the intestinal digestion. After removing protein by addition of trichloroacetic acid (20%, v/v), the digested extracts were spectrophotometrically analyzed as described for the soluble and insoluble polyphenols. For all the phenolic fractions, calibration curves were made using gallic acid as standard and the results (mean of three replicates) were expressed as mg of GAE 100 g^{-1} of breadsticks, d.m.

The free radical scavenging activity of the experimental GF samples was determined using the DPPH assay as previously described by [26], with some modifications. In brief, aliquots of 0.3 mL of organic extracts were made to react with 2.7 mL of 0.0634 µmol mL^{-1} DPPH methanol solution using a spectrophotometer (mod. Cary 3500, Agilent, Cernusco, Milan, Italy) set at 515 nm and 22 °C to obtain a decrease in absorbance by the radical DPPH. The absorbance was read at 1 min and every 5–10 min until the plateau was reached

(70 min). The test was performed in triplicate and plots of µmol DPPH vs. time (min) were drawn. The antiradical activity (AR) was calculated using the following equation:

$$AR = ((DPPH_{initial} - DPPH_{plateau}) \times 100)/DPPH_{initial} \qquad (2)$$

2.4.5. Determination of Oxidation Stability (Oxitest) and Shelf-Life Estimation

Oxitest (VELP Scientifica, Usmate Velate (MB), Italy) is an oxidation stability reactor that, according to AOCS International Standard Procedure Cd 12c-16 [30], allows rapid measurements of the stability of foods against the lipid oxidation by subjecting the sample to high temperature and pure oxygen overpressure. It is controlled through the specific OXISoft^{TM} Software, which also allows the prediction of oxidation stability for shelf-life studies. In the present research, 30 g of ground breadstick samples (10 g per plate in each oxidation chamber) were analyzed at different working temperatures (60, 70, 80, and 90 °C) and at a pressure of 6 bar. The induction period (IP) was calculated from the obtained pressure–time curves as the time required to achieve a 10% drop of the oxygen pressure inside the oxidation chambers. In case of linear dependence with the temperature, the software calculates a linear regression equation on a semi-log scale (log of the IP–temperature curve) to predict the estimated IP and, thus, the shelf-life (days) of the products at the desired storage temperature (25 °C). All measurements were made in duplicate.

2.4.6. Sensory Evaluation

Sensory analysis was conducted to assess whether there were differences in preference for the freshly prepared breadsticks by performing two different quantitative affective tests. Firstly, a ranking preference test was carried out in two different sessions to allow separate comparisons between the control and the samples enriched with OL_E and OMW_E, respectively. Subsequently, a paired comparison test was carried out to compare the two breadstick samples found to be the most preferred in the previous ranking tests, according to the preference degree.

The ranking test was performed in a laboratory and conducted in individual booths with a panel composed of 60 consumers (39 men and 21 women aged between 18 and 65 years). The samples were presented at room temperature in a randomized and balanced order and water was provided to allow appropriate cleansing of the palate between sample tastings. To establish whether there was a difference among samples for the given attribute (preference), data were analyzed using the Friedman's test.

If the F_{test} value is higher than the $F_{critical}$ value–which can be found in the χ^2 table for the defined confidence level (95%) and the degrees of freedom k-1–it can be argued that at least two samples are significantly different for the attribute analyzed. If it is established that a significant difference exists, to find out which samples differ from each other, the Fisher's least significant difference (LSD) is calculated.

The paired comparison test was conducted using a panel composed of 76 consumers (42 men and 34 women aged between 18 and 65 years). In this case, data analysis was carried out by comparing the largest number of responses for one sample with the critical value reported in the statistical table for paired difference test (two tailed) that–at the defined confidence level (95%) and with the effective number of participants–indicates the minimum number of responses required to conclude that there is a significant difference between the two samples.

2.5. *Statistical Analysis*

Statistical analysis of the results was performed using Statistica 10.0 software (StatSoft, Inc., Tulsa, OK, USA). The experimental data were submitted to one-way analysis of variance (ANOVA) followed by Fisher's least significant differences (LSD) to know the difference between each pair of means with 95% confidence. Pearson correlation analysis for relationships between some selected parameters was also used.

3. Results and Discussion

3.1. Total Phenolic Content and Antioxidant Activity of OL and OMW Extracts

As reported in Table 1, the total polyphenol content of OL and OMW samples was 134.7 ± 2.1 and 13.4 ± 0.2 mg GAE g^{-1}, respectively, with the OL$_E$ showing a phenolic concentration at about 10-fold higher than that found in OMW$_E$. These results are considerable different from those previously reported in the literature for aqueous OL$_E$, as well as for extracts obtained from OMW produced according to a three-phase centrifugation process [31,32]. However, considering that the amount of bioactive compounds accumulated in olive oil by-products may vary widely depending on many factors, the most important of which are pedoclimatic conditions, olive varieties, degree of maturity, and processing and extraction conditions, making an effective comparison of the data very difficult [33].

Table 1. Total polyphenol content and antioxidant activity of olive leaf and oil mill wastewater extracts.

Samples [1]	Total Polyphenols (mg GAE g^{-1}, As Is)	Antioxidant Activity (% per mg of Extract)
OL$_E$	134.7 ± 2.1 a	4.26 ± 0.08 a
OMW$_E$	13.4 ± 0.2 b	0.32 ± 0.01 b

[1] Mean value $\pm$ standard deviation. Within columns, values (mean of three repetitions) with the same letter do not differ significantly from each other according to Least Significant Difference (LSD) test ($p < 0.05$). OL$_E$: olive leaf extract; OMW$_E$: olive mill wastewater extract. GAE: gallic acid equivalent.

The antioxidant activity of the investigated extracts evidenced a similar trend, with the OL sample exhibiting a free radical scavenging capacity higher than that of the OMW$_E$ (4.26 ± 0.08 vs. 0.32 ± 0.01).

Moreover, another important aspect that emerged from the comparison of the two experimental extracts and that deserves to be emphasized is related to the fact that OL rather than OMW seemed to be the best option, in terms of environmental resources, to obtain polyphenol-rich extracts. In fact, the aqueous extraction of phenolic compounds from OL, besides being the easiest, efficient, and solvent-free extraction procedure, allows for a higher yield with a low environmental impact.

3.2. Moisture Content and Water Activity of GF Breadsticks

As reported in Table 2, when comparing the control breadsticks to the fortified samples significant ($p < 0.05$) differences in terms of both moisture content and a$_w$ were observed. All the enriched breadsticks showed moisture values (ranging from 9.24 to 11.50 g 100 g^{-1}) significantly higher than the control (7.66 ± 0.07 g 100 g^{-1}), the extent of the change being more pronounced at increasing level of supplementation for both the investigated extracts. In particular, the highest moisture values were observed in the sample Leaf100 (11.50 ± 0.12), closely followed by the sample Leaf50 (10.87 ± 0.14). Such an increase was probably due to the presence of fiber in the OL$_E$, enabling the absorption of more water than the basic GF ingredients, thus suggesting a potential effect on the water absorption capacity of the resulting breadsticks.

Table 2. Moisture content and a$_w$ of gluten-free (GF) breadsticks.

Characteristics	Samples [1]				
	Control	Leaf50	Leaf100	WasteW50	WasteW100
Moisture (g 100 g^{-1})	7.66 ± 0.07 e	10.87 ± 0.14 b	11.50 ± 0.12 a	9.24 ± 0.08 d	10.25 ± 0.09 c
a$_w$	0.406 ± 0.00 e	0.655 ± 0.00 b	0.667 ± 0.01 a	0.547 ± 0.00 d	0.593 ± 0.00 c

[1] Mean values $\pm$ standard deviation. Within rows, values (mean of five replicates) with the same letter do not differ significantly from each other according to LSD test ($p < 0.05$).

A similar trend was observed for the a$_w$, which significantly ($p < 0.05$) increased from a value of 0.41 observed in the control sample to a maximum value of 0.67 exhibited by the breadsticks prepared with the high level of addition of the OL_E (Table 2). As confirmation of this, values of correlation coefficients (r) revealed that higher moisture content corresponded to larger amount of available water ($r = 0.985; p < 0.001$). As expected, the data obtained in this study are considerably higher than those observed in wheat-based breadsticks by other authors, who reported moisture content ranging between 2.09 and 6.15 [23,34,35] and a_w values of about 0.14–0.38 [23,36]. Even though such differences should be not surprising given the higher amount of water usually required to obtain machinable GF dough with a proper consistency, it is worth noting that an increase in both moisture and a_w, could lead to undesirable stability issues of the GF breadsticks during storage. In fact, in starch-based system, the presence of higher amounts of available or weakly associated water, which is not able to bind to starch as strongly as it does with protein, could lead to a faster moisture migration both within and out of the products, thus shortening their shelf life [37].

3.3. Color and Textural Properties of GF Breadisticks

Visual and textural characteristics play an important role in determining the final quality of baked products and, in turn, in influencing consumer choice. This is particularly true in GF baked products, which, unlike their gluten-containing counterparts, often exhibit an overly white coloration and a too hard, dry, and grainy texture that consumers find unappealing [27].

In terms of color features, all the enriched GF breadsticks exhibited the same lightness L^* and yellowness (b^* positive), but lower red (a^* positive) values when compared to the control breadsticks, with no significant differences between samples prepared with the same supplementation level (Table 3). The only exception was observed in the sample WasteW100, which significantly differed from the control also in terms of b^* values. In particular, all the fortified breadsticks showed a significant tendency ($p < 0.05$) to change from a red to a more greenish coloration, especially at the highest level of addition, probably as a consequence of the typical green and green to yellow color of the added OL_E and OMW_E, respectively. However, as confirmed by the total color difference, which values were <1 for all the analyzed samples (Table 3), these observed differences were not obvious for the human eye, suggesting only a slight influence of both extracts on the color of the resulting breadsticks.

Table 3. Textural and color properties of GF breadsticks.

Characteristics	Samples [1]				
	Control	Leaf50	Leaf100	WasteW50	WasteW100
	Color properties				
L	60.10 ± 1.16 a	59.85 ± 0.82 a	60.55 ± 0.4 a	60.88 ± 1.09 a	60.67 ± 0.65 a
a	1.59 ± 0.05 a	1.40 ± 0.08 b	1.25 ± 0.04 c	1.46 ± 0.08 b	1.24 ± 0.08 c
b	12.93 ± 0.10 a	12.63 ± 0.37 ab	12.69 ± 0.31 ab	12.93 ± 0.17 a	12.39 ± 0.21 b
ΔE	–	0.13	0.22	0.08	0.23
	Textural properties				
Hardness (N)	51.57 ± 3.83 a	45.34 ± 5.02 c	45.07 ± 3.24 c	49.12 ± 3.05 b	48.39 ± 2.74 b
Brittleness (mm)	0.70 ± 0.08 a	0.80 ± 0.15 a	0.72 ± 0.18 a	0.65 ± 0.11 a	0.69 ± 0.14 a

[1] Mean values ± standard deviation. Within rows, values (means of 10 repetitions for color measurements and 20 repetitions for textural properties) with the same letter do not differ significantly from each other according to LSD test ($p < 0.05$).

From a textural point of view, baked snacks like breadsticks are characterized by a rigid, stiff structure with a little tendency to deform before fracture when subjected to small forces [22]. In the present study, the different mechanical behavior of the experimental breadsticks was measured on the day of baking to assess their quality in terms of hardness and brittleness.

As reported in Table 3, both OL_E and OMW_E significantly ($p < 0.05$) lowered the maximum force needed to break the experimental breadsticks, irrespective of whether they have been added at low or high level. In particular, while the control sample showed the highest values of force at break (or the maximum resistance when broken) (51.57 ± 3.83 N), the most pronounced decrease in hardness values was observed in the samples enriched with the OL_E (at about 45 N for both samples). Since in low moisture food systems, water mainly acts as plasticizer [38], a possible explanation for this softening effect could be related to the higher moisture content and a_w observed in the samples enriched with OL_E, closely followed by those enriched with the OMW_E.

Brittleness, which is a textural parameter describing the distance traveled by the blade through the sample before its breaking and, thus, how far a sample can be deformed before fracture, did not show significant differences ($p < 0.05$) among the experimental samples. However, a slight increase of this parameter was observed in the sample Leaf50, suggesting a more leathery or rubbery behavior. The authors of [38] when measuring the mechanical properties equilibrated at different a_w of a particular baked snack, called dried bread, demonstrated that when values of a_w are higher than 0.56 the stress is released on rupture in an increasingly gradual manner, making prominent the ductile behavior and, thus, deformation over brittleness. However, it is noteworthy that the effect of hydration on the textural properties of baked cereal-based snacks is quite complex and varies depending on the basic formulation of the product, so that the critical values of both a_w and water content, corresponding to changes from a crispy to a more deformable behavior, may be different [39].

3.4. Polyphenol Fractions and Antioxidant Activity of GF Breadsticks

In the present study, the polyphenol fractions and the antioxidant activity of GF breadsticks enriched with natural phenolic-rich extracts from both OL and OMW were compared with a conventional unfortified GF breadstick sample. Results are summarized in Table 4 and Figure 1.

Among the experimental breadsticks, the control formulation exhibited amounts of total polyphenols significantly lower (162.87 ± 1.15 mg of GAE 100^{-1} d.m.) than those observed when the OL_E and OMW_E were individually added to the basic formulation (from 168.40 ± 1.86 to 189.28 ± 3.55 mg of GAE 100^{-1} d.m.) (Table 4). Only the sample WasteW50, which was prepared by adding the low percentages of OMW_E, showed a total polyphenols content similar to that observed in the control, indicating that the most efficient increase was achieved by supplementing the basic recipe with the OL_E at both supplementation levels. In particular, the increment in the total phenolic content of the breadsticks enriched with low and high levels of OL_E was four-fold and two-fold higher than that observed by adding low and high levels of OMW_E, respectively (14–16% vs. 3–8%). These results were somewhat expected considering the higher polyphenol content observed in the OL_E, which was at about 10-fold higher compared to that of the OMW_E (Table 1).

Table 4. Polyphenol fractions and antioxidant activity of GF control and fortified breadsticks.

Characteristics	Samples [1]				
	Control	Leaf50	Leaf100	WasteW50	WasteW100
	Polyphenol fractions (mg GAE/100 g d.m.)				
Soluble	20.32 ± 0.97 e	35.85 ± 1.36 b	46.26 ± 2.47 a	27.05 ± 1.05 d	31.81 ± 0.90 c
Insoluble	142.45 ± 1.64 a	149.76 ± 5.56 a	143.02 ± 1.51 a	141.35 ± 1.93 a	143.71 ± 3.56 a
IP/SP	6.98 a	4.18 c	3.10 d	5.23 b	4.52 c
Total [2]	162.87 ± 1.15 c	185.61 ± 6.56 a	189.28 ± 3.55 a	168.40 ± 1.86 bc	175.52 ± 4.43 b
Bio-accessible	113.51 ± 2.30 d	129.93 ± 2.44 bc	139.68 ± 2.46 a	125.29 ± 1.09 c	130.64 ± 4.35 b
Δ bio-accessibility (%)	-	14.5	23.0	10.4	15.1
Antioxidant activity [3] (%)	32.28 ± 2.19 d	38.74 ± 0.24 b	43.84 ± 0.93 a	35.69 ± 0.17 c	37.59 ± 1.08 bc

[1] Mean values ± standard deviation. Within rows, values (mean of three repetitions) with the same letter do not differ significantly from each other according to LSD test ($p < 0.05$). [2] The total polyphenol content was calculated as the sum of the soluble and insoluble fractions. [3] Corresponding to 36 mg of breadsticks, which consumed these percentages when 0.17 μmol of 2,2-diphenyl-1-picrylhydrazyl (DPPH) are available for reaction. GAE: gallic acid equivalent; IP/PS: insoluble polyphenols/soluble polyphenols ratio.

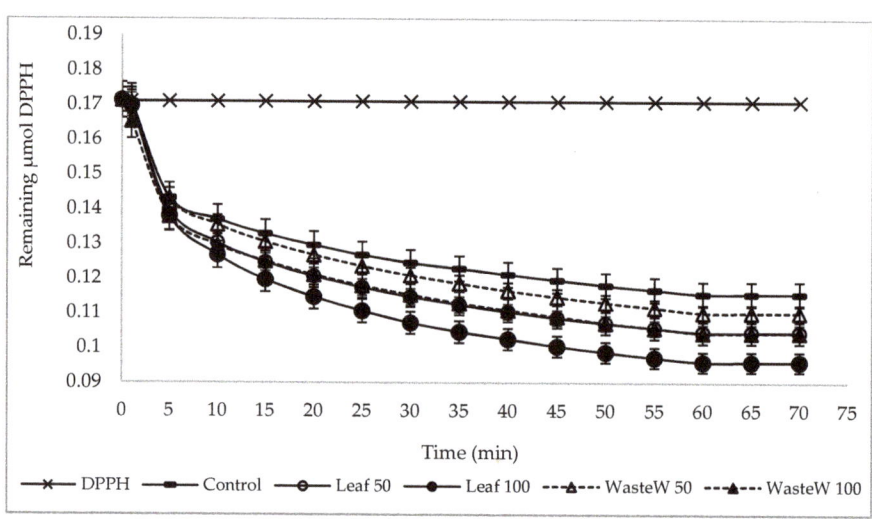

Figure 1. Time evolution of the DPPH curves in methanol of organic extracts from GF control and fortified breadsticks.

However, as it is well known, polyphenols can exist in the plant kingdom in both free and bound form. Therefore, to better evaluate the composition of the phenolic compounds in both control and fortified breadsticks, soluble and insoluble polyphenol fractions were also determined. As reported in Table 4, while the incorporation of both OL_E and OMW_E was significant ($p < 0.05$) in enhancing the soluble phenolic content of the resulting breadsticks, neither of the two affected the insoluble polyphenol fraction, which did not show significant differences among the investigated samples. In particular, the most significant changes were observed in the samples containing the OL_E–with an increment with respect to the control ranging from 76% (Leaf50) to 126% (Leaf100)–followed by those prepared with the OMW_E (+32% and +56% for WasteW50 and WasteW100, respectively). Thus, the incorporation of the investigated phenolic-rich extracts led to a significant decrease ($p < 0.05$) in the insoluble/soluble polyphenols average ratio of the resulting breadsticks, an effect more prominent at increasing levels of addition, especially in the case of OL_E supplementation (Table 4). To better understand such an improvement, it must be born in mind that fruits and vegetables, compared with cereal grains like rice and corn–in which at about 70% of the total polyphenols exists in the bound forms–have most of their phenolic compounds in the free or soluble conjugate forms [40,41]. Therefore, the addition of extracts from plant wastes can favor the accumulation of the phenolic fraction more rapidly absorbed in the grastrointestinal tract, thus leading to an effective enrichment of the final products [42].

Since there is no direct relationship between the amount of polyphenols in foods and their bioavailability, to evaluate how many of the ingested phenolic compounds could be effectively absorbed and utilized by the human body, thus exerting their biological effects [43], the bio-accessible polyphenol fraction was also determined. As reported in Table 4, although all the fortified breadsticks showed amounts of bioaccessible polyphenols significantly higher than those observed in the control, the most significant increment in polyphenol bioavailability was achieved in the sample prepared with the highest level of OL_E (+23% with respect to the control), followed by the samples Leaf50 and WasteW100, wich behaved in a similar way (+14.5% and 15.1%, respectively). This enhancement effect was in line with the same effect previously described for the soluble polyphenol fraction, as also confirmed by the highly significant ($p < 0.001$) linear correlations observed between the soluble and bioaccessible fractions (r = 0.919). In fact, the contribution of the insoluble polyphenols to the bioavailability of the final product is usually lower than that of the soluble phenolic fraction, since they have to be released from the cell structure before being absorbed [41]. Interestingly, the authors of [44] demonstrated that the absorption and, consequently, the bioavailability of hydroxytyrosol in the gastrointestinal tract could be maximize after a supplementation of the diet with its naturally precursor oleuropein–which is the most abundant bio-phenol in the OL_E–rather than with its free form (mainly present in the OMW_E) or aglycone forms. However, to the best of our knowledge, data on soluble, insoluble and bioaccessible polyphenols in GF baked products are limited to only one previous study from the same authors, who found similar results in GF breads fortified with a natural apicultural product like bee pollen [26]. For this reason, a comprehensive comparison of the obtained data with the literature is difficult.

Very often, the main source of antioxidants in food products are represented by phenolic compounds, which are able to exert several biological functions, including antioxidant and free radical scavenging activity [4]. In the present study, the antioxidant activity of both control and enriched breadsticks was evaluated by using the DPPH radical scavenging assay, which is a method based on electron donation of antioxidants to neutralize the free radicals. As reported in Figure 1, the time evolution of the DPPH concentration curves in methanol of organic extracts from both control and GF breadsticks evidenced that all the fortified samples exhibited an antioxidant activity higher than the control. In particular, the OL_E seemed to be more effective in enhancing the scavenging activity against the stable radical DPPH of the resulting breadsticks, especially at the highest supplementation level (at about 44%). These findings are similar to those reported by other authors in wheat-based baked snacks enriched with OL_E [25]. As in the case of OL_E, the OMW_E, irrespective of whether it had been added at low or high level, also increased the antioxidant activity of the enriched samples, but the extent of this increase was significantly lower than that observed for the OL_E, with the sample WasteW100 showing values similar to those observed in the sample Leaf50 (at about 38% and 39%, respectively) (Table 4). These results were in line with those previously observed for phenolic compounds. As a confirmation of this, significant positive correlations ($p < 0.001$) have been found between antiradical activity and soluble ($r = 0.958$), bioaccessible ($r = 0.899$), and total ($r = 0.868$) polyphenols.

3.5. Oxidation Stability (Oxitest) and Estimated Shelf-Life of GF Breadsticks

Lipid oxidation is one of the major causes of quality deterioration of dehydrated or low moisture bakery products–such as breadsticks and other bread substitutes–which usually require the addition of non-negligible amounts of fatty substances (from 5–15%, or even more, depending on their nature) to obtain desirable texture, appearance and flavor attributes [45]. Lipids, in fact, are susceptible to complex chemical changes that proceed through free-radical propagated chain reactions, which are triggered by unsaturated fatty acids reacting with oxygen. Other oxidation initiators, such as light or heat, certain enzymes, and metal ions are also involved in the process, by enhancing the lipid oxidation during storage [46]. The formation of free radicals and primary oxidation products, such as hydroperoxides, and their decomposition into secondary oxidation products, such as aldheydes, ketones, and hydrocarbons, are directly responsible for the formation of undesirable flavors in rancid foods. Therefore, the use of natural antioxidants–such as OL_E and OMW_E–well known for their ability to protect against free radicals, may be an effective way to promote the oxidation stability of functional GF breadsticks, thus exending their shelf-life [23]. In the present study, the effect of the investigated extracts on the oxidation stability of both GF control and fortified breadsticks was evaluated by using accelerated oxidation tests performed in the OXITEST reactor at four different working temperatures (60, 70, 80, and 90 °C) and at a constant oxygen overpressure (6 bar) to allow the estimation, in a short period of time, of the potential shelf-life of the experimental samples. Monitoring the drop in the oxygen pressure inside the oxidation chambers–which correspond to a certain level of detectable rancidity or a rapid change in the oxidation rate–evidenced that, while the control samples exhibited the lowest lipid stability to oxidation (at about 3 and 52 h at 90 and 60 °C, respectively), the addition of both extracts significantly increased the

IP of the fortified breadsticks at all the working temperatures, with the samples WasteW100 being more stable (at about 5 and 99 h at 90 and 60 °C, respectively) than the samples WasteW50 (3 and 57 h at 90 and 60 °C), Leaf100 (3 and 60 h at 90 and 60 °C), and Leaf50 (2 and 58 h at 90 and 60 °C), respectively. Starting from this point and considering that with increasing working temperature the IP of the oxidative reaction decreased, thus suggesting a linear dependence of oxidation stability and temperature, the potential shelf-life of the experimental samples at the storage temperature of 25 °C was estimated by using linear regression equations on a semi-log scale. The estimated shelf-life of the GF breadsticks based on lipid oxidation data is shown in Table 5.

Table 5. Estimated shelf-life of GF breadsticks based on lipid oxidation data (day at 25 °C).

Characteristics	Samples [1]				
	Control	Leaf50	Leaf100	WasteW50	WasteW100
Estimated shelf-life	62 ± 2 e	76 ± 3 d	82 ± 2 c	92 ± 6 b	116 ± 1 a
R^2	0.999	0.999	0.999	0.999	0.997

[1] Mean values ± standard deviation. Within rows, values (mean of two repetitions) with the same letter do not differ significantly from each other according to LSD test ($p < 0.05$).

As can be seen, the incorporation of increasing percentages of both OL_E and OMW_E was followed by a concurrent increase in the estimated shelf-life of the resulting breadsticks (Table 5), indicating an effective role of antioxidants in preserving the final products. In particular, the greatest values were observed in the sample prepared with the high level of addition of OMW_E which nearly doubled the shelf-life of the control, closely followed by the sample WasteW50 (+48% increment with respect to the control). A significant shelf-life extension was also registered in those breadsticks prepared with the addition of OL_E, but the extent of this increase was significantly lower than that observed for the OMW_E extracts at both supplementation levels (+23 and 32% for Leaf50 and Leaf100, respectively). These findings, however, were somewhat unexpected considering the opposite results observed in terms of antioxidant activity among the fortified samples. In fact, in spite of a lower (or similar) antioxidant activity (Table 4), the OMW_E-enriched breadsticks seemed to be kept fresh for longer, at least in terms of oxidation stability, than those prepared with the OL_E, irrespective of the level of substitution used. A possible explanation of this contrasting behavior may be related to the significant differences recorded in the a_w values among OL_E and OMW_E breadsticks (Table 2). In fact, it has been demonstrated that water can play both pro-oxidant and antioxidant roles in lipid oxidation, depending on whether its content in the food is within or above the monolayer moisture content [47,48]. At low levels of a_w–near to the monolayer range–water can exhibit an antioxidant role by forming a barrier that protects the sensitive sites from reactions with oxygen, but also by lowering metal catalytic activity, increasing hydration of hydroperoxides (and consequently decreasing the rate of free radicals formation), as well as by promoting recombination of free radicals. In contrast, at higher a_w values, as is the case with the experimental OL_E-enriched breadsticks, water can show a pro-oxidant role by acting as a plasticizing agent, thus promoting mobility and solubilization of catalysts, as well as by inducing matrix swelling, thus exposing new reactive sites [47,48]. Therefore, the lower oxidative stability observed in the breadsticks fortified with the OL_E in comparison to that observed in those enriched with the OMW_E might suggest that, in such a complex matrix. the effect of a_w is greater than the conservative role exerted by the added antioxidants.

However, considering that in low moisture foods the causes of lipid oxidation are still not completely understood and that both monolayer and glass transition theories have led to contrasting results when used to predict lipid oxidation rates as a function of a_w [48], further studies are needed to give consistency to the obtained results.

3.6. Sensory Evaluation of GF Breadsticks

In the first two sessions of the sensory evaluation, participants were asked to rank the overall preferences for the freshly prepared breadsticks, by comparing the control to the samples enriched with both OL_E and OMW_E separately. As reported in Table 6, the obtained results evidenced that no sample was significantly preferred to another in both comparisons.

Table 6. Results of the ranking preference tests on the freshly prepared GF breadsticks.

Samples	(a)[1]			(b)		
	Control	Leaf50	Leaf100	Control	WasteW50	WasteW100
Rank sum	131	120	109	125	122	113
Significance $p < 0.05$	a	a	a	a	a	a

[1] (a) comparisons among the control sample and the OL_E-enriched breadsticks; (b) comparisons among the control sample and the OMW_E-enriched breadsticks.

More specifically, when comparing the control sample to the OL_E-enriched breadsticks, the lowest ranking score was assigned to the sample Leaf100, followed by the Leaf50 and the control breadsticks (Table 6(a)), suggesting a clear tendency for consumers to recognize the fortified samples as the most palatable choice. However, as evidenced by the obtained F_{test} value (4.03), which was lower than the $F_{critical}$ value reported in the χ^2 table (5.99), the recorded difference in the preference degree was not significant. A similar trend was also observed when comparing the control sample to the two breadsticks fortified with OMW_E (Table 6(b)). In this case, although the differences in the preference degree assigned to the three samples were less pronounced (F_{test} value: 1.13 and $F_{critical}$ value: 5.99), a consumer's tendency to prefer the breadsticks prepared with the high level of OMW_E could also be observed.

Based on these results, Leaf100 and WasteW100 samples were then subjected to a paired comparison test to assess if a significant difference exists between them in terms of preference (Figure 2).

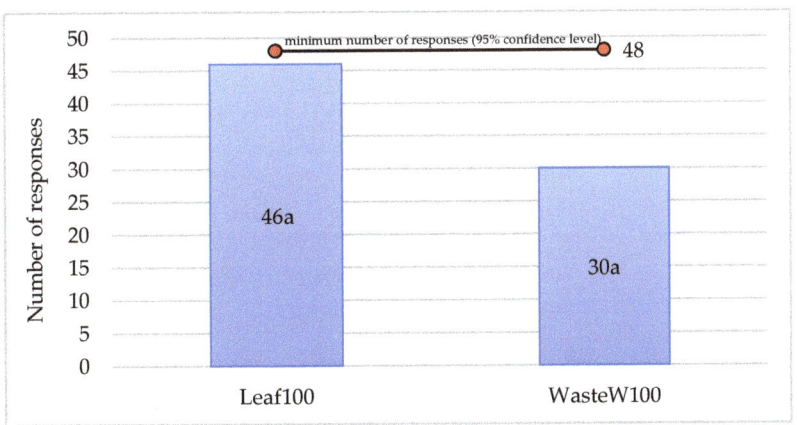

Figure 2. Differences in the number of responses accorded by consumer to the Leaf100 and WasteW100 samples in the paired comparison test.

According to the statistical table for paired difference test (two tailed), 48 is the minimum number of responses needed to conclude that a preference exists between two samples at the selected significance level (5%) and with a total number of assessors of 76. This value was not reached by either of the two experimental breadsticks, even if the sample Leaf100 came very close to this minimum number (Figure 2). However, it should be noted that, although the number of responses did not differ significantly between the two enriched breadsticks, the sample Leaf100 was preferred by the 61% of the assessors, indicating a marked, but not significant, tendency for consumers to consider it as the most appetizing choice.

4. Conclusions

Data obtained in the present study evidenced, for the first time, that the incorporation of phenolic-rich extracts from olive oil by-products in GF formulations can be considered a successful strategy in the preparation of technologically viable functional breadsticks with extended shelf-life. Although all fortified samples, especially those enriched with the OL_E, were softer, they also exhibited a similar crumbly texture and minimal color changes compared to the control, indicating an only small impairment in their technological feasibility. This was also confirmed by the sensory evaluation, which showed a marked (but not significant) tendency of consumers to consider the enriched breadsticks, especially those prepared with the high percentage of OL_E, as the most preferred. The incorporation of both extracts also resulted in improved nutritional and functional properties of the final breadsticks, as evidenced by the changes observed in the insoluble/soluble polyphenol ratio in favor of the soluble fraction, by the enhanced bioavailability of polyphenols, as well as by the higher antioxidant activity, especially in those samples prepared with the higher percentage of OL_E. Furthermore, all the fortified breadsticks exhibited higher stability against the lipid oxidation and, in turn, an extended estimated shelf-life, even if, in this case, better behavior was observed in the samples prepared with both levels of OMW_E.

In conclusion, the best results were achieved in those samples fortified with the high percentage of OL_E in which, in spite of a slightly shorter shelf-life, a more proper balance among the technological, sensory and functional properties was observed. Further experiments are needed to give consistency to these preliminary findings, especially to better clarify the influence of both antioxidant activity and a_w in the estimation of the shelf-life of GF low moisture baked snacks.

Author Contributions: Conceptualization, P.C., A.D.C., C.F., P.P.U. and A.P.; methodology, P.C., S.P., A.D.C., P.P.U.; formal analysis, P.C. and S.P.; investigation, P.C., S.P., A.D.B., G.D. and R.R.; data curation, P.C., S.P., A.D.C., P.P.U.; writing—original draft preparation, P.C., S.P., A.D.C., C.F. and A.P.; writing—review and editing, P.C., S.P., A.D.C., C.F., P.P.U., A.D.B., G.D., F.C., R.R. and A.P.; supervision, F.C. and A.P.; funding acquisition, P.C., F.C. and A.P. All authors have read and agreed to the published version of the manuscript.

Funding: This research was funded by the AGER 2 Project, grant n. 2016-0105. This work was also funded by the University of Sassari (Fondo di Ateneo per la Ricerca 2020): FAR2020CONTEP.

Institutional Review Board Statement: Ethical review and approval were waived for this study since the participation was voluntary. All data were anonymous.

Informed Consent Statement: Informed consent was obtained from all subjects involved in the study.

Data Availability Statement: Not applicable.

Conflicts of Interest: The authors declare no conflict of interest.

References

1. Galanakis, C.M. Recovery of high added-value components from food wastes: Conventional, emerging technologies and commercialized applications. *Trends Food Sci. Technol.* **2012**, *26*, 68–87. [CrossRef]
2. Faustino, M.; Veiga, M.; Sousa, P.; Costa, E.M.; Silva, S.; Pintado, M. Agro-food byproducts as a new source of natural food additives. *Molecules* **2019**, *24*, 1056. [CrossRef] [PubMed]
3. Otles, S.; Despoudi, S.; Bucatariu, C.; Kartal, C. Food waste management, valorization, and sustainability in the food industry. In *Food Waste Recovery: Processing Technologies and Industrial Techniques*, 1st ed.; Galanakis, C.M., Ed.; Elsevier Academic Press: London, UK, 2015; pp. 3–23.
4. Mulinacci, N.; Romani, A.; Galardi, C.; Pinelli, P.; Giaccherini, C.; Vincieri, F.F. Polyphenolic content in olive oil waste waters and related olive samples. *J. Agric. Food Chem.* **2001**, *49*, 3509–3514. [CrossRef]

6. Simonato, B.; Trevisan, S.; Tolve, R.; Favati, F.; Pasini, G. Pasta fortification with olive pomace: Effects on the technological characteristics and nutritional properties. *LWT Food Sci. Technol.* **2019**, *114*. [CrossRef]
7. Tabera, J.; Guinda, Á.; Ruiz-Rodríguez, A.; Señoráns, F.J.; Ibáñez, E.; Albi, T.; Reglero, G. Countercurrent supercritical fluid extraction and fractionation of high-added-value compounds from a hexane extract of olive leaves. *J. Agric. Food Chem.* **2004**, *52*, 4774–4779. [CrossRef]
8. Rahmanian, N.; Jafari, S.M.; Wani, T.A. Bioactive profile, dehydration, extraction and application of the bioactive components of olive leaves. *Trends Food Sci. Technol.* **2015**, *42*, 150–172. [CrossRef]
9. Şahin, S.; Bilgin, M. Olive tree (*Olea europaea* L.) leaf as a waste by-product of table olive and olive oil industry: A review. *J. Sci. Food Agric.* **2018**, *98*, 1271–1279. [CrossRef] [PubMed]
10. Skaltsounis, A.L.; Argyropoulou, A.; Aligiannis, N.; Xynos, N. Recovery of High Added Value Compounds from Olive Tree Products and Olive Processing Byproducts. In *Olive and Olive Oil Bioactive Constituents*; Boskou, D., Ed.; Academic Press and AOCS Press: London, UK, 2015; pp. 333–356.
11. Romero-García, J.M.; Niño, L.; Martínez-Patiño, C.; Álvarez, C.; Castro, E.; Negro, M.J. Biorefinery based on olive biomass. State of the art and future trends. *Bioresour. Technol.* **2014**, *159*, 421–432. [CrossRef]
12. Roig, A.; Cayuela, M.L.; Sánchez-Monedero, M.A. An overview on olive mill wastes and their valorisation methods. *Waste Manag.* **2006**, *26*, 960–969. [CrossRef]
13. Rahmanian, N.; Jafari, S.M.; Galanakis, C.M. Recovery and removal of phenolic compounds from olive mill wastewater. *JAOCS J. Am. Oil Chem. Soc.* **2014**, *91*, 1–18. [CrossRef]
14. Azaizeh, H.; Halahlih, F.; Najami, N.; Brunner, D.; Faulstich, M.; Tafesh, A. Antioxidant activity of phenolic fractions in olive mill wastewater. *Food Chem.* **2012**, *134*, 2226–2234. [CrossRef] [PubMed]
15. De las Hazas, M.C.L.; Rubio, L.; Macia, A.; Motilva, M.J. Hydroxytyrosol: Emerging Trends in Potential Therapeutic Applications. *Curr. Pharm. Des.* **2018**, *24*, 2157–2179. [CrossRef] [PubMed]
16. Crespo, M.C.; Tomé-Carneiro, J.; Dávalos, A.; Visioli, F. Pharma-Nutritional Properties of Olive Oil Phenols. Transfer of New Findings to Human Nutrition. *Foods* **2018**, *7*, 90. [CrossRef]
17. Obied, H.K.; Prenzler, P.D.; Robards, K. Potent antioxidant biophenols from olive mill waste. *Food Chem.* **2008**, *111*, 171–178. [CrossRef]
18. Faccioli, L.S.; Klein, M.P.; Borges, G.R.; Dalanhol, C.S.; Machado, I.C.K.; Garavaglia, J.; Dal Bosco, S.M. Development of crackers with the addition of olive leaf flour (*Olea europaea* L.): Chemical and sensory characterization. *LWT Food Sci. Technol.* **2021**, *141*. [CrossRef]
19. Ataei, F.; Hojjatoleslamy, M. Physicochemical and sensory characteristics of sponge cake made with olive leaf. *J. Food Meas. Charact.* **2017**, *11*, 2259–2264. [CrossRef]
20. Mohammadi, A.; Jafari, S.M.; Esfanjani, A.F.; Akhavan, S. Application of nano-encapsulated olive leaf extract in controlling the oxidative stability of soybean oil. *Food Chem.* **2016**, *190*, 513–519. [CrossRef] [PubMed]
21. Difonzo, G.; Squeo, G.; Calasso, M.; Pasqualone, A.; Caponio, F. Physico-chemical, microbiological and sensory evaluation of ready-to-use vegetable pâté added with olive leaf extract. *Foods* **2019**, *8*, 138. [CrossRef] [PubMed]
22. Crofton, E.C.; Scannell, A.G.M. Snack foods from brewing waste: Consumer-led approach to developing sustainable snack options. *Br. Food J.* **2020**, *122*, 3899–3916. [CrossRef]
23. Zeppa, G.; Rolle, L.; Piazza, L. Textural characteristics of typical italian "grissino stirato" and "rubatà" bread-sticks. *Ital. J. Food Sci.* **2007**, *19*, 449–459.
24. Alamprese, C.; Cappa, C.; Ratti, S.; Limbo, S.; Signorelli, M.; Fessas, D.; Lucisano, M. Shelf life extension of whole-wheat breadsticks: Formulation and packaging strategies. *Food Chem.* **2017**, *230*, 532–539. [CrossRef]
25. Uribe-Wandurraga, Z.N.; Igual, M.; García-Segovia, P.; Martínez-Monzó, J. Effect of microalgae addition on mineral content, colour and mechanical properties of breadsticks. *Food Funct.* **2019**, *10*, 4685–4692. [CrossRef]
26. Difonzo, G.; Pasqualone, A.; Silletti, R.; Cosmai, L.; Summo, C.; Paradiso, V.M.; Caponio, F. Use of olive leaf extract to reduce lipid oxidation of baked snacks. *Food Res. Int.* **2018**, *108*, 48–56. [CrossRef] [PubMed]
27. Conte, P.; Del Caro, A.; Urgeghe, P.P.; Petretto, G.L.; Montanari, L.; Piga, A.; Fadda, C. Nutritional and aroma improvement of gluten-free bread: Is bee pollen effective? *LWT Food Sci. Technol.* **2020**, *118*. [CrossRef]
28. Conte, P.; Fadda, C.; Drabińska, N.; Krupa-Kozak, U. Technological and nutritional challenges, and novelty in gluten-free breadmaking: A review. *Pol. J. Food Nutr. Sci.* **2019**, *69*, 5–21. [CrossRef]
29. Romeo, R.; De Bruno, A.; Imeneo, V.; Piscopo, A.; Poiana, M. Evaluation of enrichment with antioxidants from olive oil mill wastes in hydrophilic model system. *J. Food Process. Preserv.* **2019**, *43*. [CrossRef]
30. Singleton, V.L.; Orthofer, R.; Lamuela-Raventós, R.M. Analysis of total phenols and other oxidation substrates and antioxidants by means of folin-ciocalteu reagent. *Methods Enzymol.* **1999**, *299*, 152–178.
31. AOCS Standard Procedure Cd 12c-16. In *Official Methods and Recommended Practices of the AOCS*; AOCS Press: Champaign, IL, USA, 2017; pp. 83–85.
32. Palmeri, R.; Monteleone, J.I.; Spagna, G.; Restuccia, C.; Raffaele, M.; Vanella, L.; Li Volti, G.; Barbagallo, I. Olive leaf extract from sicilian cultivar reduced lipid accumulation by inducing thermogenic pathway during adipogenesis. *Front. Pharmacol.* **2016**, *7*. [CrossRef]
33. El-Abbassi, A.; Kiai, H.; Hafidi, A. Phenolic profile and antioxidant activities of olive mill wastewater. *Food Chem.* **2012**, *132*, 406–412. [CrossRef] [PubMed]

33. Aggoun, M.; Arhab, R.; Cornu, A.; Portelli, J.; Barkat, M.; Graulet, B. Olive mill wastewater microconstituents composition according to olive variety and extraction process. *Food Chem.* **2016**, *209*, 72–80. [CrossRef]
34. Petchoo, J.; Jittinandana, S.; Tuntipopipat, S.; Ngampeerapong, C.; Tangsuphoom, N. Effect of partial substitution of wheat flour with resistant starch on physicochemical, sensorial and nutritional properties of breadsticks. *Int. J. Food Sci. Technol.* **2020**. [CrossRef]
35. Chakraborty, P.; Bhattacharyya, D.K.; Ghosh, M. Extrusion treated meal concentrates of Brassica juncea as functionally improved ingredient in protein and fiber rich breadstick preparation. *LWT Food Sci. Technol.* **2021**, *142*. [CrossRef]
36. Ktenioudaki, A.; Chaurin, V.; Reis, S.F.; Gallagher, E. Brewer's spent grain as a functional ingredient for breadsticks. *Int. J. Food Sci. Technol.* **2012**, *47*, 1765–1771. [CrossRef]
37. Matos, M.E.; Rosell, C.M. Understanding gluten-free dough for reaching breads with physical quality and nutritional balance. *J. Sci. Food Agric.* **2015**, *95*, 653–661. [CrossRef]
38. Chang, Y.P.; Cheah, P.B.; Seow, C.C. Variations in flexural and compressive fracture behavior of a brittle cellular food (dried bread) in response to moisture sorption. *J. Texture Stud.* **2000**, *31*, 525–540. [CrossRef]
39. Roudaut, G. Water Activity and Physical Stability. In *Water Activity in Foods: Fundamentals and Applications*; Barbosa-Cánovas, G.V., Fontana, A.J., Schmidt, S.J., Labuza, T.L., Eds.; Blackwell Publishing: Oxford, UK, 2008; pp. 199–213.
40. Chan, C.L.; Gan, R.Y.; Corke, H. The phenolic composition and antioxidant capacity of soluble and bound extracts in selected dietary spices and medicinal herbs. *Int. J. Food Sci. Technol.* **2016**, *51*, 565–573. [CrossRef]
41. Acosta-Estrada, B.A.; Gutiérrez-Uribe, J.A.; Serna-Saldívar, S.O. Bound phenolics in foods, a review. *Food Chem.* **2014**, *152*, 46–55. [CrossRef]
42. Colantuono, A.; Vitaglione, P.; Ferracane, R.; Campanella, O.H.; Hamaker, B.R. Development and functional characterization of new antioxidant dietary fibers from pomegranate, olive and artichoke by-products. *Food Res. Int.* **2017**, *101*, 155–164. [CrossRef]
43. Angelino, D.; Cossu, M.; Marti, A.; Zanoletti, M.; Chiavaroli, L.; Brighenti, F.; Del Rio, D.; Martini, D. Bioaccessibility and bioavailability of phenolic compounds in bread: A review. *Food Funct.* **2017**, *8*, 2368–2393. [CrossRef]
44. López de las Hazas, M.C.; Piñol, C.; Macià, A.; Romero, M.P.; Pedret, A.; Solà, R.; Rubió, L.; Motilva, M.J. Differential absorption and metabolism of hydroxytyrosol and its precursors oleuropein and secoiridoids. *J. Funct. Foods* **2016**, *22*, 52–63. [CrossRef]
45. Gomes, T.; Delcuratolo, D.; Paradiso, V.M.; Nasti, R. The Oxidative State of Olive Oil Used in Bakery Products with Special Reference to Focaccia. In *Olives and Olive Oil in Health and Disease Prevention*; Preedy, V., Watson, R., Eds.; Academic Press: London, UK, 2010; pp. 745–753.
46. McClements, D.J.; Decker; Andrew, E. Lipids. In *Fennema's Food Chemistry*, 5th ed.; Damodaran, S., Parkin, K.L., Eds.; CRC Press: Boca Raton, FL, USA, 2017; pp. 171–233.
47. Karel, M.; Yong, S. Autoxidation-Initiated Reactions in Foods. In *Water Activity: Influences on Food Quality*; Rockland, L.B., Stewart, G.F., Eds.; Academic Press: London, UK, 1981; pp. 511–529.
48. Maltini, E.; Torreggiani, D.; Venir, E.; Bertolo, G. Water activity and the preservation of plant foods. *Food Chem.* **2003**, *82*, 79–86. [CrossRef]

Article

Consumer Attitudes towards Local and Organic Food with Upcycled Ingredients: An Italian Case Study for Olive Leaves

Maria Angela Perito [1,2,*], Silvia Coderoni [3] and Carlo Russo [4]

1. Faculty of Bioscience and Technology for Food, Agriculture and Environment, University of Teramo, 64100 Teramo, Italy
2. INRAE, ALISS, Université Paris-Saclay, 94205 Ivry-sur-Seine, France
3. Department of Agricultural and Food Economics, Catholic University of the Sacred Heart, 29122 Piacenza, Italy; silvia.coderoni@unicatt.it
4. Department of Economics and Law, University of Cassino and Southern Lazio, 03043 Cassino, Italy; carlo.russo@unicas.it
* Correspondence: maperito@unite.it

Received: 4 September 2020; Accepted: 16 September 2020; Published: 20 September 2020

Abstract: Food made with upcycled ingredients has received considerable attention in very recent years as a result of the need to both reduce waste and increase food nutritional properties. However, consumer acceptance of these novel foods is fundamental to their market uptake. This paper aims to assess the likelihood of the acceptance of food obtained from upcycled ingredients of olive oil productions and its association with some relevant recent consumption trends, such as organic food consumption and attention to food origin. In addition, particular attention is given to age group behaviors to appraise the differences between generations. Results suggest that, despite the negative influence of food technophobia, a core of sustainability-minded consumers seems to emerge that is interested in organic or local products, that could also favor the uptake of these novel food made with upcycled ingredients in the market. Results suggest that developing organic or "local" food products with upcycled ingredients can increase the probability of consumer acceptance.

Keywords: olive leaves; organic; local; consumer attitude; up-cycled ingredients; by-products; generational differences

1. Introduction

The olive tree is a central plant in the history of civilizations of the Mediterranean basin. It has historically been considered a sacred tree: in Genesis, the olive branch is a symbol of peace. Even in Greek Mythology, the olive branch was considered a symbol of peace and life. Ancient Romans rewarded valiant citizens with crowns of olive branches. In many cultures in the Mediterranean area, the olive branch is a symbol of justice. In addition, in the traditional culture of many Italian regions, olive leaves, as well as olive oil, have played the role of good luck against the evil eye [1]. Beyond the symbolic properties attributed to olive leaves, they have historically had a pharmacological role in Mediterranean countries and are widely used in traditional herbal medicine to prevent and treat various diseases [2,3].

This historical and anthropological excursus outlines the notion that olive leaves have almost never had a nutritional function in the tradition of Mediterranean areas, despite the importance of olive oil consumption [4,5]. The emergence of modern medicine and culture, however, has led to a loss of interest in olive leaves, and they have gradually become a processing waste for olive oil producers, with high disposal costs and no economic value.

Nevertheless, recent scientific literature has shown that olive leaves are rich in nutrients that could be used as by-products by the food industry to enrich food products with functional properties [6,7]. An essential point to understand the real market uptake of food enriched with food by-products that are not part of the traditional diet is to estimate consumer acceptance for such products, especially for food enriched with olive leaves.

Consumers consider food a complex good that contains both quality attributes as well as nutritional ones. Consumer acceptance is shaped by different factors, such as food habits and sociodemographic characteristics [8,9], food origin [5,10,11], the information that consumers access, and their trust in food production [12,13], the role of transparency [14], the range and prices of existing products [15–19], and the perception of health benefits and safety of food products [20,21].

In this context, the new objective of food production is no longer just food security but also the satisfaction of consumer needs and preferences with new products with both functional properties [22] and quality aspects (e.g., local, organic, etc.) [18,23]. In this context, the production of functional and nutritious food, obtained from upcycled ingredients, is a recent development [6,24–27].

Several studies referring to the use of unusual products in food production (i.e., insect products, cultured meat, etc.) report how new products can cause a strong food technophobia and neophobia in consumers [28–32]. Furthermore, studies on consumer acceptance of food derived from upcycled ingredients have found a negative impact of food technophobia and neophobia on the likelihood of acceptance [25,33]. However, interestingly, studies found that consumers may accord these products a premium status if promoted as a new food category akin to organic foods [34]. In fact, according to Bhatt et al. [34], consumers perceive "value-added surplus products" as having benefits for society and the individual.

The environmental sustainability of food production has become a matter of growing interest for consumers worldwide in recent years [35–37]. Often, environmental sustainability is understood by consumers as the preference to purchase organic food products, which are perceived as being healthier than conventional foods and better for the environment [18,38–42]. More recently, consumers have no longer been satisfied with the benefits offered by organic products only [43] but have been demanding domestic and local products as well [5,22,44–46]. In this respect, some studies suggest that consumers are interested in local food not only because they associate higher quality with these products but also for environmental friendliness, preference for their cultural roots [5,46], and support for the local economy [47] and local farmers [48].

Against this background, it seems to be of interest to analyze the possible impact of some frequent food consumption habits, such as buying organic products or giving importance to food origin in consumer acceptance of food derived from upcycled ingredients. In fact, foods obtained from upcycled ingredients of olive oil productions are new to the consumers, and their acceptance may be a problem despite their health or environmental benefits.

The main objective of the present study is to assess the association between the willingness to try food obtained from upcycled ingredients and consumer preferences for organic food and food origin. In particular, by estimating ordered probit models, we test if attributes, such as superior nutritional and/or environmental properties, of food with olive leaves are appealing to consumers. By doing so, we also investigate the possible market niches for these products.

In addition, to better understand consumer responses to food attributes, we run the analysis by age groups, as many studies have shown that elder and younger generations may actually differ in food preferences [49,50].

2. Materials and Methods

To investigate the association between the willingness to try food obtained from upcycled ingredients and the preference for organic and food origin, we surveyed a sample of 852 Italian consumers. The core of our analysis was an ordered probit regression [51] of a discrete variable measuring the willingness to try food with upcycled ingredients on a set of regressors, including demographic variables, measures of consumer environmental responsibility, technophobia, and concerns for product origin.

Data were collected through a web-based survey administered in Italy between April 2018 and April 2019 with a convenience approach. Participants were reached via different social media networks, which is becoming a more popular means of reaching participants in social sciences research for both convenience and inclusion reason [52]. In fact, the use of the Internet makes it as convenient as possible for participants to take part in the survey and allows reaching a high number of participants from all Italian regions. In particular, the information was then posted on Twitter, LinkedIn, and Facebook pages. As sampling in Internet research studies is not random and could generate selection bias, to minimize this possible problem, we posted the questionnaire on pages and online groups with a general target audience (e.g., web pages of Italian radio programs).

Before answering the questions, participants were briefly informed about the research project that motivated the survey.

Respondents were given a short, four-section questionnaire. Section 1 collected the demographic information, Section 2 assessed the respondent's attitudes toward the covariates of interest (organic food and food origin), Section 3 investigated the respondent's technophobia [29,30], and Section 4 asked the respondent's willingness to try food obtained from by-products [33]. Descriptive statistics of the sample and the questionnaire are presented in Table 1.

Table 1. Questionnaire and descriptive statistics of variables (n = 852).

Questionnaire	Label Variable	Scale	Mean				
Gender	GEN		Male 35.33			Female 64.67	
Age	AGE		37.56				
Education	EDU		Elemen. 0.35	Middle 2.35		High S. 37.09	College 60.21
Employment	EMPL		Worker 59.04	Unempl. 5.87	Student 28.29	Homem. 2.70	Retired 4.11
Frequency of the purchase of organic food	ORGANIC	5-point scale	Never 46.48	Seldom 23.59	Somet. 5.99	Often 14.44	Always 9.51
Is food Origin (e.g., local/typical product) important when shopping for food?	ORIGIN	5-point scale	Abs. Not 2.46	Not 3.76	Indiff. 7.28	Yes 38.97	Abs. Yes 47.53
There is no need for new food technologies because there are so many types of foods.	NNNT	5-point scale	Abs. Disag. 39.20	Disagr. 35.68	Indiff. 10.09	Agree 10.68	Abs. Agree 4.34
The benefits associated with innovative food technologies are often overestimated	NTOR	5-point scale	Abs. Disag. 18.31	Disagr. 35.56	Indiff. 18.19	Agree 20.07	Abs. Agree 7.86
New food technologies reduce the natural quality of foods	NTLQ	5-point scale	Abs. Disag. 23.83	Disagr. 36.97	Indiff. 15.96	Agree 15.61	Abs. Agree 7.63
I am willing to try food with upcycled ingredients of olive leaves if:							
it has superior nutritional properties	TNUT	5-point scale	Abs. Not 6.81	Not 12.56	Indiff. 13.03	Yes 45.66	Abs. Yes 21.95
it reduces the environmental impact of food production and consumption	TENV	5-point scale	Abs. Not 5.52	Not 8.80	Indiff. 10.21	Yes 33.57	Abs. Yes 41.90

To lessen collection cost and maximize the response rate, we minimized the number of items in the questionnaires and drafted questions as five-point Likert scales. The final design was extremely parsimonious and included 11 questions (see Table 1).

We measured the attitude toward organic food, asking respondents to state the frequency of their purchase of organic food on a 5-point scale (from never to always). To limit the possible self-representation bias, we referred to a specific action (buying organic food) instead of asking to report the attitude directly. This approach was possible because Italian consumers, on average, are familiar with organic products, given the sharp increase in organic consumption in the last decade (in 2017, 78% of Italian family had bought organic food at least once, and 48% had bought them at least once a week [53]).

Consumer perception of food origin was blurred mostly because of the overlapping of several different concepts, such as local food, typical food, or food safety. As a consequence, the questionnaire asked to report the importance of product origin in the food purchase decision.

Food technophobia (or food technology neophobia) is defined as consumers' fear, dislike, or avoidance of novel food technology [29]. Perito et al. [33] found that it is a key driver limiting consumer acceptance of food with olive by-products. As a consequence, we included it as a control variable in our empirical investigation.

Technophobia is a complex attitude to measure. Several contributions in the literature proposed scale measures [29,54,55]. In this paper, we adopt the approach proposed by Perito et al. [33]. The measure is based on three statements:

NNNT: There is no need for new food technologies because there are so many types of foods;
NTOR: The benefits associated with innovative food technologies are often overestimated;
NTLQ: New food technologies reduce the natural quality of foods.

Respondents were asked to agree or disagree with the statements on a 5-point Likert scale from 1 (strongly disagree) to 5 (strongly agree). Cronbach α of 0.78 confirmed that the construct items were consistent. We defined the technophobia index (TFI) as the average of the scores of the three variables.

Finally, the questionnaire asked respondents to report their willingness to try food with olive leaves as upcycled ingredients on a 5-point scale from 1 (No, Absolutely) to 5 (Yes, Absolutely). Because Perito et al. [33] identified environmental and nutritional concerns as key drivers of consumer acceptance, we conditioned the answer to two situations: A) the upcycled ingredients of olive leaves has superior nutritional properties (TNUT), and B) the upcycled ingredients reduces the environmental impact of food production and consumption (TENV). In this way, we provide useful insights to novel food researchers willing to market food with upcycled ingredients and different characteristics. Our hypothesis is that products with different attributes (nutritional or environmental) may be appealing to different consumers.

To investigate the association between consumer willingness to try food enriched with upcycled ingredients, organic food, and food origin, we ran a regression of the dependent variables TNUT and TENV on a vector of demographic variables (gender, age, education, and employment status), the technophobia index (TFI), and the attitudes regarding organic food (ORGANIC) and product origin (ORIGIN). Given the discrete nature of the dependent variables, we used an ordered probit model. In fact, in our case, the dependent variables were ordinal, but not continuous in the sense that the metric used to code the variables was substantively meaningful. For instance, the 5-point scale adopted to measure the dependent variable assigned the numerals to the categories but the metric underlying response identification was not necessarily the same as the linear metric relating the numerals. In other terms, the difference between 0 and 2 on the coded responses may be quite different from the difference between 2 and 4. A widely used approach to estimating models of this type is an ordered response model. The basic assumption of such models is that there is a latent continuous metric underlying the ordinal responses observed by the analyst.

The model estimated, assuming that the values of TNUT and TENV were the observable outcome of latent variables, is the following:

$$\hat{Y}_h = f(GENB, AGE, EDUCATION, EMPLOYMENT, TFI, ORGANIC, ORIGIN) \quad (1)$$

where h = N, E; GENB is a binary variable that is equal to 1 if the respondent is female; AGE is the respondent's age; EDUCATION is a categorical variable with four entries: elementary school, middle school, high school, and college, depending on the respondent's degree; EMPLOYMENT is a five-entry categorical variable equal to the worker (if the respondent is employed or self-employed), unemployed, student, homemaker or retired, depending on the respondent's status.

Regarding the variable AGE, a further refinement is then introduced as a separate regression for each age group has been conducted to exploit the large sample size of our survey.

3. Results

3.1. Sample Description

The sample was composed of 852 respondents aged between 18 and 90 years old, with an average of 37 years and six months. Sixty-five percent of respondents were female. Table 2 reports the social and demographic characteristics of the sample.

Table 2. Sample demographics characteristics.

Education	% Freq.	Employment Status	% Freq.	Gender	% Freq.	Age	% Freq.
Elementary school	0.35	Worker	59.04	Male	35.33	Generat. Z (18–24)	25.23
Middle School	2.35	Unemployed	5.87	Female	64.67	Millennials (25–39)	31.34
High School	37.09	Student	28.29			Generat. X (40–54)	30.99
College	60.21	Homemaker	2.70			Elder Gen. (55+)	12.44
		Retired	4.11				
Total	100.00	Total	100.00	Total	100.00	Total	100.00

Table 3 illustrates the distribution of the variables. The majority of respondents stated that they buy organic food at least "seldom". Eighty-seven percent of respondents stated that product origin was an important or very important driver of food purchase decisions (Table 3). Such a high figure is explained by the broad meaning of the term origin that in the consumers' mind, is associated with local food, Italian food, typical products, and, in a broader sense, even to food safety. The value origin and the specific reputation for very local products are extensively documented in the literature [11]. Region of origin evokes tradition, habits, culture, and so on, and these aspects directly influence preference for a regional product [56–58].

Table 3. Distribution of selected variables.

BUYBIO			ORIGIN						NNNT		NTOR		NTQL	
Scale	N	%	Scale	N	%	Scale	N	%	N	%	N	%	N	%
Never	396	46.5	Absol. not important	21	2.5	Strongly disagree	334	39.2	156	18.3	203	23.8		
Seldom	201	23.6	Not important	32	3.8	Disagree	304	35.7	303	35.6	315	37.0		
Sometimes	51	6.0	Indifferent	62	7.3	Not agree nor disagree	86	10.1	155	18.2	136	16.0		
Often	123	14.4	Important	332	39.0	Agree	91	10.7	171	20.1	133	15.6		
Always	81	9.5	Very important	405	47.5	Strongly agree	37	4.3	67	7.9	65	7.6		
Total	852	100.0	Total	852	100.0	Total	852	100.0	852	100.0	852	100.0		

Note: BUYBIO: Do you buy organic food? ORIGIN: How important is product origin when deciding your food purchase? NNT: There is no need for new food technologies because there are so many types of foods; NTQR: The benefits associated with innovative food technologies are often overestimated; NTQL: New food technologies reduce the natural quality of foods.

Table 3 reports the distribution of the three original variables (NNNT, NTOR, and NTLQ), and Figure 1 illustrates the distribution of TFI.

Our respondents reported different willingness to try food with upcycled ingredients of olive leaves based on environmental and nutritional concerns. Our finding is that products with different attributes (nutritional or environmental) can be appealing to different consumers. Figure 2 illustrates the distribution of the two variables.

Variables TNUT and TENT are not independent. A Fisher's exact test rejected the null hypothesis of independence at a 99% confidence level. Table 4 reports Pearson's standardized residuals from the contingency table, showing that the diagonal elements were positive, while the off-diagonal ones were mostly negative. (Pearson's standardized residuals are computed by subtracting the expected frequency in a given cell under the null hypothesis of independence from the actual observed frequency and then dividing by the square root of the expected frequency.) This result suggests a positive association and that the two drivers of consumer acceptance do not offset each other on average. The result is of particular importance because it shows that the two drivers may be pursued at the same time.

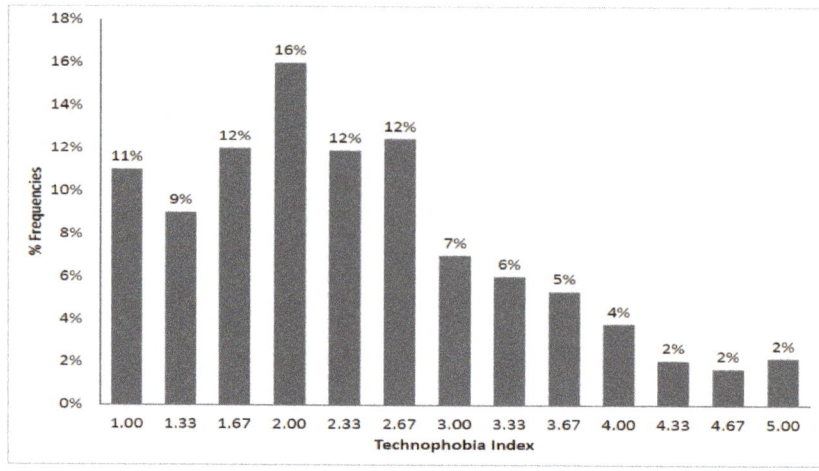

Figure 1. Distribution of the Technophobia Index (TFI).

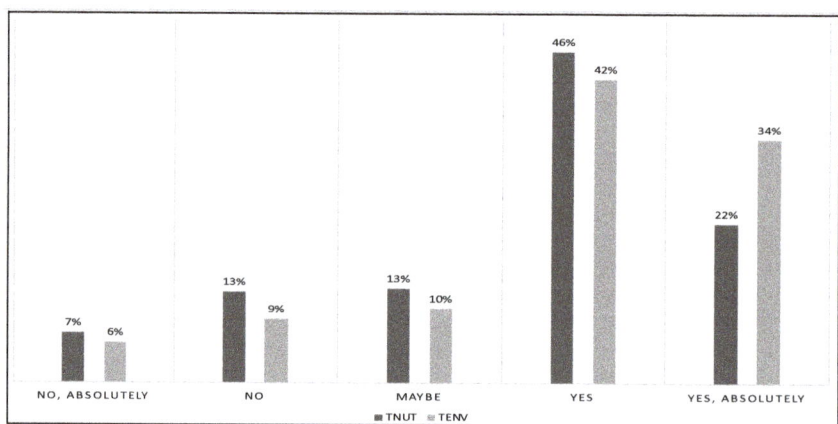

Figure 2. Willingness to try food containing upcycled ingredients with superior nutrition properties (TNUT) or lower environmental impact (TENV).

Table 4. Pearson's standardized residuals from a two-way table of superior nutrition properties (TNUT) and lower environmental impact (TENV).

	TENV				
TNUT	No, absol.	No	Maybe	Yes	Yes, absol.
No, absolutely	17.8	0.8	−2.0	−3.1	−3.1
No	0.5	12.9	−0.3	−3.0	−3.3
Maybe	−1.3	−0.6	6.1	−0.5	−2.0
Yes	−4.4	−4.1	−0.7	6.1	−2.5
Yes, absolutely	−2.9	−3.8	−2.3	−4.4	9.4

3.2. Model Results

Table 5 reports the results of the estimation of the model in Equation (1). An χ_2 test on the joint significance of the coefficients of demography, education, and working status variables failed to reject the null hypothesis that all coefficients were jointly equal to zero (p-values 0.290 and 0.597 in the TNUT and TENV regressions, respectively) (The p-values of the χ^2 on the joint significance of specific groups of variables (demography, education, employment status) are reported in Table 3.)

Table 5. Regression results.

		TNUT			TENV		
		Coeff	Std. Error	p-Value	Coeff	Std. Error	p-Value
Demography				*0.971*			*0.437*
Female respondent	Genb	−0.012	0.076	0.872	−0.035	0.080	0.658
Age of respondent	Age	0.009	0.029	0.750	0.041	0.027	0.130
Age squared	age2	0.000	0.000	0.796	0.000	0.000	0.164
Education				*0.098*			*0.857*
Elementary school							
Middle school	ed2	−0.816	0.612	0.183	0.106	0.396	0.789
High school	ed3	−1.081	0.590	0.067	0.188	0.370	0.611
College	ed4	−0.915	0.588	0.120	0.231	0.365	0.526
Working status				*0.167*			*0.358*
Worker							
Unemployed	w2	0.135	0.174	0.439	0.031	0.182	0.867
Student	w3	0.224	0.167	0.180	0.216	0.177	0.221
Homemaker	w4	0.582	0.275	0.034	0.436	0.262	0.096
Retired	w5	−0.006	0.298	0.983	−0.031	0.264	0.906
Organic				*0.145*			*0.000*
Never							
Seldom	bb2	−0.209	0.098	0.033	−0.218	0.099	0.028
Sometimes	bb3	0.132	0.169	0.436	0.583	0.147	0.000
Often	bb4	−0.105	0.104	0.314	0.504	0.120	0.000
Always	bb5	0.024	0.140	0.866	0.817	0.160	0.000
Origin				*0.006*			*0.000*
Absol. not important							
Not important	bo2	0.177	0.368	0.631	0.116	0.365	0.751
Indifferent	bo3	0.559	0.323	0.084	0.460	0.328	0.161
Important	bo4	0.725	0.310	0.019	0.812	0.315	0.010
Very important	bo5	0.769	0.310	0.013	0.897	0.317	0.005
Technophobia	TFI	−0.368	0.046	0.000	−0.301	0.045	0.000
Observations		852			852		
Pseudo R^2		0.052			0.076		
Wald chi^2		115.38		0.000	152.01		0.000

Note: Figures in Italic font report the p-values of a test on the joint significance of the regression coefficient of the corresponding group of variables (demography, education, working status, organic, origin).

Technophobia's coefficient was negative and statistically different from zero at a 95% confidence level in both regressions. High values of TFI were associated with a higher probability of being absolutely unwilling to try olive by-product food. Table 6 reports the marginal probabilities.

An χ_2 test failed to reject the hypothesis that ORGANIC did not affect TNUT. Instead, we detected a statistically significant association with TENV (95% confidence level). Holding all other variables constant, buyers of organic food are expected to be more willing to try food with olive by-products if the consumption is beneficial for the environment.

In addition, respondents considering product origin an important or very important issue in food choice are more likely to be very willing to try food with olive by-products. The results hold in both regressions.

Table 6. Marginal probabilities.

	Variables	TNUT					TENV				
		TNUT = 1	TNUT = 2	TNUT = 3	TNUT = 4	TNUT = 5	TENV = 1	TENV = 2	TENV = 3	TENV = 4	TENV = 5
Demography	Female respondent	0.001	0.002	0.001	-0.001	-0.003	0.003	0.004	0.004	0.002	-0.013
	Age of respondent	-0.001	-0.001	-0.001	0.001	0.003	-0.003	-0.005	-0.004	-0.002	0.015
	Age squared	0.000	0.000	0.000	0.000	0.000	0.000	0.000	0.000	0.000	0.000
Educational	Elementary school										
	Middle school *	0.153	0.122	0.041	-0.160	-0.155	-0.007	-0.012	-0.011	-0.008	0.039
	High school *	0.150	0.158	0.082	-0.123	-0.266	-0.014	-0.022	-0.020	-0.013	0.067
	College *	0.088	0.126	0.087	-0.029	-0.272	-0.018	-0.027	-0.024	-0.011	0.081
Employment	Worker										
	Unemployed *	-0.013	-0.020	-0.014	0.007	0.040	-0.002	-0.004	-0.003	-0.002	0.011
	Student *	-0.022	-0.033	-0.022	0.012	0.065	-0.015	-0.024	-0.023	-0.016	0.078
	Homemaker *	-0.039	-0.072	-0.061	-0.024	0.196	-0.023	-0.042	-0.044	-0.058	0.166
	Retired *	0.001	0.001	0.001	0.000	-0.002	0.002	0.004	0.003	0.002	-0.011
Organic	Never										
	Seldom *	0.024	0.032	0.019	-0.020	-0.056	0.018	0.027	0.022	0.008	-0.075
	Sometimes *	-0.013	-0.019	-0.013	0.007	0.039	-0.028	-0.053	-0.057	-0.086	0.224
	Often *	0.012	0.016	0.010	-0.009	-0.028	-0.028	-0.050	-0.051	-0.062	0.190
	Always *	-0.002	-0.004	-0.002	0.002	0.007	-0.035	-0.068	-0.076	-0.135	0.314
Origin	Absol. not important										
	Not important *	-0.016	-0.026	-0.018	0.007	0.053	-0.008	-0.013	-0.012	-0.009	0.042
	Indifferent *	-0.040	-0.072	-0.058	-0.015	0.185	-0.025	-0.045	-0.046	-0.060	0.175
	Important *	-0.070	-0.102	-0.070	0.028	0.215	-0.056	-0.087	-0.080	-0.070	0.293
	Very important *	-0.082	-0.111	-0.072	0.048	0.217	-0.070	-0.101	-0.088	-0.055	0.314
	TFI	0.039	0.055	0.036	-0.027	-0.103	0.023	0.035	0.031	0.017	-0.107

* The marginal probability is for a discrete change in a binary variable from 0 to 1.

3.3. Differences across Generations

The regressions in Table 3 failed to reject the null hypothesis that the conditional expectations of TENV and TNUT were unaffected by the respondent's age. However, several studies have shown that older generations show different food preferences to younger generations [49,50,59]. To investigate this point further, we exploited the large sample size of our survey to run separate regression for age groups.

We split the sample into four age groups: Generation Z (age between 18 and 24), Millennials (or Generation Y, age between 25 and 39), Generation X (between 40 and 54), and Baby Boomers+ (age 55 or above). Figures 3 and 4 illustrate the distribution of TNUT and TENV by age group, respectively. An χ2 test failed to reject the null hypothesis of independence between TNUT and age groups at a 95% confidence level (*p*-value 0.057). Instead, the independence of TENV and age group was rejected (*p*-value 0.01).

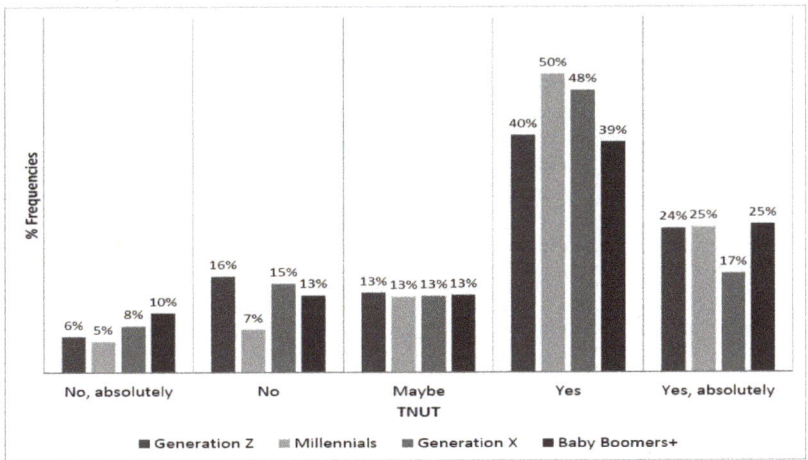

Figure 3. Distribution of TNUT by generations.

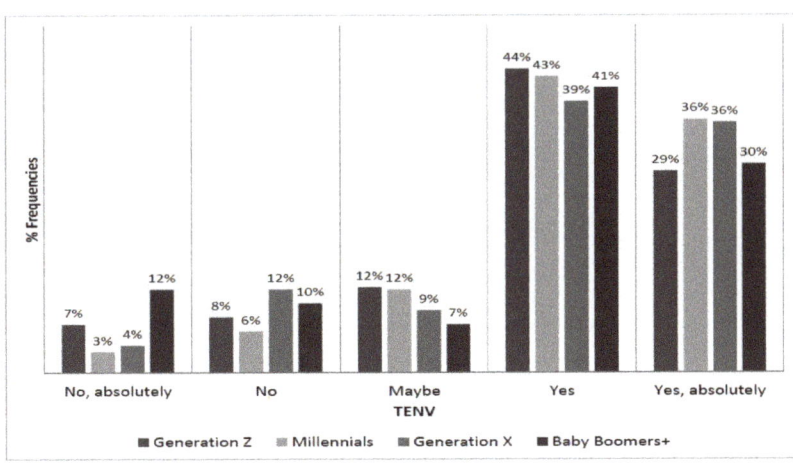

Figure 4. Distribution of TENV by generations.

Tables 7 and 8 report the outcome of the regressions of TNUT and TENV by generation groups, respectively. For the reader's convenience, the results of the χ_2 test on the joint significance of coefficients of groups of variables (Demography, Education, Employment status, Organic, and Origin) are summarized in Table 9.

Table 7. Regression of willingness to try food with upcycled ingredients and superior nutrition properties (TNUT) by generations.

		Generation Z (n = 215)		Millennials (n = 267)		Generation X (n = 264)		Baby Boomers+ (n = 106)	
		Coeff	p-Val	Coeff	p-Val	Coeff	p-Val	Coeff	p-Val
Demography			*0.015*		*0.661*		*0.651*		*0.316*
Female respondent	Genb	−0.185	0.221	0.162	0.262	0.140	0.319	−0.085	0.724
Age of respondent	Age	−2.182	0.109	0.089	0.753	−0.341	0.359	0.507	0.075
Age squared	age2	0.053	0.088	−0.001	0.732	0.004	0.362	−0.004	0.071
Education			*0.679*		*0.591*		*0.186*		*0.183*
Elementary school									
Middle school	ed2	−0.056	0.960	0.786	0.229			−1.507	0.033
High school	ed3	−0.171	0.383	−0.017	0.955	−0.775	0.071	−1.462	0.046
College	ed4			−0.030	0.918	−0.682	0.123	−1.496	0.034
Working status			*0.015*		*0.560*		*0.000*		*0.626*
Worker									
Unemployed	w2	−0.151	0.820	0.191	0.414	0.186	0.633	−0.737	0.429
Student	w3	0.571	0.351	0.318	0.253	8.978	0.000		
Homemaker	w4			−0.309	0.599	0.817	0.014	0.511	0.531
Retired	w5							−0.278	0.424
Organic			*0.300*		*0.148*		*0.064*		*0.493*
Never									
Seldom	bb2	−0.130	0.497	−0.370	0.024	−0.446	0.022	0.288	0.295
Sometimes	bb3	0.444	0.101	−0.147	0.676	0.127	0.639	0.111	0.881
Often	bb4	0.163	0.468	−0.128	0.539	−0.301	0.106	0.077	0.824
Always	bb5	0.209	0.437	−0.537	0.093	0.062	0.791	0.512	0.115
Origin			*0.874*		*0.441*		*0.000*		*0.115*
Absol. not important									
Not important	bo2	0.024	0.979	0.876	0.231	−0.601	0.207	0.129	0.932
Indifferent	bo3	−0.230	0.760	1.225	0.081	0.501	0.214	0.717	0.616
Important	bo4	−0.117	0.875	1.170	0.087	0.592	0.123	1.265	0.375
Very important	bo5	−0.288	0.698	1.110	0.109	0.870	0.027	1.327	0.337
Technophobia	TFI	−0.457	0.000	−0.525	0.000	−0.255	0.002	−0.337	0.002
R2		0.072		0.100		0.075		0.094	

Table 8. Regression of willingness to try food with upcycled ingredients and lower environmental impact (TENV) by Generations.

		Generation Z		Millennials		Generation X		Baby Boomers+	
		Coeff	p-Val	Coeff	p-Val	Coeff	p-Val	Coeff	p-Val
Demography			*0.852*		*0.624*		*0.570*		*0.425*
Female respondent	Genb	0.013	0.938	0.005	0.976	−0.041	0.802	−0.082	0.738
Age of respondent	Age	−0.760	0.585	−0.360	0.194	−0.162	0.694	0.032	0.912
Age squared	age2	0.019	0.564	0.005	0.199	0.002	0.657	−0.001	0.795
Education	(omitted)		*0.084*		*0.001*		*0.658*		*0.167*
Elementary school									
Middle school	ed2	−0.844	0.038	1.266	0.192			−0.277	0.598
High school	ed3	−0.403	0.068	0.988	0.001	−0.251	0.608	0.135	0.809
College	ed4			1.151	0.000	−0.369	0.461	−0.473	0.353
Working status			*0.007*		*0.436*		*0.000*		*0.500*
Worker									
Unemployed	w2	0.948	0.177	−0.178	0.406	0.379	0.376	−0.931	0.330
Student	w3	1.265	0.002	0.218	0.442	8.193	0.000		
Homemaker	w4			−0.529	0.239	0.827	0.012	1.177	0.297
Retired	w5							0.081	0.816
Organic			*0.004*		*0.013*		*0.000*		*0.023*
Never									
Seldom	bb2	−0.133	0.479	−0.320	0.082	−0.617	0.001	0.097	0.725
Sometimes	bb3	0.635	0.046	0.798	0.018	0.342	0.086	0.907	0.101
Often	bb4	0.602	0.007	0.261	0.240	0.564	0.025	1.560	0.001
Always	bb5	0.596	0.070	0.217	0.412	1.831	0.000	0.466	0.330
Origin			*0.057*		*0.450*		*0.000*		*0.000*
Absol. not important									
Not important	bo2	1.241	0.029	1.206	0.203	−1.585	0.004	0.706	0.486
Indifferent	bo3	0.855	0.024	1.381	0.140	−0.754	0.101	1.337	0.188
Important	bo4	1.064	0.004	1.512	0.096	−0.174	0.685	2.282	0.030
Very important	bo5	1.016	0.006	1.536	0.093	0.059	0.891	2.321	0.020
Technophobia	TFI	−0.366	0.000	−0.313	0.000	−0.274	0.001	−0.461	0.000
R2		0.082		0.077		0.161		0.179	

Table 9. Summary of the findings of the generation analysis *.

Var. Group	Full Sample		Generation Z		Millennials		Generation X		Baby Boomers+	
	TNUT	TENV	TNUT	TENV	TNUT	TENV	TNUT	TENV	TNUT	TENV
Demography	✗	✗	✓✓	✗	✗	✗	✗	✗	✗	✗
Education	✗	✗	✗	✓	✗	✓✓	✗	✗	✗	✗
Emp. Status	✗	✗	✓✓	✓✓	✗	✗	✓✓	✓✓	✗	✗
Organic	✗	✓✓	✗	✓✓	✗	✓✓	✓	✓✓	✗	✓✓
Origin	✓✓	✓✓	✗	✓	✗	✗	✓✓	✓✓	✗	✓✓
Technophobia	✓✓	✓✓	✓✓	✓✓	✓✓	✓✓	✓✓	✓✓	✓✓	✓✓

* The null hypothesis that all coefficients in the variable group are jointly equal to zero was: (✗) not rejected, (✓) rejected at 90% confidence level, (✓✓) rejected at 95% confidence level in an χ^2 test.

The analysis by age group showed that the drivers of willingness to accept food with upcycled ingredients were not monotonic with respect to the respondent's age. Each generation had distinctive characteristics that were not necessarily similar to the next age group. This result explains why the coefficients of the AGE and AGE2 variables were not statistically different from zero in the full sample regression and suggests that the ensemble of beliefs driving the behavior of a generation may be defined in contrast to the previous generation.

Generation Z's (age 18–24) acceptance of food with olive by-products and improved nutrition characteristics was driven mainly by technophobia. Working status affects decisions, with students showing a higher willingness to try than unemployed respondents (An χ^2 test rejected the null hypothesis that the coefficients w2 and w3 were equal with *p*-value 0.006.). The coefficient of the quadratic form of the variable AGE on TNUT was statistically different from zero, suggesting that preferences within Generation Z were not homogeneous. Generation Z's decisions to try food with upcycled ingredients that reduce environmental impact were affected by technophobia and working status. A positive association with organic purchase was found as well.

The willingness of Millennials (age 25–39) to try upcycled ingredients with improved nutrition attributes was driven by technophobia alone. Decisions regarding environmentally friendly by-products were driven by technophobia, education, and organic purchase. The importance of product origin was not associated with the dependent variables.

The drivers of Generation X's (40–54) decisions regarding TENV and TNUT were similar. The coefficients of Employment status variables, TFI, and ORIGIN were statistically different from zero in both regressions at a 95% confidence level. ORGANIC was statistically significant at a 95% confidence level in the TENV regression and at a 90% confidence level in the TNUT regression.

Finally, the age group Baby Boomers+ (55+) exhibited different behaviors depending on whether the upcycled ingredients were associated with improved nutrition or environmental responsibility. In the former case, technophobia was the only driver. In the latter case, ORGANIC and ORIGIN variables were associated with higher values of willingness to try.

4. Discussion

Among the greatest challenges the world faces today are how to ensure that a growing global population has access to enough healthy food and how to reduce food loss and waste. Rethinking the food system and implementing circular resource management systems will help mitigate the effects of food production on the environment and limited availability of resources [60].

Waste valorization has been defined by Arancon et al. [61] as the process of converting waste into more useful products. For example, the olive tree pruning produces 25 Kg of waste biomass for each tree annually, and approximately 25% is leaves [62]. Olive leaves are rich in phenolic compounds [6], and the food sector should use them to produce value-added products [25]. However, consumer acceptance of these products is fundamental to their market uptake.

This study aimed at answering two main research questions: What interest do consumers have in food products enriched with waste-to-value food? Which variables are important predictors of consumer willingness to buy food products enriched with waste-to-value food? Previous results suggest that, although the production of foods with up-cycled ingredients is technically feasible [6,7], carefully-designed marketing campaigns are necessary to ensure consumer acceptance and, ultimately, economic success [38,45]. Aschemann-Witzel and Peschel [27], analyzing how Danish consumers react to the use of by-products in some food products, indicated that specific brand, design, and specific quality information on these new ingredients could improve consumer attitudes towards the "waste-to-value" products.

Our study is useful to highlight what specific consumer profiles may be targeted for marketing campaigns. Environmentally responsible organic consumers, in fact, are likely to be an important niche for food with olive by-products. In the sample analyzed, buyers of organic food were expected to be more willing to try this novel food if it was more beneficial for the environment. This result is quite well known in the literature and might be explained with the special concern for the environmental aspects that organic buyers show [25,63–65].

The preference of organic consumers for such novel environmentally sustainable products is very important, considering the current market trends. In fact, nowadays, organic food is not a niche market anymore, accounting for approximately 3% of the total value of the agri-food sector [66,67]. In particular, Italy is ranked sixth in the world among the countries with the largest area cultivated with organic farming methods [68,69]. Furthermore, environmentally sustainable consumption is gaining importance in the market, with consumers showing higher interest in the impacts on natural resources of their food purchases [23].

On the contrary, nutritional attributes were not appealing for environmentalist consumers, as the lack of association between organic purchase and acceptance of upcycled food with superior nutritional properties showed. This finding seems in line with Grasso and Assioli [26], who analyzed three different groups of consumers and the group called "environmentalist", more interested in the environment, had the lowest rejection towards upcycled sun-flower flour in biscuits.

Another result deserving particular attention is one of product origin. In fact, respondents considering product origin an important or very important issue in food choice were more likely to be willing to try food with olive by-products. As product origin is a very important driver of consumer choice in the Italian market [5], this result is of interest because it suggests that there could be a marketing potential for local food made with upcycled ingredients. In fact, our results confirm that the origin of the by-product may mitigate the food technophobia, and origin information on the olive by-products can increase consumer acceptance and preference for food with upcycled ingredients.

Interesting insights on the consumer characteristics can be derived from the differences across the generations analyzed. In fact, the drivers of willingness to accept food with upcycled ingredients were not the same with respect to the respondent's age. Each generation had distinctive characteristics that were not necessarily similar to the next age group. This result explains why the coefficients of the AGE and AGE2 variables were not statistically different from zero in the full sample regression and suggests that the ensemble of beliefs driving the behavior of a generation may be defined in contrast to the previous generation.

For the youngest Z Generation, technophobia and working status were the relevant drivers in determining acceptance of food with by-products with improved nutrition characteristics or with reduced environmental impact. In this latter case, a positive association with organic purchase and a slight importance of a product's origin was found as well, confirming the general result regarding organic consumers and product origin discussed above.

For Millennials, technophobia alone seemed to drive the acceptance of food with upcycled ingredients with improved nutrition attributes, while decisions regarding environmentally friendly by-products were driven by technophobia, education, and organic purchase. Interestingly, a distinctive characteristic of Millennials, compared to other age groups, was that the importance of product origin was not associated with their willingness to accept, thus highlighting a different behavior of that generation regarding this product's feature.

Generation X seemed, instead, to show more similar and coherent preferences regarding both food with upcycled ingredients with improved nutrition attributes and lower environmental impact. For this group of consumers, technophobia, product's origin, and organic consumption were all relevant in affecting their purchase intentions.

Baby Boomers+, instead, showed different preference structures. For food with upcycled ingredients with improved nutritional properties, technophobia seemed to be the only driver. For environmentally sustainable food with up-cycled ingredients, the willingness to accept was mainly driven by product origin and organic preferences.

Differences across generations allow even better targeting of market delivery of the product, focusing the attention on the specific driver of each age group segment. In fact, results, if confirmed by further surveys in other countries and with larger and more representative samples, suggest that different age groups respond differently to product characteristics and they could be better targeted with more specific and ad-hoc campaigns. For example, the aspect of a product's origin seemed not to be relevant in determining the Millennials acceptance for food with up-cycled ingredients, and thus it could be argued that to target such consumers, the attention should be paid to better presenting the "low technological component" in the production process, rather than the local origin of the product to the consumer.

A final consideration should be given to the fact that, at the time of the survey, the proposed novel product was not yet available in the market. This aspect might represent a weakness in the proposed analysis. The fact that consumers were not able to test or see the product could have influenced their replies. However, the results presented here could be useful in the market launch of these products as they target consumers who have shown willingness to purchase such products (e.g., consumers of organic and local products) according to their specific age group.

5. Conclusions

Consumer's general perception of the use of upcycled ingredients for food production is that the food industry tries to save money with inputs obtained at a lower cost [70]. Consumers accept many by-products for pharmaceutical use because they are rich in healthy components. The acceptance of by-products used for food production is a more complex matter because of the influence of certain levels of food neophobia or technophobia that might hamper the uptake of such products.

The present study confirmed this general influence as both the complete sample and all age groups of consumers demonstrated that technophobia was negatively influenced by the probability of accepting food enriched with olive oil by-products [33]. However, two major determinants of consumption of food made with upcycled ingredients emerged: organic consumers are more likely to accept this novel food and, also, consumers who consider product origin an important or very important issue in food choice are more likely to be willing to try food with olive by-products. The impact of such aspects slightly differs when referred to different product's characteristics. When we consider products with superior nutrition properties, the association with origin attributes was stronger, while when looking at food with lower environmental impact, the consumption of organic food seemed to be highly associated with the acceptance of the novel food.

This latter aspect also emerged if we look at the results for the generation groups. In all the age groups, the consumption of organic products was positively associated with a likely acceptance of food made with upcycled ingredients, which show a lower environmental impact.

Results here presented would thus suggest that there could be a core of consumers interested in organic or local products, that could also favor the uptake of these novel foods made with upcycled ingredients in the market. Marketing policies are of great importance in that sense because indicating the benefits these foods could bring to health and the environment clearly in the label should help to deliver novel food to the greater public. According to the results of this study, developing organic or "local" varieties of food with upcycled ingredients might increase the probability of consumer acceptance.

Our study adds another piece to the puzzle of the research into upcycling or waste-to-value products in the area of food where studies are yet scarce [25–27,33].

However, this manuscript has two main limitations. First, the sample analyzed in this study is not representative of the whole Italian population. However, given the size of the sample, the relationships between the variables analyzed and the positive purchase intention eventually expressed remain valid and allow us to obtain interesting results. Second, as also reported by Grasso and Assioli [26], because upcycled ingredients for food products are not on the market yet and there is not an appropriate definition of these products, our study might suffer from hypothetical bias, which could have affected the estimation of consumer acceptance.

Future research is needed to confirm our results in other countries and using different products and/or upcycled ingredients. In particular, an experimental approach can be used to overcome the hypothetical bias.

Author Contributions: Conceptualization, M.A.P., S.C., and C.R.; Data curation, M.A.P., S.C., and C.R.; Formal analysis, M.A.P., S.C., and C.R.; Funding acquisition, M.A.P.; Investigation, M.A.P., S.C., and C.R.; Methodology, M.A.P., S.C., and C.R.; Resources, M.A.P.; Supervision, M.A.P., S.C., and C.R.; Writing—original draft, M.A.P., S.C., and C.R.; Writing—review and editing, M.A.P., S.C., and C.R. All authors have read and agreed to the published version of the manuscript.

Funding: This research was funded by the AGER 2 Project, grant number 2016-0105.

Acknowledgments: We would like to thank Carla Daniela Di Mattia for inspiring and supporting our research interests in consumer acceptance of novel food enriched with upcycled ingredients.

Conflicts of Interest: The authors declare no conflict of interest.

Data Availability: The data supporting the findings of this study are available from the corresponding author (M.A.P.), upon reasonable request.

References

1. Pazzini, A. *Storia, Tradizioni e Leggende Nella Medicina Popolare*; Ricordati: Bergamo, Italy, 1940; pp. 1–142.
2. Rossi, M. *Tinture Madri in Fitoterapia*; Studio Edizioni: Milano, Italy, 1992; pp. 1–225.
3. Acar-Tek, N.; Ağagündüz, D. Olive Leaf (Olea europaea L. folium): Potential Effects on Glycemia and Lipidemia. *Ann. Nutr. Metab.* **2020**, *76*, 63–68. [CrossRef]
4. Coderoni, S.; Perito, M.A.; Cardillo, C. Consumer behaviour in Italy. Who spends more to buy a Mediterranean Diet? *New Medit.* **2017**, *162*, 38–46.
5. Perito, M.A.; Sacchetti, G.; Di Mattia, C.D.; Chiodo, E.; Pittia, P.; Saguy, I.S.; Cohen, E. Buy local! Familiarity and preferences for extra virgin olive oil of Italian consumers. *J. Food Prod. Market.* **2019**, *25*, 462–477. [CrossRef]
6. Flamminii, F.; Di Mattia, C.D.; Difonzo, G.; Neri, L.; Faieta, M.; Caponio, F.; Pittia, P. From by-product to food ingredient: Evaluation of compositional and technological properties of olive-leaf phenolic extracts. *J. Sci. Food Agric.* **2019**, *99*, 6620–6627. [CrossRef] [PubMed]
7. Flamminii, F.; Di Mattia, C.D.; Nardella, M.; Chiarini, M.; Valbonetti, L.; Neri, L.; Difonzo, G.; Pittia, P. Structuring alginate beads with different biopolymers for the development of functional ingredients loaded with olive leaves phenolic extract. *Food Hydrocoll.* **2020**, in press. [CrossRef]
8. Lähteenmäki, L. Claiming health in food products. *Food Qual. Prefer.* **2013**, *27*, 196–201. [CrossRef]
9. Fernqvist, F.; Ekelund, L. Credence and the effect on consumer liking of food—A review. *Food Qual. Prefer.* **2014**, *32*, 340–353. [CrossRef]
10. Menapace, L.; Colson, G.; Grebitus, C.; Facendola, M. Consumers' preferences for geo-graphical origin labels: Evidence from the Canadian olive oil market. *Eur. Rev. Agric.* **2011**, *38*, 193–212. [CrossRef]
11. Van der Lans, I.A.; Van Ittersum, K.; De Cicco, A.; Loseby, M. The role of the region of origin and EU certificates of origin in consumer evaluation of food products. *Eur. Rev. Agric.* **2001**, *28*, 451–477. [CrossRef]
12. Hobbs, J.E.; Goddard, E. Consumers and trust. *Food Policy* **2015**, *52*, 71–74. [CrossRef]
13. Russo, C.; Simeone, M. The growing influence of social and digital media. *Br. Food J.* **2017**, *119*, 1766–1780. [CrossRef]

14. Peschel, A.O.; Aschemann-Witzel, J. Sell more for less or less for more? The role of transparency in consumer response to upcycled food products. *J. Clean. Prod.* **2020**, *273*, 122884. [CrossRef]
15. Asioli, D.; Næs, T.; Granli, B.S.; Lengard Almli, V. Consumer preferences for iced coffee determined by conjoint analysis: An exploratory study with Norwegian consumers. *Int. J. Food Sci. Technol.* **2014**, *49*, 1565–1571. [CrossRef]
16. Lusk, J.L.; Briggeman, B.C. Food values. *Am. J. Agric. Econ.* **2009**, *91*, 184–196. [CrossRef]
17. Steenhuis, I.H.; Waterlander, W.E.; de Mul, A. Consumer food choices: The role of price and pricing strategies. *Public Health Nutr.* **2011**, *14*, 2220–2226. [CrossRef]
18. Defrancesco, E.; Perito, M.A.; Bozzolan, I.; Cei, L.; Stefani, G. Testing consumers' preferences for environmental attributes of pasta. Insights from an ABR approach. *Sustainability* **2017**, *9*, 1701. [CrossRef]
19. Nait Mohand, N.; Hammoudi, A.; Radjef, M.S.; Hamza, O.; Perito, M.A. How do food safety regulations influence market price? A theoretical analysis. *Br. Food J.* **2017**, *119*, 1687–1704. [CrossRef]
20. De Pelsmacker, P.; Driesen, L.; Rayp, G. Are fair trade labels good business? Ethics and coffee buying intentions. *J. Consum. Aff.* **2003**, *39*, 1–20.
21. Boutouis, M.Z.; Hammoudi, A.; Benhassine, W.; Perito, M.A. Uncertainty of food contamination origin and liability rules: Implications for bargaining power. *Agribusiness* **2018**, *34*, 77–92. [CrossRef]
22. Hempel, C.; Hamm, U. Local and/or organic: A study on consumer preferences for organic food and food from different origins. *Int. J. Consum.* **2016**, *40*, 732–741. [CrossRef]
23. Canavari, M.; Coderoni, S. Consumer stated preferences for dairy products with carbon footprint labels in Italy. *Agric. Food Econ.* **2020**, *8*, 1–16. [CrossRef]
24. Galanakis, C.M. Recovery of high added-value components from food wastes: Conventional, emerging technologies and commercialized applications. *Trends Food Sci. Technol.* **2012**, *26*, 68–87. [CrossRef]
25. Coderoni, S.; Perito, M.A. Sustainable consumption in the circular economy. An analysis of consumers' purchase intentions for waste-to-value food. *J. Clean. Prod.* **2020**, *252*, 119870. [CrossRef]
26. Grasso, S.; Asioli, D. Consumer preferences for upcycled ingredients: A case study with biscuits. *Food Qual. Prefer.* **2020**, *84*, 103951. [CrossRef]
27. Aschemann-Witzel, J.; Peschel, A.O. How circular will you eat? The sustainability challenge in food and consumer reaction to either waste-to-value or yet un derused novel ingredients in food. *Food Qual. Prefer.* **2019**, *77*, 15–20. [CrossRef]
28. Pliner, P.; Hobden, K. Development of a scale to measure the trait of food neophobia in humans. *Appetite* **1992**, *19*, 105–120. [CrossRef]
29. Cox, D.N.; Evans, G. Construction and validation of a psychometric scale to measure consumers' fears of novel food technologies: The food technology neophobia scale. *Food Qual. Prefer.* **2008**, *19*, 704–710. [CrossRef]
30. Verbeke, W. Profiling consumers who are ready to adopt insects as a meat substitute in a Western society. *Food Qual. Prefer.* **2015**, *39*, 147–155. [CrossRef]
31. Palmieri, N.; Perito, M.A.; Macrì, M.C.; Lupi, C. Exploring consumers' willingness to eat insects in Italy. *Br. Food J.* **2019**, *121*, 2937–2950. [CrossRef]
32. Palmieri, N.; Perito, M.A.; Lupi, C. Consumer acceptance of cultured meat: Some hints from Italy. *Br. Food J.* **2020**, in press. [CrossRef]
33. Perito, M.A.; Di Fonzo, A.; Sansone, M.; Russo, C. Consumer acceptance of food obtained from olive by-products. *Br. Food J.* **2019**, *122*, 212–226. [CrossRef]
34. Bhatt, S.; Lee, J.; Deutsch, J.; Ayaz, H.; Fulton, B.; Suri, R. From food waste to value-added surplus products (VASP): Consumer acceptance of a novel food product category. *J. Consum. Behav.* **2018**, *17*, 57–63. [CrossRef]
35. Berghoef, N.; Dodds, R. Potential for sustainability eco-labeling in Ontario's wine industry. *Int. J. Wine Bus. Res.* **2011**, *23*, 298–317. [CrossRef]
36. Barber, N. "Green" wine packaging: Targeting environmental consumers. *Int. J. Wine Bus. Res.* **2010**, *22*, 423–444. [CrossRef]
37. Sörqvist, P.; Haga, A.; Holmgren, M.; Hansla, A. An eco-label effect in the built environment: Performance and comfort effects of labeling a light source environmentally friendly. *J. Environ. Psychol.* **2015**, *42*, 123–127. [CrossRef]

38. Ditlevsen, K.; Sandøe, P.; Lassen, J. Healthy food is nutritious, but organic food is healthy because it is pure: The negotiation of healthy food choices by Danish consumers of organic food. *Food Qual. Prefer.* **2019**, *71*, 46–53. [CrossRef]
39. Ginon, E.; Ares, G.; dos Santos Laboissière, L.H.E.; Brouard, J.; Issanchou, S.; Deliza, R. Logos indicating environmental sustainability in wine production: An exploratory study on how do Burgundy wine consumers perceive them. *Food Res. Int.* **2014**, *62*, 837–845. [CrossRef]
40. Grankvist, G.; Biel, A. The importance of beliefs and purchase criteria in the choice of eco-labeled food products. *J. Environ. Psychol.* **2001**, *21*, 405–410. [CrossRef]
41. Magnusson, M.K.; Arvola, A.; Hursti, U.K.K.; Åberg, L.; Sjödén, P.O. Choice of organic foods is related to perceived consequences for human health and to environmentally friendly behaviour. *Appetite* **2003**, *40*, 109–117. [CrossRef]
42. Hughner, R.S.; McDonagh, P.; Prothero, A.; Shultz, C.J.; Stanton, J. Who are organic food consumers? A compilation and review of why people purchase organic food. *J. Consum. Behav. Int. Res. Rev.* **2007**, *6*, 94–110. [CrossRef]
43. Aschemann-Witzel, J.; Zielke, S. Can't buy me green? A review of consumer perceptions of and behavior toward the price of organic food. *J. Cons. Affairs* **2015**, *51*, 211–251. [CrossRef]
44. Loureiro, M.L.; Umberger, W.J. Estimating consumer willingness to pay for country-of- origin labelling. *J. Agric. Resour. Econ.* **2003**, *28*, 287–301.
45. Zander, K.; Stolz, H.; Hamm, U. Promising ethical arguments for product differentiation in the organic food sector. A mixed methods research approach. *Appetite* **2013**, *62*, 133–142. [CrossRef] [PubMed]
46. Feldmann, C.; Hamm, U. Consumers' perceptions and preferences for local food: A review. *Food Qual. Prefer.* **2015**, *40*, 152–164. [CrossRef]
47. Grebitus, C.; Lusk, J.L.; Nayga, R.M., Jr. Effect of distance of transportation on willingness to pay for food. *Ecol. Econ.* **2013**, *88*, 67–75. [CrossRef]
48. Stephenson, G.; Lev, L. Common support for local agriculture in two contrasting Oregon communities. *Renew. Agric. Food Syst.* **2004**, *19*, 210–217. [CrossRef]
49. Vanhonacker, F.; Lengard, V.; Hersleth, M.; Verbeke, W. Profiling European traditional food consumers. *Br. Food J.* **2010**, *112*, 871–886. [CrossRef]
50. Fibri, D.L.N.; Frøst, M.B. Indonesian millennial consumers' perception of tempe and how it is affected by product information and consumer psychographic traits. *Food Qual. Prefer.* **2020**, *80*, 103798. [CrossRef]
51. Wooldridge, J.M. *Econometric Analysis of Cross Section and Panel Data*, 2nd ed.; MIT Press: Cambridge, MA, USA, 2010.
52. Kayam, O.; Hirsch, T. Using social media networks to conduct questionnaire based research in social studies case study: Family language policy. *J. Sociol. Res.* **2012**, *3*, 57–67. [CrossRef]
53. Osservatorio Sana. 2018: "Tutti i Numeri del Bio Italiano: I Driver del Consumatore e le Novità del Canale Specializzato". Available online: http://www.sana.it/media-room/archivio-news/osservatorio-sana-2018-tutti-i-dati-sulle-abitudini-dacquisto-di-prodotti-bio-e-sulle-strategie-del-canale-specializzato/8843.html (accessed on 6 September 2020).
54. Evans, G.; Kermarrec, C.; Sable, T.; Cox, D.N. Reliability and predictive validity of the Food Technology Neophobia Scale. *Appetite* **2010**, *54*, 390–393. [CrossRef]
55. Verneau, F.; Caracciolo, F.; Coppola, A.; Lombardi, P. Consumer fears and familiarity of processed food. The value of information provided by the FTNS. *Appetite* **2014**, *73*, 140–146. [CrossRef] [PubMed]
56. Dekhili, S.; Sirieix, L.; Cohen, E. How consumers choose olive oil: The importance of origin cues. *Food Qual. Prefer.* **2011**, *22*, 757–762. [CrossRef]
57. Newman, C.L.; Turri, A.M.; Howlett, E.; Stokes, A. Twenty years of country of origin food labeling research: A review of the literature and implications for food marketing systems. *J. Macromark.* **2014**, *34*, 505–519. [CrossRef]
58. Thøgersen, J.; Pedersen, S.; Paternoga, M.; Schwendel, E.; Aschemann-Witzel, J. How important is country-of-origin for organic food consumers? A review of the literature and suggestions for future research. *Br. Food J.* **2017**, *119*, 542–557. [CrossRef]
59. Kwak, H.S.; Jung, H.Y.; Kim, M.J.; Kim, S.S. Differences in consumer perception of Korean traditional soybean paste (Doenjang) between younger and older consumers by blind and informed tests. *J. Sens. Stud.* **2017**, *32*, 1–10. [CrossRef]

60. Van Loo, E.J.; Hoefkens, C.; Verbeke, W. Healthy, sustainable and plant-based eating: Perceived (mis)match and involvement-based consumer segments as targets for future policy. *Food Policy* **2017**, *69*, 46–57. [CrossRef]
61. Arancon, R.A.D.; Lin, C.S.K.; Chan, K.M.; Kwan, T.H.; Luque, R. Advances on waste valorization: New horizons for a more sustainable society. *Energy Sci. Eng.* **2013**, *1*, 53–71. [CrossRef]
62. Romero-García, J.M.; Niño, L.; Martínez-Patiño, C.; Álvarez, C.; Castro, E.; Negro, M.J. Biorefinery based on olive biomass. State of the art and future trends. *Bioresour. Technol.* **2014**, *159*, 421–432. [CrossRef]
63. Aertsens, J.; Mondelaers, K.; Verbeke, W.; Buysse, J.; Van Huylenbroeck, G. The influence of subjective and objective knowledge on attitude, motivations and consumption of organic food. *Br. Food J.* **2011**, *113*, 1353–1378. [CrossRef]
64. Harper, G.C.; Makatouni, A. Consumer perception of organic food production and farm animal welfare. *Br. Food J.* **2002**, *104*, 287–299. [CrossRef]
65. Padel, S.; Foster, C. Exploring the gap between attitudes and behaviour: Understanding why consumers buy or do not buy organic food. *Br. Food J.* **2005**, *107*, 606–625. [CrossRef]
66. Willer, H.; Lernoud, J. *The World of Organic Agriculture. Statistics and Emerging Trends 2017*; Research Institute of Organic Agriculture (FiBL), Frick, and IFOAM—Organics International: Bonn, Germany, 2017.
67. Thøgersen, J.; Pedersen, S.; Aschemann-Witzel, J. The impact of organic certification and country of origin on consumer food choice in developed and emerging economies. *Food Qual. Prefer.* **2019**, *72*, 10–30. [CrossRef]
68. Vega-Zamora, M.; Naspetti, S.; Zanoli, R. Principales motivaciones del consumidor de alimentos ecológicos en italia. El caso del aceite de oliva. *Agrociencia* **2020**, *54*, 327–336.
69. Palmieri, N.; Perito, M.A. Consumers' willingness to consume sustainable and local wine in Italy. *Ital. J. Food Sci.* **2020**, *32*, 222–233. [CrossRef]
70. Nitzko, S.; Spiller, A. Comparing "Leaf-to-Root", "Nose-to-Tail" and Other Efficient Food Utilization Options from a Consumer Perspective. *Sustainability* **2019**, *11*, 4779. [CrossRef]

© 2020 by the authors. Licensee MDPI, Basel, Switzerland. This article is an open access article distributed under the terms and conditions of the Creative Commons Attribution (CC BY) license (http://creativecommons.org/licenses/by/4.0/).

Article

Consumer Preferences for Origin and Organic Attributes of Extra Virgin Olive Oil: A Choice Experiment in the Italian Market

Matteo Carzedda [1], Gianluigi Gallenti [1,*], Stefania Troiano [2], Marta Cosmina [1], Francesco Marangon [2], Patrizia de Luca [1], Giovanna Pegan [1] and Federico Nassivera [3]

1. Department of Economics, Business, Mathematics and Statistics (DEAMS), University of Trieste, 34100 Trieste, Italy; matteo.carzedda@units.it (M.C.); marta.cosmina@deams.units.it (M.C.); patrizia.deluca@deams.units.it (P.d.L.); giovanna.pegan@deams.units.it (G.P.)
2. Department of Economics and Statistics (DISES), University of Udine, 33100 Udine, Italy; stefania.troiano@uniud.it (S.T.); francesco.marangon@uniud.it (F.M.)
3. Department of Agricultural, Food, Environmental and Animal Sciences (DI4A), University of Udine, 33100 Udine, Italy; federico.nassivera@uniud.it
* Correspondence: gianluigi.gallenti@deams.units.it

Citation: Carzedda, M.; Gallenti, G.; Troiano, S.; Cosmina, M.; Marangon, F.; de Luca, P.; Pegan, G.; Nassivera, F. Consumer Preferences for Origin and Organic Attributes of Extra Virgin Olive Oil: A Choice Experiment in the Italian Market. *Foods* **2021**, *10*, 994. https://doi.org/10.3390/foods10050994

Academic Editors: Cristina Alamprese, Emma Chiavaro and Francesco Caponio

Received: 12 April 2021
Accepted: 30 April 2021
Published: 2 May 2021

Publisher's Note: MDPI stays neutral with regard to jurisdictional claims in published maps and institutional affiliations.

Copyright: © 2021 by the authors. Licensee MDPI, Basel, Switzerland. This article is an open access article distributed under the terms and conditions of the Creative Commons Attribution (CC BY) license (https://creativecommons.org/licenses/by/4.0/).

Abstract: The paper investigates Italian consumers' behavior towards characteristics of extra virgin olive oil, in particular organic production methods and geographical origin. On the basis of the existing literature, the concepts of sustainability of food systems, diets, and the olive oil supply chain are analyzed. A choice experiment (CE), using a face-to-face questionnaire with over 1000 participants, was conducted to quantify the willingness to pay (WTP) for these two attributes. Findings show positive preference for origin attributes, while the organic attribute is not highly valued. The article also offers some perspectives on future research to improve the competitiveness and sustainability of the Italian olive oil supply chain.

Keywords: choice experiment (CE); extra virgin olive oil (EVOO); willingness to pay (WTP); country of origin; organic food; consumer preferences; sustainable food system

1. Introduction

The olive oil system plays a central role in the sustainability of the food system that underlies the Mediterranean diet patterns, for both environmental and socioeconomic aspects [1].

Therefore, it is particularly interesting to analyze consumer behavior for some sustainability attributes of extra virgin olive oil (EVOO), the highest-quality product in the olive oil supply chain.

In fact, such consumers' preferences are one of the main drivers for the transition from current eating patterns to a more sustainable one. Nevertheless, consumers' preferences concern not only the agricultural and production processes, but a complex set of product attributes [2]. Among these are the perceived sensorial quality of olive oil, such as taste, acidity, and fragrance, but also color, country of origin, geographical indication (GI), and the use of olives from organic farming [3].

Unlike other types of olive oil, such as ordinary, refined, lampante, and olive-pomace oils, virgin oil is exclusively obtained through mechanical extraction processes, namely, washing, decantation, centrifugation, and filtration [4]. Superior-quality virgin oils, in terms of raw materials and quality of the production process, are classified as EVOO and are particularly valued for their organoleptic and nutritional properties [5], such as low acidity levels (below 0.8%, according to European standards) and high content of monounsaturated fat and polyphenolic, antioxidant, and anti-inflammatory compounds [6]. In addition to its intrinsic, sensory, and health attributes, the value of EVOO is further enhanced by its potential to promote multifunctional and sustainable agricultural models, which is particularly true and beneficial for traditional olive-tree-growing regions [7]. In this context, the transition from current eating patterns to more sustainable ones depends on consumers' preferences for this complex of attributes [2].

This specific role related to sustainability of the olive oil system depends on the general framework in which the idea of sustainability is developed. Sustainability has become more and more used in different strategic documents, policies, and development plans at the international, national, and local levels. Among these are the so-called "Brundtland Report" [8] and "2030 Agenda" of the United Nations (UN) [9], and inside the European Union (EU), the Common Agricultural Policy (CAP), the Green Deal [10], and the "Farm to Fork" European strategy [11].

Within the agrifood sector, the concept of sustainability is historically closely related to organic production, although sustainability can refer to a wider range of agricultural elements and practices. These include precision agriculture, organic farming, agroecology, agroforestry and stricter animal welfare standards, carbon managing and storing, and adoption of circular economic models [10]. Nevertheless, organic certification remains for the consumer the main recognizable distinctive sign of the environmental sustainability of food.

The International Federation of Organic Agriculture Movements (IFOAM) defines and constantly updates the Basic Standards for Organic Production and Processing (IBS), the founding principles, definitions, and requirements on which national organic certification schemes, such as the Soil Association Standards in the UK, the USDA National Organic Program, and the Indian National Programme for Organic Production (NPOP), are based [12]. Among the broad array of national and international regulations governing organic certification schemes, Council Regulation No. 384/2007 [13] sets the legal basis for organic farming within the EU and "defines organic production as an integral system of managing and producing food products, which combines the best practices with regard to the preservation of the environment, the level of biological diversity, the preservation of natural resources, the application of high standards of proper maintenance (welfare) of animals and a method of production that corresponds to certain requirements for products manufactured using substances and processes of natural origin" [14] (p. 4). Given the spatial extent of our study, we will from now on refer to the EU organic rules, whose principles, meaning, and visual identity are generally well recognized and acknowledged by European consumers [15,16].

The overall aims of the EU action concern the transition of the European agrifood sector towards a sustainable production and consumption model, also adopting actions to help consumers choose healthy and sustainable diets [10,11].

This objective is consistent with the results of several studies [17–19] that link sustainability with healthy diet through the concept of food system, which is recalled by the "Farm to Fork" European strategy.

Sustainable food systems emphasize the role of dietary styles as core links between foods, human health, and nutrition outcomes [20–22]. A sustainable food system should generate positive outcomes related to the three dimensions of sustainability. In other words, it should be economically profitable, provide equitable benefits for society, and have a positive or neutral environmental impact [21].

Within the sustainable food system approach, the Mediterranean diet (MD) plays a primary role [23–25]. Several findings reveal that the MD pattern demands less soil, water, and energy compared with other consumption patterns, such as the Western dietary patterns and meat-based diet, characterized by high environmental impact [26,27]. Moreover, while the Mediterranean region has been a major food-producing area with a large agro-biodiversity for millennia, environmental alterations may threaten the local food system capacities to ensure food and nutrition security [28]. In fact, the Mediterranean region is facing massive environmental changes: land use and degradation, water scarcity, environment pollution, biodiversity loss, and climate change [28,29].

Therefore, the notion of MD has undergone a progressive evolution over the past decade: from a healthy dietary pattern to a sustainable diet model and to a catalyst for a resilient strategy of the Mediterranean area [30–33].

However, the transition from less sustainable, currently widespread diet in most European countries, also in the Mediterranean area, towards a more sustainable MD requires substantial changes in consumers' values, education, and choices. According to the definition of food system, consumer behavior, together with food supply and food environments, is an important driver that determines the nutrition and connection to health [2,34].

In fact, the perception among consumers of the MD as a healthy diet and the image of olive tree as a symbol of the Mediterranean lifestyle have pushed the demand for typical local foods of the Mediterranean area, EVOO in particular [35]. Olive oil, especially EVOO, has become one of the most important and recognizable symbols of the MD patterns [33], and is conventionally linked to the concept of well-being, not only in Italy, but also worldwide.

Therefore, domestic consumption of olive oil has been continuously growing for a decade and is expected to grow in nonproducing countries as well, while a strong demand in both traditional and new markets will favor an increase of exports from producer countries [36], generating positive economic impacts on the Mediterranean area, and EU countries in particular. Spain, Italy, and Greece alone produce some 70% of the global olive oil supply. In the years 2015–2019, EU production represented 69% of world production, while the provisional figure for 2019/2020 shows a share of 60%, and the forecasts for 2020/2021 indicate an increase of up to 68% of the world production of olive oil. In the same period, the Mediterranean countries of the EU consumed more than half of the world production [36]. The EU olive oil sector is expected to grow in production capacity by 1.1% per year on average, reaching 2.4 million tons in 2030 (compared with 2 million tons in 2019) [36].

Another process of market and product evolution accompanies this trend. The olive oil market has evolved from a traditionally "bulk" market, which conceived olive oil as a mere commodity, similar to other vegetable fats, to a more customized market, in which quality and sustainability claims are multiplying. Therefore, olive oil is increasingly perceived as a food specialty, similar to wine or other high-quality products [37–40].

The analysis of the literature findings highlights heterogeneity in olive oil consumption habits not only between traditional new consumer countries, but also across Mediterranean countries (see in particular [3,37–51]). For instance, Dekhili et al. [40] point out the relevance of oil color in Tunisia and France, while Ribeiro and Santos [41] focus on Portuguese consumers' preference for low-acidity oil.

Indeed, contemporary olive oil, in particular EVOO, consumers are increasingly mindful and aware of attributes, such as sustainability, supply chain ethics, and the intrinsic quality of the product; moreover, they show a growing interest in organic production processes and geographical origin of the product and raw material [52–59]. These two attributes concern the environmental and socioeconomic sustainability of the production systems and include both the quality characteristics of the products themselves and ethical aspects: environmental protection and local development [60–67]. It should be noted that organic production and geographical indication (GI) are adequately certified by community standards, hence easily recognizable by EU consumers.

In fact, several authors highlight the relevance of the organic attributes of EVOO and their importance to consumers, in particular, Tsakiridou et al. [62], Liberatore et al. [68], Roselli et al. [69], and Perito et al. [70]. Most of the themes also investigate organic and origin attributes together.

Moreover, recent studies on food choice and consumption demonstrate that consumers pay attention to the country of origin, suggesting that a certain product image reflects the image of the region or country of production [3,71].

To a broader extent, consumer preference for GI attributes of EVOO is investigated, in particular by Di Vita et al. [46], Roselli et al. [69], Tempesta et al. [71], Erraach et al. [72], Finardi et al. [73], Ballco et al. [74], Fotopoulos et al. [75], Perito et al. [76], and Menapace et al. [77]. These studies point out that higher-income consumer groups were more aware of geographical origin certification labels. According to Erraach et al. [72], for Spanish oils, attributes of origin, region of production, and quality directly affect their market potential, while Perito et al. [76] find that olive oil production region is an important driver of choice for Italian consumers. In addition to this, Italian consumers in particular show very high knowledge of and demand for extrinsic attributes, such as place of production, designation of origin, organic certification, and type of processing for extra virgin olive oils [71], all attributes that bridge production sustainability and perception of high quality.

Given these considerations, this paper investigates the attitudes of a sample of Italian consumers towards organic and origin attributes of EVOO using a choice experiment (CE). The discussion of the results provides interesting insights on consumer preference for EVOO attributes and paves the way to further advances in scientific research on competitiveness and sustainability of the Italian olive oil supply chain.

Our findings point out a preference heterogeneity in the information perceived by olive oil consumers, identifying a number of unobserved sources of heterogeneity in their decision process. The results also reveal a strong and positive preference for locally produced olive oil rather than an organic product. These consumer attitudes towards extra virgin olive oil are not in contrast with each other but fall within the framework of a sustainable development model that takes into account not only the environmental dimension but also the socioeconomic one, linked to the local development of the Mediterranean area. This model should link local development strategies with healthy diet goals, finding its keystone in the Mediterranean diet patterns.

2. Materials and Methods

Several studies, in recent years, have been carried out in the context of new and diversified trends of EVOO consumer demand [44–46,53,55,58], and different surveys have shown that consumers' choices widely differ with respect to sensory preferences, extrinsic quality signals, experience, purchase motives, perception of supply elements, and socioeconomic characteristics [47,59–62]. Moreover, the literature results indicate that organic certification, origin of both olives and olive oil, and price are the main extrinsic attributes of EVOO guiding the choice process of consumers [68–70,76,77].

A broad meta-analysis by Del Giudice et al. [45] on scientific studies on EVOO consumer preferences published between 1994 and 2014 highlights the influence of origin and its various certifications, as well as brand recognition, on consumers' choice. Although the results emerging from this literature survey are somewhat heterogeneous, it is still possible to identify common trends, namely, the importance of the country of origin of olives; a growing interest in organic certification; and the importance of trust in the brand, whether it be a traditional long-established producer or a trustworthy private label.

Among the methods used to estimate consumers' preferences for specific attributes of goods, conjoint hedonic methods, classic hedonic testing, and alternative descriptive approaches are widely used in the recent literature on consumer studies [48,63]. The basic idea behind conjoint analysis (CA) and CE is that public and private goods can be described as a bundle of product attributes; each combination of these characteristics results in a different product, and survey respondents are asked to evaluate these changes [78]. The experimental design of CA and CE allows researchers to estimate the independent effect of each product attribute on product evaluations or product choices by respondents.

In particular, CE is based on Lancaster's [79] characteristics theory of value in combination with the random utility theory [78]. Therefore, statistical analyses of the responses obtained from CE are used to estimate the marginal values of product attributes, which represent the premium price that consumers are willing to pay for the desired characteristics.

In detail, we estimated the WTP for the attribute level by dividing β coefficients by βprice.

$$WTP = -\beta/\beta price \tag{1}$$

With reference to consumer demand, it is necessary to note that this approach means the adoption of the so-called new consumer demand theory [79], and consequently, there exists the operational problem of estimating consumers' WTP for specific product attributes. As is well known, the Lancaster approach is an evolution of the traditional microeconomic theory of demand, in which the utility of goods is derived from their characteristics (and not from the goods per se); therefore, the utility of product alternatives is a latent construct that only exists in the minds of individual consumers. Researchers cannot observe this directly. Nonetheless, indirect measurement techniques can be used to explain a significant part of the latent utility construct. However, the error component determined by additional unobservable attributes, measurement errors, and variation between individual consumers remains unexplained.

First, this study used a multinomial logit model (MNL) in which consumers are assumed to be homogeneous. Moreover, considering that consumers are widely recognized as heterogeneous in their preferences [80], we used a latent class (LC) model that assumes hidden latent classes for consumers and products. Olive oil can be characterized by different attributes, such as price, origin and environmental certification, and private brand.

This approach combines insights from the characteristics theory of value that assumes that individuals do not derive utility from a product per se, but from a product's characteristics [79], as well as from the random utility theory (RUT) [81]. RUT models consumers' preferences among mutually exclusive discrete alternatives by drawing a real-valued score on each of them (typically independently) from a parameterized distribution and ranking these alternatives according to score models.

Consumers typically have only basic knowledge of EVOO, and therefore, information plays an important role. Consequently, the label information and certification logo are important means to convey and ensure the existence of the characteristics desired by consumers. The theoretical basis for this aspect is the economics of information [82,83]. In particular, Akerlof [83] was the first to show that asymmetric information, such as uncertainty about the quality of a good, can cause a market to degenerate into an exclusively low-quality product market.

Therefore, this study applies the CE methodology to the Italian EVOO market to estimate not only the ordinal ranking of preferences of consumers, but also their willingness to pay (WTP) for key product characteristics. To do this, we used data obtained from a field experiment through face-to-face interviews with household consumption decision makers conducted using a dedicated questionnaire.

The structure and contents of the questionnaire were discussed, during its preparation, with some university researchers from different disciplines (marketing, raw materials sciences, and agricultural economics) other than the authors, who teach at the University of Trieste and the University of Udine.

In order to confirm the clarity and understandability of the questionnaire and test the statistical possibilities of the gathered data, a preliminary draft was discussed in a focus group consisting of 10 consumers responsible for their own household food shopping.

Comments and observations gathered during this preliminary study allowed us to update and revise the questionnaire, whose average response time was estimated to be about 15–20 min. Prior to the actual data collection phase, the interviewers were trained in survey administration.

The interviews were conducted outside supermarkets or food shops; therefore, the random sample adopted is not representative of the population and can represent one of the limitations of the present study.

The final version of the questionnaire was organized into three sections. The first one included descriptive information on the respondents' demographic characteristics and professional background. The second section contained questions on the respondents' consumption habits. The last section presented the participants with a discrete choice of olive oil characteristics.

The scales adopted for the second section of the questionnaire included qualitative values from "never" to "always" to identify purchasing habits, consumption frequencies, and preferred purchasing channels. To verify the consumers' knowledge of MD, EVOO, organic farming, and designation of origin, multiple-choice questions were used with only one correct answer.

The first stage of developing a CE involved identifying attributes relevant to our research, and then determining the levels of each of these attributes. In our study, the attributes of interest in the hypothetical olive oil bottles were informed by reviews of literature and interviews with different stakeholders, and discussed within a focus group interview.

Price: price is the traditional economic variable that influences consumer demand in a negative way. The different price levels were chosen based on the actual prices of olive oil as assessed during a store check in January 2018 in food stores in northeastern Italy. Subsequently, three price levels were identified for a 1000 mL EVOO bottle based on a sample of bottles with attributes corresponding to those used for the choice experiment. Therefore, we considered three levels: €4.00, €8.00, €12.00 for a bottle of 1 L EVOO.

Country of origin (COO): among the mandatory EVOO attributes [84], the country-of-origin brand is probably the most recognizable. This attribute points out that the COO is a component of an EVOO brand and adds value to an EVOO purchaser. In traditional producing countries, but also in some other new consuming countries (see the literature cited above), this is a particularly relevant attribute of EVOO. In accordance with EU legislation [84], we considered three levels in hierarchical order of value: (a) 100% Italian olive oils, (b) blend of olive oils of EU origin, and (c) blend of olive oils of EU and not-EU origins.

Geographical indication (GI): The geographical indication of EVOO is another distinctive sign used to identify a product as originating from the territory of a particular country, region, or locality where its quality, reputation, or some other characteristic is linked to its geographical origin. In this context, the term is used to refer to the EU legislation. The EU legislation [85] includes two types of certification: protected designations of origin (PDO) and protected geographical indications (PGIs). These labels certificate that the product is linked to a geographical area, where "protected designation of origin" (PDO) has a stronger quality–geography link and higher qualities than "protected geographical indication" (PGI). Such certifications are important drivers of local development and can therefore be considered attributes of the social and economic dimensions of sustainability. Our CE considered three levels in hierarchical order of value: (a) PDO, (b) PGI, and none (EVOO without GI certification).

Organic: The organic characteristic of food is the main environmental attribute of sustainability. This attribute of EVOO is particularly relevant not only in traditional producing countries but also in some new olive oil-consuming countries (see the literature cited above). According to EU legislation [12], we considered the organic label logo to identify organic certification and to inform respondents about the presence of this attribute. The CE used a dichotomous variable (yes/no) corresponding to the presence (or absence) of organic labelling.

Market leader brand: private labels play a relevant role within the EVOO market, in particular in large-scale retail distribution channels [86]. In particular, in the Italian market there exist numerous famous oil producers with their own brands, even though this is not a straightforward guarantee of the Italian origin of raw materials (see COO attribute), and even the ownership of such companies is often no longer Italian. Some private brand names, leaders in the oil market, are proposed in the questionnaire as examples for the interviewees (Bertolli, Carapelli, Dante, Farchioni, Monini, and Sasso). The CE used a dichotomous variable (yes/no) corresponding to the presence (or absence) of a brand of market leader.

Table 1 provides the description of each attribute and related levels.

Table 1. Attributes and levels for olive oil in Italy.

Attribute	Description	Level
Price	The three levels of the price per bottle (1 L).	€4.00, €8.00, €12.00
Country of origin (COO)	The country where olives were produced. It appears on the label.	100% Italian olive oils, blend of olive oils of EU origin, blend of olive oils of EU origin and not of EU origin
Geographical indication (GI)	Label that indicates whether the product has a GI certification.	PDO, PGI, None
Organic	Organic certification label.	Yes, no
Market leader brand	The presence of a market-leading top brand, if any.	Yes, no

The discrete CE captured responses regarding the choice of olive oil bottles.

An orthogonal fractional factorial design was then generated using SPSS® software, with 18 alternatives (or profiles) selected. The profiles were randomly combined into six sets of choice, all presented to each respondent. The alternative bottles proposed to the respondents had no difference in any aspect (color of the bottle, year of production, acidity, etc.), with the exception of the five specific attributes described above. Each respondent was asked to compare three EVOO bottle options and choose the favorite one. In order to simulate a realistic choice context, the opt-out (no choice) alternative was included in the choice sets to grant the consumers the freedom of choice they have in real market situations, where they can also decide not to purchase any bottle at all. In addition, the interviewees were asked to consider the choice tasks as separate, individual situations and to answer each of them.

Figure 1 graphically represents an example of a choice set.

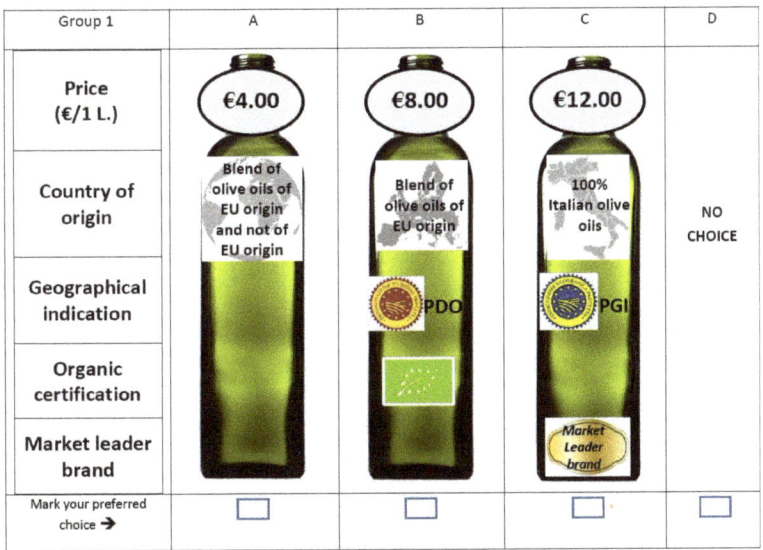

Figure 1. A choice set example for one of the six sets proposed.

A preliminary pilot study using the same characteristics was carried out in 2016 and ended on January 2017. The results of this survey [87] were used to clarify some questions, insert new ones, and remove others, and in general, to refine the overall questionnaire.

Then, a second face-to-face survey was conducted by administering a questionnaire to citizens in the northeastern part of Italy to determine their preferences for olive oil. The survey was carried out in 2018 and ended on January 2020. A total of 1024 consumers were interviewed. Participants were not offered an honorarium in exchange for their response and time.

3. Results

3.1. The General Profile of the Respondents and Their Consumption Behaviors

With respect to their socioeconomic characteristics, 60.61% of the participants were female; 45.51% were employed. All relevant age classes were represented in our sample, with 29% of the respondents being between 40 and 55 years old. The average level of education of the sample was relatively high, as 45% of the respondents held a high school diploma.

Regarding oil consumption habits, the respondents' answers show that over 90% of them mainly consume EVOO.

The analysis of purchasing channels highlights the prevalence of large-scale distribution (51.59%), followed by direct purchase from farms/olive oil producers (28.01%) and consortia or cooperatives of olive oil producers (12.05%). Purchase from retailers or specialized shops represents only 8.35% of the total answers.

Furthermore, with respect to product knowledge, over 90% of the interviewees know the characteristics of the Mediterranean diet and that olive oil is included in this diet pattern.

Approximately 73% of the interviewees stated that they know the characteristics of EVOO, but only just over 50% were able to correctly recognize in a multiple-choice question the characteristics that distinguish this product. In line with the results from previous studies on Italian EVOO consumers [88,89], this element highlights a general recognition of the high quality and wholesomeness of the product, but limited specific knowledge, despite the good level of education of the respondents.

Additionally, just over 52% of the respondents said they read the label on the EVOO bottle. However, this data could be linked to consolidated purchasing habits of the same product, to the greater attention paid to the certification logos possibly present on the bottle, or on the wording of extra virgin olive oil.

Coming to organic and geographical certifications, the results of the survey show that the interviewees generally know the characteristics of organic olive oil (69%) and PDO/PGI olive oil (80%). Moreover, 16% of the respondents buy "often" or "always" organic olive oil, and 63% of them purchase such products at least sometimes. Similarly, 31% of the respondents purchase "often" or "always" PDO/GPI olive oil, while 83% of them buy it at least sometimes.

These answers show a good knowledge and consumer attitudes towards organic and geographical indication certifications of EVOO. These results share a number of similarities with those of Polenzani et al. [90], who associate traditional consumption of EVOO and general knowledge of the product and its main attributes.

Finally, the survey results provide insights into the attributes that the consumers surveyed find most important. They were asked to express the importance of a set of attributes with values between 0 (min) and 10 (max), with the possibility to evaluate different attributes with equal scores. The results are as follows: 27% indicate the Italian origin of olives as the priority attribute in EVOO choice, 14% the regional location of the place of production of olive oil, 11% of consumers the local origin of the product, and 11% the PDO/GPI certification. Only around 7% of the respondents indicate organic certification, and 6% of them the price of the bottle. Our results are consistent with those of previous studies on the EVOO market and consumers' choice drivers [89,91,92]. Overall, almost 80% of the respondents pay attention to the national, regional, or local (including GI) origin of olives and/or olive oil. It is worth noting that the specific location of the oil mill (e.g., in one specific Italian region: Tuscany, Umbria, or Puglia) does not ensure that the raw materials (olives) come from the same territory. This may be a lack of knowledge on the part of consumers or unintentionally unclear information on the characteristics of the production chain.

3.2. Choice Experiment: Statistical Analysis

CE data were analyzed using NLogit4® software. Several empirical models were tested. In addition to the multinomial logit model (MNL) for the main effect and interactions, we further analyzed data using a latent class model (LCM). As only part of the variability in the intensity of the assessment can be associated with measurable socioeconomic characteristics, the LCM was used to reveal the component of heterogeneity associated with unobservable characteristics. This model relaxes the assumption of independence of irrelevant alternatives that result from the MNL. According to Boxall and Adamowicz [93], LCM allows for the random distribution of parameters across the population, capturing preference heterogeneity.

The LCM identifies the utility that a respondent belonging to a particular segment derives from choosing a bottle of olive oil with extrinsic attributes in different contexts. LCM determines the probability of a respondent in a segment to choose a particular alternative, and the choice probability is conditional on class probabilities. As stated by Hu et al. [94], instead of relying solely on standard demographic variables, the LCM uses information derived from respondents' choices to estimate preferences.

Taking into consideration the log-likelihood function (LL), Akaike information criterion (AIC), Bayesian information criterion (BIC), Hannan–Quinn information criterion (HQIC), and pseudo R-squared indicators (Table 2) as suggested in theory, the three-class model was chosen due to its superior performance.

Table 2. Statistical indicators for model comparison.

	LCM-2	LCM-3	LCM-4	LCM-5
LL	−5866.859	−5585.728	−5579.741	−5620.626
AIC	1.915	1.827	1.828	1.844
BIC	1.934	1.855	1.866	1.892
HQIC	1.922	1.837	1.841	1.861
McFadden pseudo R^2	0.311	0.344	0.345	0.340

Table 3 presents the results for the two models (MNL and LCM). Each coefficient (β) indicates the direction and relative importance of an attribute on utility derived by the respondents. In the base model (MNL), not all attributes were statistically significant ($p < 0.05$). Briefly, the price was significant at a 90% confidence level, while the organic attribute was not significant, indicating that this attribute was not important in determining olive oil purchase intentions among the respondents. In addition, the leading brand was significant at 94%.

According to the results of the MNL, Italian-origin olive oil had the highest preference among the respondents choosing a bottle of olive oil (β = 1.82, $p < 0.05$). Olive oil of EU origin (β = 0.83, $p < 0.05$) and PGI certification (β = 0.63, $p < 0.05$) also increased the utility perceived by the respondents, though to a lesser extent. In addition, they preferred a leading brand (β = 0.11, $p < 0.01$). Finally, in contrast to earlier findings [46,72], PDO certification decreased the utility perceived by the respondents (β = −0.28, $p < 0.05$).

Table 3. MNL and LCM results.

	MNL		LCM					
		Class 1		Class 2		Class 3		
Variable	Coeff. (S.E.)	Coeff. (S.E.)	WTP (€/l)	Coeff. (S.E.)	WTP (€/l)	Coeff. (S.E.)	WTP (€/l)
ASC	1.28	4.01	/	−1.52 (0.15) ***	/	5.97 (0.26) ***	/
Price	(0.08) ***	(0.59) ***	/	−0.07 (0.01) ***	/	0.24 (0.02) ***	/

Table 3. *Cont.*

	MNL	LCM					
		Class 1		Class 2		Class 3	
Variable	Coeff. (S.E.)	Coeff. (S.E.)	WTP (€/l)	Coeff. (S.E.)	WTP (€/l)	Coeff. (S.E.)	WTP (€/l)
COO: Italy	−0.01	−0.10	/	0.94 (0.09) ***	13.35	2.70 (0.21) ***	/
COO: EU	(0.01) **	(0.10)		0.83 (0.06) ***	11.80	0.52 (0.23) **	/
Organic	1.82	5.70	/	0.10 (0.11)	/	−0.18 (0.24)	/
Market leader brand	(0.06) ***	(0.69) ***		0.19 (0.09) **	2.67	−0.52 (0.21) **	/
GI: PGI	0.83	1.10	/	0.47 (0.08) ***	6.69	0.85 (0.24) ***	/
GI: PDO	(0.06) ***	(0.57) **		−0.26 (0.03) ***	−3.67	−0.44 (0.08) ***	/
Average probability		1.59		0.46		0.24	

*** Significant at a 95% conf. level; ** significant at a 90% conf. level.

3.3. Choice Experiment: Consumer Class Definition

The LCM showed various sources of preference heterogeneity in the information perceived by the olive oil consumers, as highlighted by the analysis we obtained for each class. The results revealed that 30% of the respondents belonged to class 1, 46% to class 2, and the remaining 24% to class 3.

- Class 1

The coefficients for the respondents belonging to class 1 were significant at a 95% confidence level apart from price, which was not significant, while EU origin and organic attribute were significant at a 90% confidence level. The participants of this group showed a strong preference for Italian olive oil, and they seemed to attribute importance to the presence of a leading brand and to PGI certification. In addition, EU origin and organic attributes were appreciated but to a lesser extent. This group disliked a PDO designation and did not consider price to be an important attribute in purchasing olive oil. Given the statistical insignificance of the price coefficient, the WTP estimation does not make sense.

- Class 2

Class 2 associates its olive oil choices with Italian and EU origins and with a PGI denomination. At a lower level, the presence of a leading brand increased the respondents' utility, while they disliked PDO certification. For this group of respondents, it was possible to look at the WTP, as coefficients were all significant at least at a 90% confidence level with the exception of the organic attribute coefficient. Specifically, the respondents declared that they were willing to pay €13.35 per liter and €11.80 per liter for Italian- and EU-origin olive oils, respectively. The estimated WTP for PGI certification and for a leading brand was €6.69 and €2.67, respectively.

- Class 3

The coefficients for group 3 were all statistically significant at 95% or 90% confidence levels apart from the organic attribute. The members of this class had a clear preference for Italian-origin olive oil and PGI certification. EU origin also increased their utility, leading brands and PDO designations. In addition, contrary to our expectations, the coefficient of the price variable for this segment was positive, implying that ceteris paribus, the higher the price, the higher will be the probability of choosing a given olive oil. Although it is plausible to think that a purchase decision could be influenced by price as a signal of quality at least to a threshold price level [95,96], this finding cannot be justified for a rational economic agent apart from Giffen goods.

These results appear coherent with the preliminary findings provided by the analysis of consumption habits (see paragraph 3.1), which show that the respondents who frequently purchase organic EVOO are about $\frac{3}{4}$ of those consuming PDO/PGI EVOO (63% and 83%). This is the same ratio of belonging to classes 1 and 2 of the LCM analysis. Moreover, these results appear coherent with the preliminary findings of the analysis of consumption habits, showing that the origin of olives and/or olive oil appears a more relevant attribute of EVOO than organic certification, given the attention consumers pay to it. Finally, the analysis reported in paragraph 3.1 also highlights consumers' limited consideration of the price of the EVOO bottle.

4. Discussion

This study aimed to investigate olive oil consumption behaviors in northeastern Italy, in particular with respect to five attributes: the country of origin (Italy, EU, or other countries), the presence (or absence) of PDO and PGI certifications, organic certification, leading brands, and price. We quantified the WTP for these attributes. Specifically, we attempted to measure the influence of various factors, such as organic certification and country of origin, on consumer purchase behavior, and to assess preference heterogeneity due to both observed and unobserved effects, as the unobserved effects could be relevant for olive oil [61].

Our findings point out a preference heterogeneity in the information perceived by olive oil consumers, identifying a number of unobserved sources of heterogeneity in their decision process. The presence of preference heterogeneity among the participants helped us to better explain underlying mechanisms driving individual choice.

This research reveals a strong and positive preference for locally produced olive oil as mainly suggested in the literature [57,64,66,67,72–74,77]; in particular, Finardi et al. [73], Casini et al. [7], and Panico et al. [50] report that Italian origin has a large positive effect on Italian EVOO buyers. Perhaps due to perceived negative or potentially negative effects on health of a number of accidents caused by contaminated food, the respondents related their preference for local products to their greater perceived safety when compared with foreign ones. This confirms the findings by Del Giudice et al. [45] on the strategic role played by knowledge, on the part of the consumer, of the oil's origin. However, this result is not obvious, as, for example, Mtimet et al. [51] demonstrate that in Tunisia, the region of origin attribute had no significant effect on respondents' purchasing decisions. In addition, Mtimet et al. [65]) state that Japanese consumers preferred olive oil of Mediterranean or Tunisian, rather than Italian, origin.

Yangui et al. [97] report that respondents did not grant superior value to EVOO organic attribute, perhaps as a consequence of the belief that olive oil is a healthy and natural product, regardless of its organic status. Similarly, our findings suggest that respondents may benefit from deeper information about organic methods of production. On the contrary, while Erraach et al. [72] demonstrate that price and PDO certification were the attributes that most affected consumer preferences, these appeared to be less relevant in our study. These attributes, which could however be appreciated by specific population segments, are not the only characteristics the respondents looked for.

The results of the LCM segmentation suggest the presence of a consumer segment who is positively impacted by the price coefficient. This is not a novelty: in fact, according to Romo-Muñoz et al. [98], respondents often consider price as a realistic and reliable quality clue.

Our study reveals useful information, which could potentially come in handy for different stakeholders. The results generally confirm expectations built on existing literature and may support the adoption of more efficient and complete marketing strategies by EVOO producers and distributors.

Indeed, a better knowledge of what olive oil consumers need and deem important and valuable is essential to both communicate salient features of existing lines of products and properly direct the selection and development of new lines according to customers' needs. At the same time, stakeholders involved in the EVOO industry can identify prejudices and misconceptions on the products and subsequently intervene and educate consumers. Better-informed customers would take more informed and rational decisions with mutual gains for them, in terms of satisfaction, and the industry as a whole, which would be pushed towards efficiency and qualitative improvement. In general, it is necessary to further reduce the information asymmetries that hinder market efficiency [82,83]. In particular, it appears important to inform consumers more about the characteristics of the products and the meaning of the certifications and to disseminate more nutritional recommendations according to international and national guidelines.

Finally, it should be noted that the sustainability of the olive oil supply chain is a key element in the context of the growing worldwide attention to the healthiness of the Mediterranean diet. Therefore, the olive oil systems can play an important role within the Mediterranean diet as "a driver of sustainable food systems within the strategies of regional development and on that of traditional local products, since quantitative food security must also be complemented by qualitative approaches" [32] (p. 40).

In this respect, the development of a sustainable food system is accompanied by local sustainable development policies that take into account different aspects of sustainability, not least the cultural heritage of rural world and the agricultural landscape [33,99] according to an endogenous development model. [100]. In relation to this last aspect, it should be pointed out that the sustainability of the local food system at the base of the Mediterranean diet must be related to the production area. Otherwise, in a global context characterized by growing international trade, the environmental impact aspects should be assessed by including transport, logistics, and distribution activities according to a "Farm to Fork" approach [11]. This perspective would require a different analysis approach for a different research scenario and highlight potential limitation of this survey focused on domestic consumption.

In accordance with the sustainable food system linked to the MD, organic certification is only one of the attributes that can be exploited together with other environmental and socioeconomic characteristics, for instance, the characteristic of a typically Italian and local product with the certification of origin of the raw material (100% Italian olive oil) and compliance with GI certification (PDO, PGI), which are particularly appreciated and demanded by consumers, according to the results of this study.

Therefore, the sustainability of an olive oil system should be analyzed by taking into account not only one dimension of sustainability, but its overall multidimensional attributes within the space of a local development model, and integrating the endogenous local development model with a healthy and sustainable diet model.

Nevertheless, this survey has a number of limitations, which suggest future research developments. First of all, the sample of respondents was characterized by a geographically limited area (mainly from northeastern Italy) and a sociocultural profile that it is not representative of the entire Italian population. Second, data collection took place in a very traditional way (face-to-face interview); hence, the adoption of other data collection methods can influence the findings.

Therefore, notwithstanding the relevance and usefulness of our findings, the need to refine results calls for further development of research and advance of knowledge on this topic. Further research should also take into account the representativeness of the sample and consider alternative data collection methods. Moreover, even though the attributes and levels used in this study were carefully selected, findings may have differed with the inclusion of other characteristics, such as carbon, water, or ecological footprint certifications, eco-packaging, and vegan certification, whose demand is growing [101]. Finally, being that Italians are traditional EVOO consumers often fond of specific products or labels, the extension of the results to less mature markets may be difficult, if not misleading. Finally, our findings might not be directly extended to foreign EVOO markets: in fact, in spite of the common ground of the IFOAM standards, organic farming regulations vary across nations, together with consumers' familiarity, understanding, and trust; therefore, further replications of our study in other contexts are highly desirable to estimate variations in consumers' preferences for and attitudes towards organic EVOO.

Author Contributions: Conceptualization, M.C. (Matteo Carzedda), G.G., F.M., and P.d.L.; resources, S.T. and F.N.; methodology, M.C. (Matteo Carzedda), G.G., F.M., S.T., and F.N.; funding acquisition, G.G., M.C. (Marta Cosmina), P.d.L., and G.P.; investigation, G.G., P.d.L., and G.P.; data curation, S.T. and M.C. (Marta Cosmina); formal analysis, M.C. (Matteo Carzedda), S.T., and F.N.; writing—original draft preparation, G.G. and S.T.; writing—review and editing, M.C. (Matteo Carzedda); supervision, G.G. and F.M.; project administration, G.G. All authors have read and agreed to the published version of the manuscript.

Funding: This research was conducted within the framework of the research project FRA (Fondo per la Ricerca di Ateneo) 2016—University of Trieste.

Data Availability Statement: The data that support the findings of this study are available from the corresponding authors upon reasonable request.

Acknowledgments: Special thanks go to the members of the focus group and to the university students who collaborated by filling out the questionnaires. The authors would also like to thank the anonymous reviewers for their insightful comments. Any remaining errors are the responsibility of the authors.

Conflicts of Interest: The authors declare no conflict of interest.

References

1. Dernini, S.; Berry, E.M. Mediterranean diet: From a healthy diet to a sustainable dietary pattern. *Front. Nutr.* **2015**, *2*, 15. [CrossRef] [PubMed]
2. Vittersø, G.; Tangeland, T. The role of consumers in transitions towards sustainable food consumption. The case of organic food in Norway. *J. Clean. Prod.* **2015**, *92*, 91–99. [CrossRef]
3. Ilak Peršurić, A.S. Segmenting Olive Oil Consumers Based on Consumption and Preferences toward Extrinsic, Intrinsic and Sensorial Attributes of Olive Oil. *Sustainability* **2020**, *12*, 6379. [CrossRef]
4. Inglese, P.; Famiani, F.; Galvano, F.; Servili, M.; Esposto, S.; Urbani, S. Factors affecting extra-virgin olive oil composition. *Hortic. Rev.* **2011**, *38*, 83–147. [CrossRef]
5. Fregapane, G.; Salvador, M.D. Production of superior quality extra virgin olive oil modulating the content and profile of its minor components. *Food Res. Int.* **2013**, *54*, 1907–1914. [CrossRef]
6. Serreli, G.; Deiana, M. Biological relevance of extra virgin olive oil polyphenols metabolites. *Antiox* **2018**, *7*, 170. [CrossRef]
7. Casini, L.; Contini, C.; Romano, C.; Scozzafava, G. New trends in food choice: What impact on sustainability of rural areas? *Agric. Agric. Sci. Proc.* **2016**, *8*, 141–147. [CrossRef]
8. World Commission on Environment and Development. *Our Common Future*; Oxford University Press: Oxford, UK, 1987.
9. United Nations. Transforming Our World: The 2030 Agenda for Sustainable Development. Resolution Adopted by the General Assembly on 25 September 2015. Available online: https://www.un.org/sustainabledevelopment/sustainable-development-goals/ (accessed on 30 March 2021).
10. European Commission. The European Green Deal COM/2019/640. 2019. Available online: https://eur-lex.europa.eu/resource.html?uri=cellar:b828d165-1c22-11ea-8c1f-01aa75ed71a1.0002.02/DOC_1&format=PDF (accessed on 4 March 2021).
11. European Commission. Farm to Fork Strategy. For a Fair, Healthy and Environmentally-Friendly Food System. 2020. Available online: https://ec.europa.eu/food/sites/food/files/safety/docs/f2f_action-plan_2020_strategy-info_en.pdf (accessed on 4 March 2021).
12. Council Regulation (EC). No 834/2007 of 28 June 2007 on Organic Production and Labelling of Organic Products and Repealing Regulation (EEC) No 2092/91. Available online: https://eur-lex.europa.eu/legal-content/EN/TXT/PDF/?uri=CELEX:32007R0834&from=EN (accessed on 4 March 2021).
13. Ward, A.; Mishra, A. Addressing Sustainability Issues with Voluntary Standards and Codes: A Closer Look at Cotton Production in India. In *Business Responsibility and Sustainability in India*; Palgrave Macmillan: Cham, Switzerland, 2019; pp. 161–193.
14. Dreval, Y.; Loboichenko, V.; Malko, A.; Morozov, A.; Zaika, S.; Kis, V. The Problem of Comprehensive Analysis of Organic Agriculture as a Factor of Environmental Safety. *Environ. Clim. Technol.* **2020**, *24*, 58–71. [CrossRef]
15. Anastasiou, C.N.; Keramitsoglou, K.M.; Kalogeras, N.; Tsagkaraki, M.I.; Kalatzi, I.; Tsagarakis, K.P. Can the "Euro-Leaf" Logo Affect Consumers' Willingness-To-Buy and Willingness-To-Pay for Organic Food and Attract Consumers' Preferences? An Empirical Study in Greece. *Sustainability* **2017**, *9*, 1450. [CrossRef]

16. Willer, H.; Schaack, D.; Lernoud, J. Organic farming and market development in Europe and the European Union. In *The World of Organic Agriculture. Statistics and Emerging Trends 2019*; Research Institute of Organic Agriculture FiBL and IFOAM-Organics International: Bonn, Germany, 2019; pp. 217–254.
17. Meybeck, A.; Gitz, V. Sustainable diets within sustainable food systems. *Proc. Nutr. Soc.* **2017**, *76*, 1–11. [CrossRef]
18. Clark, M.A.; Springmann, M.; Hill, J.; Tilman, D. Multiple health and environmental impacts of foods. *Proc. Nat. Acad. Sci. USA* **2019**, *116*, 23357–23362. [CrossRef] [PubMed]
19. EAT-Lancet Commission. Food in the Anthropocene: The EAT-Lancet Commission on healthy diets from sustainable food systems. *Lancet* **2019**, *393*, 447–492. [CrossRef]
20. Dupouy, E.; Gurinovic, M. Sustainable food systems for healthy diets in Europe and Central Asia: Introduction to the special issue. *Food Policy* **2020**, *96*. [CrossRef] [PubMed]
21. FAO. *Sustainable Food Systems: Concept and Framework*; FAO: Roma, Italy, 2018. Available online: http://www.fao.org/3/ca2079en/CA2079EN.pdf (accessed on 14 March 2021).
22. FAO. *Regional Overview of Food Security and Nutrition in Europe and Central Asia 2019: Structural Transformations of Agriculture for Improved Food Security, Nutrition and Environment*; FAO: Budapest, Hungary, 2019. Available online: http://www.fao.org/3/ca7153en/ca7153en.pdf (accessed on 14 March 2021).
23. Amiot-Carlin, M.J.; Perignon, M.; Darmon, N.; Drogue, S.; Sinfort, C.; Verger, E.; El Ati, J.; The Medina-Study Group. Promoting Sustainable Food Systems in Mediterranean Countries: A Framework to Implement Recommendations and Actions. In *Development of Voluntary Guidelines for the Sustainability of the Mediterranean Diet in the Mediterranean Region*; FAO; International Centre for Advanced Mediterranean Agronomic Studies (CIHEAM): Bari, Italy, 2017; pp. 1–141. Available online: https://hal.archives-ouvertes.fr/hal-01595254/document (accessed on 14 March 2021).
24. Berry, E.M. Sustainable Food Systems and the Mediterranean Diet. *Nutrients* **2019**, *11*, 2229. [CrossRef]
25. Serra-Majem, L.; Tomaino, L.; Dernini, S.; Berry, E.M.; Lairon, D.; Ngo de la Cruz, J.; Bach-Faig, A.; Donini, L.M.; Medina, F.-X.; Belahsen, R.; et al. Updating the Mediterranean Diet Pyramid towards Sustainability: Focus on Environmental Concerns. *Int. J. Environ. Res. Public Health* **2020**, *17*, 8758. [CrossRef]
26. Pairotti, M.B.; Cerutti, A.K.; Martini, F.; Vesce, E.; Padovan, D.; Beltramo, R. Energy consumption and GHG emission of the Mediterranean diet: A systemic assessment using a hybrid LCA-IO method. *J. Clean. Prod.* **2015**, *103*, 507–516. [CrossRef]
27. Sáez-Almendros, S.; Obrador, B.; Bach-Faig, A.; Serra-Majem, L. Environmental Footprints of Mediterranean versus Western Dietary Patterns: Beyond the Health Benefits of the Mediterranean Diet. *Environ. Health* **2013**, *12*, 1–8. [CrossRef]
28. Verger, E.O.; Perignon, M.; El Ati, J.; Darmon, N.; Dop, M.-C.; Drogué, S.; Dury, S.; Gaillard, C.; Sinfort, C.; Amiot, M.-J. A "Fork-to-Farm" Multi-Scale Approach to Promote Sustainable Food Systems for Nutrition and Health: A Perspective for the Mediterranean Region. *Front. Nutr.* **2018**, 5–30. [CrossRef]
29. García-Martín, M.; Torralba, M.; Quintas-Soriano, C.; Kahl, J.; Plieninger, T. Linking food systems and landscape sustainability in the Mediterranean region. *Landsc. Ecol.* **2020**. [CrossRef]
30. Dernini, S.; Meybeck, A.; Burlingame, B.; Gitz, V.; Lacirignola, C.; Debs, P.; El Bilali, H. Developing a Methodological Approach for Assessing the Sustainability of Diets: The Mediterranean Diet as a Case Study. *New Medit.* **2013**, *12*, 28–36. Available online: http://www.iamb.it/share/img_new_medit_articoli/949_28dernini.pdf (accessed on 15 March 2021).
31. FAO. Sustainable Diets and Biodiversity: Directions and Solutions for Policy, Research and Action. In Proceedings of the International Scientific Symposium, Biodiversity and Sustainable Diets United Against Hunger, Rome, Italy, 3–5 November 2010; FAO Headquarters: Rome, Italy, 2012. Available online: http://www.fao.org/docrep/016/i3004e/i3004e.pdf (accessed on 15 March 2021).
32. Lacirignola, C.; Dernini, S.; Capone, R.; Meybeck, A.; Burlingame, B.; Gitz, V.; El Bilali, H.; Debs, P.; Belsanti, V. (Eds.) Vers L'élaboration de Recommandations Pour Améliorer la Durabilité des Régimes et Modes de Consommation Alimentaires: La Diète Méditerranéenne Comme Étude Pilote. In *Options Méditerranéennes*; CIHEAM and FAO: Bari, Italy, 2012; p. 70. Available online: www.iamm.ciheam.org/ress_doc/opac_css/doc_num.php?explnum_id=9369 (accessed on 16 March 2021).
33. Hachem, F.; Capone, R.; Yannakoulia, M.; Dernini, S.; Hwalla, N.; Kalaitzidis, C. The Mediterranean diet: A sustainable food consumption pattern. In *Mediterra 2016. Zero Waste in the Mediterranean*; Presses de Sciences Po: Paris, France, 2016. Available online: https://www.ciheam.org/uploads/attachments/333/Mediterra2016_EN_BAT__1_.pdf (accessed on 16 March 2021).
34. Hertwich, E.G. The life cycle environmental impacts of consumption. *Econ. Syst. Res.* **2011**, *23*, 27–47. [CrossRef]
35. Xiong, B.; Sumner, D.; Matthews, W. A new market for an old food: The U.S. demand for olive oil. *Agric. Econ.* **2014**, *45*, 107–118. [CrossRef]
36. European Commission. *EU Agricultural Outlook for Markets and Income 2019–2030*; Publications Office of the European Union: Luxembourg, 2019. Available online: https://ec.europa.eu/info/sites/info/files/food-farming-fisheries/farming/documents/agricultural-outlook-2019-report_en.pdf (accessed on 16 March 2021).
37. Caporale, G.; Policastro, S.; Carlucci, A.; Monteleone, E. Consumer expectations for sensory properties in virgin olive oils. *Food Qual. Prefer.* **2006**, *17*, 116–125. [CrossRef]
38. Cacchiarelli, L.; Carbone, A.; Laureti, T.; Sorrentino, A. The Value of different Quality Clues in the Italian Olive Oil Market. *It. Rev. Agric. Econ.* **2016**, *71*, 372–379. [CrossRef]
39. Dekhili, S.; d'Hauteville, F. Effect of the region of origin on the perceived quality of olive oil: An experimental approach using a control group. *Food Qual. Prefer.* **2009**, *20*, 525–532. [CrossRef]

40. Dekhili, S.; Sirieix, L.; Cohen, E. How consumers choose olive oil: The importance of origin cues. *Food Qual. Prefer.* **2011**, *22*, 757–762. [CrossRef]
41. Ribeiro, J.C.; Santos, J.F. Portuguese olive oil and the price of regional products: Does designation of origin really matter? *Tékhne Polytech Stud. Rev.* **2005**, *2*, 61–76.
42. Piccolo, D.; Capecchi, S.; Iannario, M.; Corduas, M. Modelling consumer preferences for extra virgin olive oil: The Italian case. *Politica Agric. Internazionale Int. Agric. Policy* **2013**, *1*. [CrossRef]
43. Sandalidou, E.; Baourakis, G.; Siskos, Y. Customers' perspectives on the quality of organic olive oil in Greece: A satisfaction evaluation approach. *Br. Food J.* **2002**, *104*, 391–406. [CrossRef]
44. Cavallo, C.; Caracciolo, F.; Cicia, G.; Del Giudice, T. Extra-virgin olive oil: Are consumers provided with the sensory quality they want? A hedonic price model with sensory attributes. *J. Sci. Food Agric.* **2018**, *98*, 1591–1598. [CrossRef] [PubMed]
45. Del Giudice, T.; Cavallo, C.; Caracciolo, F.; Cicia, G. What attributes of extra virgin olive oil are really important for consumers: A meta-analysis of consumers' stated preferences. *Agric. Food Econ.* **2015**, *3*, 1–15. [CrossRef]
46. Di Vita, G.; D'Amico, M.; La Via, G.; Caniglia, E. Quality perception of PDO extra-virgin olive oil: Which attributes most influence Italian consumer. *Agric. Econ. Rev.* **2013**, *14*, 46–58. [CrossRef]
47. Barbieri, S.; Bendini, A.; Valli, E.; Toschi, T.G. Do consumers recognize the positive sensorial attributes of extra virgin olive oils related with their composition? A case study on conventional and organic products. *J. Food Comp. Anal.* **2015**, *44*, 186–195. [CrossRef]
48. Asioli, D.; Varela, P.; Hersleth, M.; Almli, V.L.; Olsen, N.V.; Naes, T. A discussion of recent methodologies for combining sensory and extrinsic product properties in consumer studies. *Food Qual. Prefer.* **2017**, *56*, 266–273. [CrossRef]
49. Casini, L.; Contini, C.; Marinelli, N.; Romano, C.; Scozzafava, G. Nutraceutical olive oil: Does it make the difference? *Nutr. Food Sci.* **2014**, *44*, 586–600. [CrossRef]
50. Panico, T.; Del Giudice, T.; Caracciolo, F. Quality dimensions and consumer preferences: A choice experiment in the Italian extra-virgin olive oil market. *Agric. Econ. Rev.* **2014**, *15*, 100–112. [CrossRef]
51. Mtimet, N.; Zaibet, L.; Zairi, C.; Hzami, H. Marketing olive oil products in the Tunisian local market: The importance of quality attributes and consumers' behavior. *J. Int. Food Agribs. Mark.* **2013**, *25*, 134–145. [CrossRef]
52. Gavruchenko, T.; Baltas, G.; Chatzitheodoridis, F.; Hadjidakis, S. Comparative marketing strategies for organic olive oil: The case of Greece and Holland. *Cahiers Opt. Mediter.* **2003**, *61*, 247–255.
53. Delgado, C.; Guinard, J.-X. How do consumer hedonic rating for extra virgin olive oil relate to quality ratings by experts and descriptive analysis ratings? *Food Qual. Prefer.* **2011**, *22*, 213–225. [CrossRef]
54. García, M.; Aragonés, Z.; Poole, N. A repositioning strategy for olive oil in the UK market. *Agribusiness* **2002**, *18*, 163–180. [CrossRef]
55. Krystallis, A.; Ness, M. Consumer preferences for quality foods from a South European perspective: A conjoint analysis implementation on Greek olive oil. *Int. Food Agric. Man. Rev.* **2005**, *8*, 62–91. [CrossRef]
56. Muñoz, R.R.; Moya, M.L.; Gil, J.M. Market values for olive oil attributes in Chile: A hedonic price function. *Br. Food J.* **2015**, *117*, 358–370. [CrossRef]
57. Roselli, L.; Carlucci, D.; De Gennaro, B.C. What Is the Value of Extrinsic Olive Oil Cues in Emerging Markets? Empirical Evidence from the US E-Commerce Retail Market. *Agribusiness* **2016**, *32*, 329–342. [CrossRef]
58. Cicia, G.; Cembalo, L.; Del Giudice, T.; Verneau, F. Il sistema agroalimentare ed il consumatore postmoderno: Nuove sfide per la ricerca e per il mercato. *Econ. Agro-Alim. Food Econ.* **2012**, *1*, 117–142. [CrossRef]
59. Gázquez-Abad, J.C. and Sánchez-Pérez, M. Factors influencing olive oil brand choice in Spain: An empirical analysis using scanner data. *Agribusiness* **2009**, *25*, 36–55. [CrossRef]
60. Nielsen, N.A.; Bech-Larsen, T.; Grunert, K.G. Consumer purchase motives and product perceptions: A laddering study on vegetable oil in three countries. *Food Qual. Prefer.* **1998**, *9*, 455–466. [CrossRef]
61. Scarpa, R.; Del Giudice, T. Market segmentation via mixed logit: Extra-virgin olive oil in urban Italy. *J. Agric. Food Ind. Organ.* **2004**, *2*, 1–18. [CrossRef]
62. Tsakiridou, E.; Mattas, K.; Tzimitra-Kalogianni, I. The influence of consumer characteristics and attitudes on the demand for organic olive oil. *J. Int. Food Agribs. Mark.* **2006**, *18*, 3–4. [CrossRef]
63. Jiménez-Guerrero, J.F.; Gázquez-Abad, J.C.; Mondéjar-Jiménez, J.A.; Huertas-García, R. Consumer Preferences for Olive-Oil. In *Constituents, Quality, Health Properties and Bioconversions*; Boskou, D., Ed.; InTech: London, UK, 2012; pp. 233–246. [CrossRef]
64. Chan-Halbrendt, C.; Zhllima, E.; Sisior, G.; Imamid, D.; Leonetti, L. Consumer preferences for olive oil in Tirana, Albania. *Int. Food Agric. Man. Rev.* **2010**, *13*, 55–74. [CrossRef]
65. Mtimet, N.; Kashiwagi, A.K.; Zaibet, L.; Masakazu, N. Exploring Japanese olive oil consumer behavior. In Proceedings of the 12th EAAE Congress People, Food and Environments: Global Trends and European Strategies, Gent, Belgium, 26–29 August 2008; pp. 1–10. Available online: https://www.academia.edu (accessed on 16 February 2021).
66. Caputo, V.; Nayga, M.R., Jr.; Sacchi, G.; Scarpa, R. Attribute non-attendance or attribute-level non-attendance? A choice experiment application on extra virgin olive oil. In Proceedings of the Agricultural and Applied Economics Association Annual Meeting, Boston, MA, USA, 31 July–2 August 2016; pp. 1–26. Available online: http://ageconsearch.umn.edu/ (accessed on 16 January 2021).

67. Bernabéu, R.; Díaz, M. Preference for olive oil consumption in the Spanish local market. *Span. J. Agric Res.* **2016**, *14*, 1–11. [CrossRef]
68. Liberatore, L.; Casolani, N.; Murmura, F. What's behind organic certification of extra-virgin olive oil? A response from Italian consumers. *J. Food Prod. Market.* **2018**, *24*, 946–959. [CrossRef]
69. Roselli, L.; Giannoccaro, G.; Carlucci, D.; De Gennaro, B. EU quality labels in the Italian olive oil market: How much overlap is there between geographical indication and organic production? *J. Food Prod. Market.* **2018**, *24*, 784–801. [CrossRef]
70. Perito, M.A.; Coderoni, S.; Russo, C. Consumer Attitudes towards Local and Organic Food with Upcycled Ingredients: An Italian Case Study for Olive Leaves. *Foods* **2020**, *9*, 1325. [CrossRef] [PubMed]
71. Tempesta, T.; Vecchiato, D. Analysis of the Factors that Influence Olive Oil Demand in the Veneto Region (Italy). *Agriculture* **2019**, *9*, 154. [CrossRef]
72. Erraach, Y.; Sayadi, S.; Gomez, A.C.; Parra-Lopez, C. Consumer stated-preferences towards Protected Designation of Origin (PDO) labels in a traditional olive-oil-producing country: The case of Spain. *New Medit.* **2014**, *13*, 11–19.
73. Finardi, C.; Giacomini, C.; Menozzi, D.; Mora, C. Consumer preferences for country-of-origin and health claim labelling of extra-virgin olive-oil. In Proceedings of the113th EAAE Seminar: A Resilient European Food Industry and Food Chain in a Challenging World, Chania, Crete, Greece, 3–6 September 2009.
74. Ballco, P.; Gracia, A.; Jurado, J. Consumer preferences for extra virgin olive oil with Protected Designation of Origin (PDO). In Proceedings of the X Congreso AEEA Alimentación y Territorios Sostenibles Desde el sur de Europa, Córdoba, Spain, 9–11 September 2015; pp. 607–612.
75. Fotopoulos, C.; Krystallis, A. Are quality labels a real marketing advantage? A conjoint application on Greek PDO protected olive oil. *J. Int. Food Agric. Mark.* **2001**, *12*, 1–22. [CrossRef]
76. Perito, M.A.; Sacchetti, G.; Di Mattia, C.D.; Chiodo, E.; Pittia, P.; Saguy, I.S.; Cohen, E. Buy local! Familiarity and preferences for extra virgin olive oil of Italian consumers. *J. Food Prod. Market.* **2019**, *25*, 462–477. [CrossRef]
77. Menapace, L.; Colson, G.; Grebitus, C.; Facendola, M. Consumers' preferences for geo-graphical origin labels: Evidence from the Canadian olive oil market. *Eur. Rev. Agric.* **2011**, *38*, 193–212. [CrossRef]
78. Hanley, N.; Wright, R.E.; Adamowicz, V. Using Choice Experiments to Value the Environment. *Environ. Res. Econ.* **1998**, *11*, 413–428. [CrossRef]
79. Lancaster, K.J. A New Approach to Consumer Theory. *J. Pol. Econ.* **1966**, *74*, 132–157. [CrossRef]
80. Wedel, M.; Kamakura, W.A. *Market Segmentation: Concepts and Methodological Foundations*; Kluwer Academic Publishers: Boston, MA, USA, 2000.
81. McFadden, D. Conditional Logit Analysis of Qualitative Choice Behaviour. In *Frontiers in Econometrics*; Zarembka, P., Ed.; Academic Press: New York, NY, USA, 1974; pp. 105–142.
82. Stigler, G.J. The economics of information. *J. Pol. Econ.* **1961**, *69*, 213–225. [CrossRef]
83. Akerlof, G.A. The market for lemons: Quality uncertainty and the market mechanism. *Q. J. Econ.* **1970**, *84*, 488–500. [CrossRef]
84. Commission Implementing Regulation (EU) No 29/2012 of 13 January 2012 on Marketing Standards for Olive Oil (Codification). Available online: https://eur-lex.europa.eu/legal-content/EN/TXT/PDF/?uri=CELEX:02012R0029-20190206&from=EN (accessed on 11 December 2020).
85. Regulation (EU) No 1151/2012 of the European Parliament and of the Council of 21 November 2012 on Quality Schemes for Agricultural Products and Food Stuffs. Available online: https://eur-lex.europa.eu/legal-content/EN/TXT/PDF/?uri=CELEX:32012R1151&from=EN (accessed on 11 December 2020).
86. Stasi, A.; Diotallevi, F.; Marchini, A.; Nardone, G. Italian Extra-Virgin Olive Oil: Impact on Demand on Being Market Leaders, Private Labels or Small Producers. *Rev. Econ. Fin.* **2018**, *13*, 39–54. Available online: https://econpapers.repec.org/scripts/redir.pf?u=http%3A%2F%2Fwww.bapress.ca%2Fref%2Fref-article%2F1923-7529-2018-03-39-16.pdf;h=repec:bap:journl:180304 (accessed on 28 December 2020).
87. de Luca, P.; Pegan, G.; Troiano, S.; Gallenti, G.; Marangon, F.; Cosmina, M. Brand e Country of Origin: Una ricerca sulle preferenze del consumatore di olio extra-vergine d'oliva. In Proceedings of the XIII Convegno Annuale della Società Italiana di Marketing SIMktg, Cassino, Italy, 20–21 October 2016.
88. Cavallo, C.; Piqueras-Fiszman, B. Visual elements of packaging shaping healthiness evaluations of consumers: The case of olive oil. *J. Sens. Stud.* **2017**, *32*, e12246. [CrossRef]
89. Spognardi, S.; Vistocco, D.; Cappelli, L.; Papetti, P. Impact of organic and "protected designation of origin" labels in the perception of olive oil sensory quality. *Br. Food J.* **2021**. ahead-of-print. [CrossRef]
90. Polenzani, B.; Riganelli, C.; Marchini, A. Sustainability Perception of Local Extra Virgin Olive Oil *and* Consumers' Attitude: A New Italian Perspective. *Sustainability* **2020**, *12*, 920. [CrossRef]
91. Ballco, P.; Gracia, A. Do market prices correspond with consumer demands? Combining market valuation and consumer utility for extra virgin olive oil quality attributes in a traditional producing country. *J. Retail. Consum. Serv.* **2020**, *53*, 101999. [CrossRef]
92. Lombardo, L.; Farolfi, C.; Capri, E. Sustainability Certification, a New Path of Value Creation in the Olive Oil Sector: The ITALIAN Case Study. *Foods* **2021**, *10*, 501. [CrossRef]
93. Boxall, P.C.; Adamowicz, W.L. Understanding heterogeneous preferences in random utility models: A latent class approach. *Environ. Res. Econ.* **2002**, *23*, 421–446. [CrossRef]

94. Hu, W.; Hünnemeyer, A.; Veeman, M.; Adamowicz, W.; Srivastava, L. Trading off health, environmental and genetic modification attributes in food. *Eur. Rev. Agric. Econ.* **2004**, *31*, 389–408. [CrossRef]
95. Troiano, S.; Marangon, F.; Tempesta, T.; Vecchiato, D. Organic vs local claims: Substitutes or complements for wine consumers? A marketing analysis with a discrete choice experiment. *New Medit.* **2016**, *15*, 14–22.
96. Cosmina, M.; Gallenti, G.; Marangon, F.; Troiano, S. Attitudes towards honey among Italian consumers: A choice experiment approach. *Appetite* **2016**, *99*, 52–58. [CrossRef] [PubMed]
97. Yangui, A.; Costa-Font, M.; Gil, J.M. Revealing additional preference heterogeneity with an extended random parameter logit model: The case of extra virgin olive oil. *Span. J. Agric. Res.* **2014**, *12*, 553–567. [CrossRef]
98. Romo-Muñoz, R.A.; Cabas-Monje, J.H.; Garrido-Henrríquez, H.M.; Gil, J.M. Heterogeneity and nonlinearity in consumers' preferences: An application to the olive oil shopping behavior in Chile. *PLoS ONE* **2017**, *12*, e0184585. [CrossRef]
99. Rodríguez-Entrena, M.; Colombo, S.; Arriaza, M. The landscape of olive groves as a driver of the rural economy. *Land Use Policy* **2017**, *65*, 164–175. [CrossRef]
100. van der Ploeg, J.D.; Renting, H.; Brunori, G.; Knickel, K.; Mannion, J.; Marsden, T.; de Roset, K.; Sevilla-Guzmán, E.; Ventura, F. Rural development: From practices and policies towards theory. In *The Rural: Critical Essays in Human Geography*, 1st ed.; Munton, R., Ed.; Routledge: London, UK, 2016. [CrossRef]
101. Pattara, C.; Russo, C.; Antrodicchia, V.; Cichelli, A. Carbon footprint as an instrument for enhancing food quality: Overview of the wine, olive oil and cereals sectors. *J. Sci. Food Agric.* **2017**, *97*, 396–410. [CrossRef]

Article

Antioxidant Properties of Olive Mill Wastewater Polyphenolic Extracts on Human Endothelial and Vascular Smooth Muscle Cells

Anna Maria Posadino [1], Annalisa Cossu [1], Roberta Giordo [2], Amalia Piscopo [3], Wael M. Abdel-Rahman [2], Antonio Piga [4,*] and Gianfranco Pintus [1,2,*]

[1] Department of Biomedical Sciences, University of Sassari, 07100 Sassari, Italy; posadino@uniss.it (A.M.P.); cossuannalisa@libero.it (A.C.)
[2] Department of Medical Laboratory Sciences, Institute for Medical Research, College of Health Sciences and Sharjah, University of Sharjah, Sharjah P.O. Box 27272, United Arab Emirates; robertagiordo2000@yahoo.it (R.G.); whassan@sharjah.ac.ae (W.M.A.-R.)
[3] Department of AGRARIA, Mediterranean University of Reggio Calabria, 89124 Vito Reggio Calabria, Italy; amalia.piscopo@unirc.it
[4] Department of Agricultural Environmental Sciences and Food Biotechnology, University of Sassari, Viale Italia 39, 07100 Sassari, Italy
* Correspondence: pigaa@uniss.it (A.P.); gpintus@sharjah.ac.ae (G.P.)

Citation: Posadino, A.M.; Cossu, A.; Giordo, R.; Piscopo, A.; Abdel-Rahman, W.M.; Piga, A.; Pintus, G. Antioxidant Properties of Olive Mill Wastewater Polyphenolic Extracts on Human Endothelial and Vascular Smooth Muscle Cells. *Foods* **2021**, *10*, 800. https://doi.org/10.3390/foods10040800

Academic Editors: Cristina Alamprese, Emma Chiavaro and Francesco Caponio

Received: 9 March 2021
Accepted: 6 April 2021
Published: 8 April 2021

Publisher's Note: MDPI stays neutral with regard to jurisdictional claims in published maps and institutional affiliations.

Copyright: © 2021 by the authors. Licensee MDPI, Basel, Switzerland. This article is an open access article distributed under the terms and conditions of the Creative Commons Attribution (CC BY) license (https://creativecommons.org/licenses/by/4.0/).

Abstract: This work aims to analyze the chemical and biological evaluation of two extracts obtained by olive mill wastewater (OMW), an olive oil processing byproduct. The exploitation of OMW is becoming an important aspect of development of the sustainable olive oil industry. Here we chemically and biologically evaluated one liquid (L) and one solid (S) extract obtained by liquid–liquid extraction followed by acidic hydrolysis (LLAC). Chemical characterization of the two extracts indicated that S has higher phenol content than L. Hydroxytyrosol and tyrosol were the more abundant phenols in both OMW extracts, with hydroxytyrosol significantly higher in S as compared to L. Both extracts failed to induce cell death when challenged with endothelial cells and vascular smooth muscle cells in cell viability experiments. On the contrary, the higher extract dosages employed significantly affected cell metabolic activity, as indicated by the MTT tests. Their ability to counteract H_2O_2-induced oxidative stress and cell death was assessed to investigate potential antioxidant activities of the extracts. Fluorescence measurements obtained with the reactive oxygen species (ROS) probe H_2DCF-DA indicated strong antioxidant activity of the two OMW extracts in both cell models, as indicated by the inhibition of H_2O_2-induced ROS generation and the counteraction of the oxidative-induced cell death. Our results indicate LLAC-obtained OMW extracts as a safe and useful source of valuable compounds harboring antioxidant activity.

Keywords: antioxidant activity; olive mill wastewaters; phenolic extract; reactive oxygen species; vascular cells

1. Introduction

The olive oil industry is a large productive sector globally, and three-quarters of world production takes place in Europe. Among different commercial categories, extra virgin olive oil (EVOO) is produced by the sole employment of mechanical processes, and its quality is tightly related to different parameters, such as agronomic practices, the olive cultivar, and the olive oil extraction technology used [1]. The final chemical composition of the EVOO is influenced by different factors, among which the olive variety is the first one to play a primary role across the supply chain. Indeed, the olive cultivars display a large range of genetic variability for several agronomic traits such as the fruit size, oil content, and the degree of adaptation to severe environmental stress [2,3]. Olive drupes possess various health-promoting bioactive substances primarily represented by biophenol secoiridoids (oleuropein, ligstroside) and their hydrolytic derivatives [4]. Although olive fruit is rich in phenolic compounds, only 2% of the entire phenolic content transfers into the oil phase, while the rest goes into the olive mill wastewater (OMW) (approximately 53%), the pomace, the olive oil filtration residue, and olive leaves (approximately 45%) [5]. As a result, the high percentage of phenolic compounds in the olive industry's byproducts is attracting great interest as a potential source of such phenolic compounds, with special attention oriented to the olive mill, which is the primary potential source of such molecules. The major components in OMW include hydroxytyrosol, tyrosol, oleuropein, ligstroside and their secoiridoids derivatives, and a variety of hydroxycinnamic and hydroxybenzoic acids [6]. Given the potential environmental impact, active molecule extraction from olive oil byproducts should embrace methodologies that employ green technologies, considering their possible exploitation as food antioxidants or nutraceuticals [7]. In fact, phenol recovery from these byproducts should be organized in order to promote their reintroduction into the food chain and coincide with greater valorization and improved olive oil industry waste management. In addition to its main implications for the local and international economy, this could reduce the environmental impact of olive oil manufacturing and contribute to this valuable production chain's sustainability. However, the quality of OMW phenolic compounds differs according to several factors, including the olive oil production technological process employed, and for this reason, it is essential to evaluate different OMW technological processes to provide promising bioactive compounds. In this regard, OMW-derived products have been tested for certain biological effects and have showed an interesting bioactivity spectrum [8]. For instance, EVOO-containing phenolic compounds have shown In Vivo and In Vitro antioxidant activity, likely due to molecules such as hydroxytyrosol, tyrosol, and secoiridoid derivatives [9,10].

Reactive oxygen species (ROS) are aerobic metabolism products and exert a pivotal role in regulating cellular functions such as proliferation, differentiation, and migration [11]. Under physiological conditions, a series of cellular antioxidant mechanisms maintain vascular ROS levels in homeostatic conditions [12]. However, an aberrant modulation of the above-mentioned mechanisms leads to critical increases in ROS levels, thus promoting different vascular-associated pathological conditions, including cancer and cardiovascular diseases [13]. In this context, endogenous ROS, released by endothelial and vascular smooth muscle cells as a result of pro-inflammatory and pro-atherosclerotic stimuli, can trigger vascular injury and blood vessel restructuring by affecting diverse intracellular signaling pathways [14]. Indeed, ROS-activated molecular machinery can modulate both endothelial cells and vascular smooth muscle cells functions including proliferation, migration, and invasion, leading to vascular pathologies such as hypertension, atherosclerosis, and cancer [15], which may be counteracted by OMW-contained compounds [16].

In this light, the present work aims to investigate whether (i) the considered OMW samples can be a source of valuable antioxidants, and (ii) whether the obtained extracts can protect human vascular cells against oxidative cell death.

2. Materials and Methods

2.1. Chemicals

Unless stated in the text, all the reagents used were from Sigma (Sigma, St. Louis, MO, USA).

2.2. Sample Collection

Ottobratica olives were sampled in November 2019 at a ripening index of 4 according to Guzmán et al. [17]. Oil extraction was performed by means of a three-phase decanter system (Alfa Laval, Monza, Italy) at 25–26 °C and 20 min of malaxation parameters in an olive oil mill located in the Calabria region (Italy). The obtained OMWs were transferred to the Food Technologies laboratory of the Mediterranea University of Reggio Calabria for the experimental project.

2.3. OMW Acquisition and Preparation

The extract was obtained following the method reported by De Marco et al., with some modifications [18]. Two liters of olive oil mill wastewater (OMW) were acidified to pH 2 with HCl and washed three times with hexane (1:1, $v{:}v$) in order to remove the lipid fraction. The mixture was vigorously shaken and centrifuged under 3000 rpm for 3 min at 10 °C. The phenolic compounds were extracted by mean of ethyl acetate three times in a separating funnel (1:4 $v{:}v$), and then the combined extracts were centrifuged for 5 min at 3000 rpm at 10 °C. The organic phase was separated and filtered through a sintered glass Buchner apparatus. Then the ethyl acetate was evaporated under vacuum using a rotary vacuum evaporator at 25 °C (headspace analysis was performed). Finally, the dry residue was again dissolved in 100 mL of water, filtered using PTFE 0.45 µm (diameter 15 mm) syringe filter, and stored at 4 °C until subsequent analyses. An aliquot of the obtained extract (called L) was freeze-dried in a VirTis lyophilizer (Gardiner, NY, USA), and the obtained sample was called S. For all the experiments, we used both types of phenolic extracts, the freeze-dried (S) and the wet (L) type.

2.4. Determination of Total and Individual Phenolic Content of OMW Polyphenolic Extracts

The total phenol content was determined spectrophotometrically as previously described by Bruno et al. with some modifications [19]. An aliquot portion (0.1 mL) of phenolic extract was placed in a 25 mL volumetric flask and mixed with 20 mL of deionized water and 0.625 mL of the Folin–Ciocalteau reagent. After 3 min, 2.5 mL of saturated solution of Na_2CO_3 (20%) was added. After that, the mixture was incubated for 12 h at room temperature and in the dark. The sample's absorbance was measured at 725 nm against a blank and compared with a gallic acid (GA) calibration curve (concentration between 1 and 10 mg L^{-1}). The results were expressed as mg of GA g^{-1} of phenolic extract.

Identification and quantification of S and L extracts' main phenolic compounds were performed by HPLC-DAD (Dionex Ultimate 3000 RSLC, Waltham, MA, USA), as previously described [20]. The phenolic determination was conducted using the Dionex Acclaim 120 C18 analytical column (3 µm, 150 × 3 mm) (Thermo Scientific, Waltham, MA, USA) set at 35 °C, a flow rate of 1 mL min^{-1}, and an injection volume of 5 µL. Water/acetic acid (98:2, v/v) (A) and acetonitrile (B) were used as mobile phases, and the applied gradient was the following: 95% A and 5% B (5 min), 80% A and 20% B (10 min), 75% A and 25% B (15 min), 65% A and 35% B (20 min), 0% A and 100% B (25 min), and 95% A and 5% B (35 min). Quantification was performed by pure standard (Sigma-Aldrich Co. LLC, St. Louis, MO, USA) and data were expressed as mg g^{-1} of phenolic extract.

2.5. Cell Culture

Human umbilical vein endothelial cells (HUVECs) were acquired from Cell Application (San Diego, CA, USA) and grown as previously described [21]. Briefly, cells were grown in endothelial cell basal medium supplemented with cell growth supplement (EGM-V2 # 213K-500) as per company instructions. When confluent, cells were sub-culture at a split ratio of 1:2 and used within three passages. Unless specified in the text, cells were plated in 96-well plates (Corning, Lowell, MA, USA) at a concentration of 10^5 cells/mL and processed for experiments in a complete medium containing the different concentrations of the extracts.

Human pulmonary artery smooth muscle cells (HPASMCs) were acquired from Cell Application (San Diego, CA, USA) and grown as previously described [22]. Briefly, cells were grown in smooth muscle cells basal medium supplemented with cell growth supplement (HSMCs-kit #311K-500) as per company instructions. When confluent, HSMCs were sub-culture at a split ratio of 1:2 and used within three passages. Unless specified in the text, cells were plated in 96-well plates (Corning, Lowell, MA, USA) at a concentration of 10^5 cells/mL and processed for experiments in a complete medium containing the different concentrations of the extracts.

According to data obtained in our previous studies [23–26], we decided to test the extracts at the doses of 10, 25, 50, and 100 µg/mL.

2.6. Cell Metabolic Assay

Cell metabolic activity was evaluated using the oxidizable and reducible colorimetric probes 3-(4,5-dimethylthiazol-2-yl)-2,5-diphenyltetrazolium bromide (MTT) [23]. Cells were treated as indicated in figure legends and then processed for the MTT assay. After treatments, cells were added with 20 µL MTT solution (5 mg/mL) in medium M199 and placed at 37 °C in a cell incubator for 4 h. After that, the medium was discarded, and the converted dye was solubilized with acidic isopropanol (0.04 N HCl in absolute isopropanol), and the multi wells were read at 570 nm using a GENios plus microplate reader (Tecan) with background subtraction at 650 nm. Results were expressed as a percent of untreated control cells. Cell viability was calculated by the following equation: Cell viability (%) = (Abs of sample/Abs of control) × 100, where Abs of sample is the absorbance of the cells incubated with the different concentrations of the two extracts, and Abs of control is the absorbance of the cells incubated with the culture medium only (positive control).

2.7. Measurement of Intracellular ROS

Intracellular ROS levels were determined by using the ROS molecular probe 2′,7′-dichlorodihydrofluorescein diacetate (H_2DCF-DA) (Molecular Probe, Eugene, OR, USA) as previously described with minor modification [24,25]. In this assay, ROS oxidize H_2DCF, producing the fluorescent compound DCF, the fluorescence levels of which are proportional to the amount of intracellular ROS. Cells were treated as indicated in figure legends and then processed for the intracellular ROS assessment. For the ROS assay, cells were incubated for 30 min with Hank's Balanced Salt Solution (HBSS) containing 5 µM H_2DCF-DA, then washed twice with HBSS, and then fluorescence was measured by using a GENios plus microplate reader (Tecan, Mannedorf, Switzerland. Excitation and emission wavelengths used for fluorescence quantification were 485 and 535 nm, respectively. All fluorescence measurements were corrected for background fluorescence and protein concentration. Using untreated cells as a reference, the antioxidant and prooxidant outcomes were evaluated by comparing five measurements and expressed as a percentage of untreated control cells.

2.8. Cell Viability Assay

Cell viability was assessed as previously described [26] by using the CytoTox-ONE™ (Promega, Madison, WI, USA) kit. The CytoTox-ONE™ homogeneous membrane integrity assay is a fluorometric method for estimating the number of nonviable cells present in multi-well plates. The assay measures the release of lactate dehydrogenase (LDH) from cells with a damaged membrane. Then the LDH released into the culture medium is measured with a 10-min coupled enzymatic assay that results in the conversion of resazurin into a fluorescent resorufin product. The amount of fluorescence produced is proportional to the number of dead cells with the lysed membrane.

Cells were treated as indicated in figure legends and then processed as per company instructions at each experimental point's end. Fluorescence was measured by using a GENios plus microplate reader (Tecan, Mannedorf, Switzerland). Excitation and emission wavelengths used for fluorescence quantification were 535 and 620 nm, respectively. All fluorescence measurements were corrected for background fluorescence and protein concentration, and results were expressed as a percentage of untreated control cells.

2.9. Statistical Analysis

Data were expressed as means ± S.D. of the indicated number of experiments. One-way analysis of variance (ANOVA) followed by a post-hoc Newman–Keuls multiple comparison test were used to detect differences of means among treatments with significance defined as $p < 0.05$. Statistical analysis was performed using GraphPad Prism version 8.00 for Windows, GraphPad Software, San Diego, CA, USA.

3. Results

In this work, two extracts derived from OMW were assessed for their phenol contents and their potential biological activity on two human vascular cells. The two extracts were obtained employing a desolvented ethyl acetate extraction followed by water recovery.

Hydroxytyrosol and tyrosol were the principal quantified compounds in both OMW extracts, with significantly greater abundance in the S extract, mainly related to the first phenolic alcohol (4.55 mg g^{-1}). The tyrosol concentration varied between 0.54 mg g^{-1} in L extract and 0.58 mg g^{-1} in L extract. The S extract also showed higher amounts of caffeic acid and oleuropein, respectively, of 0.36 and 0.15 mg g^{-1}. Finally, among the identified flavonoids, apigenin-7-O-glucoside was more abundant in L extract with a concentration of 0.15 mg g^{-1} (Figure 1).

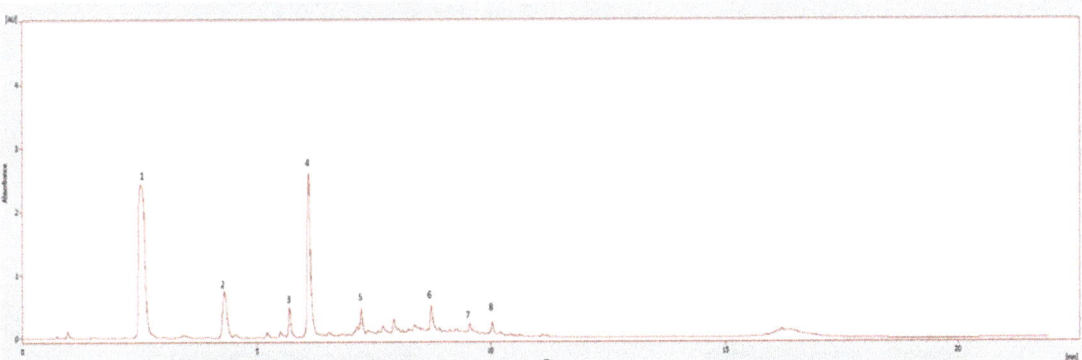

Figure 1. Chromatogram of phenolic compounds in S olive mill wastewater (OMW) extract. (1) Hydroxytyrosol; (2) tyrosol; (3) vanillic acid; (4) caffeic acid; (5) ferulic acid; (7) apigenin-7-O-glucoside; (8) oleuropein; (9) luteolin-7-O-glucoside.

Next, we proceeded with the extracts study by assessing their potential biological activities on two human vascular cells, HUVECs and HSMC, respectively, endothelial and vascular smooth muscle cells. Based on data from our previous studies [27–30], cell treatments were performed using the following extracts' concentrations: 10, 25, 50, and 100 µg/mL.

Although widely accepted as a provider of health benefits, natural antioxidants, including phenolic compounds, based on both dosages and redox environment status, have been reported to act as prooxidants, thus increasing intracellular ROS levels and inducing cell death [23,25,26,31–34]. Moreover, extracts deriving from food processing waste have been reported to be unsafe for the environment [7]. For this reason, we first investigated the potential harmful effects of OMW extracts by assessing their ability to affect cell viability, metabolic activity, and intracellular ROS production, three aspects tightly interconnected in maintaining the physiological functions of vascular cells [35,36]. As reported in Figure 2, 24-h exposition of HUVECs (Figure 2A) and HSMCs (Figure 2B) to the indicated concentrations of OMW failed to induce cell death in either cellular model. To further investigate potential harmful effects and corroborate the absence of cell toxicity, we assessed the extract's effect on cellular metabolic activity. Figure 2C,D indicate that only the highest dose tested (100 µg/mL) was able to significantly affect the cell metabolic activity, while no effects were observed at the other tested concentrations.

Since the two OMW extracts showed a remarkable quantity of antioxidant phenolic compounds, we sought to investigate their potential antioxidant activity by assessing the ability to counteract H_2O_2-elicited oxidative changes. First, we evaluated the ability of our ROS assay to reveal intracellular ROS level variations within a range of selected H_2O_2 concentrations. As reported in Figure 3A,B, the probe displayed both a significant dynamic range and linear response when challenged with increasing concentrations of H_2O_2, a well-known cellular prooxidant in these vascular cells. [37,38]. Therefore, based on these results, 75 µM was the dosage of H_2O_2 employed for the experiments concerning potential extract antioxidant activity. In this experiment, cells were treated for 4 h with the four extract concentrations followed by a 2 h-treatment with 75 µM H_2O_2. Then, the levels of intracellular ROS were measured as indicated in the Section 2.

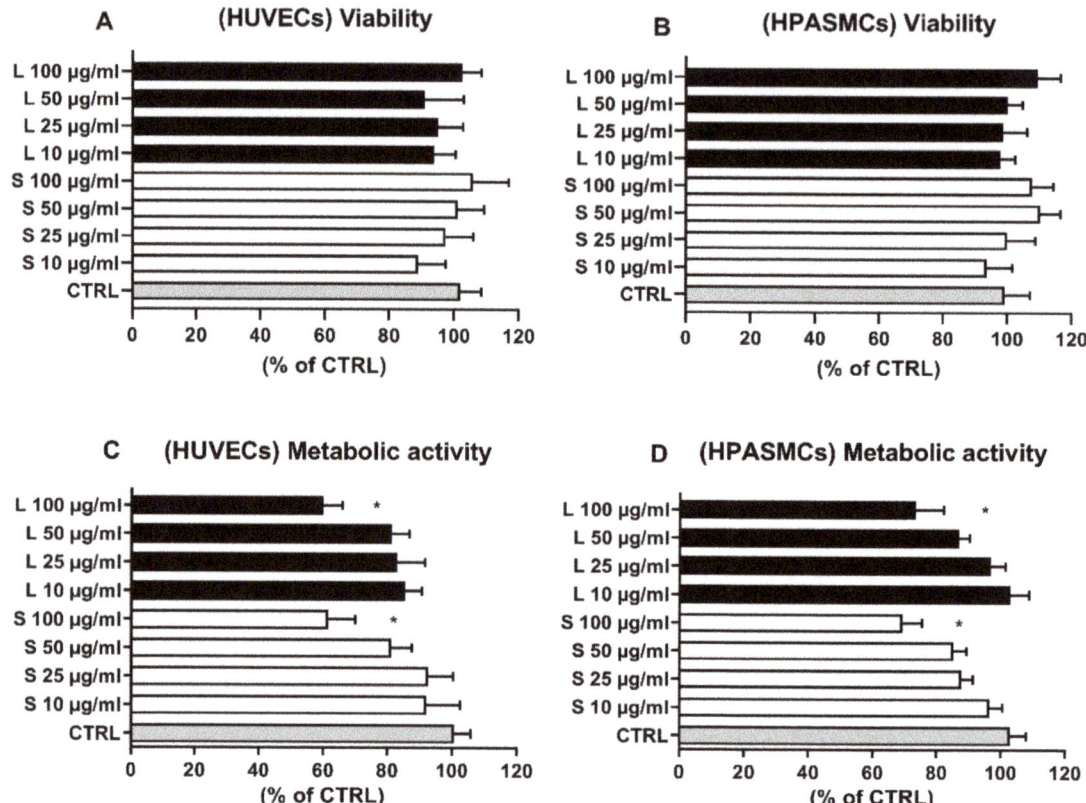

Figure 2. Effect of OMW on HUVECs and HPASMCs viability and metabolic activity. Cells were exposed for 24 h in the absence (CTRL) or presence to the indicated concentrations of OMW extracts. Cell viability (**A,B**) and metabolic activity (**C,D**) were assessed as reported in the materials and methods. HUVECs, human umbilical vein endothelial cell; HPASMCs, human pulmonary artery smooth muscle cells; L, liquid extract; S, solid extract; CTRL, untreated cells; * significantly different from CTRL; Values are shown as mean ± SD and expressed as a percentage of the CTRL. (*n* = 4).

Exposure of the HUVECs to increasing concentrations of the two extracts significantly counteracted the H_2O_2-elicited increase of ROS in all the tested concentrations (Figure 3C). Similarly, compared to H_2O_2-treated cells, exposure of HSMCs to the two extracts induced a significant antioxidant effect in the whole concentration range (Figure 3D). Since oxidative stress is recognized to cause cell injury and even death, we next hypothesized that the observed extracts' antioxidant effects could provide cellular protection against cells damaged elicited by H_2O. To this end, cells were pre-treated for 3 h with the indicated extract concentrations, and then 75 µM H_2O_2 was added for 24 h before cell viability determination. In agreement with the displayed antioxidant effect on H_2O_2-increased ROS production (Figure 3C,D), cell pre-treatments with increasing concentrations of the two OMW extracts significantly protected both cell lines from H_2O_2-elicited oxidative cell death.

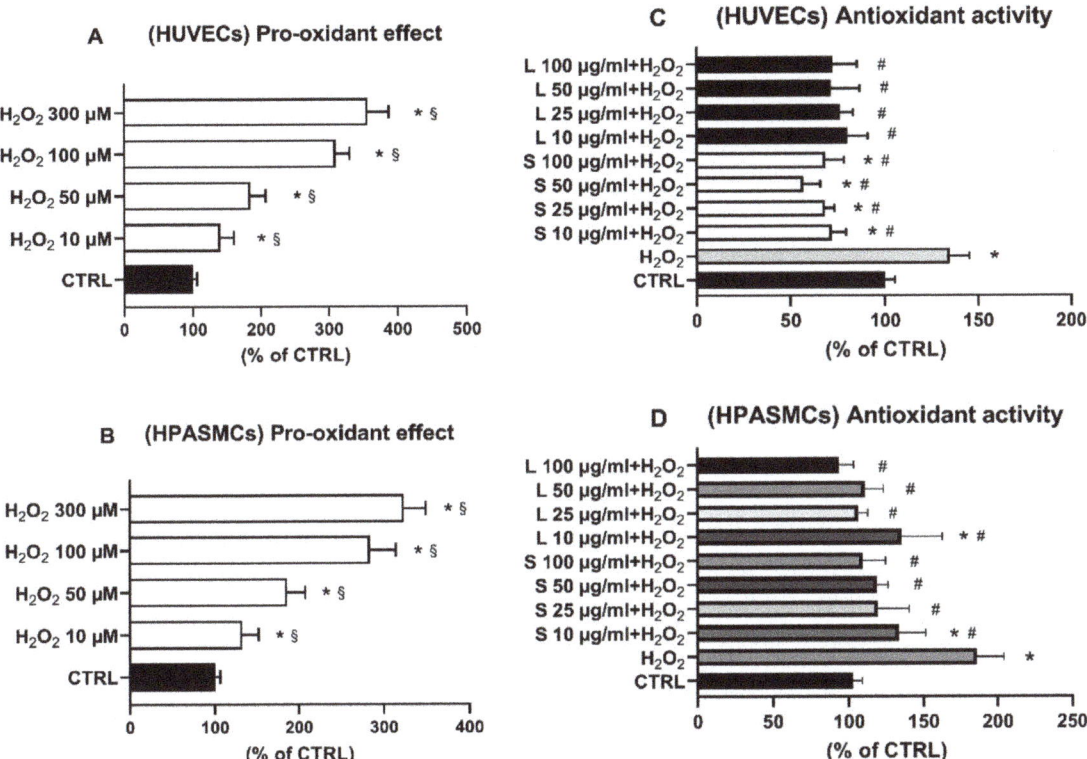

Figure 3. (A,B) Dose–response effect of H_2O_2 on intracellular ROS levels of HUVECs (A) and HPASMCs (B). Cells were exposed for 6 h to the indicated concentrations of H_2O_2. Intracellular ROS levels were assessed as reported in Materials and Methods. (C,D) OMW extracts counteract H_2O_2-induced ROS increase in both HUVECs (C) and HPASMCs (D). Cells were exposed for 3 h to the indicated concentrations of OMW extracts and then incubated for 6 h in the presence of 75 μM H_2O_2. Intracellular ROS levels were assessed, as reported in Materials and Methods. HUVECs, human umbilical vein endothelial cell; HPASMCs, human pulmonary artery smooth muscle cells; L, liquid extract; S, solid extract; CTRL, untreated cells; H_2O_2, hydrogen peroxide; * significantly different from CTRL; § significantly different from each other; # significantly different from H_2O_2. Values are shown as mean ± SD and expressed as a percentage of the CTRL. (n = 5).

4. Discussion

World globalization requires sustainable growth. In this context, the processing of different types of waste, derived from food treatment or processing, to produce value-added products is becoming a challenging reality that we must pursue. Waste processing or transformation may involve using different technologies, and the obtained products may or may not be safe for the environment and human health. However, since transformation may give rise to new, unknown, or unwanted compounds, it is useful to assess their safety, especially in terms of their impact on human health.

The olive (*Olea europaea* L.) is native to the Mediterranean area where it can be found in the wild form in the Middle East. *Olea europaea* L. is widely spread throughout the world, especially in the Mediterranean regions, where it accounts for nearly 96 percent of global olive production (FAOSTAT Food and Agriculture Data [20]). The production of olive oil forms a significant quantity of residue, such as aqueous waste called olive mill wastewater (OMW), and olive leaves. Other wastes are also produced during the filtration processes and during the storage time when the solid components migrate to the tank bottoms, generating sediments. These waste products are rich in bioactive compounds and can be therefore considered valuable byproducts to be exploited for further uses [39,40]. Although OMWs are effluents capable of degrading soil and water quality, with serious negative effects on the environment [7], studies have revealed that they contain high concentrations of antioxidants, such as phenolic molecules [41]. However, since the quality of OMW compounds differs according to the olive variety, the technological process of olive oil production, storage time, and climatic conditions [40], it is imperative to evaluate the technological processes employed in OMW processing in order to obtain valuable bioactive compounds. In this regard, phenolic compounds derived from OMW have been reported to show interesting biological properties [10,39,40]. Nonetheless, yet a large amount of crusher residues remain without actual application, since only small quantities are used as natural fertilizers, biomass fuel, additives in animal feed, and activated carbon. However, these residues are a precious starting material to produce extracts of phenolic molecules that could be used in various industrial fields [7]. Different technologies are proposed rather than classical solid–liquid extraction to enhance phenolic compounds' extraction from different olive oil byproducts.

We used a liquid–liquid extraction procedure followed by acidic hydrolysis [18], which provided two extracts, L and S, harboring a remarkable amount of total phenol content. The quantitative analysis indicated the S extract had a higher phenol content compared to the L extract. Moreover, hydroxytyrosol and tyrosol had the primary phenols in both OMW extracts, with hydroxytyrosol significantly more abundant in the S extract than the L extract (Table 1).

Table 1. Main phenolic compounds in liquid (L) and freeze dried (S) OMW extracts (mg g^{-1}).

Phenolic Compound	L	S	Sign.
Hydroxytyrosol	1.55 ± 0.01	4.55 ± 0.23	**
Tyrosol	0.54 ± 0.01	0.58 ± 0.00	*
Vanillic acid	0.15 ± 0.00	0.09 ± 0.00	**
Caffeic acid	0.22 ± 0.00	0.36 ± 0.00	**
Ferulic acid	0.02 ± 0.00	0.00 ± 0.00	**
Apigenin-7-O-glucoside	0.15 ± 0.00	0.06 ± 0.00	**
Oleuropein	0.08 ± 0.00	0.15 ± 0.00	**
Luteolin-7-O-glucoside	0.04 ± 0.00	0.05 ± 0.00	**

Data are mean (n = 2) ± standard deviation. ** Significance for $p < 0.01$; * significance for $p < 0.05$.

In the cellular experiments, we first evaluated potential extracts' harmful effects by treating endothelial and smooth muscle cells with different OMW dosages for 24 h. As reported in Figure 2A,B, the extracts were unable to induce cell death at all the concentrations tested. However, sometimes compound effects may affect cell metabolism without resulting in cell death. For this reason, we also assessed the extract effect on the two cells' metabolic activities. In this analysis, the extract showed no effect, except for the highest concentration, 100 µg/mL, where we found a slight but significant decrease in cell metabolic activity, an aspect that requires further investigations as it has been reported for other phenol compounds [23,25,26,31–34]. However, the finding that the concentration of 100 µg/mL affects only the cell metabolism without inducing cell death indicated that the extracts, even at high concentrations, appeared not to be toxic, being unable to induce cell death. Considering hydroxytyroso as the main compound present in the OMW, our findings align with In Vivo experiments reporting no toxic effects in rodents fed with HT concentrations as high as 2 g/kg b.wt. [42,43]. Similarly, no adverse clinical, biochemical, or gross necropsy effects were also reported in rats orally administered with HIDROX^{TM}, a hydrolyzed aqueous olive pulp extract with a phenol composition closer to our OMW [44]. Moreover, In Vitro experiments, performed with different cell lines, also confirmed our findings reporting lack of cytotoxicity with HT concentrations as high as 500 µM [45–47].

Although chemical tests such as TBARS, ABTS, and DPPH are often used to determine compounds' antioxidant properties, a likely better method to investigate the antioxidant activity/effect of a molecule or extract would be In Vivo or in a cellular model. Indeed, a cellular model would provide the proper environment for studying potential interactions (e.g., compound(s)–cellular signal transduction pathway(s), compound(s)–cellular receptor(s)) that would be otherwise missed in a solely chemical setting. In this regard, the intracellular oxidation of $H_2DCF-DA$ is mainly caused by H_2O_2, an oxidizing molecule physiologically present in the cell and therefore functional to induce oxidative insults and study their effects on cell functions [48,49]. Thus we created H_2O_2-induced oxidative stress and investigated whether the extracts could counteract it by following the variation of intracellular levels of DFC, the oxidized form of $H_2DCF-DA$, and assessing cell viability (Figures 3 and 4). In our experimental conditions, the different extracts' concentrations significantly inhibited the H_2O_2-induced increase of ROS in both HUVECs (Figure 3C) and in HAPVSCs (Figure 3D), bringing the oxidative stress values back to those of the controls. Consonant with observed extracts' antioxidant activity are the findings reported in Figure 4. The extracts were indeed capable of dose-dependently counteracting H_2O_2-induced cell death and restoring cell viability completely at the dosages of 50 and 100 µg/mL. Our current findings support and confirm previously performed experiments; indeed, similarly to EVOO-contained phenolic compounds [9,10,45,46,50], olive oil processing byproducts such as OMW are a source of valuable compounds harboring health benefit properties.

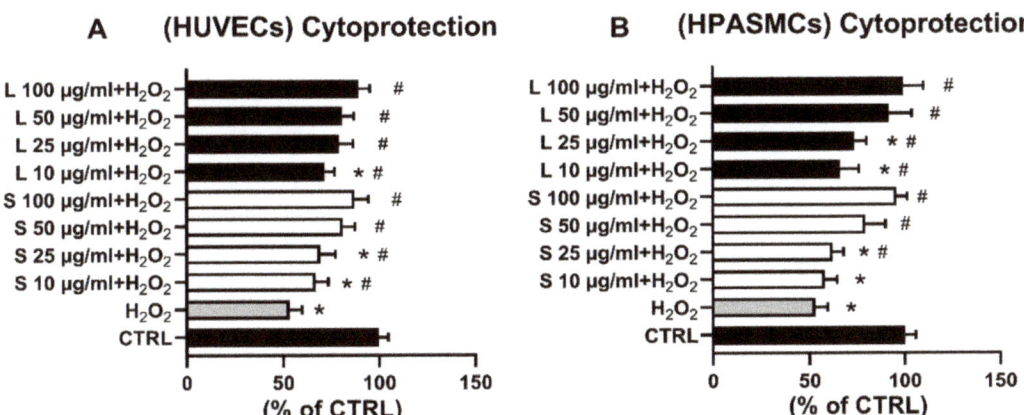

Figure 4. OMW extracts protect HUVECs (**A**) and HPASMCs (**B**) from H_2O_2-induced cell death. Cells were exposed for 3 h to the indicated concentrations of OMW extracts and then incubated in the absence (CTRL) or presence of 75 µM H_2O_2. Cell viability was assessed as reported in Materials and Methods. HUVECs, human umbilical vein endothelial cell; HPASMCs, human pulmonary artery smooth muscle cells; L, liquid extract; S, solid extract; CTRL, untreated cells; H_2O_2, hydrogen peroxide; * significantly different from CTRL; # significantly different from H_2O_2. Values are shown as mean ± SD and expressed as a percentage of the CTRL. (n = 4).

5. Conclusions

The OMW extracts obtained by the production of olive oil showed interesting antioxidant activity on both the employed cell models. The tested extracts were capable of effectively protecting cells from oxidative stress-indued cell death, failing indeed to interfere with cell viability and even with the metabolism, except for the highest tested concentration. Our results indicates a different point of view concerning the food processing residues, which should not be considered waste but precious material containing molecules capable of modulating essential functions of cellular biological models, such as the vascular model. Further studies are required to establish these substances' fates in the organism and to understand their biological functions and fine molecular mechanisms.

Author Contributions: Conceptualization, A.M.P., A.P. (Antonio Piga) and G.P.; Formal analysis, A.P. (Amalia Piscopo) and G.P.; Funding acquisition, A.P. (Antonio Piga) and G.P.; Investigation, A.M.P., A.C. and R.G.; Methodology, A.M.P., A.C., R.G. and A.P. (Amalia Piscopo); Resources, A.P. (Antonio Piga) and G.P.; Validation, G.P.; Writing—original draft, A.M.P. and A.P. (Amalia Piscopo); Writing—review & editing, A.M.P., A.P. (Amalia Piscopo), W.M.A.-R., A.P. (Antonio Piga) and G.P. All authors have read and agreed to the published version of the manuscript.

Funding: This research was funded by the AGER 2 Project, grant no. 2016-0105, and by the Fondo UNISS di Ateneo per la Ricerca 2020 to G.P. and A.P.

Institutional Review Board Statement: Not applicable.

Informed Consent Statement: Not applicable.

Data Availability Statement: All the data are reported in the article.

Acknowledgments: The authors thank the firm Olearia San Giorgio for supplying OMW.

Conflicts of Interest: The authors declare no conflict of interest.

References

1. Stillitano, T.; Falcone, G.; De Luca, A.I.; Piga, A.; Conte, P.; Strano, A.; Gulisano, G. A life cycle perspective to assess the environmental and economic impacts of innovative technologies in extra virgin olive oil extraction. *Foods* **2019**, *8*, 209. [CrossRef] [PubMed]
2. Owen, C.A.; Bita, E.-C.; Banilas, G.; Hajjar, S.E.; Sellianakis, V.; Aksoy, U.; Hepaksoy, S.; Chamoun, R.; Talhook, S.N.; Metzidakis, I. AFLP reveals structural details of genetic diversity within cultivated olive germplasm from the Eastern Mediterranean. *Theor. Appl. Genet.* **2005**, *110*, 1169–1176. [CrossRef] [PubMed]
3. Hatzopoulos, P.; Banilas, G.; Giannoulia, K.; Gazis, F.; Nikoloudakis, N.; Milioni, D.; Haralampidis, K. Breeding, molecular markers and molecular biology of the olive tree. *Eur. J. Lipid Sci. Technol.* **2002**, *104*, 574–586. [CrossRef]
4. Sivakumar, G.; Uccella, N.A.; Gentile, L. Probing downstream olive biophenol secoiridoids. *Int. J. Mol. Sci.* **2018**, *19*, 2892. [CrossRef]
5. Rodis, P.S.; Karathanos, V.T.; Mantzavinou, A. Partitioning of olive oil antioxidants between oil and water phases. *J. Agric. Food Chem.* **2002**, *50*, 596–601. [CrossRef]
6. Obied, H.K.; Allen, M.S.; Bedgood, D.R.; Prenzler, P.D.; Robards, K.; Stockmann, R. Bioactivity and analysis of biophenols recovered from olive mill waste. *J. Agric. Food Chem.* **2005**, *53*, 823–837. [CrossRef]
7. Araújo, M.; Pimentel, F.B.; Alves, R.C.; Oliveira, M.B.P. Phenolic compounds from olive mill wastes: Health effects, analytical approach and application as food antioxidants. *Trends Food Sci. Technol.* **2015**, *45*, 200–211. [CrossRef]
8. Zbakh, H.; El Abbassi, A. Potential use of olive mill wastewater in the preparation of functional beverages: A review. *J. Funct. Foods* **2012**, *4*, 53–65. [CrossRef]
9. Tripoli, E.; Giammanco, M.; Tabacchi, G.; Di Majo, D.; Giammanco, S.; La Guardia, M. The phenolic compounds of olive oil: Structure, biological activity and beneficial effects on human health. *Nutr. Res. Rev.* **2005**, *18*, 98–112. [CrossRef]
10. Giuffrè, A.; Sicari, V.; Piscopo, A.; Louadj, L. Antioxidant activity of olive oil mill wastewater obtained from different thermal treatments. *Grasas y Aceites* **2012**, *63*, 209–213. [CrossRef]
11. Zarkovic, N. Roles and Functions of ROS and RNS in Cellular Physiology and Pathology. *Cells* **2020**, *9*, 767. [CrossRef]
12. Costa, T.J.; Barros, P.R.; Arce, C.; Santos, J.D.; da Silva-Neto, J.; Egea, G.; Dantas, A.P.; Tostes, R.C.; Jimenez-Altayó, F. The homeostatic role of hydrogen peroxide, superoxide anion and nitric oxide in the vasculature. *Free Radic. Biol. Med.* **2021**, *162*, 615–635. [CrossRef]
13. Sabbatino, F.; Conti, V.; Liguori, L.; Polcaro, G.; Corbi, G.; Manzo, V.; Tortora, V.; Carlomagno, C.; Vecchione, C.; Filippelli, A. Molecules and Mechanisms to Overcome Oxidative Stress Inducing Cardiovascular Disease in Cancer Patients. *Life* **2021**, *11*, 105. [CrossRef]
14. Sauer, H.; Wartenberg, M. Reactive oxygen species as signaling molecules in cardiovascular differentiation of embryonic stem cells and tumor-induced angiogenesis. *Antioxid. Redox Signal.* **2005**, *7*, 1423–1434. [CrossRef]
15. Burtenshaw, D.; Hakimjavadi, R.; Redmond, E.M.; Cahill, P.A. Nox, reactive oxygen species and regulation of vascular cell fate. *Antioxidants* **2017**, *6*, 90. [CrossRef]
16. Gentile, L.; Uccella, N.A.; Sivakumar, G. Oleuropein: Molecular dynamics and computation. *Curr. Med. Chem.* **2017**, *24*, 4315–4328. [CrossRef]
17. Guzmán, E.; Baeten, V.; Pierna, J.A.F.; García-Mesa, J.A. Determination of the olive maturity index of intact fruits using image analysis. *J. Food Sci. Technol.* **2015**, *52*, 1462–1470. [CrossRef]
18. De Marco, E.; Savarese, M.; Paduano, A.; Sacchi, R. Characterization and fractionation of phenolic compounds extracted from olive oil mill wastewaters. *Food Chem.* **2007**, *104*, 858–867. [CrossRef]
19. De Bruno, A.; Romeo, R.; Fedele, F.L.; Sicari, A.; Piscopo, A.; Poiana, M. Antioxidant activity shown by olive pomace extracts. *J. Environ. Sci. Health Part B* **2018**, *53*, 526–533. [CrossRef]

20. Centrone, M.; D'Agostino, M.; Difonzo, G.; De Bruno, A.; Di Mise, A.; Ranieri, M.; Montemurro, C.; Valenti, G.; Poiana, M.; Caponio, F. Antioxidant Efficacy of Olive By-Product Extracts in Human Colon HCT8 Cells. *Foods* **2021**, *10*, 11. [CrossRef]
21. Zinellu, A.; Sotgia, S.; Scanu, B.; Pintus, G.; Posadino, A.M.; Cossu, A.; Deiana, L.; Sengupta, S.; Carru, C. S-homocysteinylated LDL apolipoprotein B adversely affects human endothelial cells In Vitro. *Atherosclerosis* **2009**, *206*, 40–46. [CrossRef]
22. Boin, F.; Erre, G.L.; Posadino, A.M.; Cossu, A.; Giordo, R.; Spinetti, G.; Passiu, G.; Emanueli, C.; Pintus, G. Oxidative stress-dependent activation of collagen synthesis is induced in human pulmonary smooth muscle cells by sera from patients with scleroderma-associated pulmonary hypertension. *Orphanet J. Rare Dis.* **2014**, *9*, 1–5. [CrossRef]
23. Posadino, A.M.; Phu, H.T.; Cossu, A.; Giordo, R.; Fois, M.; Thuan, D.T.B.; Piga, A.; Sotgia, S.; Zinellu, A.; Carru, C. Oxidative stress-induced Akt downregulation mediates green tea toxicity towards prostate cancer cells. *Toxicol. In Vitro* **2017**, *42*, 255–262. [CrossRef]
24. Fois, A.G.; Posadino, A.M.; Giordo, R.; Cossu, A.; Agouni, A.; Rizk, N.M.; Pirina, P.; Carru, C.; Zinellu, A.; Pintus, G. Antioxidant activity mediates pirfenidone antifibrotic effects in human pulmonary vascular smooth muscle cells exposed to sera of idiopathic pulmonary fibrosis patients. *Oxidative Med. Cell. Longev.* **2018**, *2018*, 2639081. [CrossRef]
25. Posadino, A.M.; Giordo, R.; Cossu, A.; Nasrallah, G.K.; Shaito, A.; Abou-Saleh, H.; Eid, A.H.; Pintus, G. Flavin oxidase-induced ROS generation modulates PKC biphasic effect of resveratrol on endothelial cell survival. *Biomolecules* **2019**, *9*, 209. [CrossRef]
26. Posadino, A.M.; Cossu, A.; Giordo, R.; Zinellu, A.; Sotgia, S.; Vardeu, A.; Hoa, P.T.; Deiana, L.; Carru, C.; Pintus, G. Coumaric acid induces mitochondrial damage and oxidative-mediated cell death of human endothelial cells. *Cardiovasc. Toxicol.* **2013**, *13*, 301–306. [CrossRef]
27. Posadino, A.M.; Biosa, G.; Zayed, H.; Abou-Saleh, H.; Cossu, A.; Nasrallah, G.K.; Giordo, R.; Pagnozzi, D.; Porcu, M.C.; Pretti, L. Protective effect of cyclically pressurized solid–liquid extraction polyphenols from Cagnulari grape pomace on oxidative endothelial cell death. *Molecules* **2018**, *23*, 2105. [CrossRef]
28. Posadino, A.M.; Porcu, M.C.; Marongiu, B.; Cossu, A.; Piras, A.; Porcedda, S.; Falconieri, D.; Cappuccinelli, R.; Biosa, G.; Pintus, G. Antioxidant activity of supercritical carbon dioxide extracts of Salvia desoleana on two human endothelial cell models. *Food Res. Int.* **2012**, *46*, 354–359. [CrossRef]
29. Cossu, A.; Posadino, A.M.; Giordo, R.; Emanueli, C.; Sanguinetti, A.M.; Piscopo, A.; Poiana, M.; Capobianco, G.; Piga, A.; Pintus, G. Apricot melanoidins prevent oxidative endothelial cell death by counteracting mitochondrial oxidation and membrane depolarization. *PLoS ONE* **2012**, *7*, e48817. [CrossRef]
30. Posadino, A.M.; Cossu, A.; Piga, A.; Madrau, M.A.; Del Caro, A.; Colombino, M.; Paglietti, B.; Rubino, S.; Iaccarino, C.; Crosio, C. Prune melanoidins protect against oxidative stress and endothelial cell death. *Front. Biosci.* **2011**, *3*, 1034–1041. [CrossRef]
31. Shaito, A.; Posadino, A.M.; Younes, N.; Hasan, H.; Halabi, S.; Alhababi, D.; Al-Mohannadi, A.; Abdel-Rahman, W.M.; Eid, A.H.; Nasrallah, G.K. Potential adverse effects of resveratrol: A literature review. *Int. J. Mol. Sci.* **2020**, *21*, 2084. [CrossRef] [PubMed]
32. Posadino, A.M.; Cossu, A.; Giordo, R.; Zinellu, A.; Sotgia, S.; Vardeu, A.; Hoa, P.T.; Carru, C.; Pintus, G. Resveratrol alters human endothelial cells redox state and causes mitochondrial-dependent cell death. *Food Chem. Toxicol.* **2015**, *78*, 10–16. [CrossRef] [PubMed]
33. Giordo, R.; Cossu, A.; Pasciu, V.; Hoa, P.T.; Posadino, A.M.; Pintus, G. Different redox response elicited by naturally occurring antioxidants in human endothelial cells. *Open Biochem. J.* **2013**, *7*, 44. [CrossRef] [PubMed]
34. Pasciu, V.; Posadino, A.M.; Cossu, A.; Sanna, B.; Tadolini, B.; Gaspa, L.; Marchisio, A.; Dessole, S.; Capobianco, G.; Pintus, G. Akt downregulation by flavin oxidase–induced ROS generation mediates dose-dependent endothelial cell damage elicited by natural antioxidants. *Toxicol. Sci.* **2010**, *114*, 101–112. [CrossRef] [PubMed]
35. Cai, H.; Harrison, D.G. Endothelial dysfunction in cardiovascular diseases: The role of oxidant stress. *Circ. Res.* **2000**, *87*, 840–844. [CrossRef] [PubMed]
36. Ruiz-Gines, J.; Lopez-Ongil, S.; Gonzalez-Rubio, M.; Gonzalez-Santiago, L.; Rodriguez-Puyol, M.; Rodriguez-Puyol, D. Reactive oxygen species induce proliferation of bovine aortic endothelial cells. *J. Cardiovasc. Pharmacol.* **2000**, *35*, 109–113. [CrossRef] [PubMed]
37. Haendeler, J.; Popp, R.; Goy, C.; Tischler, V.; Zeiher, A.M.; Dimmeler, S. Cathepsin D and H2O2 stimulate degradation of thioredoxin-1: Implication for endothelial cell apoptosis. *J. Biol. Chem.* **2005**, *280*, 42945–42951. [CrossRef]
38. Park, W.H. Exogenous H_2O_2 induces growth inhibition and cell death of human pulmonary artery smooth muscle cells via glutathione depletion. *Mol. Med. Rep.* **2016**, *14*, 936–942. [CrossRef]
39. Ricelli, A.; Gionfra, F.; Percario, Z.; De Angelis, M.; Primitivo, L.; Bonfantini, V.; Antonioletti, R.; Bullitta, S.M.; Saso, L.; Incerpi, S. Antioxidant and Biological Activities of Hydroxytyrosol and Homovanillic Alcohol Obtained from Olive Mill Wastewaters of Extra-Virgin Olive Oil Production. *J. Agric. Food Chem.* **2020**, *68*, 15428–15439. [CrossRef]
40. Tapia-Quirós, P.; Montenegro-Landívar, M.F.; Reig, M.; Vecino, X.; Alvarino, T.; Cortina, J.L.; Saurina, J.; Granados, M. Olive mill and winery wastes as viable sources of bioactive compounds: A study on polyphenols recovery. *Antioxidants* **2020**, *9*, 1074. [CrossRef]
41. Obied, H.K.; Allen, M.S.; Bedgood, D.R.; Prenzler, P.D.; Robards, K. Investigation of Australian olive mill waste for recovery of biophenols. *J. Agric. Food Chem.* **2005**, *53*, 9911–9920. [CrossRef]
42. D'Angelo, S.; Manna, C.; Migliardi, V.; Mazzoni, O.; Morrica, P.; Capasso, G.; Pontoni, G.; Galletti, P.; Zappia, V. Pharmacokinetics and metabolism of hydroxytyrosol, a natural antioxidant from olive oil. *Drug Metab. Dispos.* **2001**, *29*, 1492–1498.

43. Auñon-Calles, D.; Canut, L.; Visioli, F. Toxicological evaluation of pure hydroxytyrosol. *Food Chem. Toxicol.* **2013**, *55*, 498–504. [CrossRef]
44. Christian, M.S.; Sharper, V.A.; Hoberman, A.M.; Seng, J.E.; Fu, L.; Covell, D.; Diener, R.M.; Bitler, C.M.; Crea, R. The toxicity profile of hydrolyzed aqueous olive pulp extract. *Drug Chem. Toxicol.* **2004**, *27*, 309–330. [CrossRef]
45. Crupi, R.; Palma, E.; Siracusa, R.; Fusco, R.; Gugliandolo, E.; Cordaro, M.; Impellizzeri, D.; De Caro, C.; Calzetta, L.; Cuzzocrea, S. Protective effect of Hydroxytyrosol against oxidative stress induced by the Ochratoxin in kidney cells: In Vitro and In Vivo study. *Front. Vet. Sci.* **2020**, *7*, 136. [CrossRef]
46. Calahorra, J.; Martínez-Lara, E.; De Dios, C.; Siles, E. Hypoxia modulates the antioxidant effect of hydroxytyrosol in MCF-7 breast cancer cells. *PLoS ONE* **2018**, *13*, e0203892. [CrossRef]
47. Gugliandolo, E.; Fusco, R.; Licata, P.; Peritore, A.F.; D'amico, R.; Cordaro, M.; Siracusa, R.; Cuzzocrea, S.; Crupi, R. Protective Effect of Hydroxytyrosol on LPS-Induced Inflammation and Oxidative Stress in Bovine Endometrial Epithelial Cell Line. *Vet. Sci.* **2020**, *7*, 161. [CrossRef]
48. Schmidt, K.N.; Traenckner, E.B.-M.; Meier, B.; Baeuerle, P.A. Induction of oxidative stress by okadaic acid is required for activation of transcription factor NF-κB. *J. Biol. Chem.* **1995**, *270*, 27136–27142. [CrossRef]
49. Zulueta, J.J.; Sawhney, R.; Yu, F.S.; Cote, C.C.; Hassoun, P.M. Intracellular generation of reactive oxygen species in endothelial cells exposed to anoxia-reoxygenation. *Am. J. Physiol. Lung Cell. Mol. Physiol.* **1997**, *272*, L897–L902. [CrossRef]
50. Hormozi, M.; Marzijerani, A.S.; Baharvand, P. Effects of Hydroxytyrosol on Expression of Apoptotic Genes and Activity of Antioxidant Enzymes in LS180 Cells. *Cancer Manag. Res.* **2020**, *12*, 7913. [CrossRef]

Article

Antioxidant Efficacy of Olive By-Product Extracts in Human Colon HCT8 Cells

Mariangela Centrone [1], Mariagrazia D'Agostino [1], Graziana Difonzo [2], Alessandra De Bruno [3], Annarita Di Mise [1], Marianna Ranieri [1], Cinzia Montemurro [2], Giovanna Valenti [1], Marco Poiana [3], Francesco Caponio [2] and Grazia Tamma [1,*]

[1] Department of Biosciences, Biotechnologies and Biopharmaceutics, University of Bari Aldo Moro, 70125 Bari, Italy; mariangela.centrone@uniba.it (M.C.); mariagrazia.dagostino@uniba.it (M.D.); annarita.dimise@uniba.it (A.D.M.); marianna.ranieri@uniba.it (M.R.); giovanna.valenti@uniba.it (G.V.)
[2] Department of Soil, Plant and Food Sciences, University of Bari Aldo Moro, 70125 Bari, Italy; graziana.difonzo@uniba.it (G.D.); cinzia.montemurro@uniba.it (C.M.); francesco.caponio@uniba.it (F.C.)
[3] Department of AGRARIA, University Mediterranea of Reggio Calabria, Vito, 89124 Reggio Calabria, Italy; alessandra.debruno@unirc.it (A.D.B.); mpoiana@unirc.it (M.P.)
* Correspondence: grazia.tamma@uniba.it; Tel.: +39-080-5442388

Abstract: The production of olive oil is accompanied by the generation of a huge amount of waste and by-products including olive leaves, pomace, and wastewater. The latter represents a relevant environmental issue because they contain certain phytotoxic compounds that may need specific treatments before the expensive disposal. Therefore, reducing waste biomass and valorizing by-products would make olive oil production more sustainable. Here, we explore the biological actions of extracts deriving from olive by-products including olive pomace (OP), olive wastewater (OWW), and olive leaf (OLs) in human colorectal carcinoma HCT8 cells. Interestingly, with the same phenolic concentration, the extract obtained from the OWW showed higher antioxidant ability compared with the extracts derived from OP and OLs. These biological effects may be related to the differential phenolic composition of the extracts, as OWW extract contains the highest amount of hydroxytyrosol and tyrosol that are potent antioxidant compounds. Furthermore, OP extract that contains a higher level of vanillic acid than the other extracts displayed a cytotoxic action at the highest concentration. Together these findings revealed that phenols in the by-product extracts may interfere with signaling molecules that cross-link several intracellular pathways, raising the possibility to use them for beneficial health effects.

Keywords: olive by-product; reactive oxygen species (ROS); olive leaf; pomace; olive wastewater

1. Introduction

Olive cultivation is typical of Mediterranean countries and it is developed in several regions of the world having similar climate features [1]. By-products and waste from olive production and the olive oil industry represent a relevant environmental issue because they contain several compounds that may also be phytotoxic [2]. Olive wastewater (OWW) contains a huge bulk of organic compounds, metals, minerals including potassium, calcium, and phosphorous. Furthermore, OWW has high biological and chemical oxygen demand (BOD and COD) and very low pH that make this olive waste highly pollutant [3]. In developed countries, the growing demand for food increases the mass of waste and by-products in an unsustainable manner. In the last few years, novel approaches, based on the new concept of the circular economy, lead to reduce waste biomass and valorize by-products. The improvement of extraction strategies provides a better exploit of olive by-products for functional food production, the development of pharmaceutical supplements and, cosmeceuticals [4]. Along the olive oil chain, olive mills produce tons of by-products including olive pomace (OP), olive mill wastewater (OWW), and olive leaves (OLs). These latter are commonly used for animal feeding or direct combustion. Recent studies report that OLs contain certain bioactive compounds that may be useful for the preparation of nutraceuticals and supplements.

OP contains pulp, skin, stone and water. However, the chemical composition and the biological activity of OP may differ according to the oil extraction method. Pomace is mainly used to produce pomace olive oil and a more recent application includes the generation of fuel after microbiological fermentation processes [5]. In contrast, the disposal of the olive mill wastewater is still an expensive cost for the oil olive industry. Notably, recent reports propose new methods based on nanocentrifugation and ultrafiltration to revalorize the olive wastewater [6,7]. Chemical characterization studies revealed that some bioactive compounds are transferred from olives to oil and their by-products and waste during the oil production process [8]. Specifically, OP, OWW, and OLs contain bioactive compounds including secoiridoids, squalene, flavonoids, lignans, phytosterols, tocopherols, and phenols. Among them, phenols, which displayed a significant antioxidant capability, are extensively investigated in different fields of research. The highest bulk of phenols have been isolated in olive wastewater. By contrast, a low amount of phenols was detected in the oil [9]. Numerous studies investigated the functional involvement of phenols in human health and healthy aging. The biological actions of phenols may be due to their antioxidant and anti-inflammatory properties. Nevertheless, the molecular mechanism of action of these compounds, on different diseases, is still unclear because they may have multiple intracellular targets. In bronchial epithelial NCI-H292 cells, a green olive leaf extract (OLE), obtained using water as an extraction solvent, reduced the tBHP-induced reactive oxygen species (ROS) production. This extract indeed displayed a significant antioxidant capability in vegetable oil [10]. Furthermore, in renal collecting duct MCD4 cells, the green OLE counteracted the cytotoxic effects due to long exposure to low doses of cadmium [11]. Specifically, in cells exposed to cadmium, OLE reduced the frequency of micronuclei, DNA double-strand breaks, and anomalous alterations of the cytoskeleton by modulating the S-glutathionylation of actin [11]. Therefore, OLE may exert different cellular responses that may be only partially due to antioxidant features. The olive derived extracts and by-products contain numerous compounds such as hydroxytyrosol and oleuropein that can modulate different intracellular signal transduction pathways including the ones regulating autophagy and apoptosis [12,13]. Chronic and acute disturbances affecting the gastrointestinal tract such as inflammatory bowel diseases are associated with an abnormal intracellular level of reactive species due to a relevant unbalance between the ROS production systems and the antioxidant intestinal defenses [14]. Moreover, colonic mucosa of patients with chronic inflammatory diseases produced a high amount of ROS compared with normal mucosa [15]. Therefore, the administration of selective antioxidant phytocompounds may provide beneficial effects to counteract intracellular processes leading to abnormal production of ROS in the intestine. In the present study, the human colon HCT8 cells were used as an experimental model to assay the cell viability and the antioxidant activity of extracts deriving from olive by-products including OP, OWW, and OLs. HCT-8 cell line derived from the ileocecal colon displays a high proliferation and migration rate associated with high ROS production [16]. Moreover, compared with the fully differentiated HT-29 and Caco-2 colon cell lines, HCT-8 cells have poorly organized junctional complexes [16,17]. Interestingly, here, with the same phenolic concentration, the extract obtained from the olive wastewater displayed a higher antioxidant activity compared with the extracts derived from olive pomace and leaves.

2. Materials and Methods

2.1. Chemicals and Reagents

Cell culture media and FBS (fetal bovine serum) were from GIBCO (Thermo Fisher Scientific, Waltham, MA, USA). Dihydrorhodamine-123 was obtained from Invitrogen™ (Thermo Fisher Scientific, Waltham, MA, USA). The tert-Butyl hydroperoxide (tBHP) was a kind gift from A. Signorile (University of Bari Aldo Moro, Bari, Italy).

2.2. Extraction of Phenolic Compounds from By-Products

2.2.1. Olive Wastewater (OWW)

OWW was collected during the crop seasons 2019/2020 from Ottobratica olive cultivar and produced according to a three-phase centrifugation process. The phenolic extract was obtained following the method reported by Romeo et al. [18] with some modifications. Two liters of OWW was acidified to pH 2 with HCl and washed three times with hexane (1:1, v/v) to remove the lipid fraction. The mixture was vigorously shaken and centrifuged under 3000 rpm for 3 min at 10 °C. The extraction procedure was carried out using ethyl acetate three times and the solvent was recovered in a separating funnel (1:4, v/v). The two-phases (water and ethyl acetate) were separated and both were evaporated using a rotary vacuum evaporator at 25 °C. Finally, the dry residues were again dissolved in 100 mL of water, filtered using PTFE 0.45 µm (diameter 15 mm) syringe filter, and stored at 4 °C until subsequent analyses.

2.2.2. Olive Pomace (OP)

OP was collected during the crop seasons 2019/2020 from Ottobratica olive cultivar and produced according to a three-phase centrifugation process. The phenolic extract was obtained following the method reported by De Bruno et al. [19] with some modifications. 200 g of defatted olive pomace was extracted by way ultrasound system (W%, 15 OFF 5 s) with ethanol/water 80/20 for 1 h. The extract was filtered using Buchner funnel and evaporated to 80% in a rotary evaporator at 25 °C.

2.2.3. Olive Leaves (OLs)

The olive leaves were collected in the crop seasons 2019/2020 from Coratina olive cultivar, stored at 4 °C, and processed in less than 24 h. The extraction was performed according to Difonzo et al. [10]. After washing with tap water at room temperature, the olive leaves were dried at 120 °C for 8 min in a ventilated oven (Argolab, Carpi, Italy) to reach a moisture content <1%, milled with a blender (Waring-Commercial, Torrington, CT, USA). The extraction from leaves was ultrasound-assisted (CEIA, Viciomaggio, Italy) and water was added in a ratio 1/20 (w/v). Three washes were done, each one for 30 min at 35 ± 5 °C. Finally, the extracts were filtered through Whatman (GE Healthcare, Milan, Italy) filter paper, freeze-dried, and stored at −20 °C. Before the analysis, the extract was filtered by nylon filters of 0.45 µm.

2.3. Chemical Characterization of Extracts

2.3.1. Olive Wastewater (OWW) and Olive Pomace (OP)

The total phenol content (TPC) and antioxidant activity (ABTS, 2,2′-azino bis(3 ethylbenzothiazoline-6-sulfonic acid)) were determined spectrophotometrically following the method described by De Bruno [19] with some modifications. TPC, was quantified on the obtained extracts by FolinCiocalteau method, 0.1 mL of the phenolic extract (OWW and OP), were placed in a 25 mL volumetric flask and mixed with 20 mL of deionized water and 0.625 mL of the FolinCiocalteau reagent. After 3 min, 2.5 mL of a saturated solution of Na_2CO_3 (20%) were added. The content was mixed and diluted to volume with deionized water. Thereafter the mixture was incubated for 12 h at room temperature and in the dark. The absorbance of the samples was measured at 725 nm against a blank using a double-beam ultraviolet-visible spectrophotometer (Perkin-Elmer UV-Vis λ2, Waltham, MA, USA) and comparing with a gallic acid calibration curve (concentration between 1 and 10 mg·L^{-1}). The results were expressed as mg of GAE mL^{-1}. For total antioxidant activity determination, ABTS assay was used. The reaction mixture was prepared by mixing 2990 µL of ABTS and 10 µL of extracts (OWW and OP), and the absorbance was measured after 6 min at 734 nm. The quenching of initial absorbance was plotted against the Trolox concentration (from 1.5 to 24 mmol L^{-1}) and the TEAC value was expressed as µmol TE mL^{-1} of PE. Identification and determination of the main bioactive phenolic compounds were performed by UHPLC-DAD of the phenolic extract, following the method described by Romeo et al. [18] with some modifications. The UHPLC system consisted of an UHPLC PLATINblue (Knauer, Berlin, Germany) equipped with a binary pump system using a Knauer blue orchid column C18 (1.8 µm, 100 × 2 mm) coupled with a PDA-1 (Photo Diode Array Detector) PLATINblue (Knauer, Berlin, Germany). The used software was Clarity 6.2 (Clarity-DataApex, Prague, The Czech Republic). The samples were filtered with a 0.22 µm nylon syringe filters (diameter 13 mm) and then injected in the system with a volume of 5 µL. The mobile phases were (A) water acidified with acetic acid (pH 3.10) and (B) acetonitrile; the gradient elution program consisted of 0–3 min, 95% A; 3–15 min, 95%–60% A; 15–15.5 min, 60%–0% A; finally, returning to the initial conditions was achieved during analysis keeping the column at 30 °C and the injection volume 5 µL. External standards (concentration between 1 and 100 mg·L^{-1}) were used for the quantification and the results were expressed as mg·mL^{-1}.

2.3.2. Olive Leaves (OLs)

Total phenol content, antioxidant activity, and single phenolic compounds identification were performed according to Difonzo et al. [10] with some modifications. For the total phenol content determination, 20 µL of the extract was added to 980 µL of ddH$_2$O and 100 µL of FolinCiocalteu reagent. After 3 min, 5% Na$_2$CO$_3$ solution was added, following incubation at room temperature for 60 min. The absorbance was read at 750 nm using a Cary 60 spectrophotometer (Agilent, Milan, Italy). The TPC was expressed as gallic acid equivalents (GAE) in mg·L^{-1} juice.

For the antioxidant activity assay ABTS, the radical was generated by a chemical reaction with potassium persulfate (K$_2$S$_2$O$_8$). For this purpose, 25 mL of ABTS (7 mM in H$_2$O) was spiked with 440 µL of K$_2$S$_2$O$_8$ (140 mM) and allowed to stand in darkness at room temperature for 12–16 h (the time required for the formation of the radical). The working solution was prepared by taking a volume of the previous solution and diluting it in ethanol until its absorbance at λ = 734 nm was 0.70 ± 0.02. A Cary 60 spectrophotometer Agilent (Milan, Italy) was used. The reaction took place directly in the measuring cuvette: 50 µL of each sample was added to 950 µL of the final ABTS$^{·+}$ solution. After 8 min the decrease of absorbance was measured at 734 nm. The results are expressed in µmol Troloxequivalents (TE) g^{-1} dry weight. Each sample was analyzed in triplicate.

The main identified phenolic compounds were quantified by means of HPLC-DAD (Dionex Ultimate 3000 RSLC, Waltham, MA, USA). Dionex Acclaim 120 C18 analytical column (150 × 3 mm i.d.) with a particle size of 3 µm (Thermo Scientific, Waltham, MA, USA) was used. The mobile phases were (A) water/acetic acid (98:2, v/v) and (B) acetonitrile at a constant flow rate of 1 mL·min^{-1}. The column temperature was set at 35 °C. The gradient program was as follows: 5 min, 95% A; 10 min, 80% A; 15 min, 75% A; 20 min, 65% A; 25 min, 0% A; 35 min, 95% A. The identification of phenolic compounds was based on a comparison of retention times obtained by pure standard (Sigma-Aldrich Co. LLC, St. Louis, MO, USA) and literature data [10]. The quantification was carried out by means of external calibration curves with the relative standard for the selected compounds.

2.4. Cell Culture and Treatment

The human colon cancer cells HCT-8 were cultured as previously described [20]. Briefly, cells were grown in Advanced RPMI-1640 supplemented with 10% fetal bovine serum (FBS), 100 i.u. mL^{-1} penicillin, 100 µg·mL^{-1} streptomycin at 37 °C in 5% CO$_2$. Cells were left under basal condition or treated overnight with the same phenolic concentration (0.03 mg·L^{-1}, 0.06 mg·L^{-1}, and 0.12 mg·L^{-1}) obtained from extracts of olive leaves or olive pomace or olive mill wastewater.

2.5. Crystal Violet Assay

Crystal violet assay was performed as previously described [11]. Briefly, cells were grown in a 96-well plate and left under basal condition or stimulated as mentioned before. Cells were fixed with 4% paraformaldehyde for 20 min and then stained with a solution containing 0.1% crystal violet for 20 min. After washing, cells were lysed with 10% acetic acid. The optical density at 595 nm (DO595) of each well was measured with a Microplate Reader (Bio-Rad Laboratories, Inc., Hercules, CA, USA) and was used as a measurement of cell viability.

2.6. Reactive Oxygen Species (ROS) Detection

ROS were detected as already shown [11]. After treatments, cells were incubated with dihydrorhodamine-123 (10 µM) for 30 min at 37 °C with 5% CO_2 and recovered in complete medium for 30 min. Cells were lysed in RIPA buffer containing 150 mM NaCl, 10 mM Tris-HCl pH 7.2, 0.1% SDS, 1.0% Triton X-100, 1% sodium deoxycholate and 5 mM EDTA. Samples were centrifuged at $12,000 \times g$ for 10 min at 4 °C and the supernatants were used for ROS detection. As a positive control, cells were treated with tert-Butyl hydroperoxide (tBHP, 2 mM for 30 min). The fluorescence emission signal was recorded using a fluorimeter (RF-5301PC, Shimadzu Corporation, Kyoto, Japan) at excitation and emission wavelengths of 508 and 529 nm, respectively.

2.7. Statistical Analysis

All values are reported as means ± S.E.M. Statistical analysis was performed by one-way ANOVA followed by Dunnett's multiple comparisons test with * $p < 0.05$ considered statistically different.

3. Results

3.1. Characterization of Olive By-Products and Waste

The valorization of waste and by-products derived from the olive oil industry represents an attractive and sustainable challenge aiming to reduce the environmental impact and the disposal costs. Extracts derived from by-products can contain several compounds displaying bioactive activity. The phenolic compounds recovery from waste and by-products is usually carried out using organic solvent extraction. The choice of the solvent to use is made according to the purpose of extraction, the polarity of the interesting components, overall cost, safety, etc. For this reason, in this study, the recovery of phenolic compounds was realized with hydro-alcoholic solutions for OP and OLs and ethyl acetate for OWW. The latter solvent showing a high extraction efficiency for phenolic extraction from OWW, and at the end of the extraction procedure is eliminated (through concentration), and the phenolic compounds were recovered with water.

Table 1 reports the results of total phenol content, antioxidant activity, and the chemical characterization of the main phenolics in the considered olive waste and by-products. Extracts obtained from OWW showed higher total phenolic content (about 9 mg·mL^{-1}) compared to the pomace extract (4.5 mg·mL^{-1}).

Among these, the main ones were: hydroxytyrosol, tyrosol, and oleuropein in agreement with our previous work [21]. Particularly, the OWW was rich in hydroxytyrosol (1.4 mg·mL^{-1}), also if this content may vary according to several factors as oil extraction technique, OWW characteristics, cultivar, extraction method of phenolic content. Furthermore, for OP extract, the main phenolic compound was hydroxytyrosol, but in a lower amount (0.2 mg·mL^{-1}). OWW extract showed a notably higher value of antioxidant activity compared to OP extract (3579 and 451 µmol TE mL^{-1}, respectively), denoting a strong antioxidant activity against ABTS radical.

The OLE showed high total phenol content and antioxidant activity according to Ghasemi et al. [22]. Among phenols, oleuropein, verbascoside, apigenin-7-O-glucoside, luteolin-7-O-glucoside, tyrosol, and rutin were identified.

The most represented compounds were oleuropein with a concentration of 137 mg g^{-1} in line with the results of other authors [22,23], in fact, generally, the amount of this secoiridoid is in the range of 25–30% of the total polyphenols in OLE [24].

Table 1. Total phenol content (TPC, mg GAE mL^{-1} for OWW and OP; mg GAE g^{-1} for freeze-dried OLE), antioxidant activity (μmol TE mL^{-1} for OWW and OP; μmol TE g^{-1} for OLE) and phenolic profile (mg·mL^{-1} for OWW and OP; mg·g^{-1} for OLE) of phenolic extracts.

Parameter	OWW	OP	OLE
TPC	8.61	4.46	190
ABTS	3579	451	780
Hydroxytyrosol	1.43	0.23	nd
Tyrosol	0.25	0.16	3.61
Caffeic acid	0.07	0.01	nd
Apigenin-7-O-glucoside	0.03	0.03	6.92
Vanillic acid	0.01	0.03	nd
Oleuropein	0.02	0.09	137
Verbascoside	nd	nd	20
Luteolin-7-O-glucoside	0.008	0.012	3.14
Rutin	nd	nd	3.83

GAE: gallic acid equivalents; OWW: olive mill wastewater; OP: olive pomace; TE: Trolox equivalents; OLE: olive leaves extract.

3.2. Biological Characterization of Olive By-Products

An increase in systemic oxidative stress related to abnormal production of ROS is often associated with aging and several diseases including hypertension, obesity, and cancer. A growing number of bioactive phytocompounds, known for their beneficial and antioxidant actions on human health, are more frequently recommended as therapeutic adjuvants in several diseases [25]. Here, we evaluated the effects of extracts obtained from olive by-products and wastes on cell viability in human colorectal carcinoma HCT8 cells. Furthermore, the potential antioxidant efficacy of olive by-products and wastes was assayed as well. Cells were incubated overnight with extracts derived from OWW at increasing concentrations (0.03 mg·L^{-1}, 0.06 mg·L^{-1}, and 0.12 mg·L^{-1} of phenols). Compared to cells left under control conditions, treatment with OWW extracts does not alter cell viability at the concentration used (Figure 1a). Under similar conditions, ROS content was measured in HCT8 cells. Compared to untreated cells (CTR), ROS production increased in cells incubated with OWW extracts at 0.12 mg·L^{-1} (Figure 1b). Indeed, compared to the positive control (tBHP), cells co-incubated with the oxidant tBHP, and OWW extracts showed a significant decrease in ROS generation induced by tBHP (Figure 1b).

Next, HCT8 cells were treated with extracts obtained from OP at increasing concentrations (0.03 mg·L^{-1}, 0.06 mg·L^{-1} and 0.12 mg·L^{-1} of phenols). Compared to CTR cells, treatment with OP extracts at 0.12 mg·L^{-1} of phenols displayed a cytotoxic effect (Figure 2a). Fluorimetric measurements revealed that cells co-incubated with the oxidant tBHP and OP extracts at 0.12 mg·L^{-1} of phenols showed a significant decrease in ROS generation induced by tBHP. However, a relevant but not significant reduction in ROS content was measured in cells co-treated with tBHP and OP extracts at 0.03 mg·L^{-1} and 0.06 mg·L^{-1} of phenols. Incubation with the OP extracts alone, at increasing concentrations (0.03 mg·L^{-1}, 0.06 mg·L^{-1}, and 0.12 mg·L^{-1} of phenols) does not alter ROS production.

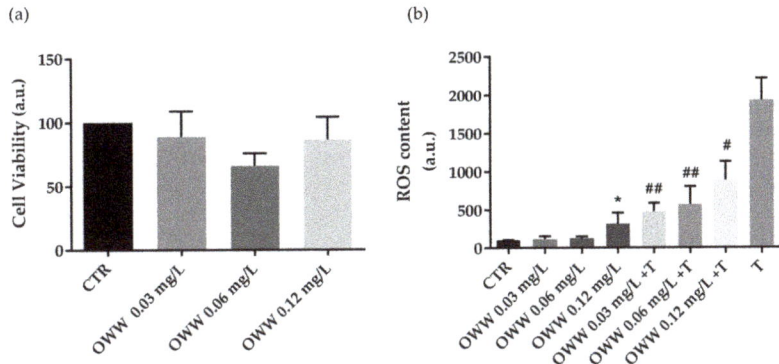

Figure 1. Cell viability (**a**) and ROS content (**b**) in HCT8 cells exposed to olive wastewater (OWW). (**a**) Cells were left under basal condition or exposed to increasing concentrations (0.03 mg·L^{-1}, 0.06 mg·L^{-1}, and 0.12 mg·L^{-1} of phenols) of OWW and were stained with crystal violet solution. Data are shown as means ± Standard Error Mean (S.E.M.) of 3 independent experiments and analyzed by one-way ANOVA followed by Dunnett's multiple comparisons test. (**b**) ROS content was measured using dihydrorhodamine-123 fluorescence in cells treated as previously described. As a positive control, cells were treated with tert-butyl hydroperoxide (tBHP). Data are shown as means ± S.E.M. of 3 independent experiments and analyzed by one-way ANOVA followed by Dunnett's multiple comparisons test. (* $p < 0.05$ vs. CTR; ## $p < 0.01$ and # $p < 0.05$ vs. tBHP).

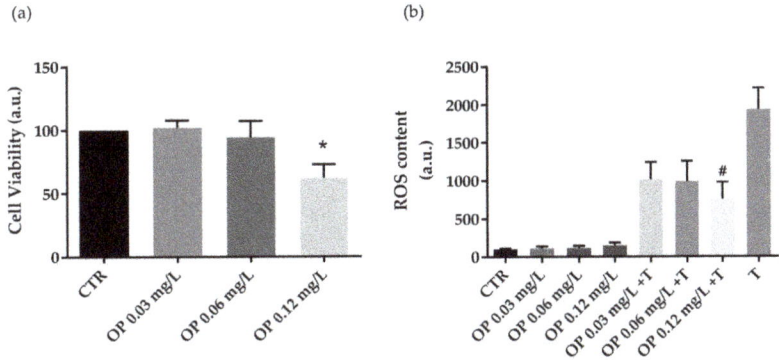

Figure 2. Cell viability (**a**) and ROS content (**b**) in HCT8 cells exposed to olive pomace (OP). (**a**) Cells were left under basal condition or exposed to increasing concentrations (0.03 mg·L^{-1}, 0.06 mg·L^{-1}, and 0.12 mg·L^{-1} of phenols) of OP and were stained with crystal violet solution. Data are shown as means ± S.E.M. of 3 independent experiments and analyzed by one-way ANOVA followed by Dunnett's multiple comparisons test (* $p < 0.05$ vs. CTR). (**b**) ROS content was measured using dihydrorhodamine-123 fluorescence in cells treated as previously described. As a positive control, cells were treated with tert-butyl hydroperoxide (tBHP). Data are shown as means ± S.E.M. of 3 independent experiments and analyzed by one-way ANOVA followed by Dunnett's multiple comparisons test. (# $p < 0.05$ vs. tBHP).

Similar evaluations were performed in cells treated with OLE. Specifically, compared to CTR, treatment with OLE does not alter cell viability at the used concentrations (Figure 3a). Moreover, compared to the positive control (tBHP), HCT8 cells coincubated with tBHP and OLE at 0.12 mg·L^{-1} of phenols displayed a significant reduction in ROS content. Incubation with the OLE alone, at increasing concentrations (0.03 mg·L^{-1}, 0.06 mg·L^{-1}, and 0.12 mg·L^{-1} of phenols) does not alter ROS content (Figure 3b).

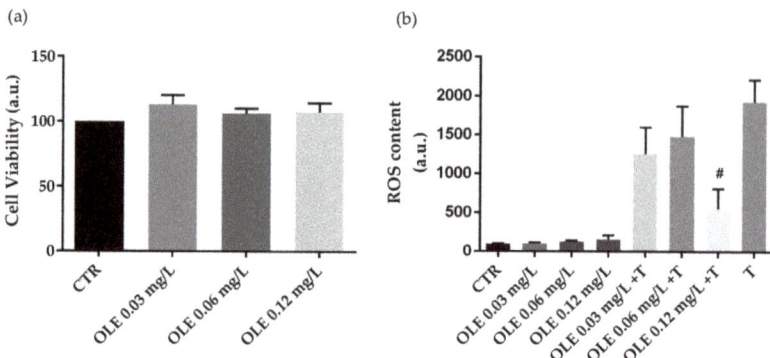

Figure 3. Cell viability (**a**) and ROS content (**b**) in HCT8 cells exposed to olive leaves extract (OLE). (**a**) Cells were left under basal condition or exposed to increasing concentrations (0.03 mg·L^{-1}, 0.06 mg·L^{-1}, and 0.12 mg·L^{-1} of phenols) of OLE and were stained with crystal violet solution. Data are shown as means ± S.E.M. of 3 independent experiments and analyzed by one-way ANOVA followed by Dunnett's multiple comparisons test. (**b**) ROS content was measured using dihydrorhodamine-123 fluorescence in cells treated as previously described. As a positive control, cells were treated with tert-butyl hydroperoxide (tBHP). Data are shown as means ± S.E.M. of 3 independent experiments and analyzed by one-way ANOVA followed by Dunnett's multiple comparisons test. (# $p < 0.05$ vs. tBHP).

4. Discussion

Extra virgin olive oil (EVOO) consumption displays protective effects on human health [26,27]. Acute intake of EVOO significantly ameliorates insulin sensitivity and glycemia in patients with metabolic syndrome by modulating the expression of genes involved in inflammation, metabolism, and cancer [28]. Furthermore, in high-fat-diet fed rats, administration of polyphenol-enriched olive oil improved insulin sensitivity, lipid profile, and interfered with inflammatory pathways [29]. Several compounds of EVOO, indeed, can modulate the NF-kB signal transduction pathway and decrease oxidative stress [30]. Importantly, the volume of wastes derived from the olive oil industry represents a worrying issue. While olive pomace can be treated to produce olive pomace oil or biomasses for fuel generation, olive mill wastewater represents a serious environmental matter and the disposal costs are very high. Notably, it has been reported that olive mill by-products and wastes can still contain numerous bioactive and valuable compounds useful in the pharmaceutical, food, and cosmetic industries. It is well established that the phenolic composition within the oil and its derivative by-products is complex and it is related to olive variety, the maturation stage, and seasons [31,32].

Oleuropein, which is found in olives, olive leaves, and oil can be degraded during olive oil production. Hydroxytyrosol and tyrosol are the main degradation products of oleuropein displaying several biological activities. Among the considered by-products, OLs are the main source of oleuropein, as reported in Table 1, since this waste is not subjected to the processes involved in oleuropein degradation during the olive oil production [33,34]. In pancreatic cancer cells, dose-dependently treatment with oleuropein or hydroxytyrosol significantly reduced cell viability and proliferation [32]. Moreover, oleuropein administration promotes apoptosis by upregulating the p53 signal pathway in breast and colon cancer cells [34,35]. Verbascoside is one of the most abundant phenolic compounds in the OLE; this compound displays beneficial effects for its antioxidant and anti-inflammatory and antitumor actions beyond the neuroprotective properties [35]. In the present report, incubation with an increasing concentration of extracts obtained from olive by-products and wastes does not alter cell viability which might be due to the low concentrations that were used. However, only the treatment with the extract obtained from the olive pomace, used at the highest concentration, caused a significant reduction in cell viability as assessed by the crystal violet assay. By comparing the extract composition of the OP with the ones of the other extracts, we found that OP extract contains higher amounts of vanillic acid and luteolin-7-glucoside. In colon HCT116 cells, exposure to vanillic acid induced G1 phase arrest and inhibited cell proliferation, possibly by inhibiting the expression of HIF-1, mTOR, and Raf/MEK/ERK signal transduction pathways [31]. Furthermore, luteolin-7-glucoside reduced keratinocyte proliferation by inhibiting IL22/STAT3 signaling [33]. Therefore, exposure to high levels of vanillic acid and luteolin-7-O-glucoside may explain the observed reduction of HCT8 cell survival exposed to the highest concentration of the OP extract.

Olive oil and olive by-products are thus a natural source of antioxidant compounds and their beneficial effects have been tested in several in vitro and in vivo experimentations [36–38]. Interestingly, it has been proposed that the concentration of polyphenols in the intestinal tract is higher than elsewhere because phenols can reach the colon directly or through the bile. However, relevant variability within volunteers has been also documented, which might be due to the difference in the expression level of metabolizing enzymes and to a different microbiota profile [39]. Recent reports revealed that selective phenols can interfere with oxidative and inflammatory pathways. Specifically, in a mouse model of acute colitis, administration of oleuropein-loaded lipid nanocarriers significantly reduced TNF-α release and intracellular ROS [40] strongly suggesting that phenols in the extracts may be considered valuable compounds for pharmaceutical applications. Here, with the same phenolic concentration, the extract derived from the olive wastewater displayed a higher antioxidant activity compared with the extracts derived from olive pomace and leaves. These responses might be due to the high concentration level of hydroxytyrosol and tyrosol detected in the olive mill wastewater [41]. Among phenolic compounds isolated from the olive tree, hydroxytyrosol is the most powerful antioxidant followed by oleuropein, caffeic acid and, tyrosol [42]. Conversely, at a high concentration level, antioxidants can react with molecular oxygen thereby acting as pro-oxidants [43]. Impairing antioxidant cellular defense and inducing the generation of reactive species may be a useful strategy to remove, and kill cancer cells. In human colon cancer DLD1 cells, hydroxytyrosol induced cell apoptosis, and the activation of PIPK3/AKT pathway through ROS production [44]. Accordingly, incubation with the only extract obtained from OWW, at the highest concentration, significantly increased the generation of ROS, which might be useful to activate specific intracellular signals leading to cell death as already shown by others [44,45]. Furthermore, we found that at the highest concentration level, incubation with the extract obtained from OP reduced cell viability. By contrast, treatment with phenols at a lower concentration does not exert a cytotoxic effect but it can reduce the ROS production induced by the synthetic oxidant compounds tBHP. Together, these findings suggest that phenols in the olive by-products can tightly modulate the intracellular ROS content that can act as signaling molecules cross-linking several intracellular pathways modulating cell viability.

5. Conclusions

This study underscores the crucial importance of the chemical and functional characterization of olive by-products. Using the ileo-carcinoma cell line HCT8 cells we found differential effects of the extracts on ROS generation and cell viability even though all extracts contain an enriched fraction of polyphenols. Therefore, further investigations would be needed to describe and identify the potential role and the possible applications and use of the bioactive compounds in the olive wastes in the pharmacological and food industry.

To conclude, according to the guiding principles of the circular economy, this study highlights and underlines the possibility of valorizing olive by-products and wastewater as they contain numerous bioactive compounds that individually or synergistically can bring beneficial effects to human health.

Author Contributions: Conceptualization, M.C., C.M., G.T. and M.R.; methodology, M.C., M.D., G.D., A.D.B. and A.D.M.; data curation, G.T., G.V. and F.C.; writing—original draft preparation, M.C. and G.T.; writing—review and editing, G.T., F.C. and M.P.; funding acquisition, F.C. All authors have read and agreed to the published version of the manuscript.

Funding: This research was funded by the AGER 2 Project, grant no.2016-0105.

Acknowledgments: The financial support of Marianna Ranieri is supported by "Intervento cofinanziato dal Fondo di Sviluppo e Coesione 2007–2013–APQ Ricerca Regione Puglia, Programma Regionale a Sostegno della Specializzazione Intelligente e della Sostenibilità Sociale ed Ambientale–FutureInResearch" (code CHVNKZ4). Annarita Di Mise is supported by "Attrazione e Mobilità dei Ricercatori, PON "R&I" 2014–2020, Azione I.2" (code AIM1893457).

Conflicts of Interest: The authors declare no conflict of interest.

References

1. Kouka, P.; Tsakiri, G.; Tzortzi, D.; Dimopoulou, S.; Sarikaki, G.; Stathopoulos, P.; Veskoukis, A.S.; Halabalaki, M.; Skaltsounis, A.L.; Kouretas, D. The Polyphenolic Composition of Extracts Derived from Different Greek Extra Virgin Olive Oils Is Correlated with Their Antioxidant Potency. *Oxidative Med. Cell. Longev.* **2019**, *2019*, 1870965. [CrossRef] [PubMed]
2. Saez, J.A.; Perez-Murcia, M.D.; Vico, A.; Martinez-Gallardo, M.R.; Andreu-Rodriguez, F.J.; Lopez, M.J.; Bustamante, M.A.; Sanchez-Hernandez, J.C.; Moreno, J.; Moral, R. Olive mill wastewater-evaporation ponds long term stored: Integrated assessment of in situ bioremediation strategies based on composting and vermicomposting. *J. Hazard. Mater.* **2020**, *402*, 123481. [CrossRef] [PubMed]
3. Babic, S.; Malev, O.; Pflieger, M.; Lebedev, A.T.; Mazur, D.M.; Kuzic, A.; Coz-Rakovac, R.; Trebse, P. Toxicity evaluation of olive oil mill wastewater and its polar fraction using multiple whole-organism bioassays. *Sci. Total Environ.* **2019**, *686*, 903–914. [CrossRef] [PubMed]
4. Sahin, S.; Bilgin, M. Olive tree (*Olea europaea* L.) leaf as a waste by-product of table olive and olive oil industry: A review. *J. Sci. Food Agric.* **2018**, *98*, 1271–1279. [CrossRef] [PubMed]
5. Russo, M.; Bonaccorsi, I.L.; Cacciola, F.; Dugo, L.; De Gara, L.; Dugo, P.; Mondello, L. Distribution of bioactives in entire mill chain from the drupe to the oil and wastes. *Nat. Prod. Res.* **2020**, 1–6. [CrossRef] [PubMed]
6. Ochando-Pulido, J.M.; Martinez-Ferez, A. Novel micro/ultra/nanocentrifugation membrane process assessment for revalorization and reclamation of agricultural wastewater. *J. Environ. Manag.* **2018**, *222*, 447–453. [CrossRef]
7. Sygouni, V.; Pantziaros, A.G.; Iakovides, I.C.; Sfetsa, E.; Bogdou, P.I.; Christoforou, E.A.; Paraskeva, C.A. Treatment of Two-Phase Olive Mill Wastewater and Recovery of Phenolic Compounds Using Membrane Technology. *Membranes* **2019**, *9*, 27. [CrossRef]
8. De Bruno, A.; Romeo, R.; Piscopo, A.; Poiana, M. Antioxidant quantification in different portions obtained during olive oil extraction process in an olive oil press mill. *J. Sci. Food Agric.* **2020**. [CrossRef]
9. Romani, A.; Ieri, F.; Urciuoli, S.; Noce, A.; Marrone, G.; Nediani, C.; Bernini, R. Health Effects of Phenolic Compounds Found in Extra-Virgin Olive Oil, By-Products, and Leaf of *Olea europaea* L. *Nutrients* **2019**, *11*, 1776. [CrossRef]
10. Difonzo, G.; Russo, A.; Trani, A.; Paradiso, V.M.; Ranieri, M.; Pasqualone, A.; Summo, C.; Tamma, G.; Silletti, R.; Caponio, F. Green extracts from Coratina olive cultivar leaves: Antioxidant characterization and biological activity. *J. Funct. Foods* **2017**, *31*, 63–70. [CrossRef]
11. Ranieri, M.; Di Mise, A.; Difonzo, G.; Centrone, M.; Venneri, M.; Pellegrino, T.; Russo, A.; Mastrodonato, M.; Caponio, F.; Valenti, G.; et al. Green olive leaf extract (OLE) provides cytoprotection in renal cells exposed to low doses of cadmium. *PLoS ONE* **2019**, *14*, e0214159. [CrossRef] [PubMed]
12. Asgharzade, S.; Sheikhshabani, S.H.; Ghasempour, E.; Heidari, R.; Rahmati, S.; Mohammadi, M.; Jazaeri, A.; Amini-Farsani, Z. The effect of oleuropein on apoptotic pathway regulators in breast cancer cells. *Eur. J. Pharmacol.* **2020**, *886*, 173509. [CrossRef] [PubMed]
13. Dong, Y.Z.; Li, L.; Espe, M.; Lu, K.L.; Rahimnejad, S. Hydroxytyrosol Attenuates Hepatic Fat Accumulation via Activating Mitochondrial Biogenesis and Autophagy through the AMPK Pathway. *J. Agric. Food Chem.* **2020**, *68*, 9377–9386. [CrossRef]
14. D'Evoli, L.; Morroni, F.; Lombardi-Boccia, G.; Lucarini, M.; Hrelia, P.; Cantelli-Forti, G.; Tarozzi, A. Red chicory (*Cichorium intybus* L. cultivar) as a potential source of antioxidant anthocyanins for intestinal health. *Oxidative Med. Cell. Longev.* **2013**, *2013*, 704310. [CrossRef]
15. Owen, R.W.; Giacosa, A.; Hull, W.E.; Haubner, R.; Spiegelhalder, B.; Bartsch, H. The antioxidant/anticancer potential of phenolic compounds isolated from olive oil. *Eur. J. Cancer* **2000**, *36*, 1235–1247. [CrossRef]
16. Tang, X.; Kuhlenschmidt, T.B.; Li, Q.; Ali, S.; Lezmi, S.; Chen, H.; Pires-Alves, M.; Laegreid, W.W.; Saif, T.A.; Kuhlenschmidt, M.S. A mechanically-induced colon cancer cell population shows increased metastatic potential. *Mol. Cancer* **2014**, *13*, 131. [CrossRef] [PubMed]
17. Lievin-Le Moal, V.; Servin, A.L. Pathogenesis of human enterovirulent bacteria: Lessons from cultured, fully differentiated human colon cancer cell lines. *Microbiol. Mol. Biol. Rev.* **2013**, *77*, 380–439. [CrossRef] [PubMed]

18. Romeo, R.; De Bruno, A.; Imeneo, V.; Piscopo, A.; Poiana, M. Evaluation of enrichment with antioxidants from olive oil mill wastes in hydrophilic model system. *J. Food Process. Preserv.* **2019**, *43*, e14211. [CrossRef]
19. De Bruno, A.; Romeo, R.; Fedele, F.L.; Sicari, A.; Piscopo, A.; Poiana, M. Antioxidant activity shown by olive pomace extracts. *J. Environ. Sci. Health* **2018**, *53*, 526–533. [CrossRef]
20. Marroncini, G.; Fibbi, B.; Errico, A.; Grappone, C.; Maggi, M.; Peri, A. Effects of low extracellular sodium on proliferation and invasive activity of cancer cells in vitro. *Endocrine* **2020**, *67*, 473–484. [CrossRef]
21. Romeo, R.; De Bruno, A.; Imeneo, V.; Piscopo, A.; Poiana, M. Impact of Stability of Enriched Oil with Phenolic Extract from Olive Mill Wastewaters. *Foods* **2020**, *9*, 856. [CrossRef] [PubMed]
22. Ghasemi, S.; Koohi, D.E.; Emmamzadehhashemi, M.S.B.; Khamas, S.S.; Moazen, M.; Hashemi, A.K.; Amin, G.; Golfakhrabadi, F.; Yousefi, Z.; Yousefbeyk, F. Investigation of phenolic compounds and antioxidant activity of leaves extracts from seventeen cultivars of Iranian olive (*Olea europaea* L.). *J. Food Sci. Technol.* **2018**, *55*, 4600–4607. [CrossRef] [PubMed]
23. Hayes, J.; Allen, P.; Brunton, N.; O'Grady, M.; Kerry, J.P. Phenolic composition and in vitro antioxidant capacity of four commercial phytochemical products: Olive leaf extract (*Olea europaea* L.), lutein, sesamol and ellagic acid. *Food Chem.* **2011**, *126*, 948–955. [CrossRef]
24. Benavente-Garcia, O.; Castillo, J.; Lorente, J.; Ortuño, A.D.R.J.; Del Rio, J.A. Antioxidant activity of phenolics extracted from *Olea europaea* L. leaves. *Food Chem.* **2000**, *68*, 457–462. [CrossRef]
25. Tun, S.; Spainhower, C.J.; Cottrill, C.L.; Lakhani, H.V.; Pillai, S.S.; Dilip, A.; Chaudhry, H.; Shapiro, J.I.; Sodhi, K. Therapeutic Efficacy of Antioxidants in Ameliorating Obesity Phenotype and Associated Comorbidities. *Front. Pharmacol.* **2020**, *11*, 1234. [CrossRef] [PubMed]
26. Ordovas, J.M.; Kaput, J.; Corella, D. Nutrition in the genomics era: Cardiovascular disease risk and the Mediterranean diet. *Mol. Nutr. Food Res.* **2007**, *51*, 1293–1299. [CrossRef]
27. Waterman, E.; Lockwood, B. Active components and clinical applications of olive oil. *Altern. Med. Rev. A J. Clin. Ther.* **2007**, *12*, 331–342.
28. D'Amore, S.; Vacca, M.; Cariello, M.; Graziano, G.; D'Orazio, A.; Salvia, R.; Sasso, R.C.; Sabba, C.; Palasciano, G.; Moschetta, A. Genes and miRNA expression signatures in peripheral blood mononuclear cells in healthy subjects and patients with metabolic syndrome after acute intake of extra virgin olive oil. *Biochim. Et Biophys. Acta* **2016**, *1861*, 1671–1680. [CrossRef]
29. Lama, A.; Pirozzi, C.; Mollica, M.P.; Trinchese, G.; Di Guida, F.; Cavaliere, G.; Calignano, A.; Mattace Raso, G.; Berni Canani, R.; Meli, R. Polyphenol-rich virgin olive oil reduces insulin resistance and liver inflammation and improves mitochondrial dysfunction in high-fat diet fed rats. *Mol. Nutr. Food Res.* **2017**, *61*, 1600418. [CrossRef]
30. Cariello, M.; Contursi, A.; Gadaleta, R.M.; Piccinin, E.; De Santis, S.; Piglionica, M.; Spaziante, A.F.; Sabba, C.; Villani, G.; Moschetta, A. Extra-Virgin Olive Oil from Apulian Cultivars and Intestinal Inflammation. *Nutrients* **2020**, *12*, 1084. [CrossRef]
31. Gong, J.; Zhou, S.; Yang, S. Vanillic Acid Suppresses HIF-1alpha Expression via Inhibition of mTOR/p70S6K/4E-BP1 and Raf/MEK/ERK Pathways in Human Colon Cancer HCT116 Cells. *Int. J. Mol. Sci.* **2019**, *20*, 456. [CrossRef] [PubMed]
32. Goldsmith, C.D.; Bond, D.R.; Jankowski, H.; Weidenhofer, J.; Stathopoulos, C.E.; Roach, P.D.; Scarlett, C.J. The Olive Biophenols Oleuropein and Hydroxytyrosol Selectively Reduce Proliferation, Influence the Cell Cycle, and Induce Apoptosis in Pancreatic Cancer Cells. *Int. J. Mol. Sci.* **2018**, *19*, 1937. [CrossRef] [PubMed]
33. Palombo, R.; Savini, I.; Avigliano, L.; Madonna, S.; Cavani, A.; Albanesi, C.; Mauriello, A.; Melino, G.; Terrinoni, A. Luteolin-7-glucoside inhibits IL-22/STAT3 pathway, reducing proliferation, acanthosis, and inflammation in keratinocytes and in mouse psoriatic model. *Cell Death Dis.* **2016**, *7*, e2344. [CrossRef] [PubMed]
34. Hassan, Z.K.; Elamin, M.H.; Omer, S.A.; Daghestani, M.H.; Al-Olayan, E.S.; Elobeid, M.A.; Virk, P. Oleuropein induces apoptosis via the p53 pathway in breast cancer cells. *Asian Pac. J. Cancer Prev.* **2013**, *14*, 6739–6742. [CrossRef] [PubMed]
35. Cardeno, A.; Sanchez-Hidalgo, M.; Rosillo, M.A.; Alarcon de la Lastra, C. Oleuropein, a secoiridoid derived from olive tree, inhibits the proliferation of human colorectal cancer cell through downregulation of HIF-1alpha. *Nutr. Cancer* **2013**, *65*, 147–156. [CrossRef] [PubMed]
36. Karanovic, D.; Mihailovic-Stanojevic, N.; Miloradovic, Z.; Ivanov, M.; Vajic, U.J.; Grujic-Milanovic, J.; Markovic-Lipkovski, J.; Dekanski, D.; Jovovic, D. Olive leaf extract attenuates adriamycin-induced focal segmental glomerulosclerosis in spontaneously hypertensive rats via suppression of oxidative stress, hyperlipidemia, and fibrosis. *Phytother. Res.* **2020**. [CrossRef] [PubMed]
37. Lins, P.G.; Marina Piccoli Pugine, S.; Scatolini, A.M.; de Melo, M.P. In vitro antioxidant activity of olive leaf extract (*Olea europaea* L.) and its protective effect on oxidative damage in human erythrocytes. *Heliyon* **2018**, *4*, e00805. [CrossRef]
38. Visioli, F.; Bellomo, G.; Galli, C. Free radical-scavenging properties of olive oil polyphenols. *Biochem. Biophys. Res. Commun.* **1998**, *247*, 60–64. [CrossRef]
39. Deiana, M.; Serra, G.; Corona, G. Modulation of intestinal epithelium homeostasis by extra virgin olive oil phenolic compounds. *Food Funct.* **2018**, *9*, 4085–4099. [CrossRef]
40. Huguet-Casquero, A.; Xu, Y.; Gainza, E.; Pedraz, J.L.; Beloqui, A. Oral delivery of oleuropein-loaded lipid nanocarriers alleviates inflammation and oxidative stress in acute colitis. *Int. J. Pharm.* **2020**, *586*, 119515. [CrossRef]
41. Azaizeh, H.; Halahlih, F.; Najami, N.; Brunner, D.; Faulstich, M.; Tafesh, A. Antioxidant activity of phenolic fractions in olive mill wastewater. *Food Chem.* **2012**, *134*, 2226–2234. [CrossRef] [PubMed]
42. Martinez, L.; Ros, G.; Nieto, G. Hydroxytyrosol: Health Benefits and Use as Functional Ingredient in Meat. *Medicines* **2018**, *5*, 13. [CrossRef] [PubMed]

43. Sotler, R.; Poljsak, B.; Dahmane, R.; Jukic, T.; Pavan Jukic, D.; Rotim, C.; Trebse, P.; Starc, A. Prooxidant Activities of Antioxidants and Their Impact on Health. *Acta Clin. Croat.* **2019**, *58*, 726–736. [CrossRef] [PubMed]
44. Sun, L.; Luo, C.; Liu, J. Hydroxytyrosol induces apoptosis in human colon cancer cells through ROS generation. *Food Funct.* **2014**, *5*, 1909–1914. [CrossRef]
45. Toric, J.; Markovic, A.K.; Brala, C.J.; Barbaric, M. Anticancer effects of olive oil polyphenols and their combinations with anticancer drugs. *Acta Pharm.* **2019**, *69*, 461–482. [CrossRef]

Article

Increasing the Content of Olive Mill Wastewater in Biogas Reactors for a Sustainable Recovery: Methane Productivity and Life Cycle Analyses of the Process

Souraya Benalia, Giacomo Falcone, Teodora Stillitano *, Anna Irene De Luca, Alfio Strano, Giovanni Gulisano, Giuseppe Zimbalatti and Bruno Bernardi

Dipartimento di Agraria, Università degli Studi Mediterranea di Reggio Calabria, Località Feo di Vito, 89122 Reggio Calabria, Italy; soraya.benalia@unirc.it (S.B.); giacomo.falcone@unirc.it (G.F.); anna.deluca@unirc.it (A.I.D.L.); astrano@unirc.it (A.S.); ggulisano@unirc.it (G.G.); gzimbalatti@unirc.it (G.Z.); bruno.bernardi@unirc.it (B.B.)
* Correspondence: teodora.stillitano@unirc.it

Abstract: Anaerobic codigestion of olive mill wastewater for renewable energy production constitutes a promising process to overcome management and environmental issues due to their conventional disposal. The present study aims at assessing biogas and biomethane production from olive mill wastewater by performing biochemical methane potential tests. Hence, mixtures containing 0% (blank), 20% and 30% olive mill wastewater, in volume, were experimented on under mesophilic conditions. In addition, life cycle assessment and life cycle costing were performed for sustainability analysis. Particularly, life cycle assessment allowed assessing the potential environmental impact resulting from the tested process, while life cycle costing in conjunction with specific economic indicators allowed performing the economic feasibility analysis. The research highlighted reliable outcomes: higher amounts of biogas (80.22 ± 24.49 $NL.kg_{SV}^{-1}$) and methane (47.68 ± 17.55 $NL.kg_{SV}^{-1}$) were obtained when implementing a higher amount of olive mill wastewater (30%) (v/v) in the batch reactors. According to life cycle assessment, the biogas ecoprofile was better when using 20% (v/v) olive mill wastewater. Similarly, the economic results demonstrated the profitability of the process, with better performances when using 20% (v/v) olive mill wastewater. These findings confirm the advantages from using farm and food industry by-products for the production of renewable energy as well as organic fertilizers, which could be used in situ to enhance farm sustainability.

Keywords: anaerobic codigestion; biomethane; life cycle assessment (LCA); life cycle costing (LCC); olive mill by-products

1. Introduction

The Euro-Mediterranean region boasts an olive oil production exceeding 2950 thousand tons, representing 92.41% of the worldwide production [1] and Italy in its own produced more than 211 thousand tons in 2020. Calabria, in Southern Italy, is the second largest olive oil producer, with over 184,000 ha of olive groves and a production of more than 82,262.7 t of olive oil in 2020 according to ISTAT data [2]. Among the 4480 active mills in Italy, this commodity is mainly produced in small and medium ones with production capacities of up to 500 tons of olive in 76% of cases [3]. This indicates, as for the olive groves, how fragmented the olive processing sector is. Most of these mills adopt three-phase extraction system, generating two kinds of by-products—i.e., olive mill wastewater (OMWW) and olive mill solid waste (OMSW) or olive cake, in addition to olive oil. Messineo et al. [4] estimated that one ton of olives may generate up to 1.6 cubic meters of olive mill wastewater using a three-phase extraction system. This wastewater mainly derives from olive washing as well as from the addition of water during centrifugation. Up until recently, the most common and implemented routines for OMWW management were its storage in evaporation ponds or its controlled disposal on agricultural terrains [5]. However, this practice presents several negative aspects [6]. The concentration of olive milling from both spatial and temporal points of view and the low biodegradability that characterizes OMWW limit its disposal on agricultural lands according to the regulation in vigor and therefore create management problems as well as environmental impacts. Hence, it becomes crucial to look for an eco-friendly way to manage this kind of effluent. In this sense, its recovery for energy production through anaerobic digestion process may constitute a reliable solution.

1.1. Application of Anaerobic Codigestion Process to Olive Mill Wastewater

Due to the physical-chemical features that characterize OMWW it is difficult to expect acceptable yields in terms of biogas and methane [7], without applying previous treatments [8].

Bearing in mind the objective of the European Union in terms of energy strategy, which aims at increasing the share of renewable energy up to 32% by 2030 [9], and considering the fact that only 18% of the renewable energy produced in Italy comes from biomass and organic waste, anaerobic codigestion of OMWW with other farm, livestock or food industry by-products constitutes a suitable and a sustainable way to buffer OMWW properties in view of biogas and biomethane production, such as carried out by Kougias et al. [10]. These authors performed batch and continuous trials under mesophilic conditions mixing up to 40% OMWW with swine manure and obtained up to 373 $mL.g_{VS}^{-1}$ of methane. Battista et al. [11] tested mixtures of olive mill wastewater and olive mill solid waste coming from both three-phase and two-phase extraction systems, with milk whey. They obtained better results (1.23 $L_{CH4}.Lreactor^{-1}.d^{-1}$) with three-phase solid waste rather than two-phase solid waste. Thanos et al. [12] obtained an increase in biogas yield ranging from 0.7 ± 0.4 $L.Lreactor^{-1}.d^{-1}$ to 1.2 ± 0.3 $L.Lreactor^{-1}.d^{-1}$ using 40% v/v of OMWW mixed with liquid pig manure and cheese whey.

1.2. Life Cycle Analysis

Life cycle-based methodologies are increasingly approved as very powerful and reliable tools to quantify the impact generated from a product/service along the entire production process, which is explored in all its phases and constituents, and throughout its whole duration. In this context, Life Cycle Assessment (LCA) and Life Cycle Costing (LCC) methodologies were developed within the so-called Life Cycle Management (LCM) framework and validated by means of standardization processes. LCA has been defined as "[...] an objective process to evaluate the environmental burdens associated to a product, a process, or an activity by identifying energy and materials usage and environmental releases, and to evaluate opportunities to achieve environmental improvements" [13] and it has been standardized with the International Organization for Standardization (ISO) norms [14,15]. Hence, a correct implementation of the life cycle assessment (LCA), to account for potential direct or indirect loads, considers specific and consecutive steps (Figure 1). The first step aims at defining the goal and scope of the study, the product system and its functions, in terms of specific parameters, such as the Functional Unit (FU), i.e., the measurement unit to which all inputs and outputs data are related, and system boundaries, i.e., the size of the life cycle, which characterize the object of analysis. The description of data quality and allocation procedures is also considered in this step. The second step deals with life cycle inventory (LCI), e.g., a qualitative and quantitative data collection, and then the quantification of incoming and outgoing flows (energy, materials, and emissions) and validation. The life cycle inventory assessment (LCIA) represents the third step and allows relating all data previously considered to specific impact categories, indicators, and characterization models. With the last step, results are interpreted by highlighting potential critical points of the production process and suggesting improvement strategies for production process performances. Over time, LCA has gained increasing interest and many applications have been carried out in different productive sectors and services.

Currently, LCA represents the paramount tool to be adopted for achieving environmental certifications, such as carbon footprint, water footprint, and ecological footprint.

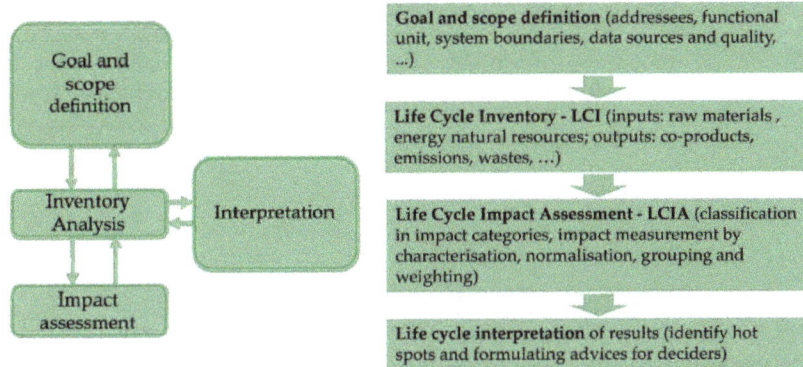

Figure 1. Methodological steps of Life Cycle Assessment (LCA). Source: ISO 14040:2006 [14].

In this framework, LCC, representing the alter-ego of LCA for economic analyses, allows considering both the initial and operating costs by suggesting alternatives for optimizing budget allocation during the system/product lifetime [16]. LCC has been developed in the context of management accounting as an investment analysis tool by using the discounting technique and cash flow models, which represent, up to date, the most widespread approaches. However, several methods and standards for performing LCC have grown over time. In this context, environmental LCC assesses internal costs including externalities that are planned to comprise the monetized effects of environmental impacts not directly accounted for in the firms [17]. In these terms, one of the great potentials of life cycle approaches is to properly analyze the whole life cycle of the object under study considering its interactions with the environmental context, upstream and downstream, in terms of supplying inputs, land use, and load or benefit generated by the products, coproducts, by-products or wastes, as well with the economic dimensions, such as production costs, revenues, cash flows, etc. The multiple direct and indirect connections existing between natural contexts and agri-food systems make the latter particularly interesting for sustainability evaluations. Indeed, the accountability of agri-food production processes in generating negative externalities confirms the need for effective tools to quantify environmental impacts consistently with economic analysis, which aims at evaluating firm performances related to cost reduction, income stabilization, productivity, and competitiveness in the markets. From a life cycle perspective, one of the most investigated agricultural systems is olive growing, which is very representative of Mediterranean countries, as well as olive oil industry, which represents, nowadays, a fast-growing sector worldwide [18]. For over ten years, scholars have applied life cycle tools (more or less methodologically integrated) to analyze olive groves by comparing different production systems (traditional vs. innovative; conventional vs. organic), different agricultural areas, different technological solutions (e.g., irrigation systems), with the objective to measure farm sustainability performances. In recent years, the attention of the scientific research has increasingly shifted towards the so-called "circularity evaluation" or, in other terms, the measurement of alternative systems that not only reduce generated environmental impacts, but also make the whole process more efficient by reducing the consumption of raw materials and avoiding waste [19]. In this sense, and in the case of life cycle studies applied to olive oil production, the most challenging direction is to investigate which production methods can represent viable alternatives to optimize a functioning circular economic system by evaluating the way to convert agricultural by-products into energy or into valuable material fractions. For example, Palmieri et al. [20] analyzed the economic and environmental sustainability of an agri-energy chain from pruning residues of olive groves in nine municipalities in southern Italy; Uceda-Rodríguez et al. [21] evaluated the environmental benefits linked to the production of artificial inert materials created with olive pomace as an alternative to the final disposal of this waste in a landfill; Moreno et al. [22] quantified environmental and economic indexes related to different innovative processes of the conversion of biomass coming from olive pruning residues into energy; finally, Batuecasa et al. [23] conducted an LCA of olive oil production by-products by analyzing both anaerobic digestion and conventional disposal on the soil (Table 1).

Table 1. Analysis of the main literature dealing with life cycle studies applied to agricultural by-products recovery. Source: Our elaboration.

Authors	Year	Title	Journal	Field of Application	Applied Methodologies
Palmieri, N., Suardi, A., Alfano, V., Pari, L.	2020	Circular Economy Model: Insights from a Case Study in South Italy.	Sustainability	Electricity production from pruning residues of olive groves	Profitability and efficiency ratios; Greenhouse gas emissions
Uceda-Rodríguez, M., López-García, A.,B., Moreno-Maroto, J.,M., Cobo-Ceacero, C., J., Cotes-Palomino, M.,T., Martínez García, C.	2020	Evaluation of the Environmental Benefits Associated with the Addition of Olive Pomace in the Manufacture of Lightweight Aggregates.	Materials	Olive pomace recycling as a substitute for clay	Life Cycle Assessment
Moreno, V.C., Iervolino, G., Tugnoli, A., Cozzani, V.	2020	Techno-economic and environmental sustainability of biomass waste conversion based on thermocatalytic reforming.	Waste Management	Biomass waste (olive wood pruning and digestate) to energy conversion process	Mass and energy balances
Batuecasa, E., Tommasi, T., Battista, F., Negro, V., Sonetti, G.	2019	Life Cycle Assessment of waste disposal from olive oil produion: Anaerobic digestion and conventional disposal on soil.	Journal of Environmental Management	Management of by-products from olive oil production: solid–liquid olive pomace and olive mill wastewater	Life Cycle Assessment

Considering the above, the present study aims at assessing the production of biogas and biomethane from the codigestion of olive mill wastewater with digestate. Particularly, different percentages of olive mill wastewater in the reactor contents were experimented on under mesophilic conditions in order to evaluate the eventual threshold of using this by-product in the anaerobic codigestion process. In addition, taking into account that each innovative processes should be evaluated in order to verify its economic feasibility and potentially to prevent its impacts or enhance its benefits, this work aims at analyzing the sustainability of the above-mentioned processes by quantifying the environmental loads and economic implications by applying LCA and LCC methodologies in conjunction with specific economic indicators. Therefore, data input was provided by experimental trials carried out in Calabria (Southern Italy). Particularly, global warming, depletion of the ozone layer, eutrophication, acidification, human and ecosystem toxicity, depletion of natural resources, energy consumption, land use, and water use are the environmental impacts categories considered in LCA implementation, while in LCC analysis, operating costs of the production system were accounted for by monetizing inputs and outputs values.

For this purpose, the Material and Methods section provides the methodological approaches used for laboratory experimental trials as well as LCA and LCC methodology implementation. The study outputs are reported in the Results and Discussion section. Finally, suggestions about practical utilization of the study outcomes are reported in the Conclusions.

2. Materials and Methods

2.1. Biochemical Methane Potential (BMP) of Olive Mill Wastewater

Experimental trials have been conducted at the laboratory scale to assess biochemical methane potential (BMP) of olive wastewater through anaerobic codigestion process.

2.1.1. Anaerobic Codigestion Experiments

Olive mill wastewater (OMWW) was withdrawn during the 2020/2021 campaign from a private mill situated in the Province of Reggio Calabria (Southern Italy, 38°23′28.70″ N;

16°04′31.10″ E), which implements a three-phase extraction system. Digestate (Dig) was withdrawn from a biogas production plant situated in the same province and which already implements olive mill by-products among the feedstock.

Biochemical methane potential (BMP) tests were performed using 2000 mL DURAN® GL 45 laboratory glass bottles as reactors. These were later half-filled with mixtures containing 0% (blank), 20% and 30% (v/v) olive mill wastewater, the remaining content consisted in the digestate. The experimental design is reported in Table 2. Batch reactors were sealed and connected hermetically to the gasbags for biogas sampling. Each thesis was performed in triplicate. Once filled and before sealing, batch reactors were blown through with pure nitrogen (N_2) to remove atmospheric air at the beginning of the fermentation and favor anaerobic conditions. Then, they were incubated for 30 days at 37 °C in a laboratory forced air oven (AgroLab, Italy, TCF 200) to guarantee mesophilic conditions, as shown in Figure 2.

Table 2. Experimental setup of biochemical methane potential (BMP) tests. Source: Our elaboration.

	Thesis 1 (Blank)	Thesis 2	Thesis 3
Olive mill wastewater content (v/v)	0%	20%	30%
Digestate (v/v)	100%	80%	70%

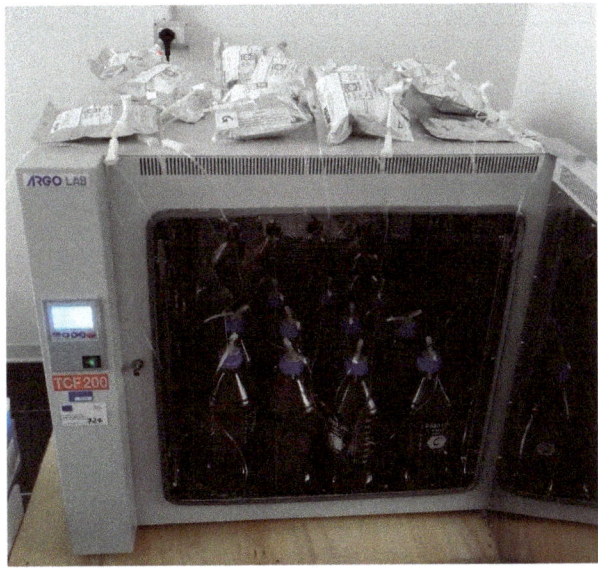

Figure 2. Biochemical methane potential (BMP) tests of olive mill wastewater under mesophilic conditions (37 °C). Source: Picture acquired in our own laboratory.

2.1.2. Substrate and Matrix Characterization

Before each experiment, physical-chemical features of the employed matrices (olive mill wastewater and digestate) as well as the considered mixtures were characterized. This included pH using pH probe (Cryson, GLP 21+), dry content (DC) (%) at 105 °C using a moister analyzer (Ohaus, MB120) and volatile solids VS (%) after ignition at 550 °C using a muffle furnace (Fabber, FBL 70) [24]. Chemical oxygen demand (COD) (g.L^{-1}) was measured using a bench photometer after reaction with Hanna high rate COD reagents. In addition, total polyphenols (PPs) were measured for OMWW according to Folin Ciocalteu method [25], total carbon (TC), total nitrogen (TN) and C/N ratio were determined with an elemental analyzer (Leco, CN628).

2.1.3. Biogas Characterization

The produced biogas was sampled in 1 L multilayer foil gas sampling bags (Restek S.r.l.). Biogas content was analyzed using a dual channel micro gas chromatograph (Agilent, 490MicroGC), implementing Molesieve 5A and PoraPLOT Q columns, both running with helium and a micromachined Thermal Conductivity Detector (TCD), while biogas volume was determined according to the water displacement method.

2.2. Environmental and Economic Analyses

Environmental and economic analyses of biogas and biomethane production from olive processing by-products through an anaerobic codigestion process were carried out using, respectively, LCA and LCC methodologies. As previously mentioned, LCA follows ISO standards, [14] and [15], which define the principles, framework, and requirements of handling a LCA study. Therefore, the procedure followed in this study includes the following four methodological steps: goal and scope definition; life cycle inventory; life cycle impact assessment (LCIA); and interpretation. The LCC methodology applied in this work was based on the approaches described by Ciroth et al. [26] and Moreau and Weidema [27] and is congruent with and complementary to the LCA methodology. Therefore, the system boundary and the functional unit were similar to those of the LCA (Figure 3). The LCC was also implemented in conjunction with specific economic indicators to assess the economic profitability of biogas production.

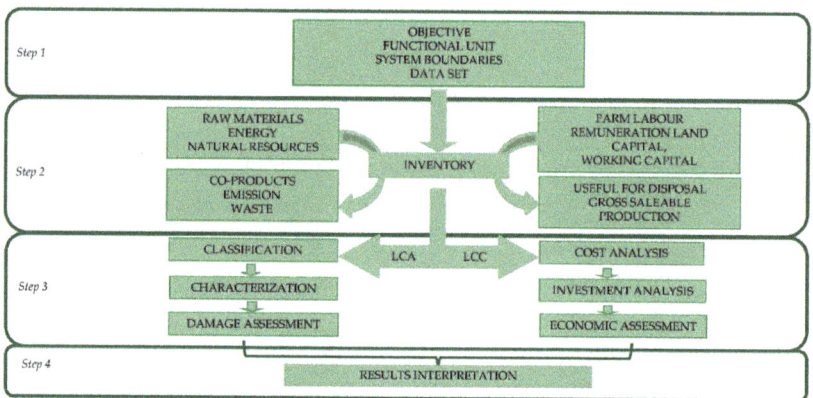

Figure 3. Methodological implementation of Life Cycle Assessment (LCA) and Life Cycle Costing (LCC). Source: Our elaboration.

2.2.1. Scenarios Description

The production of biogas from anaerobic codigestion was evaluated both from environmental and economic points of view, considering the mixtures previously tested at the laboratory scale. Hence, a scaling up from the laboratory level to the industrial level was performed. Specifically, an anaerobic reactor that produces biogas and generates an electrical power of 200 kW was considered for the analyses.

Environmental and economic impact assessments were then performed considering only the mixtures containing OMWW (the control thesis was excluded). Particularly, the following was evaluated: biogas production from the mix containing 20% (v/v) OMWW and 80% (v/v) digestate with a retention time of 16 days (Thesis 2); biogas production from the mix containing 30% (v/v) OMWW and 70% (v/v) digestate with a retention time of 29 days (Thesis 3).

The scaling up was based on the results obtained in the laboratory experiments, modelling the size of the plant into an industrial one considering matrix availability. It has been assumed that the plant is located in the vicinity of another anaerobic digestion plant with an electrical power of 998 kW that produces an annual quantity of digestate equal to 37,000 t, enough to satisfy the feeding needs of the plant fed with OMWW, which, according to the different retention times, requires annual quantities of matrices equal to 45,625.00 (Thesis 2) and 25,172.41 t (Thesis 3).

Specific assumptions for environmental and economic assessments are discussed in the following paragraph.

2.2.2. Goal and Scope Definition, Functional Unit and System Boundaries

For both LCA and LCC, the same assumptions were used for life cycle modelling so that the life cycle of biogas production from OMWW could be assessed according to common criteria from both environmental and economic perspectives. In particular, the function to be analyzed is OMWW recovery to energy; therefore, "1 m^3 of normalized biogas" has been defined as the Functional Unit (FU). Since wastewater is normally considered a waste product of the olive milling process, it has been chosen to limit the system boundaries "from digester gate to the biogas production", considering OMWW as a residual product with zero impact (Figure 4).

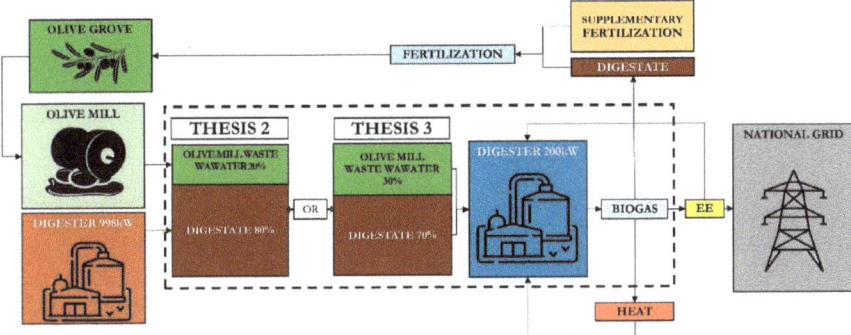

Figure 4. Flowchart of the system boundaries considered in the two scenarios. Source: Our elaboration.

2.2.3. Specific LCA Implementation

Data on quantities of matrices, transport, quantities of biogas and generated heat as well as produced digestate were taken from laboratory trials and scaled up to the 200 kW electrical power plant. According to current regulatory requirements, the plant electricity consumption has been set at 11% for self-consumption. Secondary data on fuel production for transport and digester construction were taken from the Ecoinvent 3.5 database. The methane and ammonia emissions from digestate storage were assessed according to Lovarelli et al. [28]. Fugitive methane losses from digesters and post-digesters and losses during biogas treatment and combustion were considered equal to 2% following Dressler et al. [29].

The inventory data (Table 3) were processed through Simapro 8.5 software using the ILCD 2011 midpoint impact assessment method [30], through which the following impact categories were assessed: climate change; ozone depletion; human toxicity, non cancer effects; human toxicity, cancer effects; particulate matter; ionizing radiation HH; ionizing radiation E (interim); photochemical ozone formation; acidification; terrestrial eutrophication; freshwater eutrophication; marine eutrophication; freshwater ecotoxicity; land use; water resource depletion; mineral, fossil and ren resource depletion.

Table 3. Inventory data. Source: Our elaboration.

	Unit	Thesis 2	Thesis 3
Products			
Biogas	m^3	1.00	1.00
Primary inputs			
Carbon Dioxide	g	428.40	417.69
Inputs			
Transports	$t.km^{-1}$	0.20	0.26
Electricity	kWh	0.67	0.75
Power plant	p	2.24×10^{-7}	3.42×10^{-7}
Emissions			
Carbon dioxide	g	71.97	71.97
Methane	g	12.23	13.71
Ammonia	g	1.41	1.58
Heat	MJ	0.52	0.52

Cut-off criteria were set ignoring all inventory data that would have an impact of less than 1%, such as energy for plant control computers.

The main limitation of the study lies in the scaling operation, whereby productions are directly proportional to those obtained in laboratory trials. Since a decrease in plant efficiency at full-scale is possible, a sensitivity analysis was carried out, reducing the biogas production by 10% and 20%.

2.2.4. Specific LCC Implementation and Profitability Analysis

The LCC analysis aimed at evaluating the overall cost of the two scenarios of biogas production under study (Thesis 2 and Thesis 3). Data collection was conducted in parallel with the inventory phase of LCA to estimate costs related to plant acquisition, operation, and disposal in accordance with Gonzalez et al. [31]. As pointed out by Herbes et al. [32], the site-specific conditions in which the process takes place should be considered. Therefore, in performing the economic analysis, site-specific cost drivers are taken into account.

The initial investment cost for the plant acquisition was EUR 900,000 according to the current market prices, corresponding to a specific cost of EUR 4500 per kW.

Operating costs were split into three categories: materials and services, labor, quota, and other attributions. In the first category, only transport costs for matrix handling were considered, assuming an average distance of 500 km per year. The diesel average price was taken as EUR 0.92 per liter, taking into account an average consumption of 0.05 $L.t^{-1}.km^{-1}$.

In this work, the purchase price of both OMWW and digestate was assumed to be EUR 0 per ton. In the first case, we assumed that the transport cost is covered by olive mills, which avoid the traditional disposal of the wastes on the soil. In the second case, the cost of digestate was considered for free.

Within the labor costs, human labor cost based on local current wage (EUR 8 per hour) and administrative overheads (EUR 10.3 per hour) were included.

In the quota and other attributions category, all those cost items not directly attributable to specific biogas production process stage, represented by quotas (i.e., depreciation, maintenance and insurance), interests in advance capital and capital goods, land rent and levies, were considered.

The expected revenues were estimated considering only the sale of electricity after internal consumption. The electricity produced was assumed to be fed into the national grid. A FiT tariff of EUR 0.233 per kWh was considered [33].

Estimation of end of life costs for the biogas plant disposal was obtained from the literature [31]. The plant disposal was estimated by subtracting from disposal cost the used equipment revenue.

The following assumptions were made to carry out the economic analysis of the two scenarios:

- All of the costs and revenues were discounted for the entire life cycle of 15 years (plant lifetime).
- To select a discount rate, the opportunity cost approach in terms of alternative investments with similar risks and times was used [34]. Here, a discount rate set to 5% was assumed, as in other studies [31,35].
- Constant prices by excluding adjustments for inflation [36] were taken into account.

In order to evaluate the investment feasibility of the biogas production scenarios, specific economic indicators were identified—i.e., Discounted Gross Margin (DGM), Net Present Value (NPV), Internal Rate of Return (IRR) and Discounted Payback Period (DPP). These represent the most common indicators used to compare investment options, which are based on the cash flow model [32].

The DGM indicator provides information on project profitability, as advised by Mel et al. [37] and Stillitano et al. [38], defined in Equation (1):

$$DGM = \sum_{t=1}^{n} \frac{TR_t}{(1+r)^t} - \frac{VC_t}{(1+r)^t} \quad (1)$$

where TR_t is the total revenue in the t-th year; VC_t is the variable cost in the t-th year; t is the time of the cash flow (year); n is the plant lifetime (15 years) and r is the discount rate (5%).

The NPV and IRR indicators were calculated according to Equations (2) and (3), respectively, as suggested by Moreno et al. [39]:

$$NPV = \sum_{t=1}^{n} \frac{CF_t}{(1+r)^t} - I_0 \quad (2)$$

where t is the time of the cash flow (year); n is the plant lifetime (15 years); CF_t is the net cash flow in the t-th year; r is the discount rate (5%) and I_0 is the initial investment, which equals the total facility investment.

$$\sum_{t=1}^{n} \frac{CF_t}{(1+IRR)^t} - I_0 = 0 \quad (3)$$

where IRR is the discount rate, which will make the NPV equal to zero.

When the conditions NPV > 0 and IRR > r occur, the investment is profitable; otherwise, it should be rejected [40].

The formula for calculating the DPP indicator is presented in Equation (4), as suggested by Tse et al. [41]:

$$DPP = LNC \frac{ADC}{DCA} \quad (4)$$

where LNC is the last period with a negative discount cumulative cash flow; ADC is the absolute value of discount cumulative cash flow at the end of the period LNC; DCA is the discount cash flow during the period after LNC.

As argued by Ong and Chun [42], the payback period, defined as the expected number of years required to recover the initial investment, is often used as an indicator of a project's riskiness. In any case, the payback period must be shorter than the time horizon considered.

Lastly, each indicator value has been defined for the FU of 1 m^3 of normalized biogas.

As a final step, a sensitivity analysis was performed for the two scenarios to examine the influence of varying specific parameters over the economic indicators under study [43]. The variables independently evaluated were discount rate (r) set to be floated with ± 20% and biogas yields floated with −10% and −20%.

3. Results and Discussion
3.1. Results of Biochemical Methane Potential (BMP) of Olive Mill Wastewater
3.1.1. Matrix and Substrate Characterization

The results of the initial characterization of the matrices and the substrates are, respectively, reported in Table 4.

Table 4. Matrix and substrate preliminary characterization. Values are expressed as mean ± St. Dev of minimum three replicates for each parameter and each matrix/substrate. Source: Our elaboration.

	Unit	OMWW	Dig/Blank	Thesis 2	Thesis 3
pH		4.65 ± 0.05	7.97 ± 0.16	7.20 ± 0.01	6.93 ± 0.03
DC	%	8.18 ± 0.15	9.31 ± 0.52	9.46 ± 0.74	8.99 ± 0.56
VS $_{\text{dry matter}}$	%	82.08 ± 0.34	79.84 ± 0.72	80.59 ± 0.12	80.47 ± 0.42
COD	$g \cdot L^{-1}$	125.39 ± 3.57	70.35 ± 4.47	80.82 ± 1.59	79.56 ± 1.27
TC	$g \cdot kg^{-1}$	/	481.57 ± 0.77	487.53 ± 3.15	491.13 ± 2.40
TN	$g \cdot kg^{-1}$	/	26.65 ± 0.48	27.83 ± 0.11	29.84 ± 0.27
C/N		/	18.08 ± 0.30	17.52 ± 0.17	16.46 ± 0.20
PPs	$g \cdot L^{-1}$	4.60	/	/	/

According to the obtained data, the pH value of OMWW is, as expected, very low and similar to values reported by other authors [44], while values inherent to theses subjected to anaerobic digestion process are between 6.93 ± 0.03 and 7.97 ± 0.16, with optimal values for both mixes, meaning that the comatrix, i.e., the digestate, exerted a good buffering effect. Dry content (DC) in all cases does not exceed 10%, indicating that the process runs in wet conditions, which consists of the operating mode of most of large-scale reactors worldwide [45,46]. Volatile solid (VS) or organic substance content also represents an important parameter for the anaerobic digestion process as it refers to the susceptible content to be decomposed [47]. In addition, the chemical oxygen demand (COD), whose values are between 70.35 ± 4.47 and 80.82 ± 1.59 $g \cdot L^{-1}$ for the substrates subjected to AcoD, measures the content of oxidizable compounds in the substrate [48] and theoretically enable predicting methane production as 1 g of converted COD corresponds to a maximum of 350 mL of methane [49]. In the three theses, the carbon/nitrogen ratio (C/N) was below the recommended value that should be comprised between 20 and 30, with an optimal value of 30. However, Guarino et al. [50] investigated the effect of a wider C/N interval ranging from 9 to 50 on anaerobic digestion of buffalo manure and obtained a high biomethane productivity (around 60–70%) even with lower values than those obtained in our experiments. C/N values decreased in favor of nitrogen as OMWW content in the reactor increased.

As the trials aim at assessing the BMP of the OMWW, it was important to quantify polyphenol (PP) contents since they represent inhibiting compounds of the bacterial pool, particularly methanogens. The analysis revealed an amount of 4.60 $g \cdot L^{-1}$.

3.1.2. Biogas and Methane Yields

The biogas volume recorded in each sampling date as well as methane content were normalized to normal liters (dry gas, at temperature = 0 °C and pressure = 1013 hPa), according to the standard procedures described in the VDI 4630 [48], as carried out in [44]. Cumulative biogas production of the tested theses during the AcoD period is represented in Figure 5. Higher biogas production was registered in Thesis 2, which contains 20% v/v olive mill wastewater, until day 20. After that, this tendency changed in favor of Thesis 3, which registered a total amount of biogas equal to 5.80 ± 1.77 $NL \cdot L^{-1}$ of substrate, corresponding to 80.22 ± 24.49 $NL \cdot kg_{SV}^{-1}$.

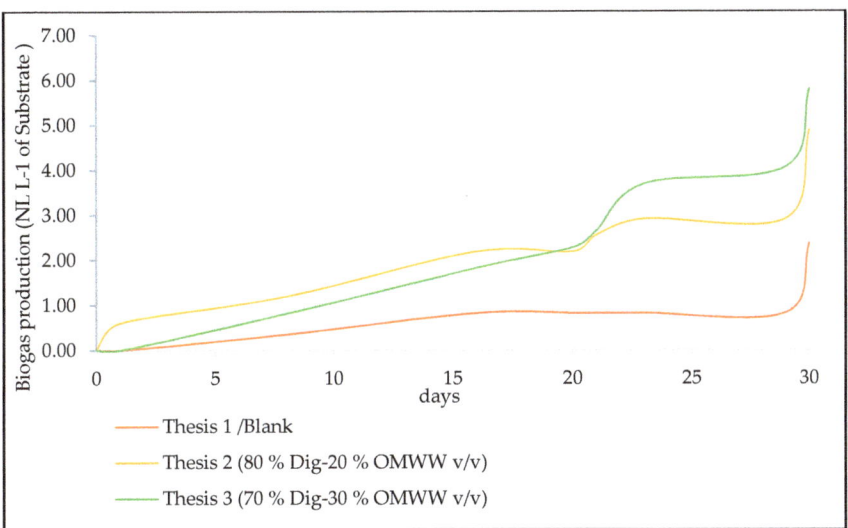

Figure 5. Cumulative biogas production for 30-day AcoD of olive mill wastewater. Values are the mean production values obtained from the three replicates of each thesis at different sampling time. Source: Our elaboration.

Considering biogas total amount, Thesis 2 (with 4.88 ± 2.03 $NL.L^{-1}$ of substrate corresponding to 64.06 ± 26.64 $NL.kg_{SV}^{-1}$) and Thesis 3, respectively, recorded 2- and 2.5-times higher productions than that of the blank equal to 2.37 ± 0.37 $NL.L^{-1}$ of substrate corresponding to 31.89 ± 4.98 $NL.kg_{SV}^{-1}$ (Figure 6), meaning that reactor content in OMWW favored biogas production. Nevertheless, statistical analysis by performing one-way ANOVA did not show any significant difference (F = 4.082; df (2; 6), Pr (>F) = 0.076).

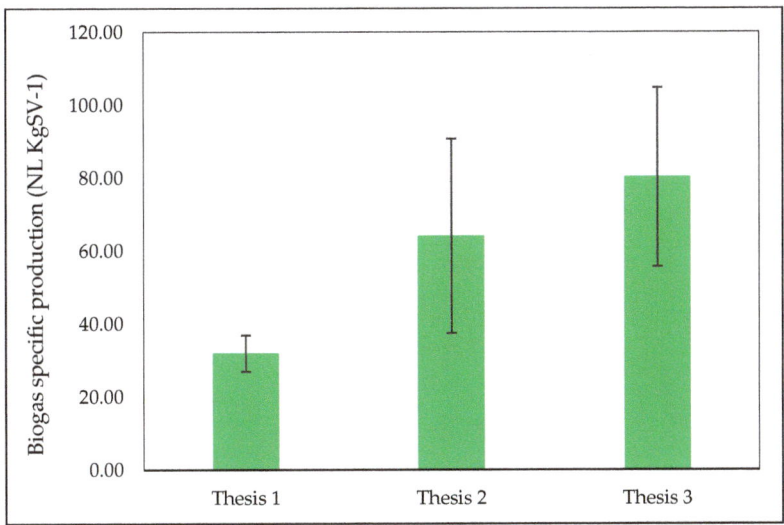

Figure 6. Mean values ± St. Dev. of total biogas specific production for 30 days AcoD of olive mill wastewater. Source: Our elaboration.

Regarding biogas composition, the highest methane percentage, 75.48%, was obtained by Thesis 3. Methane content had the same tendency as biogas production, with higher amount in Thesis 2 at the beginning of AcoD process until the 16th day, after which a decline was observed in both Theses 1 and 2, whereas higher amounts were found in Thesis 3 until day 29, after which they decreased by 58.12% (Figure 7). Total amount of methane during the whole process was equal to 0.99 ± 0.06 NL.L^{-1} of substrate, corresponding to 13.29 ± 0.79 NL.kg$_{SV}$$^{-1}$, 2.55 ± 1.15 NL.L^{-1} of substrate, corresponding to 33.53 ± 15.15 NL.kg$_{SV}$$^{-1}$, and 3.45 ± 1.27 NL.L^{-1} of substrate, corresponding to 47.68 ± 17.55 NL.kg$_{SV}$$^{-1}$, respectively, for Theses 1, 2 and 3. Additionally, for methane content, no significant difference was found (F = 4.997; df (2; 6), Pr (>F) = 0.0528). Figure 8 illustrates the biogas composition considering the overall process for the three tested theses.

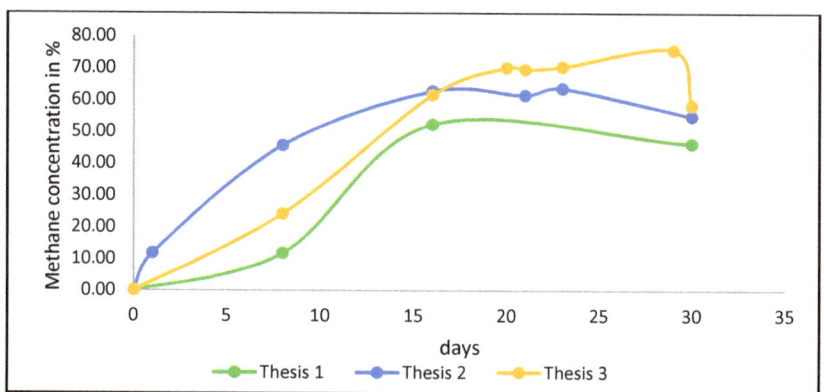

Figure 7. Methane content in the biogas expressed as percentage. Source: Our elaboration.

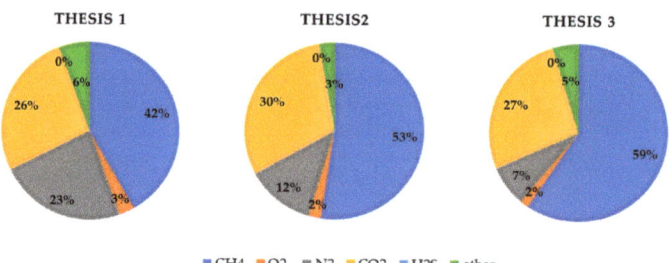

Figure 8. Biogas composition considering the whole process of OMWW AcoD. Source: Our elaboration.

Biogas and biomethane results obtained in this study are lower than those reported by other authors, who implemented OMWW in an anaerobic codigestion process. Indeed, experimental trials performed by Zema et al. [44] provided 0.362 Nm^3.kg$_{TVS}$$^{-1}$ of biogas (0.187 Nm^3.kg$_{TVS}$$^{-1}$ of methane) and 0.176 Nm^3 kg$_{TVS}$$^{-1}$ of biogas (0.067 Nm^3.kg$_{TVS}$$^{-1}$ of methane) after 25 days using, respectively, 20% and 30% OMWW with a polyphenol concentration of 2.8 g.kg^{-1} in a mesophilic AcoD process with digested liquid manure. In contrast with the results presented here, they yielded higher amounts of biogas and methane with lower quantities of OMWW. Bovina et al. [51] reported an increase in biogas and methane yield with the increasing of OMWW content instead of sewage sludge, obtaining the best performances with 25% OMWW (with 1.01 ± 40 g.L^{-1} of polyphenols)—i.e., 116 NL.kg$_{VS}$$^{-1}$ of methane. Calabrò et al. [52] used raw and concentrated OMWW with polyphenols values ranging between 1.1 ± 0.12 and 4.4 ± 0.03 g.L^{-1} in order to obtain up to 2 g.L^{-1} PPs in the blends they tested in batch under mesophilic conditions. They obtained 0.419 NL.g$_{TVS}$$^{-1}$ with a PP concentration of 0.5 g.L^{-1}. The blend with 2 g.L^{-1} PPs provided better results due to the adaptation of the inoculum to polyphenols (170 against 45 NL.g$_{TVS}$$^{-1}$ when using non acclimated inoculum).

Accordingly, it can be stated that the lower amounts obtained in this study are mainly due to the high polyphenol (PP) contents, equal to 4.60 g.L^{-1}. Regarding this aspect, Borja et al. [53] and Fedorak et al. [54] suggest not exceeding a phenol concentration of 2 g.L^{-1} to avoid an inhibiting effect on the methanation process.

3.2. Environmental Results

Environmental impacts related to the production of 1 m^3 of biogas are presented in Table 5. The ecoprofile of biogas from "Thesis 2" shows better results than "Thesis 3"; however, they are fully comparable, due to the higher biogas production that could be achieved with "Thesis 2" considering an annual duration, as it has shorter retention times and therefore allows more matrices to be processed. The shorter retention times can be attributed to the lower amount of OMWW in the mix, which favors a faster start-up of the anaerobic digestion and methanation processes.

Table 5. Characterization of impacts linked to 1 m^3 of biogas production. Source: Our elaboration.

Impact Categories	Unit	Thesis 2	Thesis 3
Climate change	kg CO$_2$ eq	2.22×10^{-1}	3.12×10^{-1}
Ozone depletion	kg CFC-11 eq	1.02×10^{-8}	1.26×10^{-8}
Human toxicity, noncancer effects	CTUh	2.10×10^{-8}	2.63×10^{-8}
Human toxicity, cancer effects	CTUh	9.25×10^{-9}	1.19×10^{-8}
Particulate matter	kg PM2.5 eq	2.78×10^{-4}	3.17×10^{-4}
Ionizing radiation HH	kBq U235 eq	2.14×10^{-2}	2.51×10^{-2}
Ionizing radiation E (interim)	CTUe	6.51×10^{-8}	7.62×10^{-8}
Photochemical ozone formation	kg NMVOC eq	9.89×10^{-4}	1.18×10^{-3}
Acidification	molc H+ eq	1.11×10^{-2}	1.25×10^{-2}
Terrestrial eutrophication	molc N eq	4.86×10^{-2}	5.48×10^{-2}
Freshwater eutrophication	kg P eq	1.49×10^{-5}	1.84×10^{-5}
Marine eutrophication	kg N eq	5.67×10^{-4}	6.59×10^{-4}
Freshwater ecotoxicity	CTUe	4.64×10^{-1}	5.83×10^{-1}
Land use	kg C deficit	1.41×10^{-1}	1.73×10^{-1}
Water resource depletion	m3 water eq	1.43×10^{-4}	1.78×10^{-4}
Mineral, fossil and ren resource depletion	kg Sb eq	2.63×10^{-6}	3.37×10^{-6}

The impacts are coherent with those obtained in other works such as [28,55,56], but in order to compare them, some clarifications are needed. Indeed, the above-mentioned studies used the production of electricity from cogeneration as a functional unit, whereas in the present study we have limited the study to the production of biogas. Therefore, the impacts of cogeneration should be added to the impacts presented in Table 4 and should be attributed to the production of electricity. The impacts could be even more favorable if the avoided emissions from the storage of input digestate and wastewater had been considered in the calculation of emissions. In this regard, the emissions from storage could be considered zero due to the balance between avoided and generated emissions. In addition, the assumptions made by Lovarelli et al. [28] were used to estimate ammonia and methane emissions from digestate management, although they refer to digestate from livestock manure. In the case of the two tested theses, the mixes are largely made from digestate, so the experimented anaerobic digestion process allows the recovery of all biogas still obtainable from this matrix, leaving, as an output, a rather exhausted "second generation" digestate.

Therefore, methane emission resulting from the storage of digestate could also be considered zero, which together with the avoided impacts for the management of the incoming digestate could generate a strong reduction in impacts.

The analysis of the contribution of individual processes was carried out for both theses and it is shown in Figures 9 and 10. It emerged that the electricity is the main hotspot in both analyzed scenarios and for all impact categories. Electricity has been considered as an external input but it is modelled on the basis of 11% legally required self-consumption. The use of electricity as an input purchased on the market is a consequence of the system boundaries being set for biogas production and not for the combined production of heat and power (CHP). Using self-produced energy would probably result in significantly lower impacts. Transport is, on average, the second hotspot. The impacts of this process are related to the movement of matrices. This element is a critical factor in the production of biogas from agricultural waste since the need for matrices to feed the plant often requires them to be supplied from long distances. In the specific case study, the amount of wastewater needed to operate the plant for one year is generated from about 1100 ha (11 km^2) of olive grove, while the digestate is all produced by a single 998 kW anaerobic digestion plant. Therefore, assuming that the OMWW digestion plant is built close to the 998 kW plant, transport is only related to wastewater and therefore an average supply distance of 5 km has been estimated.

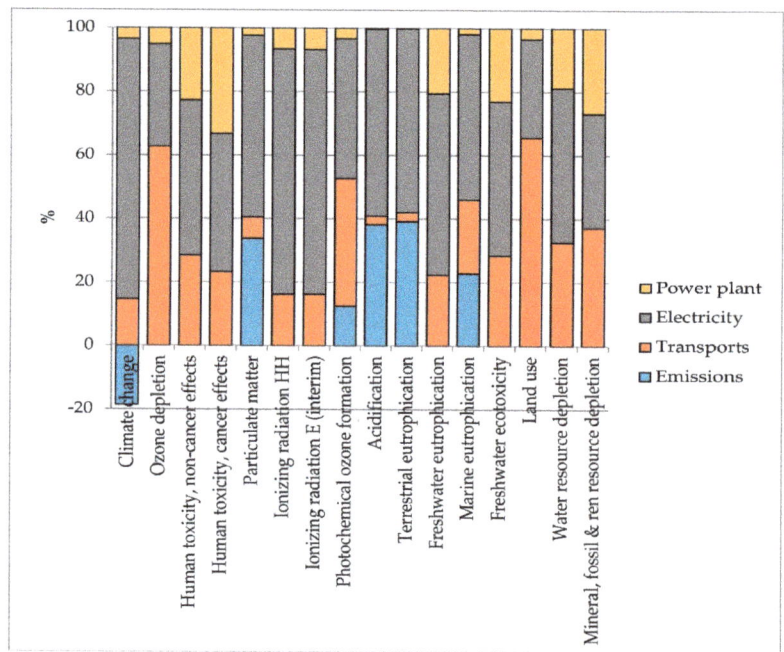

Figure 9. Contribution analysis in Thesis 2. Source: Our elaboration.

The positive impacts in the Climate Change category are due to the fixation of CO_2 during the anaerobic digestion process. In Thesis 2, the balance between fixation and emissions is positive, while in Thesis 3, emissions almost equal CO_2 fixation, so the positive impact of this phenomenon is almost reduced to zero. The impacts from the plant refer to a plant of 200 kW electrical power with a lifespan of 15 years.

Since the scaling up operation could lead to lower yields than those found in the laboratory experiments, due to a lower control of the anaerobic digestion process caused by the full-scale dimensions of the 200 kW plant, the effect of a reduction in the production capacity of the two theses was analyzed, assuming yield decreases of 10% and 20%, respectively, in both tested theses. The results of the sensitivity analysis show a linear increase in impacts in almost all impact categories (on average +6% in the hypothesis of a 10% yield reduction and +13% in the hypothesis of a 20% yield reduction), except for the Climate Change category. Impacts in terms of GHG emissions increase exponentially (+137.68% and +243.03% for Thesis 2; +103.12% and +112.79% for Thesis 3) and these results are attributed to the lower production efficiency linked to the use of inputs. Indeed, for the same quantity of implemented inputs, the sensitivity scenarios predict lower yields; therefore, the incidence of impacts per unit of product increases (Table 6). Given the high influence of electricity use on Climate Change impacts (see Figures 9 and 10), a reduction in yields has very clear consequences on this impact category. Thesis 2 is more sensitive than Thesis 3 as the amounts of implemented inputs in the process are larger given the shorter retention times in Thesis 2.

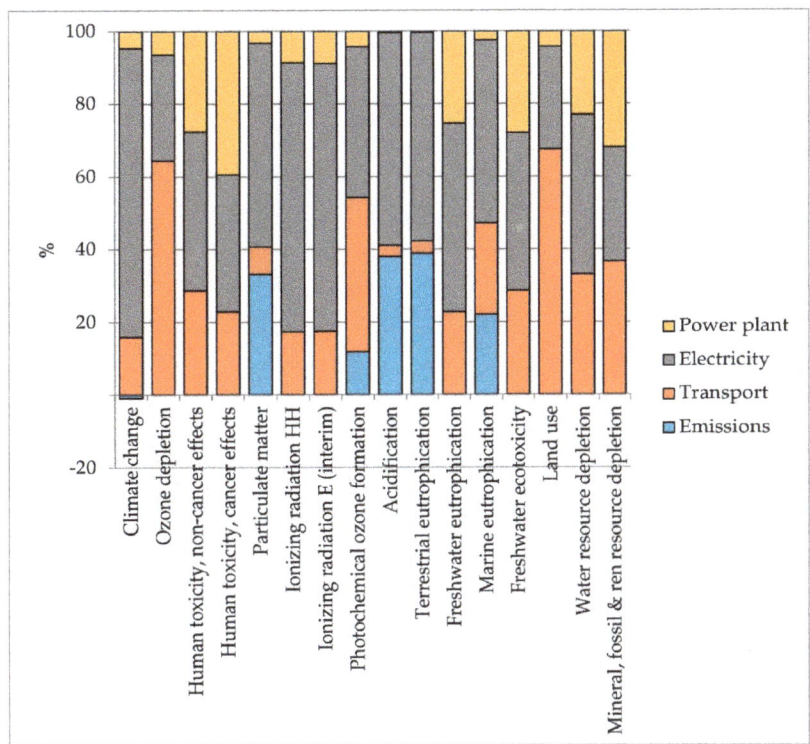

Figure 10. Contribution analysis in Thesis 3. Source: Our elaboration.

Different modelling of the two production processes with self-consumption of the energy produced instead of energy purchased on the market could lead to a flattening of the effects of yield reductions on climate change.

The methane yield of the two tested theses is in favor of Thesis 3, which had an average peak of 68.6% compared to 61.2% for Thesis 2. Extending the boundaries of the system to cogeneration and using the amount of energy produced from biogas, the results could change in favor of Thesis 3.

Table 6. Sensitivity analysis of results with reductions in biogas yield, respectively, of −10% and −20%. Impact deviations from the baseline scenario. Source: Our elaboration.

	Thesis 2		Thesis 3	
Impact category	−10%	−20%	−10%	−20%
Climate change	+137.68%	+243.03%	+103.12%	+112.79%
Ozone depletion	+7.55%	+16.99%	+7.87%	+17.70%
Human toxicity, noncancer effects	+5.69%	+12.81%	+6.26%	+14.09%
Human toxicity, cancer effects	+6.30%	+14.18%	+6.93%	+15.59%
Particulate matter	+4.78%	+10.75%	+4.89%	+10.99%
Ionizing radiation HH	+2.54%	+5.72%	+2.91%	+6.54%
Ionizing radiation E (interim)	+2.56%	+5.76%	+2.93%	+6.59%
Photochemical ozone formation	+11.02%	+17.83%	+11.03%	+18.25%
Acidification	+4.58%	+10.32%	+4.62%	+10.38%
Terrestrial eutrophication	+4.70%	+10.57%	+4.73%	+10.63%
Freshwater eutrophication	+4.77%	+10.73%	+5.36%	+12.06%
Marine eutrophication	+5.31%	+11.96%	+5.51%	+12.41%
Freshwater ecotoxicity	+5.72%	+12.88%	+6.30%	+14.17%
Land use	+7.68%	+17.29%	+7.98%	+17.96%
Water resource depletion	+5.71%	+12.85%	+6.24%	+14.05%
Mineral, fossil and ren resource depletion	+7.12%	+16.02%	+7.62%	+17.14%

3.3. Economic Results

The main results of the economic evaluation are presented in Table 7. It is worth noting that the findings are clearly influenced by the biogas production yield, which was greater in Thesis 2 than in Thesis 3 considering an annual duration. Therefore, the best scenario in terms of total life cycle cost was Thesis 2, with a value of EUR 4.55 per m^3 per year vs. EUR 6.94 per m^3 per year (achieved by the Thesis 3). In both scenarios, the cost driver was initial investment, contributing, in overall, with 88.8% of the total LCC.

Table 7. Life cycle costs of the biogas plant under two scenarios (EUR.m^{-3}.year^{-1} of biogas). Source: Our elaboration.

Cost Item	Thesis 2	Thesis 3
Initial investment cost	4.04	6.16
Operating costs	0.34	0.53
-Materials and Services	0.004	0.01
-Labor	0.03	0.05
-Quotas and other duties	0.31	0.47
End of life disposal costs	0.17	0.25

The analysis of operating costs showed that, in terms of FU, the quota and other attributions category was the greatest contributor to the total operating costs (89.8%). This is due to the higher costs of maintenance and depreciation incurred for plant investment.

Within the material and services category, only transport cost for matrix handling was included and was estimated at 9% of the total operating cost in both scenarios. Since we assumed that the supply of raw materials was free (see Section 2.2.4), no raw material cost was calculated.

The results obtained from the feasibility analysis of the two scenarios under study are shown in Table 8. The findings indicated that both scenarios were profitable. In fact, under assumptions considered for each economic indicator, it was found that:

- NPV was greater than zero, being EUR 0.37 per m^3 and EUR 0.20 per m^3 for Thesis 2 and Thesis 3, respectively;
- IRR was higher than discount rate, being 21.64% and 11.71% for Thesis 2 and Thesis 3, respectively;
- DPP was shorter than the time horizon considered, being 5.05 and 8.62 years for Thesis 2 and Thesis 3, respectively.

Table 8. Comparison of the economic feasibility for the two scenarios under study. Source: Our elaboration.

Economic Indicator	Unit	Thesis 2	Thesis 3
Discounted Gross Margin (DGM)	EUR.m^{-3}	0.88	0.98
Net Present Value (NPV)	EUR.m^{-3}	0.37	0.20
Internal Rate of Return (IRR)	%	21.64	11.71
Discounted Payback Period (DPP)	years	5.05	8.62

However, the Thesis 2 showed the best performance in most of the examined indicators. This is largely due to the bigger revenues, which were estimated considering only the sale of electricity after internal consumption, in Thesis 2 compared to Thesis 3.

Figure 11 shows the sensitivity analysis carried out by changing the discount rate with a ±20% variation and biogas yield floated with −10% and −20%. The results indicated the biogas yield was the most important parameter in the profitability variation. This factor had a remarkable impact on NPV, IRR and DPP indicators. Its decrease led to the worst economic configuration of the plant in both scenarios, and has shown the following:

- Considering a −10% variation in biogas yield, −15.61%, −16.56% and +18.44% variations were observed in the NPV, IRR and DPP, respectively, for Thesis 2, and −43.03%, −26.57% and 24.64% variations for Thesis 3.
- Considering a −20% variation in biogas yield, −35.11%, −33.95% and 45.43% variations were observed in the NPV, IRR and DPP, respectively, for the Thesis 2, and −96.81%, −55.67% and 68.75% variations for the Thesis 3.

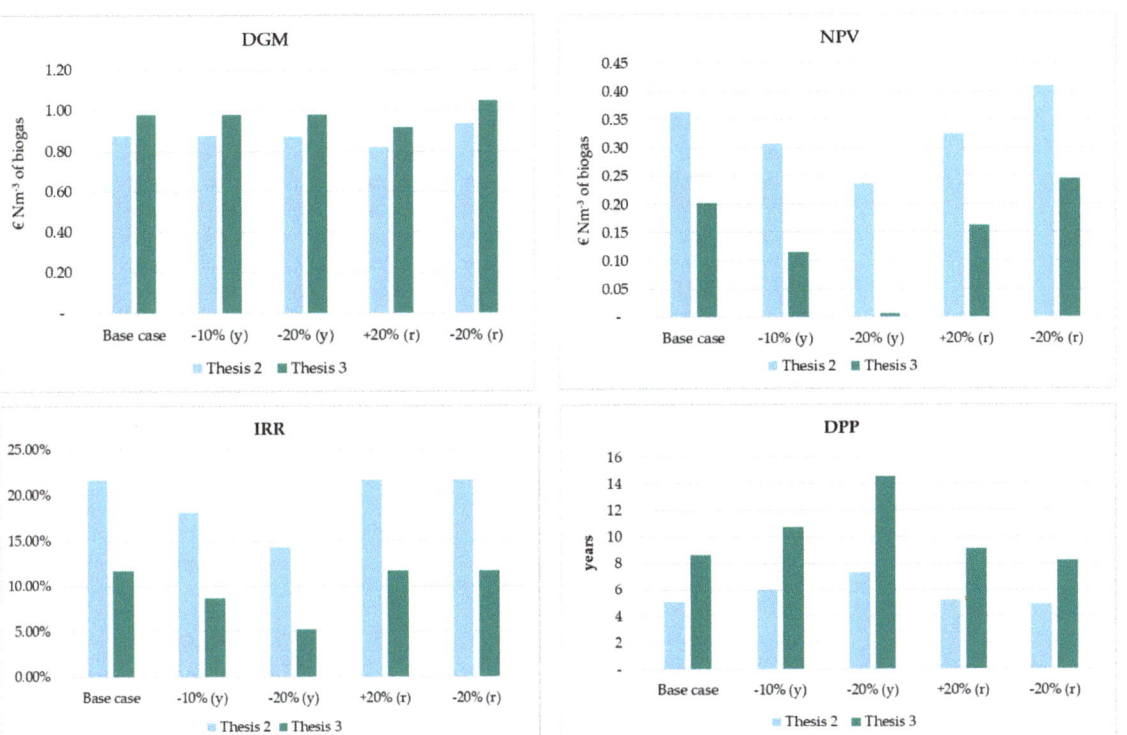

Figure 11. Sensitivity analysis for the two scenarios under study: −10% and −20% represent a decrease in biogas yield (y); +20 and −20% represent, respectively, an increase and decrease in discount rate (r) (DGM = Discounted Gross Margin; NPV = Net Present Value; IRR = Internal Rate of Return; DPP = Discounted Payback Period). Source: Our elaboration.

These findings were consistent with results reported by Li et al. [43]. Less significant variations were achieved for the DGM indicator, ranging from −0.04% for Thesis 2 to −0.05% for Thesis 3, decreasing the biogas yield by 10%, and −0.08% and −0.11%, respectively, with a decrease of 20%.

The sensitivity results also showed that changes in the discount rate affect the magnitude of NPV in both scenarios, in accordance with the studies by Herbes et al. [32] and Hamedani et al. [57]. NPV decreased by more than 11% for Thesis 2 and 19% for Thesis 3 when the discount rate increased by 20%. When the discount rate decreased by 20%, NPVs rose accordingly by 12.24% in Thesis 2 and 21.25% in Thesis 3. While no change was recorded for the IRR indicator, weak changes were achieved for DGM and DPP.

4. Conclusions

The present study illustrates the intermediate results conducted in the framework of the Sustainability of the Olive oil System (S.O.S.) project, with a particular interest in the recovery of olive mill wastes for energy purposes. The findings up to now considering technical, economic, and environmental aspects are very promising, since laboratory experimental trials showed higher amount of biogas (80.22 ± 24.49 NL.kg$_{SV}^{-1}$) and methane (47.68 ± 17.55 NL.kg$_{SV}^{-1}$) when implementing a higher content of olive mill wastewater (30% of v/v in our case). Our results are lower than those obtained in other studies, due to the high content of OMWW in polyphenols; however, they suggest the possibility to explore furthermore AcoD of OMWW considering other farm, livestock, or food industry by-products to enhance biogas and biomethane production, and consequently increase the positive effects on the environment by limiting the impacts, as supported by LCA. In this regard, the analysis of environmental impacts has shown that the production of biogas from OMWW is advantageous from several points of view. In fact, through anaerobic digestion, multiple benefits are obtained such as: the management of a critical waste to be managed from an environmental point of view; the valorization of the biogas present in the digestate that, otherwise, would be dispersed in the environment, representing an environmental load; the production of renewable energy produced exclusively from waste; the production of an exhausted digestate that can be used as organic fertilizer and avoid the impacts generated by the production and distribution of synthetic fertilizers. This represents a strategy for implementing circular economy models, with benefits that go beyond just improving environmental performances. In fact, the economic analysis has demonstrated the profitability of this solution, which, due to the feed-in tariff guaranteed for 15 years, allows dealing with the investment for the plant without the risk of the aleatory energy selling price. Moreover, the recovery and valorization of thermic energy for district heating would allow a further advantage, both for the producer and for the community—using a source of heating with a low environmental impact and that is economic and independent from the public network and therefore free from fixed costs in the bill. Current policies and subsidies encourage the use of by-products instead of energy crops for renewable energy production, so this could be a further stimulus to the spread of small plants dedicated to the digestion of OMWW. In addition, global policies now focus on the adoption of sustainable and circular development models, so it is desirable that the force research should be concentrated on the valorization of waste, transforming it into by-products capable of producing benefits for the whole community.

Author Contributions: Conceptualization, S.B., G.F., T.S., A.I.D.L. and B.B.; data curation, S.B., G.F., T.S., A.I.D.L. and B.B.; formal analysis, S.B., G.F., T.S., A.I.D.L. and B.B.; funding acquisition, B.B., A.S., G.G. and G.Z.; investigation, S.B., G.F., T.S., A.I.D.L. and B.B.; methodology, S.B., G.F., T.S. and A.I.D.L.; project administration, A.S., G.G. and G.Z.; resources, A.S., G.G. and G.Z.; supervision, A.S., G.G. and G.Z.; validation, S.B., G.F., T.S., A.I.D.L., A.S., G.G., G.Z. and B.B.; visualization, S.B., G.F., T.S., A.I.D.L., A.S., G.G., G.Z. and B.B.; writing—original draft, S.B., G.F., T.S., A.I.D.L. and B.B.; writing—review and editing, S.B., G.F., T.S., A.I.D.L. and B.B. Please turn to the CRediT taxonomy for explanation of the term. All authors have read and agreed to the published version of the manuscript.

Funding: This research was funded by the AGER 2 Project (grant no. 2016-0105).

Data Availability Statement: Data is contained within the article. Raw data of the study results may be available on request from the corresponding author.

Acknowledgments: Raw material implemented in experimental trials was kindly provided by Olearia San Giorgio olive mill free of charge.

Conflicts of Interest: The authors declare no conflict of interest. The funders had no role in the design of the study; in the collection, analyses, or interpretation of data; in the writing of the manuscript, or in the decision to publish the results.

Abbreviations

%	percent
°C	degree Celsius
AcoD	anaerobic codigestion
BMP	biochemical methane potential
C/N	carbon/nitrogen ratio
CHP	combined heat and power
COD	chemical oxygen demand
DC	dry content
DGM	discounted gross margin
Dig	digestate
DPP	discounted payback period
E	Ecosystem
EUR	Euro
$EUR.m^{-3}$	Euro per cubic meter
$EUR.m^{-3}.year^{-1}$	Euro per cubic meter per year
FU	functional unit
$g.kg^{-1}$	gram per kilogram
$g.L^{-1}$	gram per liter
g	gram
GC	gas chromatograph
GHGs	greenhouse gases
ha	hectare
HH	Human health
IRR	internal rate of return
ISO	International Organization for Standardization
ISTAT	Istituto Nazionale di Statistica/Italian National Institute of Statistics
km	kilometer
km^2	kilometer square
kW	kilowatt
kWh	kilowatt-hour
$L.L_{reactor}^{-1}.d^{-1}$	liter (of biogas or methane) per liter of or reactor content per day
$L.t^{-1}.km^{-1}$	liter per ton per kilometer
L	liter
LCA	life cycle assessment
LCC	life cycle costing
$L_{CH4}.L_{reactor}^{-1}.d^{-1}$	liter of methane per liter of reactor content per day
LCI	life cycle inventory
LCIA	life cycle impact assessment
LCM	life cycle management
m^3	cubic meter
MJ	megajoule
$mL.g_{VS}^{-1}$	milliliter per gram of volatile solids

mL	milliliter
N_2	nitrogen
$NL.g_{VS}^{-1}$	normal liter per gram of volatile solids
$NL.kg_{SV}^{-1}$	normal liter per kilogram of volatile solids
$NL.L^{-1}$	normal liter per liter
$Nm^3.kg_{VS}^{-1}$	normal cubic meter per kilogram of volatile solids
NPV	Net Present Value
OMSW	olive mill solid waste
OMWW	olive mill wastewater
p	power
pH	potential of hydrogen
PPs	polyphenols
r	discount rate
St. Dev	standard deviation
t	ton
TC	total carbon
TCD	thermal conductivity detector
$t^{-1}km^{-1}$	ton per kilometer
TN	total nitrogen
v/v	volume per volume
VS	volatile solid

References

1. IOC. *International Olive Oil Council. World Olive Oil Figures—PRODUCTION*; IOC: Madrid, Spain, 2021. Available online: https://www.internationaloliveoil.org/wp-content/uploads/2020/12/HO-W901-23-11-2020-P.pdf (accessed on 30 March 2021).
2. ISTAT. Istituto Nazionale di Statistica. Coltivazioni: Uva, Vino, Olive, Olio. Available online: http://dati.istat.it/Index.aspx?QueryId=33706 (accessed on 30 March 2021).
3. ISMEA. *Istituto di Servizi per il Mercato Agricolo Alimentare. Scheda di Settore: Olio di Olva*; ISMEA: Rome, Italy, 2020.
4. Antonio, M.; Maniscalco, M.P.; Roberto, V. Biomethane recovery from olive mill residues through anaerobic digestion: A review of the state of the art technology. *Sci. Total Environ.* **2020**, *703*, 135508.
5. Kapellakis, I.; Tzanakakis, V.A.; Angelakis, A.N. Land Application-Based Olive Mill Wastewater Management. *Water* **2015**, *7*, 362–376. [CrossRef]
6. Dutournié, P.; Jeguirim, M.; Khiari, B.; Goddard, M.-L.; Jellali, S. Olive Mill Wastewater: From a Pollutant to Green Fuels, Agricultural Water Source, and Bio-Fertilizer. Part 2: Water Recovery. *Water* **2019**, *11*, 768. [CrossRef]
7. Mouftahi, M.; Tlili, N.; Hidouri, N.; Bartocci, P.; Alrawashdeh, K.A.B.; Gul, E.; Liberti, F.; Fantozzi, F. Biomethanation Potential (BMP) Study of Mesophilic Anaerobic Co-Digestion of Abundant Bio-Wastes in Southern Regions of Tunisia. *Processes* **2020**, *9*, 48. [CrossRef]
8. Bernardi, B.; Benalia, S.; Zema, D.A.; Tamburino, V.; Zimbalatti, G. An automated medium scale prototype foranaerobic co-digestion of olive mill wastewater. *Inf. Process. Agric.* **2017**, *4*, 316–320.
9. European Parliament and the Council of the European Union. *Directive (EU) 2018/2001 of the European Parliament and of the Council on the Promotion of the Use of Energy from Renewable Sources*; European Parliament and the Council of the European Union: Strasbourg, France, 2018; pp. 82–209.
10. Kougias, P.; Kotsopoulos, T.A.; Martzopoulos, G.G. Effect of feedstock composition and organic loading rate during the mesophilic co-digestion of olive mill wastewater and swine manure. *Renew. Energy* **2014**, *69*, 202–207. [CrossRef]
11. Battista, F.; Fino, D.; Erriquens, F.; Mancini, G.; Ruggeri, B. Scaled-up experimental biogas production from two agro-food wastemixtures having high inhibitory compound concentrations. *Renew. Energy* **2015**, *81*, 71–77. [CrossRef]
12. Thanos, D.; Maragkaki, A.; Venieri, D.; Fountoulakis, M.; Manios, T. Enhanced Biogas Production in Pilot Digesters Treating a Mixture of Olive Mill Wastewater and Agro industrial or Agro livestock By Products in Greece. *Waste Biomass Valorization* **2021**, *12*, 135–143. [CrossRef]
13. Fava, J.; Denison, R.; Jones, B.; Curran, M.A.; Vigon, B.; Selke, S.; Barnum, J. *A Technical Framework for Life-Cycle Assessments*; Society of Environmental Toxicology and Chemistry (SETAC): Washington, DC, USA, 1991.
14. ISO. *ISO 14040:2006 Environmental Management—Life Cycle Assessment—Principles and Framework*; ISO: Geneve, Switzerland, 2006.
15. ISO. *ISO 14044:2006 Environmental Management—Life Cycle Assessment—Requirements and Guidelines*; ISO: Geneve, Switzerland, 2006.
16. Dhillon, B.S. *Life Cycle Costing: Techniques, Models and Application*; Gordon and Breach Science Publishers: New York, NY, USA, 1989.
17. De Luca, A.I.; Falcone, G.; Iofrida, N.; Stillitano, T.; Strano, A.; Gulisano, G. Life Cycle methodologies to improve agri-food systems sustainability. *Riv. Studi Sulla Sostenibilità* **2015**, *1*, 135–150. [CrossRef]

18. Alonso-Farinas, B.; Oliva, A.; Rodriguez-Galan, M.; Esposito, G.; Gracia-Martin, J.F.; Rodrigues-Gutierres, G.; Serrano, A.; Fermoso, F.G. Environmental Assessment of Olive Mill Solid Waste Valorization via Anaerobic Digestion Versus Olive Pomace Oil Extraction. *Processes* **2020**, *8*, 626. [CrossRef]
19. Stillitano, T.; Spada, E.; Iofrida, N.; Falcone, G.; De Luca, A. Sustainable Agri-Food Processes and Circular Economy Pathways in a Life Cycle Perspective: State of the Art of Applicative Research. *Sustainability* **2021**, *13*, 2472. [CrossRef]
20. Palmieri, N.; Suardi, A.; Alfano, V.; Pari, L. Circular Economy Model: Insights from a Case Study in South Italy. *Sustainability* **2020**, *12*, 3466. [CrossRef]
21. Uceda-Rodríguez, M.; López-García, A.B.; Moreno-Maroto, J.M.; Cobo-Ceacero, C.J.; Cotes-Palomino, M.T.; Martínez-García, C. Evaluation of the Environmental Benefits Associated with the Addition of Olive Pomace in the Manufacture of Lightweight Aggregates. *Materials* **2020**, *13*, 2351. [CrossRef] [PubMed]
22. Moreno, V.C.; Iervolino, G.; Tugnoli, A.; Cozzani, V. Techno-economic and environmental sustainability of biomass waste conversion based on thermocatalytic reforming. *Waste Manag* **2020**, *201*, 106–115. [CrossRef]
23. Batuecasa, E.; Tommasi, T.; Battista, F.; Negro, V.; Sonetti, G. Life Cycle Assessment of waste disposal from olive oil production: Anaerobic digestion and conventional disposal on soil. *J. Environ. Manag.* **2019**, *237*, 94–102. [CrossRef]
24. *Method 1684—Total, Fixed, and Volatile Solids in Water, Solids, and Biosolids*; U.S. Environmental Protection Agency: Washington, DC, USA, 2001.
25. Singleton, V.L.; Orthofer, R.; Lamuela-Raventós, R.M. Analysis of total phenols and other oxidation substrates and antioxidants by means of folin-ciocalteu reagent. *Methods Enzymol.* **1999**, *299*, 152–178.
26. Ciroth, A.; Hildenbrand, J.; Steen, B. Life Cycle Costing. In *Sustainability Assessment of Renewables-Based Products*; John Wiley and the Sons: Hoboken, NJ, USA, 2015; pp. 215–228.
27. Moreau, V.; Weidema, B.P. The computational structure of environmental life cycle costing. *Int. J. Life Cycle Assess.* **2015**, *20*, 1359–1363. [CrossRef]
28. Lovarelli, D.; Falcone, G.; Orsi, L.; Bacenetti, J. Agricultural small anaerobic digestion plants: Combining economic and environmental assessment. *Biomass Bioenergy* **2019**, *128*, 105302. [CrossRef]
29. Dressler, D.; Loewen, A.; Nelles, M. Life cycle assessment of the supply and use of bioenergy: Impact of regional factors on biogas production. *Int. J. Life Cycle Assess.* **2012**, *17*, 1104–1115. [CrossRef]
30. EC-JRC–European Commission, Joint Research Centre. *Characterisation Factors of the ILCD Recommended Life Cycle Impact Assessment Methods. Database and Supporting Information*; Institute For Environment and Sustinability: Luxembourg, 2012.
31. González, R.; Rosas, J.G.; Blanco, D.; Smith, R.; Martínez, E.J.; Pastor-Bueis, R.; Gómez, X. Anaerobic digestion of fourth range fruit and vegetable products: Comparison of three different scenarios for its valorisation by life cycle assessment and life cycle costing. *Environ. Monit. Assess.* **2020**, *192*, 1–19. [CrossRef] [PubMed]
32. Herbes, C.; Roth, U.; Wulf, S.; Dahlin, J. Economic assessment of different biogas digestate processing technologies: A scenario-based analysis. *J. Clean. Prod.* **2020**, *255*, 120282. [CrossRef]
33. Dm. Incentivazione Dell'energia Elettrica Prodotta da Fonti Rinnovabili Diverse dal Fotovoltaico. Gazzetta Ufficiale della Re-pubblica Italiana Serie generale, n. 150, Rome, Italy, 23 June 2016. Available online: https://www.gse.it/documenti_site/Documenti%20GSE/Servizi%20per%20te/FER%20ELETTRICHE/NORMATIVE/DM%2023%20giugno%202016.PDF (accessed on 30 March 2021).
34. De Luca, A.I.; Falcone, G.; Stillitano, T.; Iofrida, N.; Strano, A.; Gulisano, G. Evaluation of sustainable innovations in olive growing systems: A Life Cycle Sustainability Assessment case study in southern Italy. *J. Clean. Prod.* **2018**, *171*, 1187–1202. [CrossRef]
35. Arias, A.; Feijoo, G.; Moreira, M.T. Benchmarking environmental and economic indicators of sludge management alternatives aimed at enhanced energy efficiency and nutrient recovery. *J. Environ. Manag.* **2021**, *279*, 111594. [CrossRef] [PubMed]
36. Hussain, M.; Mumma, G.; Saboor, A. Discount rate for investments: Some basic considerations in selecting a discount rate. *Pak. J. Life Soc. Sci.* **2005**, *3*, 1–5.
37. Mel, M.; Yong, A.S.H.; Avicenna; Ihsan, S.I.; Setyobudi, R.H. Simulation Study for Economic Analysis of Biogas Production from Agricultural Biomass. *Energy Procedia* **2015**, *65*, 204–214. [CrossRef]
38. Stillitano, T.; Falcone, G.; De Luca, A.I.; Piga, A.; Conte, P.; Strano, A.; Gulisano, G. A life cycle perspective to assess the environmental and economic impacts of innovative technologies in extra virgin olive oil extraction. *Foods* **2019**, *8*, 209. [CrossRef] [PubMed]
39. Moreno, L.; González, A.; Cuadros-Salcedo, F.; Cuadros-Blázquez, F. Feasibility of a novel use for agroindustrial biogas. *J. Clean. Prod.* **2017**, *144*, 48–56. [CrossRef]
40. Orive, M.; Cebràn, M.; Zufía, J. Echno-economic anaerobic co-digestion feasibility study for two phase olive oil mill pomace and pig slurry. *Renew. Energy* **2016**, *97*, 532–540. [CrossRef]
41. Tse, K.K.; Chow, T.T.; Su, Y. Performance evaluation and economic analysis of a full scale water-based photo-voltaic/thermal (PV/T) system in an office building. *Energy Build.* **2016**, *122*, 42–52. [CrossRef]
42. Ong, T.S.; Chun, H.T. Net present value and payback period for building integrated photovoltaic projects in Malaysia. *Int. J. Acad. Res. Bus. Soc. Sci.* **2013**, *3*, 153–171.
43. Li, Y.; Lu, J.; Xu, F.; Li, Y.; Li, D.; Wang, G.; Li, S.; Zhang, H.; Wu, Y.; Shah, A.; et al. Reactor performance and economic evaluation of anaerobic co-digestion of dairy manure with corn stover and tomato residues under liquid, hemisolid, and solid state conditions. *Bioresour. Technol.* **2018**, *270*, 103–112. [CrossRef] [PubMed]

44. Zema, D.; Zappia, G.; Benalia, S.; Perri, E.; Urso, E.; Tamburino, V.; Bernadi, B. Limiting factors for anaerobic digestion of olive mill wastewater blends under mesophilic and thermophilic conditions. *J. Agric. Eng.* **2018**, *792*, 130–137. [CrossRef]
45. Kang, A.J.; Yuan, Q. Enhanced Anaerobic Digestion of Organic Waste. In *Solid Waste Management in Rural Areas*; IntechOpen: London, UK, 2017. [CrossRef]
46. Nsair, A.; Cinar, S.O.; Alassali, A.; Abu-Qdais, H.; Kuchta, K. Operational Parameters of Biogas Plants: A Review and Evalutaion Study. *Energies* **2020**, *13*, 3761. [CrossRef]
47. Lagrange, B. *Il Biogas I Rifiuti Animali e Umani Come Fonte di Energia: Principi e Tecniche di Utilizzazione*; Longanesi: Milano, Italy, 1981.
48. VDI 4630. *Fermentation of Organic Materials. Characterisation of the Substrate, Sampling, Collection of Material Data, Fermentation Tests*; VDI, Gesellschaft: Düsseldorf, Germany, 2006.
49. Weinrich, S.; Schäfer, F.; Liebetrau, J. *Value of Batch Tests for Biogas Potential Analysis, Method Comparison*; Murphy, J.D., Ed.; IEA Bioenergy: Paris, France, 2018.
50. Guarino, G.; Carotenuto, C.; Di Cristofaro, F.; Papa, S.; Morrone, B.; Minale, M. Does the C/N ratio really affect the Biomethane Yield? A three years investigation of Buffalo Manure Digestion. *Chem. Eng. Trans.* **2016**, *49*, 463–468.
51. Bovina, S.; Frascari, D.; Ragini, A.; Avolio, F.; Scarcella, G.; Pinellia, D. Development of a continuous-flow anaerobic co-digestion process of olive mill waste water and municipal sewage sludge. *J. Chem. Technol. Biotechnol.* **2021**, *96*, 532–543. [CrossRef]
52. Calabrò, P.; Fòlino, A.; Tamburino, V.; Zappia, G.; Zema, D. Increasing the tolerance to polyphenols of the anaerobic digestion of olive wastewater through microbial adaptation. *Biosyst. Eng.* **2018**, *172*, 19–28. [CrossRef]
53. Borja, R.; Banks, C.J.; Alba, J.; Maestro, R. The effect of the most important phenolic constituents of OMW on batch anaerobic methanogenis. *Environ. Technol.* **1996**, *17*, 167–174. [CrossRef]
54. Fedorak, P.M.; Hrudey, S.E. The effects of phenols and some alkil phenolics on batch anaerobic methanogenesis. *Water Resour.* **1984**, *18*, 361–367.
55. Bacenetti, J.; Sala, C.; Fusi, A.; Fiala, M. Agricultural anaerobic digestion plants: What LCA studies pointed out and what can be done to make them more environmentally sustainable. *Appl. Energy* **2016**, *179*, 669–686. [CrossRef]
56. Falcone, G.; Lovarelli, D.; Bacenetti, J. Electricity Generation from Anaerobic Digestion in Italy: Environmental Consequences Related to the Changing of Economic Subsidies. *Chem. Eng. Trans.* **2018**, *67*, 475–480.
57. Hamedani, S.R.; Villarini, M.; Colantoni, A.; Carlini, M.; Cecchini, M.; Santoro, F.; Pantaleo, A. Environmental and economic analysis of an anaerobic co-digest'ion power plant integrated with a compost plant. *Energies* **2020**, *13*, 2724. [CrossRef]

Article

Measuring Circularity in Food Supply Chain Using Life Cycle Assessment; Refining Oil from Olive Kernel

Amin Nikkhah [1,2,*], Saeed Firouzi [3], Keyvan Dadaei [3] and Sam Van Haute [1,2]

1. Department of Food Technology, Safety and Health, Faculty of Bioscience Engineering, Ghent University, Coupure Links 653, 9000 Ghent, Belgium; Sam.vanhaute@ghent.ac.kr
2. Department of Environmental Technology, Food Technology and Molecular Biotechnology, Ghent University Global Campus, Incheon 21985, Korea
3. Department of Agronomy, College of Agriculture, Rasht Branch, Islamic Azad University, Rasht 41476-54919, Iran; firoozi@iaurasht.ac.ir (S.F.); dadaee.keyvan@yahoo.com (K.D.)
* Correspondence: Amin.Nikkhah@ugent.be

Abstract: Valorization of food waste is a potential strategy toward a circular food supply chain. In this regard, measuring the circularity of food waste valorization systems is highly important to better understand multiple environmental impacts. Therefore, this study investigated the circularity of a food waste valorization system (refining oil from olive kernel) using a life cycle assessment methodology. An inventory of an industrial-based olive kernel oil production system is also provided in this study. The system boundary was the cradle to the factory gate of the production system. The results indicated that natural gas consumption was the highest contributor to most of the investigated impact categories. The global warming potential of one kg of oil produced from olive kernel was calculated to be 1.37 kg CO_2eq. Moreover, the calculated damages of 1 kg oil production from olive kernel to human health, ecosystem quality, and resource depletion were 5.29×10^{-7} DALY, 0.12 PDF·m^2·yr., and 24.40 MJ, respectively.

Keywords: circular economy; environmental impact; global warming; valorization of waste

1. Introduction

The circular economy concept is gaining growing attention as an alternative to the linear economy—"take, make, waste,"—which exists now [1,2]. In a linear economy, natural resources are transformed into goods that provide economic value; however, they come with a limited life span, and are disposed of in the environment with minimum recovery of resources [3,4]. This system puts enormous stress on the carrying capacity of the planet [5]. The circular economy describes a system with minimum loss of resources by reusing, recycling, and recovering materials and energy [6–9]. Various strategies have been suggested for moving from a linear economy to a circular one, including R-based frameworks, such as the 3Rs strategy (reduce–reuse–recycle), the 4Rs (introducing "recover" as the fourth R), the 6Rs, and even the 9Rs [10,11].

The measurement of circularity is the first step in moving toward a circular system, as quoted by Peter Ducker: "what gets measured gets managed" [1]. There is not a unique approach for measuring a circular economy, since the understanding of a circular economy is still being explored [12]. To date, some assessment indices have been applied to measure circularity, such as the material circularity indicator [13], the circular economy index [14], material flow analysis [15], food loss and waste [16], and life cycle assessment (LCA) methodology [17,18]. Corona et al. (2019) [19] reviewed the applied approaches for measuring circularity and found three assessment frameworks, seven measurement indices, and nine assessment indicators. In this regard, LCA has been used for decades for the evaluation of the environmental impact of products and services but, more recently, it has shown to be a promising method to measure circularity. It is an appropriate method to investigate the environmental consequences of circular product designs and large-scale changes to move toward a more circular economy [8]. However, in recent years, LCA has been applied to measure circularity in various sectors, such as bio-based materials [20,21], tourism [5], and concrete production [22]. There are also some published documents addressing the connection between LCA and the circular economy concept in the food supply chain [16,23,24].

In the case of the food supply chain, approximately one-third of the total global food production, which is equal to 1.3 billion tons per year, is wasted in the food production/consumption chain [25]. This includes food loss (such as losses and spoilage at the producer level before the market) or waste (such as losses at retailer and consumer levels) [26]. In fact, food loss and waste (FLW) refer to a certain amount of food, nutrients, or calories that intentionally/unintentionally disappear from food systems [27]. A large part of FLW is avoidable, and could be decreased by implementing different strategies at each level of the life cycle of the production system [28,29]. Although food waste has been understood as a critical global issue [30], food waste has high potential for reuse or recovery in a circular economy prospective [31].

In this regard, the olive-based products industry is an interesting case, as it is an economically important industry [32]. As a globally energy-intensive sector, the olive processing industry faces sustainability challenges [33]. Espadas-Aldana et al. (2019) [34] studied 23 published papers on the LCA of olives and olive oil and concluded that the global warming potential (GWP) of one liter of olive oil production is equal to 1.6 kg CO_2eq. The olive oil production supply chain also faces crucial challenges regarding waste management. For example, 80% of olive mass is composed of olive pulp and stones; therefore, waste production is four times higher than that produced within the extraction process [34]. In this regard, the by-products and residues generated in olive processing are not commonly used and end up as waste [33]. Thus, valorization of food waste could be considered as an effective strategy to make the supply chain of olive-based products more circular. In this regard, measuring the circularity of food waste valorization systems is highly important for improving understanding of multiple environmental impacts. One of these wastes is olive kernel (stone). Olive kernel is an important by-product generated in the pitted table olive industry [35]. The characterization and application of olive kernel are described in Figure 1. However, the current and main use of olive kernel is as direct solid feedstock for biofuel generation for domestic application [36]. However, this currently may not be a realistic option for an oil-rich country. In this regard, establishing an environmentally efficient approach for olive kernel utilization could actually improve the overall sustainability of olive-based product supply chains. This paper is the first report on the LCA of industrial-scale refining oil from olive kernel (as an olive processing waste valorization approach) system.

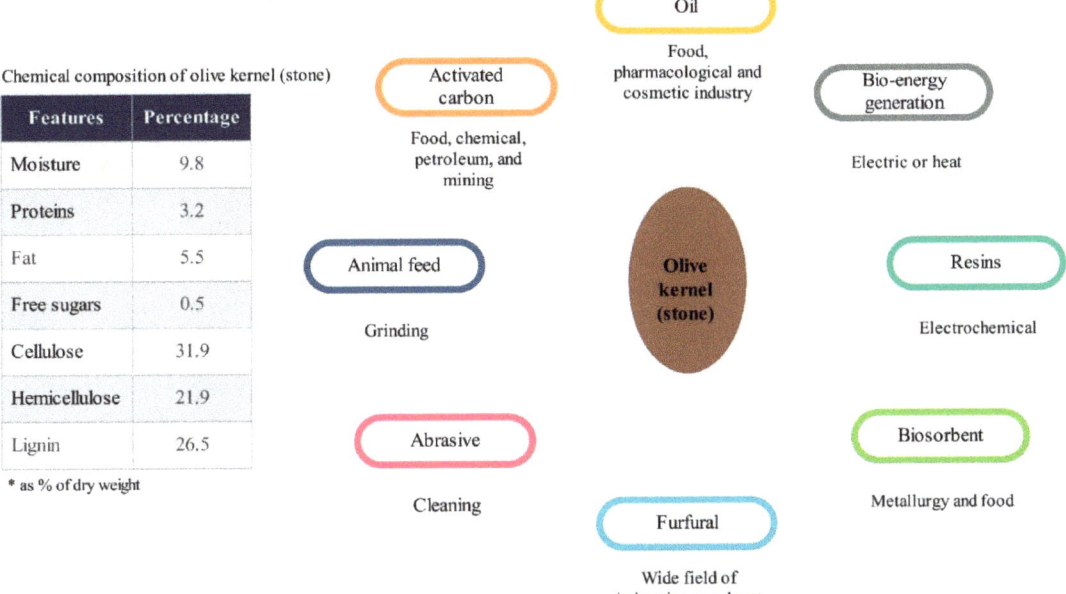

Figure 1. Characterization and application of olive kernel [37].

2. Materials and Methods

2.1. Refining Oil from Olive Kernel

The olive kernel oil production company investigated in this study is located in Iran. Olives (*Olea europaea* L.) were grown in Iran, and the uses of olive fruits in the studied region are: (i) raw material for extra virgin olive oil production, (ii) inside dishes (pitted table olive), and (iii) raw material for pickling. Solid–liquid olive pomace and olive mill wastewater are the two major by-products of the extra virgin olive oil production system [38]. The kernel must be separated from the olive fruit in the second and third abovementioned olive fruit applications. Therefore, olive kernel is a common source of waste in olive fruit processing systems (Figure 2). Olive kernel can be used to produce oil. The characterization of olive kernel oil was described by Moghaddam et al. (2012) [39].

The industrial olive kernel oil production process is shown in Figure 3. Olive kernels are transported to the processing plant, and the factory is located in the olive oil production/processing area. A small amount of olive pulp is stuck to kernels because it cannot be completely separated from the olive kernels in olive processing (Figure 2b). The received kernels are washed to remove impurities. Then, they are crushed to ease the release of the oil and are subsequently mixed. Afterward, the liquid (including oil and water) is separated through a decanter. In the next step, the oil is separated from the water and the olive kernel oil is extracted by a separator. Natural gas is consumed to heat the water in the boiler at a working temperature of 60 to 70 °C. Its circulation in the decanter's double-walled jacket heats the dough (the crushed olive kernel, oil, and water). Mixer blades of the decanter provide a uniform spread of heat throughout the dough. Heating the dough contributes more efficiently to separating the three phases of oil, water, and pulp through centrifuging at 4000–4200 rpm. Moreover, the remaining pulp needs to be warmed to flow easily through the discharge mono pump of the decanter. Warm water is also added to the oil entering the separator in order to maximize the oil extraction rate. At the final step, a centrifuge rotating at a high rotational speed of 7000–7200 rpm separates the warm water and the olive kernel oil.

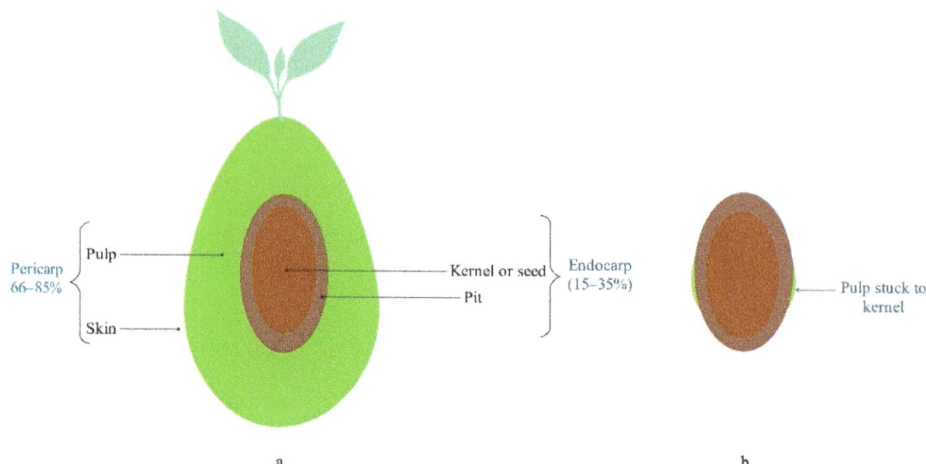

Figure 2. (**a**) different parts of an olive fruit [40,41], (**b**) the parts of the olive which were used to produce oil in the case study. Note: In this study, the term "kernel" refers to "kernel, pit, and pulp stuck to kernel".

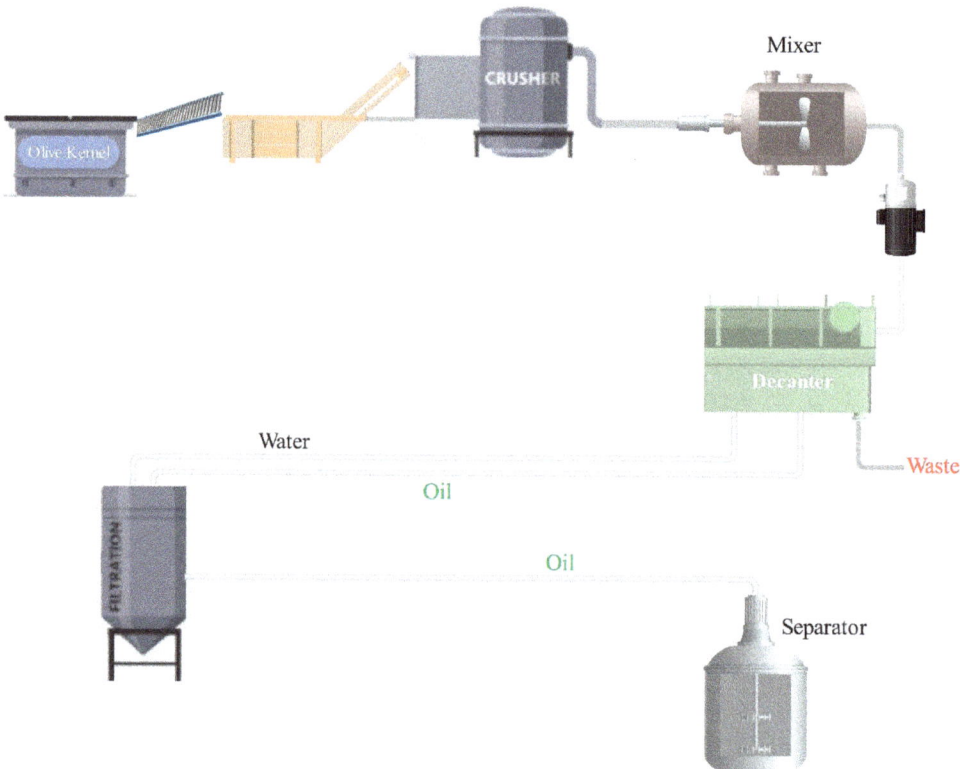

Figure 3. Olive oil production line with a daily production capacity of 100 kg.

2.2. Measuring Circularity Using LCA

In this study, LCA methodology was applied to measure the circularity of an olive waste valorization system. Based on the review by Espadas-Aldana et al. (2019) [34] on olive oil LCA studies, the functional units (FUs) are defined as the quantity of olive oil produced in kg, and the energy content of the produced oil in MJ. Therefore, 1 kg of oil produced from olive kernel and 100 MJ energy generation were selected as the FUs for this LCA analysis to compare the environmental consequences associated with oil production from olive kernel with other vegetable-based oil production systems. The heating value (energy equivalent) of olive kernel oil was considered to be the same as for olive oil at 34.5 MJ/L [42]. The system boundary of this LCA study was the cradle-to-factory gate olive kernel oil production.

The cradle-to-gate environmental impact for olive kernel oil production was evaluated using LCA. In this regard, the main primary inventory data for olive kernel production in the considered factory are shown in Table 1. It should be noted that the input and output amounts were measured and recorded at the factory level, not through surveys or interviews. The experiments were performed in 2019. Experiments were conducted three times and the mean values are reported (Table 1). The olive kernel was not included as an input for the LCA analysis, and it was considered as a burden-free input; this is because it is a by-product/waste of a food production system. Therefore, the environmental impacts of wastewater from the olive kernel production system are not included in the system boundary. In other words, the assumption of not considering the wastewater treatment is due to the fact that olive kernel was not included as an input for the LCA analysis, as it is actually a by-product/waste of an olive processing system. Packaging is also excluded from the system boundary. The emitted pollutants were divided into off-site and on-site emissions. The emitted pollutants in the off-site phase (production of input materials) were adapted from the Ecoinvent 3.0 database using SimaPro 9.0.0.49 software. The datasets applied for calculation of emissions from off-site operations are shown in Appendix A, Table 1. The on-site emissions of natural gas consumption (see Appendix A, Table 2) were calculated and added to the on-site emission section using SimaPro software. The inventory of emissions for refining one kilogram oil from olive kernel is provided as Supplementary 1.

Table 1. Main primary inventory data for the olive kernel oil production system.

Inputs and Outputs	Unit	Quantity	
		Per 1 kg Produced Oil	Per 100 MJ Produced Oil
Inputs			
—Olive kernel	kg	47.72	150.83
—Water consumption	m^3	0.05	0.14
—Natural gas	m^3	0.52	1.65
—Electricity	kWh	0.10	0.31
—Human labor	h	0.16	0.50
—Transportation of the produced oil	ton × km	0.34	1.07
Outputs			
—Olive kernel oil		1 kg	100 MJ
—Efficiency (oil/olive kernel)	%	2.09	2.09

In the third step of the LCA study, IMPACT 2002+ was employed as the impact assessment (IA) methodology due to its hybrid application of IMPACT 2002, Eco–Indicator 99, CML, and IPCC, all of which cover various impact and damage categories. This impact assessment (IA) methodology evaluates the environmental impacts based on the 15 impact categories, and also divides the impact categories into four damage categories: human health, ecosystem quality, GWP, and resource depletion. The human health damage category is represented as disability-adjusted life year (DALY), ecosystem quality as PDF·m^2·yr., GWP as kg CO_2eq, and resource depletion as MJ. The fourth and last step of conducting an

LCA study is the interpretation of the LCA results, which are explained in the Results and Discussion section. The fourth step of an LCA study includes the determination of hotspots, the specification of areas with a potential for improvement, and recommendations [43]. A detailed explanation of this IA methodology can be found in Jolliet et al. (2003) [44]. A flowchart of the utilization of LCA for measuring the circularity of olive kernel oil production systems is demonstrated in Figure 4. Edraw Max (ver. 9.1, 2018; Sheung Wan, Hong Kong, China) software was used for the representation of graphical items.

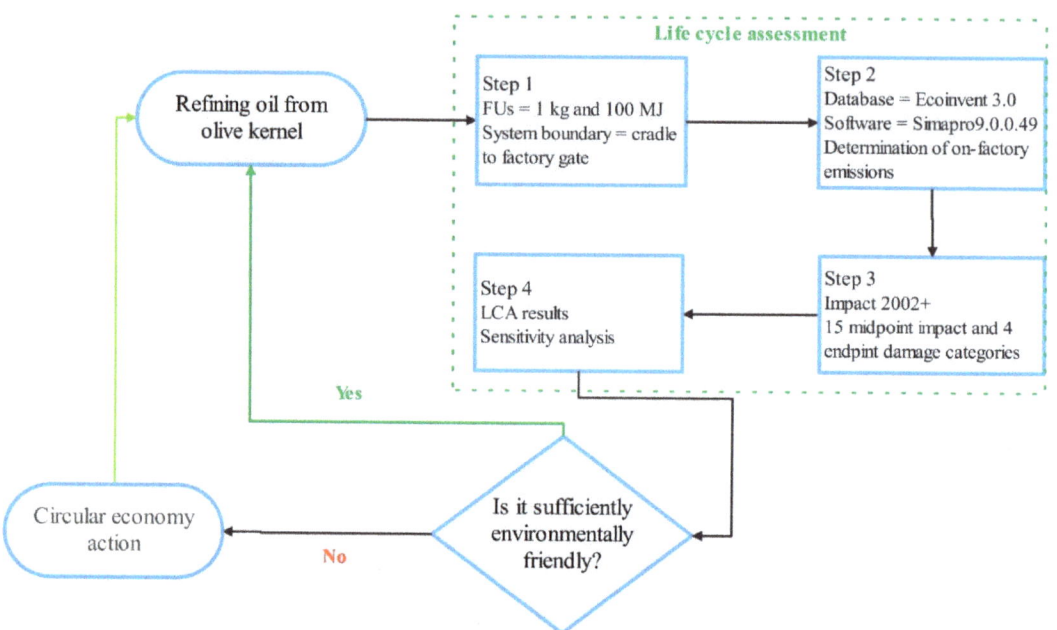

Figure 4. Measuring the circularity of olive kernel oil production using life cycle assessment (LCA) methodology.

There are several sources of uncertainty in an LCA study [45]. It is important to take into account the uncertainty of LCA results which can be due to a lack of accuracy in the collected data [46], the initial assumption [47], the allocation method [48], the selection of functional units [49], the determination of on-site emissions [50], and the type of IA methodology [51]. Therefore, to test the consistency of the LCA results obtained in this study, a quantitative uncertainty analysis was conducted to evaluate the effect of IA methodology selection on the final LCA results. In this regard, the GWP impact category was selected for comparison, as it is the mutual impact category among the investigated IA methodologies. The selected impact assessment methodologies were IMPACT 2002+, ReCiPe 2016 [52], CML–IA baseline [53], EDIP 2003 [54], EF [55], EN 15804 +A2 [56], Environmental Prices [57], EPD [58] and ILCD [59]. The IA methodologies evaluated by this study are available on SimaPro 9.0.0.49—a widely used software for LCA analysis [60,61].

One limitation of this research comes from the limited number (one industrial-scale company refining oil from olive kernel) of investigated olive kernel oil factories. Another limitation of this study is that, compared with studies on the LCA of extra virgin olive oils—the highest quality of olive oils—this olive kernel study does not consider the olive quality indices, as suggested by Salomone et al. (2015) [62].

3. Results and Discussion

3.1. Interpretation of Mid-Point LCA Results

The quantified amounts of the environmental impacts of olive kernel oil production, based on different impact categories, are presented in Table 2. The environmental impact of 1 kg of olive kernel oil production on GWP was calculated as 1.37 kg CO_2eq. This value of the abovementioned impact category for 100 MJ energy generation was 4.32 kg CO_2eq. The GWP of one liter of olive oil production is equal to 1.6 kg CO_2eq, according to Espadas-Aldana et al. (2019) [34]. They also indicated that the agricultural phase is the most impactful phase in the olive oil supply chain, responsible for 0.46 kg CO_2eq/kg of GWP. In the case of refining oil from olive kernel, olives are not used in the system, and the by-product of the olive postharvest waste was used to produce oil. Then, the environmental impacts of the agricultural phase are not accounted for in this system. Therefore, the environmental impacts of oil production from olive kernel are low, making the conventional olive processing systems more circular through waste valorization (olive kernel) of the system. If it is assumed that the production of oil from kernels means less olive oil needs to be produced elsewhere for cosmetic and pharmacological purposes (Figure 1), then this process can save around 0.23 kg CO_2 per each kg of oil produced from olive kernel in the investigated system.

Table 2. Environmental consequence of the olive kernel oil production system based on different impact categories.

Impact Category	Unit	Quantity	
		Per 1 kg Produced Oil	Per 100 MJ Produced Energy
Global warming	kg CO_2eq	1.37	4.32
Non-renewable energy	MJ primary	26.40	83.44
Mineral extraction	MJ surplus	0.005	0.02
Ozone layer depletion	kg CFC-11eq	1.57×10^{-7}	4.97×10^{-7}
Non-carcinogens	kg C_2H_3Cl eq	0.02	0.05
Carcinogens	kg C_2H_3Cl eq	0.08	0.24
Ionizing radiation	Bq C-14 eq	1.94	6.14
Respiratory organics	kg C_2H_4 eq	3.84×10^{-4}	1.39×10^{-3}
Respiratory inorganics	kg PM2.5eq	4.40×10^{-4}	1.21×10^{-3}
Aquatic ecotoxicity	kg TEG water	68.07	215.16
Terrestrial ecotoxicity	kg TEG soil	12.71	40.16
Aquatic eutrophication	kg PO_4 P-lim	3.28×10^{-5}	1.04×10^{-4}
Terrestrial acid/nutri	kg SO_2eq	0.007	0.02
Land occupation	m^2org.arable	0.009	0.03
Aquatic acidification	kg SO_2eq	0.002	0.01

Figure 5 demonstrates the proportion of inputs to the environmental effects of the olive kernel chain. Natural gas consumption was the highest contributor to the most investigated impact categories. The share of natural gas from off-farm emissions on GWP in olive kernel oil was 83%. On-site emissions contributed to the impact categories of GWP, respiratory organics, respiratory inorganics, and terrestrial ecotoxicity; contributions of on-site emissions to these impact categories were 73, 12, 16, and 2%, respectively.

3.2. Interpretation of End-Point Damage Assessment

Figure 6 displays the damage assessment of the olive kernel oil chain. According to the results, the production of 1 kg of oil from olive kernel led to 5.29×10^{-7} DALY damage to human health. The current study investigated the environmental impacts associated with the olive kernel oil production system, and the chemical and microbiological health risk of the final produced oil was not involved in the LCA analysis. Further research is required to examine the human health risk caused by the consumption of oil produced from olive kernel, as well as to gain a proper understanding of the sustainability of the olive kernel oil production system throughout its life cycle. The damages of 1 kg of oil production from olive kernel to ecosystem quality, GWP, and resource depletion were 1.21×10^{-2} PDF·m²·yr., 1.37 kg CO₂eq, and 24.40 MJ, respectively (Figure 6). The environmental indices of the investigated olive kernel oil production system based on different phases of the production system are shown in Table 3.

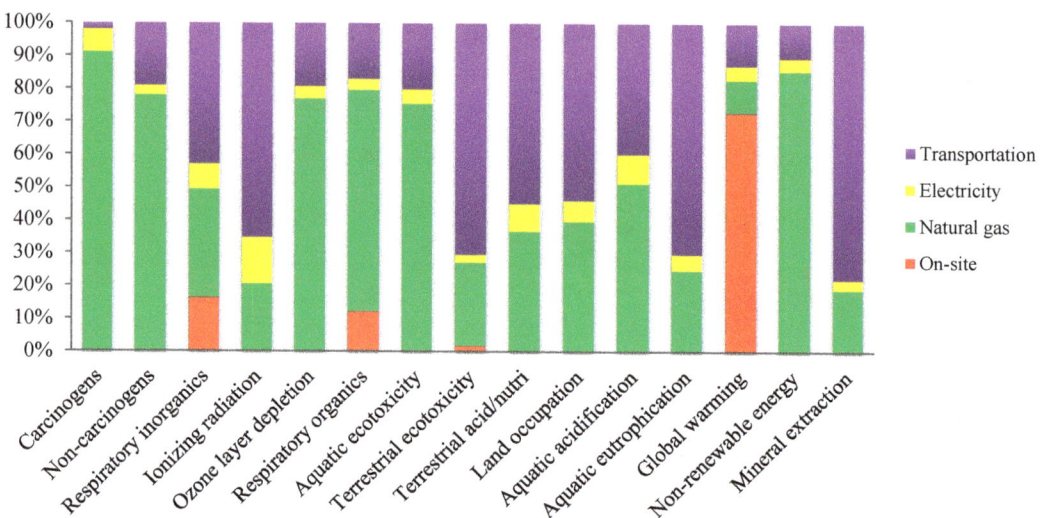

Figure 5. Contribution of consumed inputs to the environmental impact of the olive kernel oil production system.

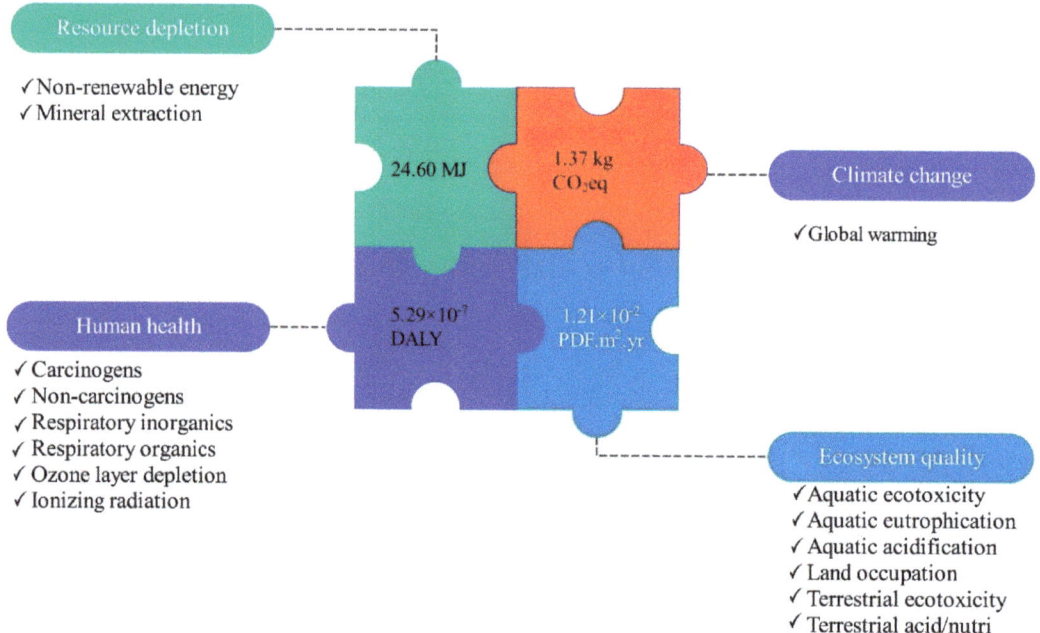

Figure 6. Damage assessment of olive kernel oil production (per 1 kg produced oil).

Table 3. Environmental impacts of the olive kernel oil production system based on different damage categories.

	Off-Site									On-Site		
	Electricity			Natural Gas			Transportation			On-Factory		
Damage Category	Damage Assessment	Normalized	Weighted	Damage Assessment	Normalized	Weighted	Damage Assessment	Normalized	Weighted	Damage Assessment	Normalized	Weighted
Human health	3.78×10^{-8} DALY	2.53×10^{-6}	5.23	3.20×10^{-7} DALY	4.51×10^{-5}	45.07	1.28×10^{-7} DALY	1.80×10^{-5}	17.98	4.46×10^{-8} DALY	6.29×10^{-6}	6.29
Ecosystem quality	3.95×10^{-3} PDF.m^2.yr	2.8×10^{-7}	0.29	3.46×10^{-2} PDF.m^2.yr	2.53×10^{-6}	2.53	8.08×10^{-2} PDF.m^2.yr	5.89×10^{-6}	5.89	2.02×10^{-3} PDF.m^2.yr	1.47×10^{-7}	0.15
Climate change	0.06 kg CO$_2$eq	6.28×10^{-6}	6.28	0.13 kg CO$_2$eq	1.35×10^{-5}	13.48	0.17 kg CO$_2$eq	1.75×10^{-5}	17.52	1.00 kg CO$_2$eq	1.01×10^{-4}	100.91
Resource depletion	1.08 MJ	7.12×10^{-6}	7.12	22.57 MJ	1.49×10^{-4}	148.52	2.75 MJ	1.81×10^{-5}	18.09	0	0	0

The single scores of the damage categories in olive kernel oil are shown in Figure 7. The total weighted environmental damage from refining oil from olive kernel was calculated as 395 µPt/FU. Natural gas was the largest contributor to the total environmental impacts of the studied system, with a share of 80% (including its background and on-site emissions). As explained in the Materials and Methods section, natural gas is used in olive kernel processing to heat water in order to (a) facilitate the separation of water and oil from pulp in the decanter, (b) contribute to the flow of the remaining pulp in the decanter, and (c) separate water from olive kernel oil in the separator. A preliminary experiment shows that the optimum dough temperature to extract purer oil is around 35 °C, which is associated with the circulating water temperature at around 60–70 °C in the studied system. Moreover, a temperature of around 35 °C for the water added to the separator might be efficient. Nevertheless, further research is needed to determine the optimum water temperature to avoid extra natural gas/heat consumption.

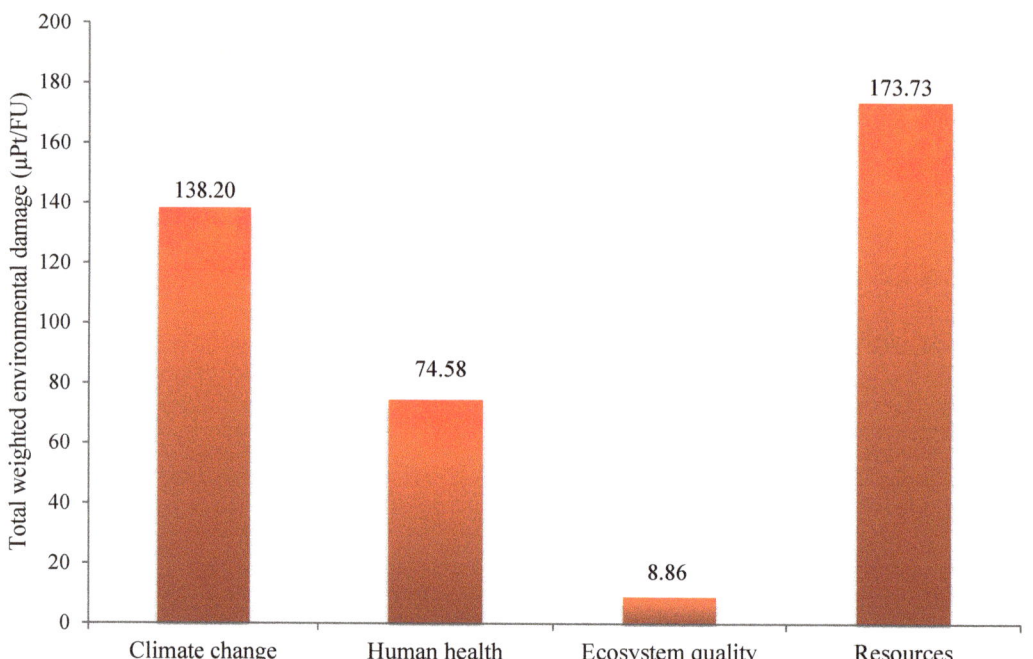

Figure 7. Single scores of damage categories in the olive kernel oil production chain (unit = µPt).

Khounani et al. (2021) [33] showed that the total environmental burdens of olive oil systems can be reduced by 12% by applying agro-biorefinery strategies based on olive cultivation and the extraction of fruit and pomace oil. There are also other ways to valorize olive processing wastes, such as energy generation. One LCA study highlighted significant greenhouse gas emission savings through olive husk (the solid portion remaining after pressing olives) application in a mobile pyrolysis process [63]. In another LCA study, Intini et al. (2012) [64] reported the environmental advantages of using de-oiled pomace and waste wood as feedstock for biofuel production, in terms of greenhouse gas emissions reduction. Multiple environmental measures could be applied in order to improve the sustainability of olive processing. In this regard, Martinez-Hernandez et al. (2014) [65] indicated how integrated process schemes can be used to develop a sustainable Jatropha-based biorefinery system.

3.3. Uncertainty of GWP's Results

A unique IA methodology for LCA analysis in food systems does not exist. Different IA methodologies may apply for characterization, and the selection of IA methodology can therefore affect the final LCA results. The characterization index of GWP for refining oil from olive kernel, based on the application of various IA methodologies, is illustrated in Figure 8. These results may correspond with the findings of relevant LCA studies on the same topic. The results revealed that the GWP of the production of 1 kg of olive kernel oil ranges from 1.37 to 1.47 kgCO$_2$eq. The lowest estimation of GWP belonged to IMPACT 2002+, and the highest to the EDPI and EF methodologies. The results are in agreement with the reports by Fathollahi et al. (2018) [47] and Paramesh et al. (2018) [51], which indicated that selection of IA methodology can slightly affect LCA results in some impact categories in the food system. Therefore, it is recommended to consider the impact of IA selection as a source of uncertainty in future research concerning LCA in the food supply chain.

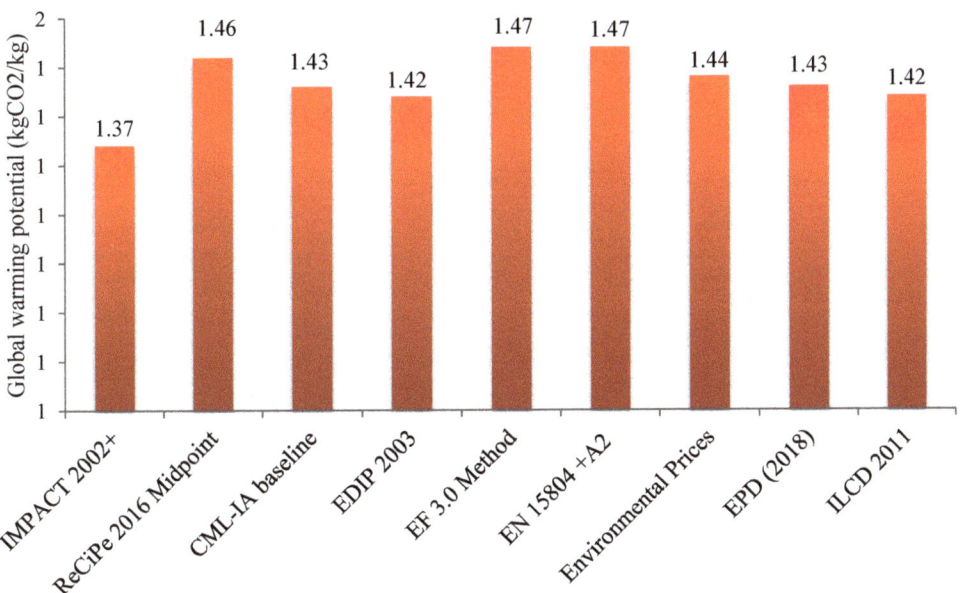

Figure 8. Effect of impact assessment methodology selection on the global warming potential (GWP) results.

4. Conclusions

This study applied LCA to measure the circularity of a food waste system—in this case, refining oil from olive kernel. The damages of 1 kg of oil production from olive kernel to GWP, human health, ecosystem quality, and resource depletion were 1.37 kg CO$_2$eq, 5.29×10^{-7} DALY, 1.21×10^{-2} PDF·m^2·yr., and 24.40 MJ, respectively. The results highlighted that the studied system is relatively circular, resulting in low environmental impacts. Olive processing systems could be made more circular through food waste valorization. By managing the consumption of energy sources, such as natural gas, olive kernel oil production systems can become environmentally efficient. In this regard, further research is needed to determine the optimum temperature of olive kernel dough in the decanter, as well as the optimum temperature for warm water when added to the separator in the process of oil extraction from olive kernel. Further research is also required to explore the human health and microbiological risks of the oil produced from olive kernels, which were not considered in this study.

Supplementary Materials: The following are available online at https://www.mdpi.com/2304-8158/10/3/590/s1, Supplementary 1.

Author Contributions: A.N. developed the initial concept of the research and contributed to the analysis and writing of the first draft of the manuscript. S.F. contributed to data collection and manuscript writing and editing. K.D. contributed to data collection and technical support. S.V.H. contributed to the concept, supervision, and editing of the article. All authors have read and agreed to the published version of the manuscript.

Funding: This research was funded by the Ghent University Global Campus.

Institutional Review Board Statement: Not applicable.

Informed Consent Statement: Not applicable.

Data Availability Statement: The inventory of the industrial-based olive kernel oil production system is provided in the paper. In addition, the inventory of emissions for refining one kilogram oil from olive kernel is also provided as Supplementary 1.

Acknowledgments: The authors would like to acknowledge the support provided by the Ghent University Global Campus.

Conflicts of Interest: The authors declare that they have no conflicts of interest.

Appendix A

Table 1. Datasets applied for calculation of emissions from off-site operations.

Activity	Database	Category	Unit	Activity Uuid/Source
Electricity	Ecoinvent 3	IR	kWh	668baf12-38db-47b7-8517-c0a18aa122f4
Natural gas	Ecoinvent 3	GLO	M^3	65f221cf-b821-4da1-a2ed-a67994b10f42
Transport	Ecoinvent 3	GLO	t.km	413c356e-677d-4676-b816-0c0b20768d7a_03bf1369-1eec-49d0-bc4b-8b29efa826b9.spold
Water-Unspecified origins	Input from nature	IR	M^3	-

Table 2. Coefficients for calculation of emissions from natural gas combustion (gram).

Emission	Coefficients [66,67]	Quantity (per on kg Produced Olive Kernel Oil)
Carbon dioxide (CO_2)	38.70	996.14
Methane (CH_4)	7.40×10^{-4}	1.90×10^{-2}
Dinitrogen monoxide (N_2O)	7.11×10^{-4}	1.83×10^{-2}
Sulfur dioxide (SO_2)	1.93×10^{-4}	4.97×10^{-3}
Nickel (Ni)	6.78×10^{-7}	1.75×10^{-5}
Lead	1.61×10^{-7}	4.14×10^{-6}
Zinc (Zn)	9.37×10^{-6}	2.41×10^{-4}
Benzo(a)pyrene	3.87×10^{-9}	9.96×10^{-8}
Selenium (Se)	7.75×10^{-9}	1.99×10^{-7}
Organic compound	3.50×10^{-3}	9.01×10^{-2}
Volatile organic compound (VOC)	1.70×10^{-3}	4.38×10^{-2}
Particulates (<2.5 μm)	2.45×10^{-3}	6.31×10^{-2}

References

1. Haupt, M.; Hellweg, S. Measuring the environmental sustainability of a circular economy. *Environ. Sustain. Indic.* **2019**, *1*, 100005. [CrossRef]
2. Secondi, L. A Regression-Adjustment Approach with Control-Function for Estimating Economic Benefits of Targeted Circular Economy Practices: Evidence from European SMEs. *Stud. Appl. Econ.* **2020**, 39. [CrossRef]
3. Ruggieri, A.; Braccini, A.M.; Poponi, S.; Mosconi, E.M. A meta-model of inter-organisational cooperation for the transition to a circular economy. *Sustainability* **2016**, *8*, 1153. [CrossRef]
4. Babbitt, C.W.; Gaustad, G.; Fisher, A.; Chen, W.Q.; Liu, G. Closing the loop on circular economy research: From theory to practice and back again. *Resour. Conserv. Recycl.* **2018**, *135*, 1–2. [CrossRef]

8. Scheepens, A.E.; Vogtländer, J.G.; Brezet, J.C. Two life cycle assessment (LCA) based methods to analyse and design complex (regional) circular economy systems. Case: Making water tourism more sustainable. *J. Clean. Prod.* **2016**, *114*, 257–268. [CrossRef]
9. Ellen MacArthur Foundation. *Towards the Circular Economy: Economic and Business Rationale for Accelerated Transition*; Ellen MacArthur Foundation: Cowes, UK, 2013.
10. European Environment Agency. *Signals: Well-Being and the Environment—Building a Resource-Efficient and Circular Economy in Europe Luxembourg*; Publications Office of the European Union: Copenhagen, Denmark, 2014.
11. Haupt, M.; Zschokke, M. How can LCA support the circular economy?—63rd discussion forum on life cycle assessment, Zurich, Switzerland, November 30, 2016. *Int. J. Life Cycle Assess.* **2017**, *22*, 832–837. [CrossRef]
12. Patwa, N.; Sivarajah, U.; Seetharaman, A.; Sarkar, S.; Maiti, K.; Hingorani, K. Towards a circular economy: An emerging economies context. *J. Bus. Res.* **2021**, *122*, 725–735. [CrossRef]
13. Kirchherr, J.; Reike, D.; Hekkert, M. Conceptualizing the circular economy: An analysis of 114 definitions. *Resour. Conserv. Recycl.* **2017**, *127*, 221–232. [CrossRef]
14. Bassi, F.; Dias, J.G. The use of circular economy practices in SMEs across the EU. *Resour. Conserv. Recycl.* **2019**, *146*, 523–533. [CrossRef]
15. Korhonen, J.; Honkasalo, A.; Seppälä, J. Circular economy: The concept and its limitations. *Ecol. Econ.* **2018**, *143*, 37–46. [CrossRef]
16. Ellen MacArthur Foundation; Granta Design. *An Approach Tomeasuring Circularity—Methodology*; Ellen MacArthur Foundation: Cowes, UK, 2015.
17. Di Maio, F.; Rem, P.C. A robust indicator for promoting circular economy through recycling. *J. Environ. Protect.* **2015**, *6*, 1095. [CrossRef]
18. Linder, M.; Sarasini, S.; van Loon, P. A metric for quantifying product-level circularity. *J. Ind. Ecol.* **2017**, *21*, 545–558. [CrossRef]
19. Vilariño, M.V.; Franco, C.; Quarrington, C. Food loss and waste reduction as an integral part of a circular economy. *Front. Environ. Sci.* **2017**, *5*, 21. [CrossRef]
20. ISO (International Organization for Standardization). *Environmental Management: Life Cycle Assessmente*; Principles and Framework; British Standards Institution: London, UK, 2006; Volume 14040.
21. ISO (International Organization for Standardization). *Environmental Management: Life Cycle Assessment*; Requirements and Guidelines; International Standard Organisation: Geneva, Switzerland, 2006.
22. Corona, B.; Shen, L.; Reike, D.; Carreón, J.R.; Worrell, E. Towards sustainable development through the circular economy—A review and critical assessment on current circularity metrics. *Resour. Conserv. Recycl.* **2019**, *151*, 104498. [CrossRef]
23. Lokesh, K.; Matharu, A.S.; Kookos, I.K.; Ladakis, D.; Koutinas, A.; Morone, P.; Clark, J. Hybridised sustainability metrics for use in life cycle assessment of bio-based products: Resource efficiency and circularity. *Green Chem.* **2020**, *22*, 803–813. [CrossRef]
24. Dahiya, S.; Katakojwala, R.; Ramakrishna, S.; Mohan, S.V. Biobased Products and Life Cycle Assessment in the Context of Circular Economy and Sustainability. *Mater. Circ. Econ.* **2020**, *2*, 1–28. [CrossRef]
25. Colangelo, F.; Navarro, T.G.; Farina, I.; Petrillo, A. Comparative LCA of concrete with recycled aggregates: A circular economy mindset in Europe. *Int. J. Life Cycle Assess.* **2020**, *25*, 1790–1804. [CrossRef]
26. Pauer, E.; Wohner, B.; Heinrich, V.; Tacker, M. Assessing the environmental sustainability of food packaging: An extended life cycle assessment including packaging-related food losses and waste and circularity assessment. *Sustainability* **2019**, *11*, 925. [CrossRef]
27. Sadhukhan, J.; Dugmore, T.I.; Matharu, A.; Martinez-Hernandez, E.; Aburto, J.; Rahman, P.K.; Lynch, J. Perspectives on "game changer" global challenges for sustainable 21st century: Plant-based diet, unavoidable food waste biorefining, and circular economy. *Sustainability* **2020**, *12*, 1976. [CrossRef]
28. Dahiya, S.; Kumar, A.N.; Sravan, J.S.; Chatterjee, S.; Sarkar, O.; Mohan, S.V. Food waste biorefinery: Sustainable strategy for circular bioeconomy. *Bioresour. Technol.* **2018**, *248*, 2–12. [CrossRef] [PubMed]
29. Shafiee-Jood, M.; Cai, X. Reducing food loss and waste to enhance food security and environmental sustainability. *Environ. Sci. Technol.* **2016**, *50*, 8432–8443. [CrossRef] [PubMed]
30. Wesana, J.; Gellynck, X.; Dora, M.K.; Pearce, D.; De Steur, H. Measuring Food Losses in the Supply Chain through Value Stream Mapping: A Case Study in the Dairy Sector. In *Saving Food*; Academic Press: New York, NY, USA, 2019; pp. 249–277.
31. Strotmann, C.; Göbel, C.; Friedrich, S.; Kreyenschmidt, J.; Ritter, G.; Teitscheid, P. A participatory approach to minimizing food waste in the food industry—A manual for managers. *Sustainability* **2017**, *9*, 66. [CrossRef]
32. Principato, L.; Ruini, L.; Guidi, M.; Secondi, L. Adopting the circular economy approach on food loss and waste: The case of Italian pasta production. *Resour. Conserv. Recycl.* **2019**, *144*, 82–89. [CrossRef]
33. Muhammad, N.I.S.; Rosentrater, K.A. Comparison of global-warming potential impact of food waste fermentation to landfill disposal. *SN Appl. Sci.* **2020**, *2*, 261. [CrossRef]
34. Ingrao, C.; Faccilongo, N.; Di Gioia, L.; Messineo, A. Food waste recovery into energy in a circular economy perspective: A comprehensive review of aspects related to plant operation and environmental assessment. *J. Clean. Prod.* **2018**, *184*, 869–892. [CrossRef]
35. Negro, M.J.; Manzanares, P.; Ruiz, E.; Castro, E.; Ballesteros, M. The biorefinery concept for the industrial valorization of residues from olive oil industry. In *Olive Mill Waste*; Academic Press: New York, NY, USA, 2017; pp. 57–78.

33. Khounani, Z.; Hosseinzadeh-Bandbafha, H.; Moustakas, K.; Talebi, A.F.; Goli, S.A.H.; Rajaeifar, M.A.; Khoshnevisan, B.; Jouzani, G.S.; Peng, W.; Kim, K.H.; et al. Environmental life cycle assessment of different biorefinery platforms valorizing olive wastes to biofuel, phosphate salts, natural antioxidant, and an oxygenated fuel additive (triacetin). *J. Clean. Prod.* **2021**, *278*, 123916. [CrossRef]
34. Espadas-Aldana, G.; Vialle, C.; Belaud, J.P.; Vaca-Garcia, C.; Sablayrolles, C. Analysis and trends for Life Cycle Assessment of olive oil production. *Sustain. Prod. Consum.* **2019**, *19*, 216–230. [CrossRef]
35. Nunes, M.A.; Pimentel, F.B.; Costa, A.S.; Alves, R.C.; Oliveira, M.B.P. Olive by-products for functional and food applications: Challenging opportunities to face environmental constraints. *Innov. Food Sci. Emerg. Technol.* **2016**, *35*, 139–148. [CrossRef]
36. Gomez-Martin, A.; Chacartegui, R.; Ramirez-Rico, J.; Martinez-Fernandez, J. Performance improvement in olive stone's combustion from a previous carbonization transformation. *Fuel* **2018**, *228*, 254–262. [CrossRef]
37. Rodríguez, G.; Lama, A.; Rodríguez, R.; Jiménez, A.; Guillén, R.; Fernández-Bolanos, J. Olive stone an attractive source of bioactive and valuable compounds. *Bioresour. Technol.* **2008**, *99*, 5261–5269. [CrossRef] [PubMed]
38. Batuecas, E.; Tommasi, T.; Battista, F.; Negro, V.; Sonetti, G.; Viotti, P.; Fino, D.; Mancini, G. Life Cycle Assessment of waste disposal from olive oil production: Anaerobic digestion and conventional disposal on soil. *J. Environ. Manag.* **2019**, *237*, 94–102. [CrossRef] [PubMed]
39. Moghaddam, G.; Vander Heyden, Y.; Rabiei, Z.; Sadeghi, N.; Oveisi, M.R.; Jannat, B.; Araghi, V.; Hassani, S.; Behzad, M.; Hajimahmoodi, M. Characterization of different olive pulp and kernel oils. *J. Food Compos. Anal.* **2012**, *28*, 54–60. [CrossRef]
40. Martins, F.P.; Kiritsakis, A. *Olives and Olive Oil as Functional Foods: Bioactivity, Chemistry and Processing*, 1st ed.; Kiritsakis, A., Shahidi, F., Eds.; Wiley: Hoboken, NJ, USA, 2017; pp. 81–105.
41. Kiritsakis, A.; Turkan, K.M.; Kiritsakis, K. *Bailey's Industrial Oil and Fat Products*, 7th ed.; Shahidi, F., Ed.; Wiley: Hoboken, NJ, USA, 2020. [CrossRef]
42. Cappelletti, G.M.; Ioppolo, G.; Nicoletti, G.M.; Russo, C. Energy requirement of extra virgin olive oil production. *Sustainability* **2014**, *6*, 4966–4974. [CrossRef]
43. Sundaram, S.; Siew, K.; Martinez-Hernandez, E. Biorefineries and chemical processes: Design, integration and sustainability analysis. *Green Process. Synth.* **2015**, *4*, 65–66. [CrossRef]
44. Jolliet, O.; Margni, M.; Charles, R.; Humbert, S.; Payet, J.; Rebitzer, G.; Rosenbaum, R. IMPACT 2002+: A new life cycle impact assessment methodology. *Int. J. Life Cycle Assess.* **2003**, *8*, 324. [CrossRef]
45. Zargar-Ershadi, S.Z.; Heidari, M.D.; Dutta, B.; Dias, G.; Pelletier, N. Comparative life cycle assessment of technologies and strategies to improve nitrogen use efficiency in egg supply chains. *Resour. Conserv. Recycl.* **2021**, *166*, 105275. [CrossRef]
46. Vásquez-Ibarra, L.; Rebolledo-Leiva, R.; Angulo-Meza, L.; González-Araya, M.C.; Iriarte, A. The joint use of life cycle assessment and data envelopment analysis methodologies for eco-efficiency assessment: A critical review, taxonomy and future research. *Sci. Total Environ.* **2020**, 139538. [CrossRef]
47. Fathollahi, H.; Mousavi-Avval, S.H.; Akram, A.; Rafiee, S. Comparative energy, economic and environmental analyses of forage production systems for dairy farming. *J. Clean. Prod.* **2018**, *182*, 852–862. [CrossRef]
48. Cherubini, E.; Franco, D.; Zanghelini, G.M.; Soares, S.R. Uncertainty in LCA case study due to allocation approaches and life cycle impact assessment methods. *Int. J. Life Cycle Assess.* **2018**, *23*, 2055–2070. [CrossRef]
49. Noya, I.; González-García, S.; Bacenetti, J.; Fiala, M.; Moreira, M.T. Environmental impacts of the cultivation-phase associated with agricultural crops for feed production. *J. Clean. Prod.* **2018**, *172*, 3721–3733. [CrossRef]
50. Rezaei, M.; Soheilifard, F.; Keshvari, A. Impact of agrochemical emission models on the environmental assessment of paddy rice production using life cycle assessment approach. *Energy Sources* **2021**, 1–16. [CrossRef]
51. Paramesh, V.; Arunachalam, V.; Nikkhah, A.; Das, B.; Ghnimi, S. Optimization of energy consumption and environmental impacts of arecanut production through coupled data envelopment analysis and life cycle assessment. *J. Clean. Prod.* **2018**, *203*, 674–684. [CrossRef]
52. Dekker, E.; Zijp, M.C.; van de Kamp, M.E.; Temme, E.H.; van Zelm, R. A taste of the new ReCiPe for life cycle assessment: Consequences of the updated impact assessment method on food product LCAs. *Int. J. Life Cycle Assess.* **2019**, 1–10. [CrossRef]
53. Guinée, J.B.; Lindeijer, E. Handbook on Life Cycle Assessment—Operational Guide to the ISO Standards. In *Handbook on Life Cycle Assessment: Operational Guide to the ISO Standards Series: Eco-Efficiency in Industry and Science*; Guinée, J.B., Ed.; Springer: Dordrecht, The Netherlands, 2002.
54. Hauschild, M.Z.; Wenzel, H. *Environmental Assessment of Products, Volume 2: Scientific Background*; Springer: New York, NY, USA, 1998.
55. Huijbregts, M.A.; Hellweg, S.; Frischknecht, R.; Hungerbühler, K.; Hendriks, A.J. Ecological footprint accounting in the life cycle assessment of products. *Ecol. Econ.* **2008**, *64*, 798–807. [CrossRef]
56. EN 15804, BS EN 15804:2012. *Standards Publication Sustainability of Construction Works-Environmental Product Declarations-Core Rules for the Product Category of Construction Products*; European Committee for Standardization: Brussels, Belgium, 2014; Volume 70.
57. De Bruyn, S.; Ahdour, S.; Bijleveld, M.; De Graaff, L.; Schep, E.; Schroten, A.; Vergeer, R. Environmental Prices Handbook 2017-Methods and Numbers for Valuation of Environmental Impacts. 2018. Available online: https://www.cedelft.eu/en/publications/2113/envionmental-prices-handbook-2017 (accessed on 12 January 2021).
58. PRE. SimaPro Database Manual Methods Library. 2019. Available online: https://simapro.com/ (accessed on 28 January 2021).

49. Chomkhamsri, K.; Wolf, M.A.; Pant, R. International reference life cycle data system (ILCD) handbook: Review schemes for life cycle assessment. *Towards Life Cycle Sustain. Manag.* **2011**, 107–117. [CrossRef]
50. Herrmann, I.T.; Moltesen, A. Does it matter which Life Cycle Assessment (LCA) tool you choose?—A comparative assessment of SimaPro and GaBi. *J. Clean. Prod.* **2015**, *86*, 163–169. [CrossRef]
51. Hiloidhari, M.; Banerjee, R.; Rao, A.B. Life cycle assessment of sugar and electricity production under different sugarcane cultivation and cogeneration scenarios in India. *J. Clean. Prod.* **2020**, *290*, 125170. [CrossRef]
52. Salomone, R.; Cappelletti, G.M.; Malandrino, O.; Mistretta, M.; Neri, E.; Nicoletti, G.M.; Notarnicola, B.; Pattara, C.; Russo, C.; Saija, G. Life Cycle Assessment in the Olive Oil Sector. In *Life Cycle Assessment in the Agri-food Sector*; Springer: Cham, Switzerland, 2015; pp. 57–121.
53. El Hanandeh, A. Carbon abatement via treating the solid waste from the Australian olive industry in mobile pyrolysis units: LCA with uncertainty analysis. *Waste Manag. Res.* **2013**, *31*, 341–352. [CrossRef]
54. Intini, F.; Kuhtz, S.; Rospi, G. Life cycle assessment (LCA) of an energy recovery plant in the olive oil industries. *Int. J. Energy Environ.* **2012**, *3*, 541–552.
55. Martinez-Hernandez, E.; Martinez-Herrera, J.; Campbell, G.M.; Sadhukhan, J. Process integration, energy and GHG emission analyses of Jatropha-based biorefinery systems. *Biomass Convers. Biorefinery* **2014**, *4*, 105–124. [CrossRef]
56. EPA-Environmental Protection Agency Emission Factor Documentation for AP-42 Section 1.4-Natural Gas Combustion, Technical Support Division, Office of Air Quality Planning and Standards, Research Triangle Park, NC, USA, 1998. Available online: http://www3.epa.gov/ttnchie1/ap42/ch01/bgdocs/b01s04.pdf (accessed on 8 January 2021).
57. Farahani, S.S.; Soheilifard, F.; Raini, M.G.N.; Kokei, D. Comparison of different tomato puree production phases from an environmental point of view. *Int. J. Life Cycle Assess.* **2019**, *24*, 1817–1827. [CrossRef]

Article

Environmental Impact of Food Preparations Enriched with Phenolic Extracts from Olive Oil Mill Waste

Alessia Pampuri [1], Andrea Casson [1,*], Cristina Alamprese [2], Carla Daniela Di Mattia [3], Amalia Piscopo [4], Graziana Difonzo [5], Paola Conte [6], Maria Paciulli [7], Alessio Tugnolo [1], Roberto Beghi [1], Ernestina Casiraghi [2], Riccardo Guidetti [1] and Valentina Giovenzana [1]

[1] Department of Agricultural and Environmental Sciences (DiSAA), Università degli Studi di Milano, 20133 Milan, Italy; alessia.pampuri@unimi.it (A.P.); alessio.tugnolo@unimi.it (A.T.); roberto.beghi@unimi.it (R.B.); riccardo.guidetti@unimi.it (R.G.); valentina.giovenzana@unimi.it (V.G.)

[2] Department of Food, Environmental, and Nutritional Sciences (DeFENS), Università degli Studi di Milano, 20133 Milan, Italy; cristina.alamprese@unimi.it (C.A.); ernestina.casiraghi@unimi.it (E.C.)

[3] Faculty of Bioscience and Technology for Agriculture, Food and Environment, University of Teramo, 64100 Teramo, Italy; cdimattia@unite.it

[4] Department of Agraria, University Mediterranea of Reggio Calabria, 89124 Reggio Calabria, Italy; amalia.piscopo@unirc.it

[5] Department of Soil, Plant and Food Sciences, Food Science and Technology Unit, University of Bari Aldo Moro, 70126 Bari, Italy; graziana.difonzo@uniba.it

[6] Department of Agricultural Sciences, Università degli Studi di Sassari, 07100 Sassari, Italy; pconte@uniss.it

[7] Department of Food and Drug, University of Parma, 43124 Parma, Italy; maria.paciulli@unipr.it

* Correspondence: andrea.casson@unimi.it; Tel.: +39-(025)-0316873

Abstract: Reducing food waste as well as converting waste products into second-life products are global challenges to promote the circular economy business model. In this context, the aim of this study is to quantify the environmental impact of lab-scale food preparations enriched with phenolic extracts from olive oil mill waste, i.e., wastewater and olive leaves. Technological (oxidation induction time) and nutritional (total phenols content) parameters were considered to assess the environmental performance based on benefits deriving by adding the extracts in vegan mayonnaise, salad dressing, biscuits, and gluten-free breadsticks. Phenolic extraction, encapsulation, and addiction to the four food preparations were analyzed, and the input and output processes were identified in order to apply the life cycle assessment to quantify the potential environmental impact of the system analyzed. Extraction and encapsulation processes characterized by low production yields, energy-intensive and complex operations, and the partial use of chemical reagents have a non-negligible environmental impact contribution on the food preparation, ranging from 0.71% to 73.51%. Considering technological and nutritional aspects, the extraction/encapsulation process contributions tend to cancel out. Impacts could be reduced approaching to a scale-up process.

Keywords: life cycle assessment; biocompounds; shelf life; environmental sustainability; biscuits; gluten-free breadsticks; salad dressing; vegan mayonnaise; circular economy; waste recovery

1. Introduction

According to the International Olive Council, the global production of olive oil has been constantly increasing from 1.8 million tons per year in the 1990s, and the production currently amounts to more than 3 million tons per year. Olive oil production represents a significant sector all over the world and in the European Union economy. EU countries contributed almost 70% of all olive oil produced in the world in the 2018–2019 harvest year campaign, and the resultant revenue was about five billion euro [1].

Italy is the second largest producer of olive oil after Spain. Most of the production (about 80%) is concentrated in Apulia, Sicily, and Calabria. On the other hand, in the northernmost regions, the climatic conditions have allowed the cultivation of olives since ancient times thanks to microclimates; however, the cultivation is much less widespread than in southern Italy [2].

Olive drupes and olive oil are potential sources of several bioactive compounds as phenolic compounds, tocopherols, and other antioxidants. During oil extraction, many of these secondary metabolites can be destroyed or degraded or transferred in olive oil mill waste. At the end of the extraction process, olive oil contains 1–2% of the total phenol content (TPC) of drupes, so the residual antioxidant compounds would be lost in olive mill wastewater and pomace [3].

In particular, the traditional press extraction method as well as the continuous three-phase decanter process, which is most widely used for the production of olive oil, generate three principal products: olive oil (20%) and two streams of waste, a wet solid waste (30%) called "crude olive cake" and an aqueous waste called "olive mill wastewater" (50%). The solid waste (crude olive cake) is the residue that remains after the first pressing of the olives and is a mixture of pomace, stones, leaves, and dust [4].

Despite the economic and nutritional importance of this food product in many countries, the olive oil industry causes diverse environmental impacts in terms of resource depletion, land degradation, air emissions, and waste generation. These impacts may vary as a result of the practices and techniques employed in olive cultivation and oil production, and they can also vary from one country to another and also from one region to another within the same geographical area [5].

Many olive mill waste products could be exploited as by-products to be used as fuels, fertilizers, or other intermediate products for the food, nutraceutical, cosmetic, and pharmaceutical industries. Olive pomace and olive oil mill wastewater could be considered a low-cost and renewable source of high-added value compounds [6].

These by-products are still undervalued even if they have a good potential as sources of bioactive components [7]: the phenolic compounds deriving from olive possess antimicrobial, anti-inflammatory, and chemopreventive properties [8,9]. Olive polyphenols have been proven to exert important technological functionality [10] such as a water/oil-holding capacity and emulsifying activity and can represent a useful ingredient that can help in the production and stabilization of complex food products such as emulsions [11].

However, the quantity and the specific characteristics of these by-products depends on climatic conditions and production practices. For these reasons nowadays, many studies are focusing on the management of by-products of the olive oil industry to try to further enhance this supply chain.

From a circular economy perspective, it is possible to assess, and in the future reduce, the environmental impact of these processes and waste products using a life cycle assessment analysis (LCA) [12]. There are several studies in the literature, mostly conducted in Mediterranean countries, in which the LCA of producing olive oil via different methods have been performed [5,13–23].

The aim of this study is to quantify the environmental impact of the extraction of bioactive phenolic compounds from olive oil mill wastewater and olive leaves. To evaluate the impact of phenols' extraction and encapsulation processes within the entire supply chain, lab-scale food preparations were analyzed. Moreover, technological and nutritional parameters (oxidation induction time and total phenols content) were used to assess the environmental performance of production process based on benefits deriving from adding the phenolic extract in four food preparations: vegan mayonnaise, salad dressing, biscuits, and gluten-free breadsticks.

2. Materials and Methods

The phenolic extracts (PE) were obtained following two extraction methods, olive oil mill wastewater (OMWW) and from olive leaves (OL), to obtain olive leaves extract (OLE). PE were added to different food preparations, as free extracts of OMWW (OMWW PE) in vegan mayonnaise (mayo) and gluten-free breadsticks (GFB), as free extracts of OL (OLE) in gluten-free breadsticks, and as encapsulated OLE (eOLE) in salad dressing and biscuits.

Life cycle assessment is a standardized method aimed at evaluating the environmental impacts studying the whole cycle of a product or a service. It considers all the inputs and outputs from raw materials extraction until end-of-life scenarios of the product or service analyzed. The following LCA study was developed in compliance with the international standards of series ISO 14040 and ISO 14044 [12]. According to ISO standards [12], the analysis was articulated in four stages: goal and scope definition, life cycle inventory (LCI), life cycle impact assessment (LCIA), and interpretation of the results (proposed in the Section 4).

2.1. Goal and Scope Definition

The goal of this study is to quantify the potential environmental impact of the use of phenolic extracts (PE) from olive mill wastewater (OMWW) and from olive leaves (OL) in particular case studies of food processes to improve nutritional and technological parameters.

The functional unit (FU) defines the reference unit of the system under analysis (ISO 14040 and 14044) [12]. To evaluate the environmental impact of the polyphenol's extraction processes and encapsulating process, 1 g of TPC was defined as FU.

Differently, a commercial unit was used to evaluate the weight of the PE as ingredients in four food formulations: (i) 350 g for vegan mayonnaise, (ii) 135 mL for salad dressing, (iii) 160 g for biscuits, and (iv) 300 g for gluten free breadsticks.

Moreover, the commercial FU for each product was normalized considering technological and nutritional parameters.

The system under study follows an approach called "from cradle to grave" where all the factors were considered from the olive oil extraction process to the formulation of the four food preparations. In detail, as reported in Figure 1 (detailed system boundaries are reported in supplementary data Figure S1), every input (extraction of raw materials, energy and water consumption, chemicals) and output (hazardous waste and food production waste) were considered. Regarding the food preparations, the consumption and the packaging were neglected.

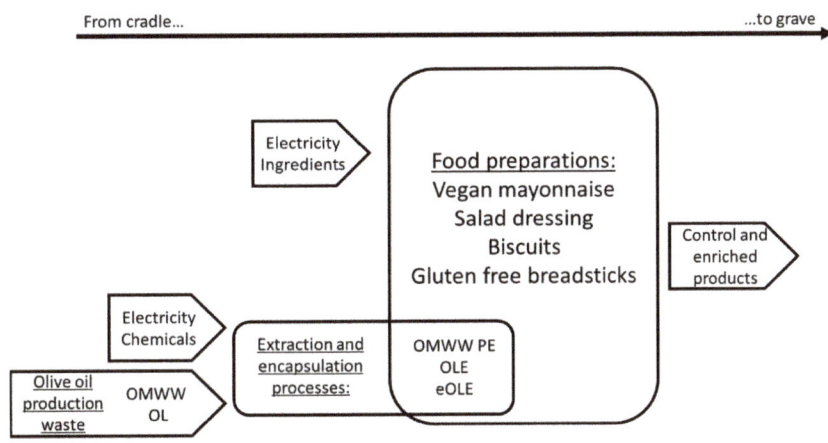

Figure 1. Simplified system boundaries.

2.1.1. Normalization Factors

Since the PE addiction implies advantages in term of technological and nutritional aspects, the environmental impact calculated considering commercial FU was normalized based on Total Phenolic Content (TPC) and oxidative stability. In order to quantify (i) the TPC, Singleton and Rossi's method [13] was performed and (ii) the oxidative stability method used by Paciulli et al. [14] was carried out. Gallic acid equivalent expressed as TPC (mg/g) and oxidative stability expressed in oxidation induction days represent the units describing nutritional and technological parameters respectively.

In this work different combinations of extracts (OMWW PE, OLE, and eOLE) and food preparations were considered and only the best performing enriched food preparations were studied (higher values of induction days and mg of TPC/g of product) and reported in Table 1. To evaluate the technological and nutritional performance of enriched food preparations, a conventional production process (control) was also analyzed.

Table 1. Technological and nutritional characterization of food preparations.

Products	Oxidative Stability	Polyphenols Content
	Induction days	mg TPC/g product
Mayo control	0.5	0.001
Mayo + OMWW PE	1	0.413
Biscuits control	4.1	1.1
Biscuits + eOLE	5.5	1.6
Salad dressing control	/	0.001
Salad dressing + eOLE	/	0.162
GFB control	84	29.39
GFB + OLE	109	34.08
GFB + OMWW PE	152	31.74

2.1.2. Allocation Criteria

Considering the olive oil mill process and according to Parascanu et al. [20], olive oil has a much higher economic value compared to the other by-products, not only representing the higher output in terms of product amount (mass allocation in Table 2). To better evaluate the environmental impact of every olive oil mill output, an economic allocation was used. In particular, according to Tsagaraki et al. [4], in addition to the conventional subdivision of the by-products, it was possible to identify in a more specific way the average composition of the different outputs: (i) pomace, (ii) wastewater, (iii) stone, and (iv) leaves and dust (Table 2).

Table 2. Economic values and average percentage of mass and economic allocation for the olive oil and by-products of the milling process.

Output	Mass Allocation (%)	Economic Value (€/kg)	Economic Allocation (%)
Olive oil	18.0	3.65	98.33
Pomace	23.0	0.015	0.52
Wastewater (OMWW)	50.0	0.001	0.07
Stone	8.0	0.09	1.08
Leaves (OL) and dust	1.0	0.001	0.00

Details about the transition from the mass allocation to the economic one was reported in Table 2. Overall, OMWW and OL values represent a cost for the milling process instead of a revenue, not having economic allocation (waste products). In this work, in order to calculate the relative environmental impact, a minimum economic value attribution was chosen also for these products (by-products). Considering the four food preparations, a mass allocation was used.

2.2. Life Cycle Inventory (LCI)

Regarding the olive oil milling process, secondary data from the WFLD (World Food LCA Database) were used and modeled in to obtain the five different outputs as argued in the allocation criteria paragraph. OMWW and OL were used as inputs for the following extraction processes. The inventory of OMWW PE and OLE extraction phases, encapsulation phase (eOLE), and the related applications on food preparations, on the contrary, were performed separately and reported below using primary data.

2.2.1. OMWW PE Extraction Phase LCI

Following the method used by Romeo et al. [24], 2 L of olive mill wastewater were acidified with 1 mL of HCl to obtain a pH 2 mixture. The mixture passed three cycles which required 2 L of hexane and 3 min of centrifuge (0.5 kW) for each cycle. Then, 0.625 L of ethyl acetate were added, and using (i) the centrifuge (3000 rpm 0.8 kW, 18 min) and (ii) the evaporator for 2 h, 3 g of dry residue were obtained. Then, 100 mL of water were added to the dry residue to obtain a solution, which was filtered to obtain 103 mL of OMWW PE. Hexane and ethyl acetate were recovered after use for 75%, and the remaining 25% were treated as hazardous waste. Input and output data related to this analysis and allocated to the FU (equivalent to 34.33 mL) are reported in Table 3.

Table 3. Input, output data and allocated quantity per FU, related to the OMWW PE process.

Input	Description	Quantity	Units	Allocation Factor	Allocated Quantity per FU
OMWW	Olive Mill Wastewater	2	L	0.333	0.667
HCl	Acid	1	mL	0.333	0.333
Hexane	2 L × 3 times	6	L	0.004	0.025
Centrifuge	Power load (0.5 kW); use time (9 min)	0.075	kWh	0.333	0.025
Ethyl-acetate	0.625 mL × 3 times	1.875	mL	0.004	0.008
Centrifuge	Power load (0.8 kW); use time (18 min)	0.24	kWh	0.333	0.080
Evaporator	Power load (1 kW); use time (2 h)	2	kWh	0.333	0.667
Water		100	mL	0.333	33.333
Filter		3	g	0.333	1.000
Waste	Description	Quantity	Units	Allocation Factor	Allocated Quantity per FU
Hazardous waste	Hexane	6	L	0.004	0.025
Hazardous waste	Ethyl-acetate	1.875	mL	0.004	0.008

2.2.2. OLE Extraction Phase LCI

According to Difonzo et al. [25], 200 g of olive leaves were washed using 1 L of water, dried firstly with paper (2–3 pieces) and then using an oven (0.53 kW, 8 min, 120 °C). Then, 100 g of hot air-dried leaves (3–4% water content) were milled for 30 s using a mill (0.175 kW). The powder obtained passed three cycles, which required (i) 2L of water, (ii) the use of ultrasound (0.2 kW, 30 min), and (iii) filtering to obtain 6 L of filtered aqueous extract. Using a freeze dryer (1.4 kW, 24 h), 10 g of OLE were obtained. Then, 100 g of exhausted leaves were treated as biowaste. Input and output data related to this analysis and allocated to the FU (equivalent to 6.67 g) are reported in Table 4.

Table 4. Input, output data, and allocated quantity per FU, related to the OLE process.

Input	Description	Quantity	Units	Allocation Factor	Allocated Quantity per FU
Olive leaves	Waste from olive oil production	400	g	0.167	66.667
Water	1 L for the washing activity	1	L	0.167	0.167
Oven	Power load (0.530 kW); use time (8 min); Temperature (120 °C); Capacity (400 g)	0.0707	kWh	0.167	0.012
Paper	2–3 pieces	12.5	g	0.167	2.083
Mill	Power load (0.175 kW); use time (30 s)	0.0015	kWh	0.667	0.001
Water	6 L for the extraction	6	L	0.667	4.002
Ultrasound	Power load (0.200 kW); use time (90 min)	0.3000	kWh	0.667	0.200
Filters	1 filter	3	g	0.667	2.001
Freeze dryer	Power load (1.100 kW); use time (8 h)	8.8	kWh	0.667	5.870
Waste	Description	Quantity	Units	Allocation Factor	Allocated Quantity per FU
Biowaste	Exhausted leaves	100	g	0.667	66.700

2.2.3. eOLE Encapsulation Phase LCI

According to Flamminii et al. [26], 0.4 g of OLE, 0.4 g of pectin, 1.64 g of calcium citrate, 0.4 g of alginate, and 17.16 mL of water were mixed to 98 g of sunflower oil and 2 g of Span 80 (emulsifier) and agitated using a stirring plate (0.4 kW, 15 min). Then, 20 g of sunflower oil and 0.5 g of glacial acetic acid were added and agitated using a stirring plate (0.4 kW, 30 min). Then, 3 g of OLE, 145.42 mL of water, 0.83 g of calcium chloride, and 0.75 g of Tween20 (surfactant) were added to the mixture and agitated again using a stirring plate (0.4 kW, 30 min). To obtain the beads (92% water content and 8% dried matter), a centrifuge was used for 5 min (0.800 kW). Then, 30 mL of ethanol solution (21 mL ethanol, 8.94 mL water, and 0.6 g OLE) was added to obtain 20 g of cleaned beads. Lastly, the beads were lyophilized using a freeze drier for 24 h (1.4 kW) to obtain 1.60 g of eOLE. Then, 300.5 g of exhausted oil were treated as waste oil. Input and output data related to this analysis and allocated to the FU (1 g of TPC equivalent to 5 g of eOLE) are reported in Table 5.

Table 5. Input, output data, and allocated quantity per FU, related to the eOLE process.

Input Eole	Description	Quantity	Units	Allocation Factor	Allocated Quantity per FU
OLE	Olive Leaf Extract	4	g	3.125	12.500
Alginate	Polymer	0.4	g	3.125	1.250
Pectin	Polymer	0.4	g	3.125	1.250
Water		168.92	mL	3.125	527.875
Calcium citrate		1.638	g	3.125	5.119
Sunflower oil	Oil	118	g	3.125	368.750
Span 80	Emulsifier	2	g	3.125	6.250
Stirring plate	Power load (0.4 kW); use time (75 min)	0.5	kWh	3.125	1.563
Glacial acetic acid	Acid	0.5	g	3.125	1.563
Calcium chloride	Gelling agent	0.832	g	3.125	2.600
Tween 20	Surfactant	0.75	g	3.125	2.344
Centrifuge	Power load (0.8 kW); use time (5 min)	0.07	kWh	3.125	0.208
Ethanol		21.00	mL	3.125	65.625
Freeze dryer	Power load (1.100 kW); use time (18 min)	0.33	kWh	3.125	1.031

Waste	Description	Quantity	Units	Allocation Factor	Allocated Quantity Per FU
Hazardous waste	Exhausted oil	300.5	mL	3.125	939.063

2.2.4. Vegan Mayonnaise LCI

For the traditional formulation (Mayo) used as control, 150 g of soy milk were mixed with 1 g of lemon juice and 199 g of sunflower oil using a blender (1 kW) for 4 min. The final product obtained weighed 350 g. Differently, for the product enriched with polyphenols (Mayo + OMWW PE), 100 g of soy milk were mixed with 1 g of lemon juice, 199 g of sunflower oil, and 50 g of OMWW PE using a blender (1 kW) for 4 min. In this case, the final product obtained weighted 350 g. Input and output data related to this preparation are reported in Table 6.

Table 6. Input, output data, and allocated quantity per FU (350 g), related to the vegan mayonnaise process.

Product	Input	Description	Quantity	Units	Allocation Factor per 350 g Mayo	Allocated Value per 350 g Mayo
Mayo control	Soy milk	Ingredient	150	g	1	150
	Sunflower oil	Ingredient	199	g	1	199
	Lemon juice	Ingredient	1	g	1	1
	Blending	Power load (1 kW); use time (4 min)	0.0667	kWh	1	0.0667
Mayo + OMWW PE	Soy milk	Ingredient	100	g	1	100
	OMWW	Ingredient	50	g	1	50
	Sunflower oil	Ingredient	199	g	1	199
	Lemon juice	Ingredient	1	g	1	1
	Blending	Power load (1 kW); use time (4 min)	0.0667	kWh	1	0.0667

2.2.5. Salad Dressing LCI

According to Jolayemi et al. [27], for the traditional formulation of salad dressing (salad dressing control), 99.55 mL of water were homogenized with 0.80 g of xanthan gum, 0.50 g of citric acid, 0.40 g of salt, and 35.75 mL of corn oil using a blender (0.75 kW) for 1 min to obtain 135 mL of salad dressing. Differently, for the product enriched with polyphenols (salad dressing + eOLE), 99.25 mL of water was mixed with 0.80 g of xanthan gum, 0.5 g of citric acid, and 0.4 g of salt. Then, 0.3 g of eOLE were added to the aqueous phase; finally, 33.75 mL of corn oil were added to obtain the salad dressing composed of 25% oil and 75% aqueous phase. Two blendings were carried out using a blender (0.75 kW) for 30 s each to obtain 135 mL of salad dressing. Input and output data related to this preparation are reported in Table 7.

Table 7. Input, output data, and allocated quantity per FU (135 mL) related to the salad dressing process.

	Input	Description	Quantity	Units	Allocation Factor per 135 mL Salad Dressing	Allocated Quantity per 135 mL Salad Dressing
Salad dressing control	Water	Ingredient	99.55	mL	1	99.55
	Salt	Ingredient	0.40	g	1	0.40
	Xanthan Gum	Ingredient	0.80	g	1	0.80
	Citric acid	Ingredient	0.50	g	1	0.50
	Corn oil	Ingredient	33.75	g	1	33.75
	Blending	Power load (0.75 kW); use time (1 min)	0.0125	kWh	1	0.0125
Salad dressing + eOLE	Water	Ingredient	99.25	mL	1	99.25
	Salt	Ingredient	0.40	g	1	0.40
	Xanthan Gum	Ingredient	0.80	g	1	0.80
	Citric acid	Ingredient	0.50	g	1	0.50
	eOLE	Ingredient	0.30	g	1	0.30
	Corn oil	Ingredient	33.75	g	1	33.75
	Blending	Power load (0.75 kW); use time (1 min)	0.0125	kWh	1	0.0125

2.2.6. Biscuits LCI

According to Paciulli et al. [28], for the biscuits, 17 g of water, 1 g of salt, and 1 g baking powder were mixed with 100 g of soft wheat flour using a blender for 15 min (0.3 kW). Then, 30 g of butter and 50 g of icing sugar were blended for 3 min and then added to the dough and blended for another 6 min. The final dough (200 g), after a chilling phase of 10 min in a refrigerator, was formed manually to obtain 12 biscuits. After the cooking phase (180 °C; 20 min; 0.530 kW), 12 biscuits (13.5 g each) were obtained. For the enriched formulation (Biscuits + PE), 99.45 g of soft wheat flour and 0.55 g of eOLE were blended for 15 min (kW); the blended mixture was added to the solution of 17 g of water, 1 g of salt, and 1 g of baking powder. The other steps that follow are the same as for the traditional formulation. Input and output data related to this preparation are reported in Table 8.

Table 8. Input, output data, and allocated quantity per FU (160 g), related to the biscuits process.

Product	Input	Description	Quantity	Units	Allocation Factor per 160 g Biscuits	Allocated Quantity per 160 g Biscuits
Biscuits control	Soft wheat flour	Ingredient	100	g	1	100
	Water	Ingredient	17	g	1	17
	Salt	Ingredient	1	g	1	1
	Baking powder	Ingredient	1	g	1	1
	Sugar	Ingredient	50	g	1	50
	Butter	Ingredient	30	g	1	30
	Blending	Power load (0.3 kW); use time (24 min)	0.12	kWh	1	0.12
	Chilling	Energy consumption (250 kWh/year); use time (10 min)	0.0048	kWh	0.01	4.8×10^{-5}
	Baking	Power load (0.53 kW); use time (20 min)	0.1767	kWh	0.25	4.4×10^{-2}
Biscuits + eOLE	Soft wheat flour	Ingredient	99.45	g	1	99.45
	eOLE	Ingredient	0.55	g	1	0.55
	Water	Ingredient	17	g	1	17
	Salt	Ingredient	1	g	1	1
	Baking powder	Ingredient	1	g	1	1
	Sugar	Ingredient	50	g	1	50
	Butter	Ingredient	30	g	1	30
	Blending	Power load (0.300 kW); use time (24 min)	0.12	kWh	1	0.12
	Chilling	Energy consumption (250 kWh/year); use time (10 min)	0.005	kWh	0.01	4.8×10^{-5}
	Baking	Power load (0.530 kW); use time (20 min)	0.177	kWh	0.25	4.4×10^{-2}
Biscuits control Biscuits + eOLE	Food waste	Production waste	15	g	1	15

2.2.7. Gluten-Free Breadsticks LCI

To obtain the experimental gluten-free breadsticks, 500 g of rice flour, 500 g of corn starch, 550 mL of warm water (26 °C), 100 g of sunflower oil, 15 g of guar gum, 15 g of psyllium fiber, 30 g of sugar, 18 g of salt, and 40 g of compressed yeast were used as the basic formulation. For the preparation of the control samples, the dry ingredients were pre-blended (0.3 kW) for 2 min to ensure a proper homogenization and then mixed with sugar, salt, and yeast—previously dissolved in aliquots of water—and 100 g of sunflower oil for 13 min by using a professional mixer (0.3 kW). The resulting dough, after a leavening phase of 30 min (33 °C), was divided in 62 pieces of 28 g each and subjected to a second leavening (30 min; 33 °C). Then, the breadstick samples were baked (0.53 kW; 180 °C) for 13 min, rested for 30 min, and baked again for 22 min. Differently from the control breadsticks, the enriched ones were formulated using 1 g of PE from OMWW or OL (GFB + OMWW PE, GFB + OLE). Input and output data related to this preparation are reported in Table 9.

Table 9. Input, output data, and allocated quantity per FU (300 g), related to gluten-free breadsticks process.

Product	Input	Description	Quantity	Units	Allocation Factor per 300 g GFB	Allocated Quantity per 300 g GFB
GFB control; GFB + OLE; GFB + OMWW PE	Rice flour	Ingredient	500	g	0.247	123.457
	Corn starch	Ingredient	500	g	0.247	123.457
	Guar gum	Ingredient	15	g	0.247	3.704
	Psyllium fiber	Ingredient	15	g	0.247	3.704
	Sugar	Ingredient	30	g	0.247	7.407
	Salt	Ingredient	15	g	0.247	3.704
	Compressed yeast	Ingredient	40	g	0.247	9.877
	Sunflower oil	Ingredient	100	g	0.247	24.691
	Water	Ingredient	550	mL	0.247	135.802
	Blending	Power load (0.3 kW); use time (15 min)	0.075	kWh	0.247	0.019
	Leavening chamber	Power load (1.5 kW); (T 33 °C; RH 90%); use time (60 min)	1.5	kWh	0.012	0.019
	Baking	Power load (0.53 kW); use time (35 min)	0.31	kWh	0.062	0.019
GFB + OLE	OLE	OLE	1	g	0.247	0.247
GFB + OMWW PE	OMWW	OMWW	1	g	0.247	0.247

Product	Waste	Description	Quantity	Units	Allocation Factor per 300 g GFB	Allocated Quantity per 300 g GFB
GFB control; GFB + OLE; GFB + OMWW PE	Food waste	Production waste of raw dough	0.467	g	0.247	0.115

2.3. Life Cycle Impact Assessment (LCIA)

Life cycle impact assessment (LCIA) translates emissions and resource extractions into a limited number of environmental impact scores by means of characterization factors. These factors convert the data from LCI to the common unit of category indicator. According to Goedkoop et al. [29], the Recipe 2016 Midpoint (H) v1.04 method was used to assess the potential environmental impact. The LCIA data results were proposed using the following impact categories: Global warming (GW), Stratospheric ozone depletion (OD), Ionizing radiation (IR), Ozone formation, human health (OF-HH), Fine particulate matter formation (PM), Ozone formation, terrestrial ecosystems (OF-TE), Terrestrial acidification (TA), Freshwater eutrophication (FE), Marine eutrophication (ME), Terrestrial ecotoxicity (TE), Freshwater ecotoxicity (FRE), Marine ecotoxicity (MECO), Human carcinogenic toxicity (HCT), Human non-carcinogenic toxicity (HNCT), Land use (LU), Mineral resource scarcity (MRS), Fossil resource scarcity (FRS), and Water consumption (WC). Ecoinvent 3.6 (allocation, cut-off by classification) and World Food LCA Database 3.5 were used as databases for the inventory phase, SimaPro version 9.1.1. (PRè Sustainability, Amersfoort, The Netherlands) was used to assess the environmental impacts of PE extractions, PE encapsulation, and food preparations.

3. Results

The PE extractions, PE encapsulation process, and their use in four food formulation will be analyzed separately in the following sections.

3.1. PE Extraction and OLE Encapsulation Impact Assessment

Table 10 represents the overall potential environmental impacts while Figures 2–4 report the contribution analysis related to the different PE extractions and encapsulation. According to the FU used for the extraction and encapsulation processes, OMWW PE represents the least environmental impactful extraction technique (6.69 times less). The encapsulation of OLE, to obtain eOLE, represents a 10 times more impactful process than the environmental impact of the extraction of 1 g TPC contained in OLE.

Table 10. Environmental impacts of PE extractions and encapsulation processes.

Impact Category	Unit	OMWW PE	OLE	eOLE
GW [1]	kg CO_2 eq	4.10×10^{-1}	2.83	7.84
OD [2]	kg CFC11 eq	3.32×10^{-7}	2.34×10^{-6}	1.63×10^{-5}
IR [3]	kBq ^{60}Co eq	4.11×10^{-2}	3.17×10^{-1}	7.73×10^{-1}
OF-HH [4]	kg NO_x eq	7.11×10^{-4}	4.92×10^{-3}	1.57×10^{-2}
PM [5]	kg $PM_{2.5}$ eq	4.52×10^{-4}	3.29×10^{-3}	9.97×10^{-3}
OF-TE [6]	kg NO_x eq	7.40×10^{-4}	5.00×10^{-3}	1.60×10^{-2}
TA [7]	kg SO_2 eq	1.38×10^{-3}	1.01×10^{-2}	3.29×10^{-2}
FE [8]	kg P eq	9.98×10^{-5}	7.16×10^{-4}	2.08×10^{-3}
ME [9]	kg N eq	9.38×10^{-6}	5.48×10^{-5}	4.75×10^{-3}
TE [10]	kg 1,4-DCB	4.44×10^{-1}	2.88	11.1
FRE [11]	kg 1,4-DCB	1.00×10^{-2}	6.98×10^{-2}	3.43×10^{-1}
MECO [12]	kg 1,4-DCB	1.30×10^{-2}	9.05×10^{-2}	2.79×10^{-1}
HCT [13]	kg 1,4-DCB	8.38×10^{-3}	6.11×10^{-2}	1.71×10^{-1}
HNCT [14]	kg 1,4-DCB	2.02×10^{-1}	1.43	5.98
LU [15]	m^2a crop eq	1.58×10^{-2}	1.08×10^{-1}	3.61
MRS [16]	kg Cu eq	4.16×10^{-4}	2.56×10^{-3}	2.00×10^{-2}
FRS [17]	kg oil eq	1.29×10^{-1}	8.37×10^{-1}	2.25
WC [18]	m^3	5.80×10^{-3}	4.72×10^{-2}	1.20×10^{-1}

[1] Global warming. [2] Stratospheric ozone depletion. [3] Ionizing radiation. [4] Ozone formation, human health. [5] Fine particulate matter formation. [6] Ozone formation, terrestrial ecosystems. [7] Terrestrial acidification. [8] Freshwater eutrophication. [9] Marine eutrophication. [10] Terrestrial ecotoxicity. [11] Freshwater ecotoxicity. [12] Marine ecotoxicity. [13] Human carcinogenic toxicity. [14] Human non-carcinogenic toxicity. [15] Land use. [16] Mineral resource scarcity. [17] Fossil resource scarcity. [18] Water consumption.

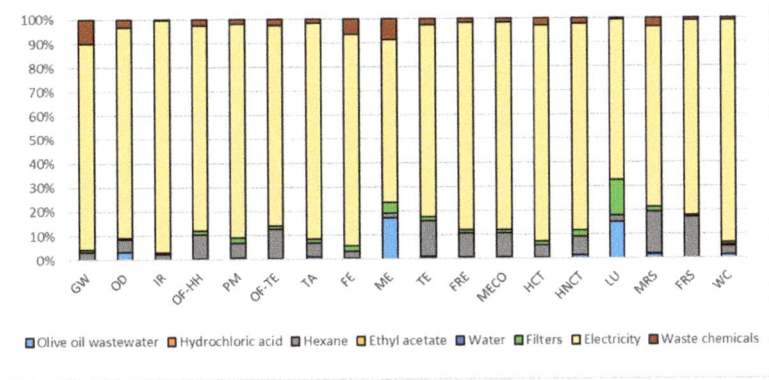

Figure 2. Hotspot deriving from contribution analysis related to OMWW PE. Global warming (GW), Stratospheric ozone depletion (OD), Ionizing radiation (IR), Ozone formation, human health (OF-HH), Fine particulate matter formation (PM), Ozone formation, terrestrial ecosystems (OF-TE), Terrestrial acidification (TA), Freshwater eutrophication (FE), Marine eutrophication (ME), Terrestrial ecotoxicity (TE), Freshwater ecotoxicity (FRE), Marine ecotoxicity (MECO), Human carcinogenic toxicity (HCT), Human non-carcinogenic toxicity (HNCT), Land use (LU), Mineral resource scarcity (MRS), Fossil resource scarcity (FRS), and Water consumption (WC).

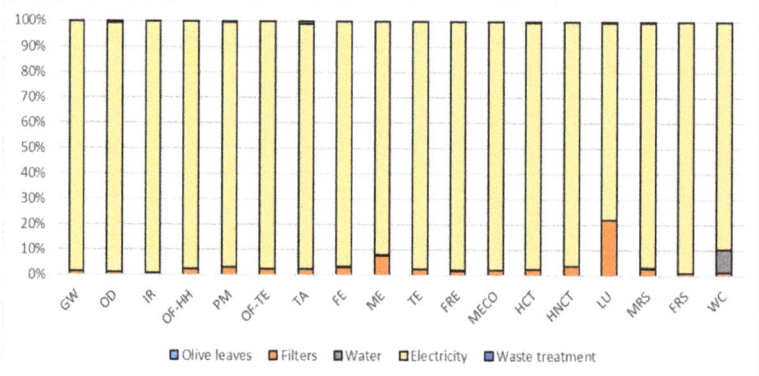

Figure 3. Hotspot deriving from contribution analysis related to OLE. Global warming (GW), Stratospheric ozone depletion (OD), Ionizing radiation (IR), Ozone formation, human health (OF-HH), Fine particulate matter formation (PM), Ozone formation, terrestrial ecosystems (OF-TE), Terrestrial acidification (TA), Freshwater eutrophication (FE), Marine eutrophication (ME), Terrestrial ecotoxicity (TE), Freshwater ecotoxicity (FRE), Marine ecotoxicity (MECO), Human carcinogenic toxicity (HCT), Human non-carcinogenic toxicity (HNCT), Land use (LU), Mineral resource scarcity (MRS), Fossil resource scarcity (FRS), and Water consumption (WC).

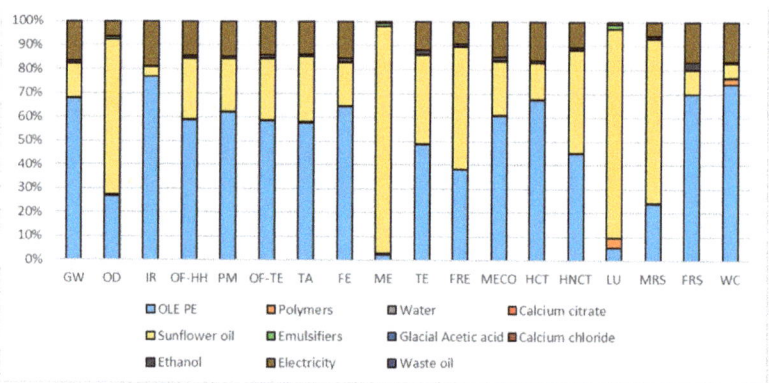

Figure 4. Hotspot deriving from contribution analysis related to eOLE. Global warming (GW), Stratospheric ozone depletion (OD), Ionizing radiation (IR), Ozone formation, human health (OF-HH), Fine particulate matter formation (PM), Ozone formation, terrestrial ecosystems (OF-TE), Terrestrial acidification (TA), Freshwater eutrophication (FE), Marine eutrophication (ME), Terrestrial ecotoxicity (TE), Freshwater ecotoxicity (FRE), Marine ecotoxicity (MECO), Human carcinogenic toxicity (HCT), Human non-carcinogenic toxicity (HNCT), Land use (LU), Mineral resource scarcity (MRS), Fossil resource scarcity (FRS), and Water consumption (WC).

To highlight hotspots, details regarding the contribution analysis of the different factors influencing extraction and encapsulation processes were reported.

3.1.1. Contribution Analysis of OMWW PE Impact Assessment

Factors influencing the OMWW PE extraction were reported in Figure 2. The factor electricity reached the highest values for all the impact categories from the lowest value reached in LU (67%) to the highest one reached in IR impact category (97%). This large contribution depends on the energy demand from the equipment necessary for the extraction of phenols from olive oil wastewater; the major contribution comes from the evaporator, which absorbs 86% of the overall energy consumption. Despite the tiny quantity used and allocated to this extraction process, the contribution of hexane covers an average weight among impact categories equal to 8% and thus represents the second contribution factor. The waste chemicals factor, which refers to the waste management of chemicals, represents the third contribution of this extraction technique, contributing with an average weight among impact categories of about 3%.

As for the OMWW PE, also for the OLE process (Figure 3), the electricity contributes mainly to all the impact categories with an average contribution of about 97%. This high level of contribution is directly linked to the energy consumption of the freeze dryer (97% of the total energy consumption). The second contribution factor is identified in the paper filters used, even if this factor has a contribution of about 4% among impact categories

Differently from the two extraction processes previously analyzed, Figure 4 shows a fragmented contribution analysis of the eOLE reporting different contribution factors. The highest contribution of the whole process comes from the OLE production process, which quantified the related weight of this factor equal to 50% among impact categories. The second contribution factor is the sunflower oil with an average weight among impact categories equal to 35%. This factor shows higher contribution in those impact categories related to the cultivation phases (ME 97%; LU 85%). The factor electricity contributes less compared to the other processes analyzed, but it anyway covers an average contribution of about 12%. In this case, the stirring plate requires 56% of the total energy consumption due to the time of usage, while the freeze dryer requires only 37% of the total energy consumption.

3.1.2. Vegan Mayonnaises Impact Assessment

The environmental impact comparison related to the formulation of 350 g of vegan mayonnaise and 350 g of enriched vegan mayonnaise is reported in Figure 5, while the comparison of environmental impact of the two formulations normalized to the shelf-life parameters is reported in Figure 6. The environmental impact results from these two analyses and TPC normalization factor was reported also in Table S1.

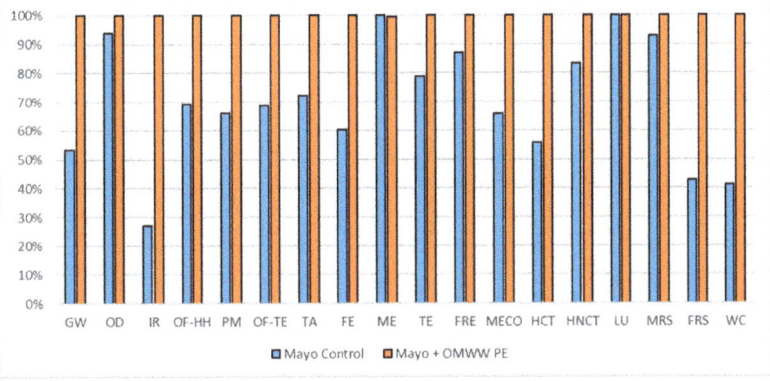

Figure 5. Environmental impact comparison of a vegan mayonnaise (350 g) vs. PE enriched product. Global warming (GW), Stratospheric ozone depletion (OD), Ionizing radiation (IR), Ozone formation, human health (OF-HH), Fine particulate matter formation (PM), Ozone formation, terrestrial ecosystems (OF-TE), Terrestrial acidification (TA), Freshwater eutrophication (FE), Marine eutrophication (ME), Terrestrial ecotoxicity (TE), Freshwater ecotoxicity (FRE), Marine ecotoxicity (MECO), Human carcinogenic toxicity (HCT), Human non-carcinogenic toxicity (HNCT), Land use (LU), Mineral resource scarcity (MRS), Fossil resource scarcity (FRS), and Water consumption (WC).

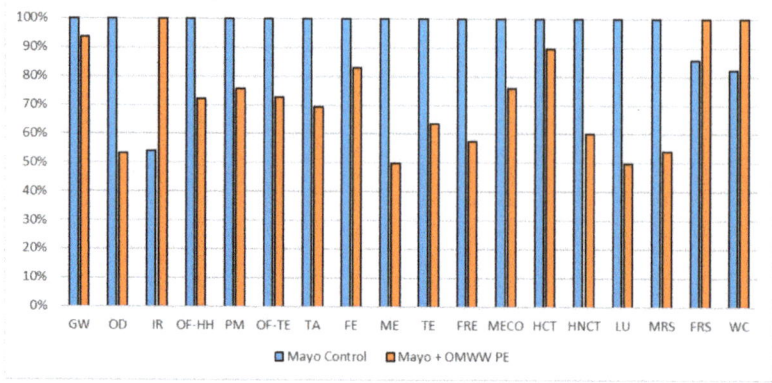

Figure 6. Environmental impact comparison of a vegan mayonnaise (350 g) vs. PE enriched product normalized to oxidation induction time. Global warming (GW), Stratospheric ozone depletion (OD), Ionizing radiation (IR), Ozone formation, human health (OF-HH), Fine particulate matter formation (PM), Ozone formation, terrestrial ecosystems (OF-TE), Terrestrial acidification (TA), Freshwater eutrophication (FE), Marine eutrophication (ME), Terrestrial ecotoxicity (TE), Freshwater ecotoxicity (FRE), Marine ecotoxicity (MECO), Human carcinogenic toxicity (HCT), Human non-carcinogenic toxicity (HNCT), Land use (LU), Mineral resource scarcity (MRS), Fossil resource scarcity (FRS), and Water consumption (WC).

According to the comparison of the commercial unit (350 g of product for each formulation), the enriched product reports higher environmental impacts in most of the impact categories. As reported in Figure 2, the high demand of electricity required for the OMWW PE process can be identified in the differences between the traditional mayonnaise and the enriched one. The average percentage of responsibility related to the OMWW PE in the enriched formulation weights of about 30% (as reported in Table 11). The gap between the two products goes from a lower value equal to 0% in ME and LU (related mainly to the field activities of sunflower oil) to the higher value reached in IR, which counts 73% less in traditional formulation. Considering only the commercial unit, the traditional mayonnaise shows better environmental impact with respect to the enriched one (30% more convenient).

Table 11. Mean and standard deviation among the impact categories of extraction and encapsulation processes impact on the whole food chain.

Products	Mean	SD
	%	%
Mayo + OMWW PE	30.90	20.77
Salad dressing + eOLE	73.51	17.22
Biscuits + eOLE	56.61	14.97
GFB + OLE	17.72	14.29
GFB + OMWW PE	0.71	0.75

Different considerations must be done according to the estimated shelf-life values quantified in Table 1. The traditional mayonnaise, which claims the worst technological performances (half induction period) with respect to the enriched one, reports different results in Figure 6.

The induction days parameter was used to compare the potential environmental impact of the two preparations. The technological characteristics of the enriched mayonnaise (1 induction day) counted double values with respect to the traditional mayonnaise (0.5 induction days). Considering the worst-case scenario, the induction day cannot be identified as a representative parameter for a normalization of the environmental impact respect to the oxidation induction time. Despite this, potential food waste has been identified as a parameter for the technological performance of the two food preparations. Considering the potential shelf life, the enriched mayonnaise, which claims a double induction period with respect to the traditional one, reports an overall impact benefit avoiding food loss of about 23%.

No considerations shall be done for the nutritional parameter, the gap of TPC between the two products was quantified in 413 times (Table 1). Then, the convenience of choosing the enriched mayonnaise is directly quantified in 413 times.

3.1.3. Salad Dressing Impact Assessment

The salad dressing shows a simple formulation, which does not require particular ingredients or transformation, highlighting a large benefit in choosing the traditional salad dressing rather than the enriched one if considering the commercial unit (135 g) (Figure 7 and Table S2).

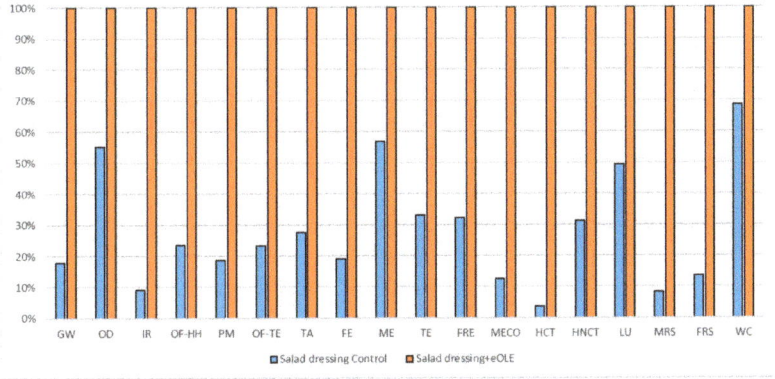

Figure 7. Environmental impact comparison of salad dressing (135 g) vs. PE enriched product. Global warming (GW), Stratospheric ozone depletion (OD), Ionizing radiation (IR), Ozone formation, human health (OF-HH), Fine particulate matter formation (PM), Ozone formation, terrestrial ecosystems (OF-TE), Terrestrial acidification (TA), Freshwater eutrophication (FE), Marine eutrophication (ME), Terrestrial ecotoxicity (TE), Freshwater ecotoxicity (FRE), Marine ecotoxicity (MECO), Human carcinogenic toxicity (HCT), Human non-carcinogenic toxicity (HNCT), Land use (LU), Mineral resource scarcity (MRS), Fossil resource scarcity (FRS), and Water consumption (WC).

The use of eOLE in the enriched salad dressing formulation represents a high risk for the environment if compared to the traditional salad dressing due to the simpleness of the formulation. The eOLE as an ingredient (as reported in Table 11) shows an overall impact among impact categories equal to 73.51%. The high level of contribution of the eOLE in the enriched product gives the salad dressing + eOLE 72% more impact than the traditional product. Opposite consideration shall be done if the TPC is considered (Table 1): in this case, the traditional formulation does not result in convenient with respect to the enriched one, showing a higher environmental impact of about 96%.

3.1.4. Biscuits Impact Assessment

The comparisons of the two biscuits formulations are reported in Figure 8 (considering 160 g commercial unit), in Figure 9 (considering oxidation induction time parameter), and in Figure 10 (considering TPC parameter). All the environmental impacts related to the three normalization parameters are reported together in Table S3.

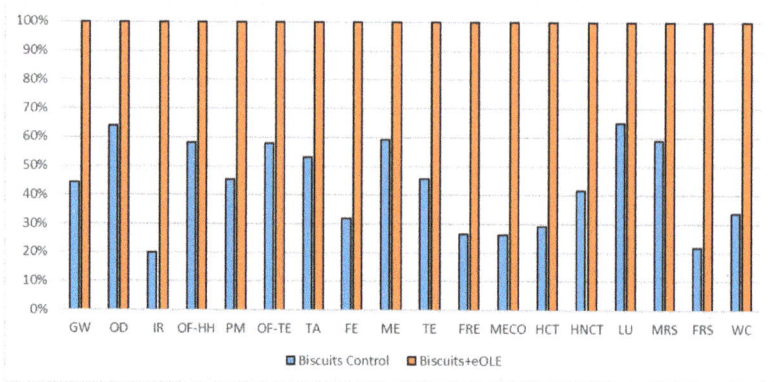

Figure 8. Environmental impact comparison of biscuits (160 g) vs. PE enriched product. Global warming (GW), Stratospheric ozone depletion (OD), Ionizing radiation (IR), Ozone formation, human health (OF-HH), Fine particulate matter formation (PM), Ozone formation, terrestrial ecosystems (OF-TE), Terrestrial acidification (TA), Freshwater eutrophication (FE), Marine eutrophication (ME), Terrestrial ecotoxicity (TE), Freshwater ecotoxicity (FRE), Marine ecotoxicity (MECO), Human carcinogenic toxicity (HCT), Human non-carcinogenic toxicity (HNCT), Land use (LU), Mineral resource scarcity (MRS), Fossil resource scarcity (FRS), and Water consumption (WC).

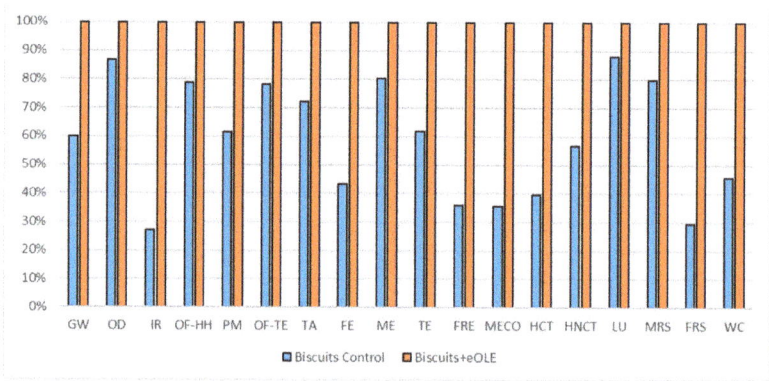

Figure 9. Environmental impact comparison of biscuits vs. PE enriched product normalized to potential shelf-life. Global warming (GW), Stratospheric ozone depletion (OD), Ionizing radiation (IR), Ozone formation, human health (OF-HH), Fine particulate matter formation (PM), Ozone formation, terrestrial ecosystems (OF-TE), Terrestrial acidification (TA), Freshwater eutrophication (FE), Marine eutrophication (ME), Terrestrial ecotoxicity (TE), Freshwater ecotoxicity (FRE), Marine ecotoxicity (MECO), Human carcinogenic toxicity (HCT), Human non-carcinogenic toxicity (HNCT), Land use (LU), Mineral resource scarcity (MRS), Fossil resource scarcity (FRS), and Water consumption (WC).

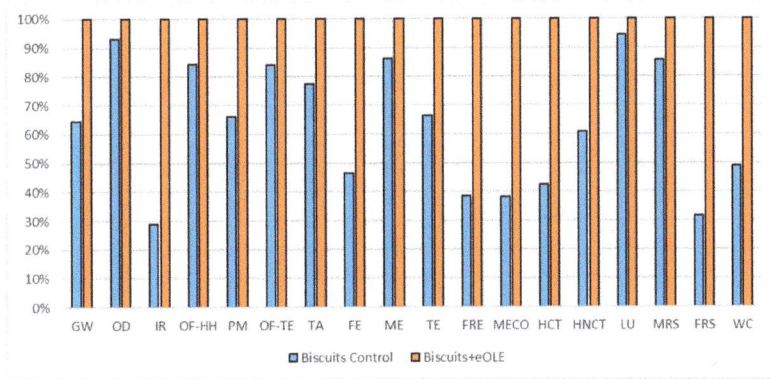

Figure 10. Environmental impact comparison of a vegan mayonnaise vs. PE enriched product normalized to TPC. Global warming (GW), Stratospheric ozone depletion (OD), Ionizing radiation (IR), Ozone formation, human health (OF-HH), Fine particulate matter formation (PM), Ozone formation, terrestrial ecosystems (OF-TE), Terrestrial acidification (TA), Freshwater eutrophication (FE), Marine eutrophication (ME), Terrestrial ecotoxicity (TE), Freshwater ecotoxicity (FRE), Marine ecotoxicity (MECO), Human carcinogenic toxicity (HCT), Human non-carcinogenic toxicity (HNCT), Land use (LU), Mineral resource scarcity (MRS), Fossil resource scarcity (FRS), and Water consumption (WC).

As for the salad dressing, also for the biscuits, which represent a more complex formulation, the high energy demand coming from the OLE extraction and encapsulation process shows higher environmental impacts for the enriched formulations in all the impact categories. The eOLE ingredient required for the commercial unit is about 0.55 g, which represents a very tiny quantity but reports a high level of contribution as reported in Table 12 (56% among impact categories).

Table 12. Strengths, weaknesses, opportunities, and threats related to use of phenolic extract deriving from the olive oil milling process in food chain.

	Internal		
	Strengths	Weaknesses	
	Reduce waste of olive oil milling process	High environmental impact of extraction/encapsulation processes	
	Transform waste products into a second-life product	Energy-intensive operations to freeze drying operation and encapsulation solvents	
		Chemicals for extraction	
Positive	Opportunities	Threats	Negative
	Promote circular economy business model	Research efforts in optimizing extraction and encapsulation processes in terms of yield extraction	
	Reduce food waste		
	Add values to olive oil sector		
	Design new packaging concept considering the use of bio-compounds to extend shelf-life		
	External		

The comparison of the two biscuits formulations (reported in Figure 8) considering a 160 g commercial unit showed that the enriched biscuits formulation impact 56% more with respect to the traditional one (average value among impact categories).

Considering the oxidation induction time parameter (Figure 9), which implies a delta between the two formulations equal to 1.3559, the traditional formulation reports again better environmental impact in all the impact categories (41% less with respect to the enriched one). The better behavior of the traditional biscuits is largely highlighted in those impact categories that are directly linked to the energy consumption as IR, HNCT, HCT, and FRS (62% average benefit), while the lowest advantages in choosing the traditional biscuits can be seen in those impact categories linked to the agricultural activities as OD, ME, and LU (15% average benefit).

As reported in Table 6, also considering the TPC normalization parameter, the benefit of the traditional biscuits is confirmed again even if it decreased with respect to the commercial functional unit comparison (37% less). Even if an increase of the relative environmental impact can be registered due to a 1.45 gap (Table 1) between traditional and enriched formulation, the enriched product represents the worst product.

3.1.5. Gluten-Free Breadsticks Impact Assessment

The gluten-free breadsticks formulation comparisons are reported in Figure 11 (considering a 300 g commercial unit), in Figure 12 (considering oxidation induction time parameter) and in Figure 13 (considering TPC parameter). All the environmental impacts related to the three normalization parameters are reported together in Table S4.

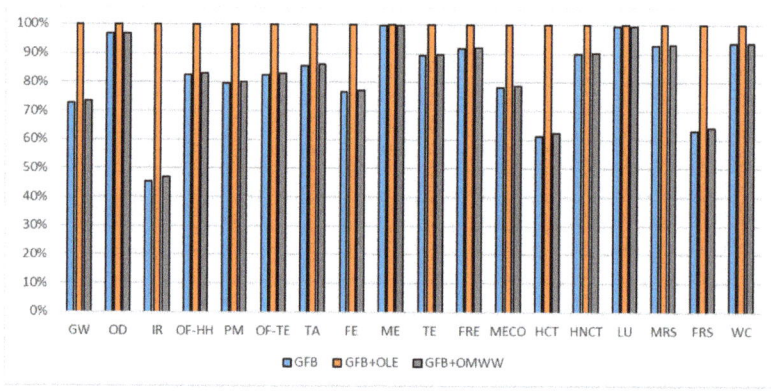

Figure 11. Environmental impact comparison of gluten-free breadsticks (300 g) vs. PE enriched product. Global warming (GW), Stratospheric ozone depletion (OD), Ionizing radiation (IR), Ozone formation, human health (OF-HH), Fine particulate matter formation (PM), Ozone formation, terrestrial ecosystems (OF-TE), Terrestrial acidification (TA), Freshwater eutrophication (FE), Marine eutrophication (ME), Terrestrial ecotoxicity (TE), Freshwater ecotoxicity (FRE), Marine ecotoxicity (MECO), Human carcinogenic toxicity (HCT), Human non-carcinogenic toxicity (HNCT), Land use (LU), Mineral resource scarcity (MRS), Fossil resource scarcity (FRS), and Water consumption (WC).

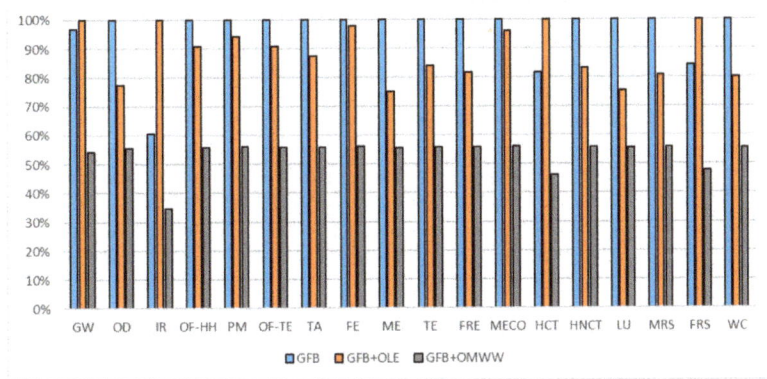

Figure 12. Environmental impact comparison of gluten-free breadsticks vs. PE-enriched product normalized to potential shelf life. Global warming (GW), Stratospheric ozone depletion (OD), Ionizing radiation (IR), Ozone formation, human health (OF-HH), Fine particulate matter formation (PM), Ozone formation, terrestrial ecosystems (OF-TE), Terrestrial acidification (TA), Freshwater eutrophication (FE), Marine eutrophication (ME), Terrestrial ecotoxicity (TE), Freshwater ecotoxicity (FRE), Marine ecotoxicity (MECO), Human carcinogenic toxicity (HCT), Human non-carcinogenic toxicity (HNCT), Land use (LU), Mineral resource scarcity (MRS), Fossil resource scarcity (FRS), and Water consumption (WC).

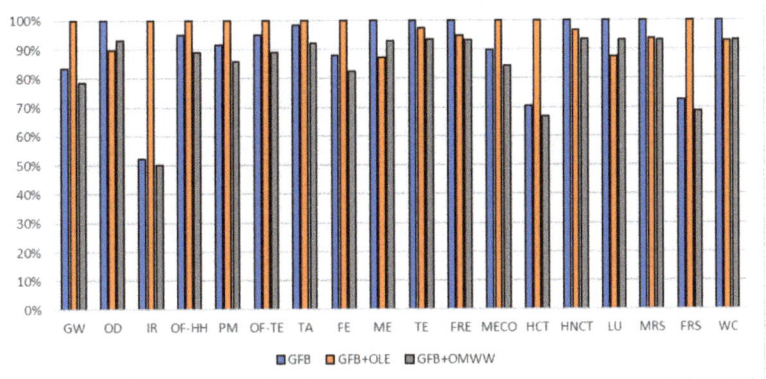

Figure 13. Environmental impact comparison of gluten-free breadsticks vs. PE-enriched product normalized to TPC. Global warming (GW), Stratospheric ozone depletion (OD), Ionizing radiation (IR), Ozone formation, human health (OF-HH), Fine particulate matter formation (PM), Ozone formation, terrestrial ecosystems (OF-TE), Terrestrial acidification (TA), Freshwater eutrophication (FE), Marine eutrophication (ME), Terrestrial ecotoxicity (TE), Freshwater ecotoxicity (FRE), Marine ecotoxicity (MECO), Human carcinogenic toxicity (HCT), Human non-carcinogenic toxicity (HNCT), Land use (LU), Mineral resource scarcity (MRS), Fossil resource scarcity (FRS), and Water consumption (WC).

According to the comparison of the commercial unit (300 g of product for each formulation), the OLE-enriched product reports higher environmental impacts in all the impact categories. As reported in Figure 10, the OMWW PE-enriched product reports lower environmental impact with respect to the OLE-enriched one (17% less on average) but higher with respect to the traditional formulation (1% more on average). The average percentage of responsibility related to the OLE in the enriched formulation was a weight of about 17% (as reported in Table 11) while the responsibility related to the OMWW PE in the enriched formulation weight was less than 1% (as reported in Table 11). The gap among the three formulations is highlighted majorly in impact categories directly related to the energy consumption as IR, HNCT, HCT, and FRS where the OLE ingredient has a higher contribution (as reported in Table 11). A consideration regarding only the commercial unit puts in first place the traditional GFB, in second place GFB + OMWWPE, and in third place the OLE enriched one.

Different results are deductible according to the estimated potential shelf life and TPC parameters as reported in Table 7. The overall benefit of potential shelf-life extension enriching the GFB control with OMWW PE can be identified in a potential environmental impact reduction of about 46%. The traditional GFB and the GFB + OLE cannot be defined as the worst solution, both report in some impact categories the higher environmental impact.

Considering the TPC normalization parameter and taking into consideration the average impact among the different impact categories, the differentiation made for the commercial unit cannot be carried out; in some impact categories, the highest level of impact is reached by the GFB control, and in others, the highest level is reached by GFB + OLE. A consideration that should be carried out is that the GFB + OLE on average represents the higher environmental impact formulation, representing the worst choice in most of the impact categories. The choice between the GFB control and GFB + OMWW PE should be linked to the impact category taken into consideration. An overview of the results obtained, considering all three parameters, identifies as the best choice the GFB + OMWW PE, even if in some cases, it means that it is not the best option among the three.

Overall, the results obtained also showed the impact of phenols extracts on the food preparations process considering mean and standard deviation among the impact categories (Table 11). The impact of PE extraction/encapsulation on the whole food chain ranged from 0.71% to 73.51%. This wide range is due to the ingredients and operations provided in the preparation processes. Since the production process consists of a simple formulation in term of (i) ingredients (low quantity and low processed products) and/or (ii) process (few and low energy demand operations), the impact of polyphenols extraction process and encapsulation reached on the whole production process impact a percentage of 73.51%. On the contrary, for those complex preparations, as for the gluten-free breadsticks, the operations relating to phenols extraction process and/or encapsulation weight on the whole process only for 17.72% and 0.71% for GFB + OLE and GFB + OMWW PE respectively, while the impact of phenols extracts on vegan mayonnaise preparation reaches 30.90%.

4. Discussion

The results obtained show that the extraction and encapsulation processes, characterized by low production yields, energy-intensive operations, and the partial use of chemical reagents has a non-negligible environmental impact. In detail, to contextualize the results obtained, it is important to analyze the whole supply chain up to the finished product. Even in other critical sectors, such as the plastic packaging field, analyzing only the impact of the material and the production process, the environmental sustainability results are very low [30].

If other aspects were also considered in evaluating the environmental impact, such as the extension of the potential shelf life and therefore the reduction of food waste rather than the entire supply chain [31], the packaging environmental impact would have another weight on the whole supply chain. In this work, after calculating the impact of the polyphenols extraction and encapsulation process from olive oil mill waste, the impact of the extraction process in the food chain was considered, in particular for the production of vegan mayonnaise, biscuits, salad dressing, and gluten-free breadsticks.

The impact of polyphenols extraction/encapsulation on the whole food chain presents very different results based on the operations provided in the formulations process. Considering the advantages in terms of technological and nutritional aspects in the use of enriched formulations, the weight of the polyphenols extraction process and/or encapsulation falls exponentially.

Considering that the LCA evaluation in this work was carried out based on lab-scale data, the impact of the polyphenols extraction process and encapsulation could be reduced in a view of a scale-up process. In fact, the development of pilot plants for the polyphenols extraction and encapsulation within a real chain of reuse of olive oil mill waste would allow the use of more efficient systems and therefore reduce the environmental impact as well as the development of a circular economy model. The environmental impact of the polyphenol extraction and encapsulation process, which for some food preparation showed high contribution, if transferred to a wide context, would allow economic advantages in the valorization of the olive oil supply chain, cancelling out the environmental impact of the polyphenols extraction and encapsulation process from waste.

Nowadays, the olive oil mill by-products are treated with a high energy demand process to transform these products into a second life product (i.e., from pomace to pomace oil, from stone to heat) identifying in the waste product as a high-level product. Other activities are simply catalogued as waste management processes, the output of the oil mill as wastewater, and leaves and dust are treated for composting or fertilizing fields, representing in any case a cost and not a profit for the mill. The revalorization of the waste products coming from the oil milling activities can rearrange all the outputs' quality level, identifying a profit in waste.

This work should be the basis for future research focused on the environmental impact comparison between the use of phenolic extract deriving from olive oil mill waste and the packaging operations to improve shelf life performance in food chain. The analysis could include different scenarios:

- If the impact on the product of the use of phenolic extract from olive oil milling process wastes results in more sustainability with respect to the use of packaging, it should promote the concept of circular economy. In fact, in this case, there is high probability of a reduction of packaging worldwide re-using olive oil mill waste and adding value to the whole olive oil sector;
- If the use of packaging results is more sustainable than the use of phenolic extracts from olive oil mill wastes, it should be anyway interesting. In this case, the utilization of biocompounds of synthesis instead of phenolic extract from wastes could be the winning choice, reducing the benefits around the food supply chain and having a lower appeal for consumers. It should open further studies to identify the best combo choice between biocompounds and packaging design.

In order to summarize the strengths, weaknesses, opportunities, and threats related to the use of polyphenols extract deriving from olive oil milling waste in the food chain promoting the circular economy, a SWOT table was created (Table 12).

5. Conclusions

Nowadays, olive mill waste products could be exploited as by-products to be used as fuels, fertilizers, or other intermediate products for the food, nutraceutical, cosmetic, and pharmaceutical industries. Olive leaf and olive oil mill wastewater have a good potential as sources of bioactive component and could be considered renewable source of high-added value. In this context, the aim of this study is to quantify the environmental impact of the extraction of phenolic compounds from olive oil mill waste. The phenolic extraction and encapsulation obtained from wastewater and olive leaves was characterized by low production yields, energy-intensive operations, and the partial use of chemical reagents. The addition of phenolic extract to food products (vegan mayonnaise, salad dressing, biscuits, and gluten-free breadsticks) leads to enhancing the environmental impact of production process but also implies an improvement of technological and nutritional performance. The potential shelf life of enriched food preparations induces an increase up to two times with respect to control due to the presence of TPC added. This is a crucial aspect to consider in the normalization of environmental impact based on technological and nutritional parameters. This work should be the basis for future research focused on the environmental impact comparison between the use of phenolic extract deriving from olive oil mill waste and the packaging operations to improve shelf life performance in food chain. Moreover, this research can open further studies to identify the best combo choice between biocompounds and packaging design to reduce food waste, reduce packaging, and promote the circular economy business model.

Supplementary Materials: The following are available online at https://www.mdpi.com/article/10.3390/foods10050980/s1, Figure S1: System boundary, Table S1: Environmental impact comparison of mayonnaise and enriched mayonnaise considering commercial unit, potential shelf-life and TPC parameters, Table S2: Environmental impact comparison of salad dressing and enriched salad dressing considering commercial unit and TPC parameters, Table S3: Environmental impact comparison of biscuits and enriched biscuits considering commercial unit, technological and nutritional parameters, Table S4: Environmental impact comparison of gluten free breadsticks and enriched GFB considering commercial unit, technological and nutritional parameters.

Author Contributions: Author Contributions: Conceptualization, A.P. (Alessia Pampuri), A.C., A.T., R.B., R.G., and V.G.; methodology, A.P. (Alessia Pampuri), A.C., A.T. and R.B.; formal analysis, A.P. (Alessia Pampuri), A.C., and V.G.; investigation, C.A, C.D.D.M., A.P. (Amalia Piscopo), G.D., P.C., and M.P.; resources, C.A, C.D.D.M., A.P. (Amalia Piscopo), G.D., P.C., and M.P.; data curation, A.P. (Alessia Pampuri), A.C., and V.G.; writing—original draft preparation, A.P. (Alessia Pampuri), A.C. and V.G.; writing—review and editing: A.T., R.B., C.A., C.D.D.M., A.P. (Amalia Piscopo), G.D., P.C., M.P., R.G. and E.C.; supervision, R.G., and V.G.; funding acquisition, E.C. All authors have read and agreed to the published version of the manuscript.

Funding: This research was funded by AGER 2 Project, grant no. 2016-0105.

Institutional Review Board Statement: Not applicable.

Informed Consent Statement: Not applicable.

Data Availability Statement: Data is contained within the article or supplementary material.

Acknowledgments: The authors wish to thank Francesco Caponio for the supervision of the whole research project.

Conflicts of Interest: The authors declare no conflict of interest. The funder had no role in the design of the study; in the collection, analyses, or interpretation of data; in the writing of the manuscript, or in the decision to publish the results.

References

1. Casson, A.; Beghi, R.; Giovenzana, V.; Fiorindo, I.; Tugnolo, A.; Guidetti, R. Visible near infrared spectroscopy as a green technology: An environmental impact comparative study on olive oil analyses. *Sustainability* **2019**, *11*, 2611. [CrossRef]
2. Tempesta, T.; Vecchiato, D. Analysis of the factors that influence olive oil demand in the Veneto Region (Italy). *Agriculture* **2019**, *9*, 154. [CrossRef]
3. De Bruno, A.; Romeo, R.; Piscopo, A.; Poiana, M. Antioxidant quantification in different portions obtained during olive oil extraction process in an olive oil press mill. *J. Sci. Food Agric.* **2021**, *101*, 1119–1126. [CrossRef]
4. Tsagaraki, E.; Lazarides, H.N.; Petrotos, K.B. Olive mill wastewater treatment. In *Utilization of By-Products and Treatment of Waste in the Food Industry*; Springer: Boston, MA, USA, 2007; pp. 133–157.
5. Salomone, R.; Ioppolo, G. Environmental impacts of olive oil production: A Life Cycle Assessment case study in the province of Messina (Sicily). *J. Clean. Prod.* **2012**, *28*, 88–100. [CrossRef]
6. Aliakbarian, B.; Paini, M.; Adami, R.; Perego, P.; Reverchon, E. Use of Supercritical Assisted Atomization to produce nanoparticles from olive pomace extract. *Innov. Food Sci. Emerg. Technol.* **2017**, *40*, 2–9. [CrossRef]
7. Difonzo, G.; Squeo, G.; Calasso, M.; Pasqualone, A.; Caponio, F. Physico-chemical, microbiological and sensory evaluation of ready-to-use vegetable pâté added with olive leaf extract. *Foods* **2019**, *8*, 138. [CrossRef] [PubMed]
8. Ranieri, M.; Di Mise, A.; Difonzo, G.; Centrone, M.; Venneri, M.; Pellegrino, T.; Tamma, G. Green olive leaf extract (OLE) provides cytoprotection in renal cells exposed to low doses of cadmium. *PLoS ONE* **2019**, *14*, e0214159.
9. Ranieri, M.; Di Mise, A.; Centrone, M.; D'Agostino, M.; Tingskov, S.J.; Venneri, M.; Tamma, G. Olive Leaf Extract (OLE) impaired vasopressin-induced aquaporin-2 trafficking through the activation of the calcium-sensing receptor. *Sci. Rep.* **2021**, *11*, 1–13. [CrossRef] [PubMed]

10. Flamminii, F.; Di Mattia, C.D.; Nardella, M.; Chiarini, M.; Valbonetti, L.; Neri, L.; Pittia, P. Structuring alginate beads with different biopolymers for the development of functional ingredients loaded with olive leaves phenolic extract. *Food Hydrocoll.* **2020**, *108*, 105849. [CrossRef]
11. Flamminii, F.; Di Mattia, C.D.; Difonzo, G.; Neri, L.; Faieta, M.; Caponio, F.; Pittia, P. From by-product to food ingredient: Evaluation of compositional and technological properties of olive-leaf phenolic extracts. *J. Sci. Food Agric.* **2019**, *99*, 6620–6627. [CrossRef] [PubMed]
12. International Organization for Standardization. *Environmental Management: Life Cycle Assessment; Principles and Framework*, 2nd ed.; ISO: 14040-14044:2006; International Organization for Standardization: Geneva, Switzerland, 2006.
13. Singleton, V.L.; Rossi, J.A. Colorimetry of total phenolics with phosphomolybdic-phosphotungstic acid reagents. *AJEV* **1965**, *16*, 144–158.
14. Paciulli, M.; Rinaldi, M.; Cavazza, A.; Ganino, T.; Rodolfi, M.; Chiancone, B.; Chiavaro, E. Effect of chestnut flour supplementation on physico-chemical properties and oxidative stability of gluten-free biscuits during storage. *LWT* **2018**, *98*, 451–457. [CrossRef]
15. Figueiredo, F.; Marques, P.; Castanheira, É.G.; Kulay, L.; Freire, F. Greenhouse gas assessment of olive oil in Portugal addressing the valorization of olive mill waste. In Proceedings of the Symbiosis International Conference, Athens, Greece, 19–21 June 2014; pp. 19–21.
16. Rajaeifar, M.A.; Akram, A.; Ghobadian, B.; Rafiee, S.; Heidari, M.D. Energy-economic life cycle assessment (LCA) and greenhouse gas emissions analysis of olive oil production in Iran. *Energy* **2014**, *66*, 139–149. [CrossRef]
17. Tsarouhas, P.; Achillas, C.; Aidonis, D.; Folinas, D.; Maslis, V. Life Cycle Assessment of olive oil production in Greece. *J. Clean. Prod.* **2015**, *93*, 75–83. [CrossRef]
18. El Hanandeh, A.; Gharaibeh, M.A. Environmental efficiency of olive oil production by small and micro-scale farmers in northern Jordan: Life cycle assessment. *Agric. Syst.* **2016**, *148*, 169–177. [CrossRef]
19. Banias, G.; Achillas, C.; Vlachokostas, C.; Moussiopoulos, N.; Stefanou, M. Environmental impacts in the life cycle of olive oil: A literature review. *J. Sci. Food Agric.* **2017**, *97*, 1686–1697. [CrossRef] [PubMed]
20. Parascanu, M.M.; Sánchez, P.; Soreanu, G.; Valverde, J.L.; Sanchez-Silva, L. Environmental assessment of olive pomace valorization through two different thermochemical processes for energy production. *J. Clean Prod.* **2018**, *186*, 771–781. [CrossRef]
21. Espadas-Aldana, G.; Vialle, C.; Belaud, J.P.; Vaca-Garcia, C.; Sablayrolles, C. Analysis and trends for Life Cycle Assessment of olive oil production. *Sustain. Prod. Consum.* **2019**, *19*, 216–230. [CrossRef]
22. Batuecas, E.; Tommasi, T.; Battista, F.; Negro, V.; Sonetti, G.; Viotti, P.; Fino, D.; Mancini, G. Life Cycle Assessment of waste disposal from olive oil production: Anaerobic digestion and conventional disposal on soil. *J. Environ. Manag.* **2019**, *237*, 94–102. [CrossRef] [PubMed]
23. Guarino, F.; Falcone, G.; Stillitano, T.; De Luca, A.I.; Gulisano, G.; Mistretta, M.; Strano, A. Life cycle assessment of olive oil: A case study in southern Italy. *J. Environ. Manag.* **2019**, *238*, 396–407. [CrossRef]
24. Romeo, R.; De Bruno, A.; Imeneo, V.; Piscopo, A.; Poiana, M. Evaluation of enrichment with antioxidants from olive oil mill wastes in hydrophilic model system. *J. Food Proc. Pres.* **2019**, *43*, e14211. [CrossRef]
25. Difonzo, G.; Russo, A.; Trani, A.; Paradiso, V.M.; Ranieri, M.; Pasqualone, A.; Caponio, F. Green extracts from Coratina olive cultivar leaves: Antioxidant characterization and biological activity. *J. Funct. Foods* **2017**, *31*, 63–70. [CrossRef]
26. Flamminii, F.; Di Mattia, C.D.; Sacchetti, G.; Neri, L.; Mastrocola, D.; Pittia, P. Physical and sensory properties of mayonnaise enriched with encapsulated olive leaf phenolic extracts. *Foods* **2020**, *9*, 997. [CrossRef]
27. Jolayemi, O.S.; Stranges, N.; Flamminii, F.; Casiraghi, E.; Alamprese, C. Influence of free and encapsulated olive leaf phenolic extract on the storage stability of single and double emulsion salad dressings. *Food Bioprocess. Technol.* **2021**, 1–13. [CrossRef]
28. Paciulli, M.; Grimaldi, M.; Flamminii, F.; Littardi, P.; Di Mattia, C.; Cavazza, A.; Rinaldi, M.; Carini, E.; Pittia, P.; Ornaghi, P.; et al. Olive leaves microencapsulated polyphenols as functional ingredient to prolong oxidative stability of biscuits. In Proceedings of the 17th Euro Fed Lipid, Seville, Spain, 20–23 October 2019.
29. Goedkoop, M.; Heijungs, R.; Huijbregts, M.; De Schryver, A.; Struijs, J.; Van Zelm, R. A life cycle impact assessment method which comprises harmonised category indicators at the midpoint and the endpoint level. *ReCiPe 2008* **2009**, *1*, 1–126.
30. Lombardi, M.; Rana, R.; Fellner, J. Material flow analysis and sustainability of the Italian plastic packaging management. *J. Clean. Prod.* **2021**, *287*, 125573. [CrossRef]
31. Pauer, E.; Wohner, B.; Heinrich, V.; Tacker, M. Assessing the environmental sustainability of food packaging: An extended life cycle assessment including packaging-related food losses and waste and circularity assessment. *Sustainability* **2019**, *11*, 925. [CrossRef]

MDPI
St. Alban-Anlage 66
4052 Basel
Switzerland
Tel. +41 61 683 77 34
Fax +41 61 302 89 18
www.mdpi.com

Foods Editorial Office
E-mail: foods@mdpi.com
www.mdpi.com/journal/foods

www.ingramcontent.com/pod-product-compliance
Lightning Source LLC
LaVergne TN
LVHW070051120526
838202LV00102B/2046